INTERPRETING THE STRATIGRAPHIC RECORD

INTERPRETING THE STRATIGRAPHIC RECORD

DONALD R. PROTHERO

Occidental College

W. H. FREEMAN & COMPANY · NEW YORK

The W. H. Freeman Series in the Geological Sciences
Raymond Siever, Consulting Editor

Cover image: The Grand Canyon from Toroweap Point; photograph by Josef Muench.

Quotations: p. xiii: Gilluly quote with permission from the *Annual Review of Earth Sciences*, 5, © 1977 by Annual Reviews, Inc.;

Billings quote from Billings, *Geol. Soc. Amer. Bull.*, 61, p. 435, reprinted with permission of the Geological Society of America;

Weller quote from Weller, *J. Paleont*, 21, pp. 570–575;

Love quote from J. McPhee, *Rising from the Plains*, 1986, p. 153, by permission of Farrar, Straus & Giroux, Inc. p. 22;

McPhee quote from J. McPhee, *Basin and Range*, 1980, pp. 128–129, by permission of Farrar, Straus & Giroux, Inc.

Library of Congress Cataloging-in-Publication Data

Prothero, Donald R.
    Interpreting the stratigraphic record / Donald R. Prothero.
      p. cm.
    Bibliography: p.
    Includes index.
    ISBN 0-7167-1854-5
    1. Geology, Stratigraphic. I. Title.
QE651P8   1989   88-35587
551.7—dc19   CIP

Printed in the United States of America

Fourth printing 1997, KP

This book is affectionately dedicated to Dewey Moore and to Larry DeMott, whose inspiration and dedication to their many students and colleagues will never be diminished by anyone.

Contents

Preface

The discipline of stratigraphy has changed tremendously in the last few years. In the 1950s and 1960s, stratigraphy classes typically emphasized the detailed study of local stratigraphic columns and often had a reputation for memorization and drudgery. The principles of stratigraphy, as shown by Dunbar and Rodgers' classic text of the same title, were mostly principles of what we now call *sedimentology*, which emphasizes the processes that form sediments and includes the interpretation of sedimentary rocks on the microscopic and macroscopic scale. A few chapters were devoted to the theoretical basis behind rock units and their correlation, but there was little discussion of the pitfalls inherent in interpreting the rock record. More than half of even the most theoretical textbook of that generation, Krumbein and Sloss' *Stratigraphy and Sedimentation*, was on sedimentology. Its theoretical discussion of stratigraphy was the most sophisticated of its time and placed strong emphasis on tectonic processes and sedimentary packages of the continental interior. But the last edition of this book was published in 1963, just before plate tectonics radically changed the way geologists look at the earth.

The 1960s and 1970s brought many fundamental changes to geology. Foremost among these changes was the scientific revolution brought about by plate tectonics. With an understanding of the mechanism that formed mountain belts and sedimentary basins, many stratigraphers were forced to reinterpret their life's work. A new generation of earth scientists began busily reinterpreting classic areas in terms of the new paradigm. Plate tectonics also shifted attention from the relatively quiet continental interior to the mountain belts, with their attendant complexity. In active mountain belts, the word *stratigraphy* has taken on new meaning. A modern stratigrapher now must deal with igneous and metamorphic, as well as sedimentary rocks, as part of the stratigraphic history of a region. A modern stratigraphy course now must incorporate a discussion of plate tectonic models as explanations for *why* mountains go up and basins go down.

My own training in stratigraphy during the mid-seventies reflected this radical change. As an undergraduate, I was taught the classic Krumbein and Sloss emphasis on sedimentology and the stable continental interior. As a graduate student, I encountered an undergraduate-level stratigraphy course that concentrated on plate tectonic models.

Stratigraphy has also been changed by another phenomenon: the explosive growth of sedimentology as a discipline in its own right. Most colleges devote a full 10 or 15 weeks to sedimentology, and the classical principles of stratigraphy are often treated as orphans, squeezed into wherever they fit or, all too often, neglected altogether.

In most colleges and universities, stratigraphy has taken on a new form. In courses devoted to stratigraphy, the classical principles of lithostratigraphy and biostratigraphy are still the core of the discipline. But a practicing stratigrapher must be conversant in many subjects that were unknown, or were in their infancy, in the 1950s and 1960s. Foremost among these is geochronology, the use of radioactive decay to determine numerical age. No active geologist can afford to be ignorant of the strengths and limitations of the various dating techniques. Geochronology is increasingly becoming a part of stratigraphy courses.

Many new techniques (particularly seismic stratigraphy, magnetic stratigraphy, and isotope stratigraphy) have developed tremendously in the last decade, so that most modern stratigraphic studies make use of at least one of these techniques. Even the classical core of stratigraphic theory has grown. The seminal works of Shaw (1964) and Ager (1973) and the development of quantitative methods of biostratigraphy have challenged many traditional methods of interpreting the rock record. Finally, the interpretation of depositional environments has become the key to nearly every stratigraphic study as more and more modern analogues for ancient stratigraphic sequences are studied.

This textbook grew out of my own stratigraphy course, which emphasizes most of these developments. Talks with friends and colleagues who teach similar courses made it apparent to me that there is a need for a textbook on stratigraphy as it is done today. This book is designed for a 10- to 15-week stratigraphy course, which follows a separate sedimentology course in most curricula. The book assumes at least a course in physical and historical geology and some knowledge of sedimentary rocks. However, Chapter 2 includes some basics of sedimentology, so that students without an extensive background in the subject will not be left out. (For a course that covers both stratigraphy and sedimentology in 15 weeks, this book could be combined with a few chapters from a sedimentology book.) I have tried to write for an audience of upper-level undergraduate geology majors, though some of the material could be discussed in greater detail in a graduate course. As a consequence, I have tried to avoid making this book too technical or detailed, or too specialized for a particular industry. This book does not emphasize well-logging or subsurface analysis because I feel that for most undergraduates, it is much more practical and important to master the principles of stratigraphy as seen in surface outcrops. To this end, I have set aside a classic outcrop sequence of each of the depositional systems in Part 2 as boxed text. Greater understanding of subsurface data can be developed once the basic concepts have been mastered from the surface where they can actually be seen.

Acknowledgments. I thank Nicholas Christie-Blick, Charles Byers, Emily CoBabe, Jim Collinson, George Engelmann, Susan Kidwell, Steve King, Ron Lewis, Norman MacLeod, Peter Sadler, Raymond Siever, Roger Walker, and several anonymous reviewers for reading the entire text and correcting many problems. Edwin Anderson, Bill Berggren, John Crowell, Brent Dalrymple, Lucy Edwards, Frank Ethridge, John Ferm, Peter Goodwin, Leo Laporte, Emiliano Mutti, Chuck Naeser, John Obradovich, Phillip Playford, Martin Rudwick, James Secord, Norman Sohl, Ron Steel, and Tony Tankard read and critized portions of the text. I take full responsibility for any remaining errors in fact or interpretation, of course.

Many friends and colleagues have been exceedingly generous in providing photographs or illustrations for the book. Their contributions are acknowledged in the appropriate places. In particular, my father, Clifford Prothero, spent much time helping me with photographs and artwork. His artistic taste has improved the visual appeal of this book immensely, both by his direct advice and also by what he taught me over the years. I may not have followed in his footsteps as an artist, but his lessons were not wasted. Steve King helped with many tasks, including the bibliography and index.

My editor, Jeremiah Lyons, has been a helpful and supportive advisor through the years of working on this project. Many other people at W. H. Freeman have contributed greatly to this book: Georgia Lee Hadler, project editor; Glenn Cochran and Yvonne Howell, copy editors; Martha Rago and Lynn Pieroni, designers; Bill Page, illustration coordinator; Norm Nason, line artist; John Hatzakis, page make-up; Julia DeRosa, production coordinator; and, Christine Rickoff, assistant to Jerry Lyons.

Finally, I thank the many people whose advice and support kept me going during the long gestation of this book. Mike Murphy and Mike Woodburne taught me the importance of stratigraphy and gave me the sound backing in the fundamentals that so few undergraduate geologists get today. If this book can convey some of what I learned from them to a new generation of geologists, then a small part of my debt will be repaid. Rich Schweikert opened my eyes to the impact of tectonics on stratigraphy. John Sanders supported me during my first sedimentology class and at many other times in my career. My first professional colleagues, John Johnsen and Jerome Regnier, were very helpful at a time when only sedimentology was employing vertebrate paleontologists. Dewey Moore and the late Larry DeMott were my inspirations in my first full-time teaching post, and their faith in me helped me survive trying times. Meg Zepp put up with my moods and helped me in many ways during the writing of this book. My current colleagues, Jim Woodhead, Margi Rusmore, and Scott Bogue, have been supportive in many ways. Finally, I thank my parents, Clifford and Shirley Prothero, for their long years of financial and emotional support during the trials of graduate school and jobs changes while I tried to maintain a research career, write books, and teach double overloads for six straight years. Without their encouragement, this book would never had been possible.

— *Donald R. Prothero*

Stratigraphy is the backbone of the science. Without sound stratigraphy, structural studies are impossible, and all historical and most economic geology depend upon it.

— JAMES GILLULY, 1977

The student of metamorphic rocks should be familiar with modern concepts of stratigraphy and sedimentation.

— MARLAND BILLINGS, 1950

Stratigraphy is the single great unifying agency of geology. Without it the findings of other branches could not be knit into a single historical whole. Perhaps the structural history of the Appalachians and that of the Rocky Mountains could be deciphered in acceptable detail without the aid of stratigraphy (I doubt it) but their histories could never be related to each other without stratigraphy. Likewise the petrogenesis of any number of rock bodies might be worked out individually but without stratigraphy these would remain so many unconnected histories and geology would be nothing more than a hodge-podge of unrelated facts and generalizations, and certainly not a historical science . . . . Stratigraphy makes possible the synthesis of a unified geological science from its component parts. Stratigraphy is the heart of geology.

— J. MARVIN WELLER, 1947

In order to know the anatomy of each mountain range, you have to know details of sedimentary history. To know the details of sedimentary history, you have to know stratigraphy . . . . Many schools don't teach it anymore. To me, that's writing the story without knowing the alphabet. The geologic literature is a graveyard of skeletons who worked the structure of mountain ranges without knowing the stratigraphy.

— J. DAVID LOVE (in McPhee, 1986)

I

INTRODUCTION

Aerial view of entrenched meanders cutting through Permian sediments at Goosenecks of the San Juan River, Utah. Road in upper left corner indicates scale. Photo courtesy Donald L. Baars.

At the beginning of any new course, most students ask themselves: why should I take this class? Why is this field of study interesting and important? What benefits will I gain for the time and effort invested? One of the first tasks of the instructor is to justify the course in these terms.

Although stratigraphy is usually defined as "the study and interpretation of layered rock sequences" (or some similar definition), it is much more than that. Layered rock sequences are the primary sources of our knowledge of the history of the earth, which was first deciphered by stratigraphers. Fossils come exclusively from stratified rocks, so much information about the many fascinating and bizarre life forms that once lived on this planet is provided by stratigraphers. Stratigraphy synthesizes the techniques and information of several geologic disciplines.

At the core of stratigraphy is the study of sedimentary rocks. At one time stratigraphy and sedimentology were almost interchangeable, but now the meaning of the term *stratigraphy* has broadened so much that nearly any technique for interpreting geologic history is covered by it. Stratigraphy is no longer earthbound; we have lunar and Martian stratigraphy. Many geologists who interpret the geologic history of igneous and metamorphic bodies (which are seldom layered) describe their work as stratigraphy. The current North American Stratigraphic Code (Appendix A) now has categories for unstratified igneous and metamorphic rocks to reflect this change. Because stratigraphy deals directly with geologic time, stratigraphers must understand the work of geochemists and nuclear physicists to obtain numerical ages for earth events (*geochronology*). For many years geophysical techniques have been important in interpreting layered sequences of rock. A whole new class of geophysical and geochemical stratigraphy has arisen, employing the techniques of seismic stratigraphy, magnetic polarity stratigraphy, isotope and trace element stratigraphy, and the physical properties of rocks as measured by well-logging. These techniques have opened new vistas on geologic history and forced reassessment of traditional assumptions. As phrased by an international commission of stratigraphers, "Stratigraphy is the study of rocks and their distribution in space and time with the object of reconstructing the history of the earth and eventually of extraterrestrial bodies" (Laffitte et al., 1972).

In this sense, historical geology and stratigraphy are intimately related. Historical geology is the sum of our knowledge of the geologic history of the earth, and stratigraphy is the method by which we decipher this history. No geologist can afford to be ignorant of the basic principles of stratigraphy and geochronology. They are the "bread and butter" of geologic research.

Stratigraphy is a bread-and-butter science for another reason. Most of the world's important economic mineral deposits lie in stratified sequences. A large percentage of the world's geologists are practicing stratigraphers, employed by the major mineral industries to find more such deposits. Oil, gas, uranium, coal, and many other economic rocks and minerals (gypsum, rock salt, sand, limestone, and cement) are found primarily in layered sedimentary rocks. There are also stratified ore deposits and placer gold deposits, both of which are found in stratified sediments. As mineral resources become scarcer, the demand for geologists trained in stratigraphy can only increase. Other branches of geology also employ stratigraphy. Groundwater geologists must know stratigraphy to locate aquifers and determine the patterns of groundwater flow. Many problems in engineering geology and geologic hazards require some understanding of the stratigraphy of the rocks that are critical to a project. Environmental problems such as the siting of hazardous waste dumps and the disposal of industrial waste require stratigraphy to determine whether a waste product or pollutant can be stored safely. Stratigraphy is a daily tool of many professional geologists.

Finally, stratigraphy is at the core of many of the most exciting fields of geological research today. In the study of plate tectonics, stratigraphic data are often the key to tracing past plate configurations. Our knowledge of ancient climates has increased tremendously in recent years, especially with the development of isotope stratigraphy. We are now capable of reconstructing not only ancient land configurations but also climatic patterns, oceanographic circulations, and surface temperatures, and are well on our way to ascertaining the causes of climatic change. In the last few years, stratigraphic studies have suggested that extraterrestrial events may have greatly affected life and climates in the past, possibly causing mass extinctions. This is an area of controversy and active research that promises to continue to attract interest in the future.

The theme of this book is the question, "How does the geologist go about interpreting the rock record?" Pure description of rock sequences is only a beginning. The determination of the *meaning* of those sequences depends on the theoretical framework of geologists' interpretation. In Part I, we look at the data base from which all stratigraphic interpretation emerges. Chapter 1 examines the theoretical assumptions of early geologists as they began to describe the rock record around the world. Many of them saw this record as instantly created in its present form or as a product of Noah's flood. In Chapter 2, we look at the descriptive aspects of the outcrops from a modern perspective.

Part II deals with the interpretation of these outcrops in terms of their mode of deposition. In Chapters 3 through 6, we examine the current framework of interpretation employed by modern geologists: depositional environments.

Part III examines the rock record on the next level above the environmental interpretation of local outcrops. The first step is the process of correlating and interpreting rock bodies on a regional scale (*lithostratigraphy*). Fossils are the best time markers in most layered rocks, and this use is discussed in Chapter 10. Other methods of correlation, more geophysical or geochemical, are covered in Chapter 11. In Chapter 12, we examine radiometric dating, the only widely applicable method of obtaining the numerical age of rocks. Chapter 13 demonstrates how all of the various techniques are tightly integrated to achieve the most reliable interpretation. In the last chapter we examine the tectonic processes that control the accumulation of rock sequences. We try to answer the question, "What are the forces that determined *how* these rocks accumulated *here when* they did?"

Why stratigraphy? Whatever the geologic problem—geologic history; reconstruction of ancient plate positions or environments; determining the age of rocks or geologic events; finding oil, gas, coal, and uranium; and deciphering the clues to the causes of mass extinction— stratigraphy is the key.

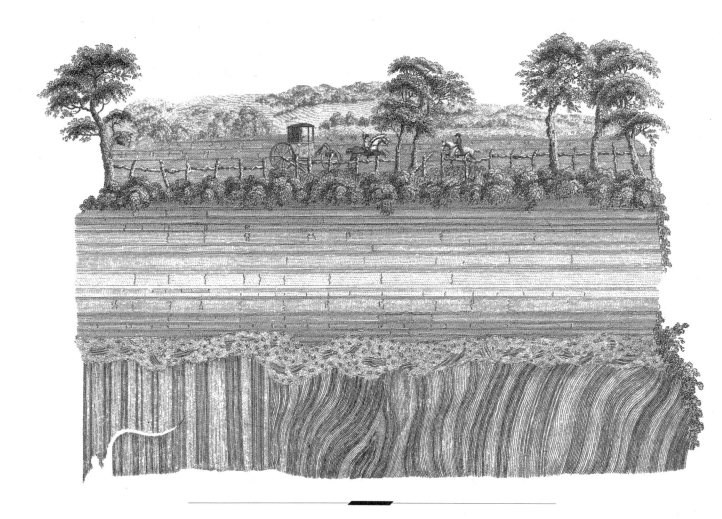

The unconformity at Jedburgh, Scotland, showing near-vertical beds of
Silurian sandstone and shale overlain by flat-lying Devonian sandstones.
Between the two layers is angular detritus composed of fragments of the
older unit. This is the same unconformity as the one exposed at Siccar
Point. From Hutton (1795).

1

The Concept of Geologic Time

DEVELOPMENT OF THE RELATIVE TIME SCALE

Ideas about the Earth

From our perspective in the late twentieth century, it is hard for us to understand the viewpoints of cultures many centuries in the past. This is particularly striking when we examine ancient concepts of the age of the earth. Certain non-Western cultures (notably those of China and India) thought of the earth as eternal and unchanging or as changing cyclically. The ancient Greeks viewed the earth as continuously changing, with the changes governed by natural laws. Heraclitus, in the sixth century BC, taught that everything on the face of the earth is continuously changing. This naturalistic view of the earth was carried on by the Romans, especially by the philosopher-poet Lucretius (98–55 BC) and by Pliny the Elder (AD 23–79), who lost his life making scientific observations of the eruption of Vesuvius as it destroyed Pompeii.

After the fall of Rome, Western culture was dominated by the biblical idea of an earth no more than a few thousand years old. According to the book of Genesis,

mankind had been a part of earth history from the beginning and was the purpose of the earth's existence. Before the nineteenth century, people viewed the changes on the earth's surface (such as erosion) as evidence that the earth was rapidly decaying from its pristine primordial state due to the Fall of Man. The widespread assumption that rocks were very young and were created in their present form delayed for a long time any questions about *how* the rocks were formed.

Early miners were certainly aware of the nature of the materials on which they based their livelihoods. They recognized rock units of distinctive character, gave them names ("chalk," "stinking vein," "peacock vein," "clunch"), and occasionally mapped their distribution and speculated about their origin. Speculation was usually restricted to the mineral being mined and seldom suggested that the host rock around it was anything but permanent and unchanging.

This static view of the earth was first seriously challenged when people began to speculate about the origin of the fossils of seashells and other sea creatures found high in the mountains and far from the sea. From our perspective, it seems obvious that if an object looks like a shell, it must be a shell formed by organic processes. As early as the sixth century BC, the Greek philosopher

Xenophanes of Colophon saw shells high on the cliffs of the island of Malta and suggested that the land had periodically been covered by the sea. The Greek historian Herodotus (fifth century BC) also described shells found far inland and suggested that the position of the shoreline had changed. By the Middle Ages, this naturalistic view of the earth had been rejected. The mysterious objects found embedded in the rocks were called *fossils*, from the Latin word meaning "dug up." Such fossils included not only the objects of organic origin that we would call fossils today but other features found in rock such as crystals and concretions. The occurrence of these objects in permanent, solid rock was not easily explained. Many scholars thought that they had been produced by mysterious "plastic forces" percolating through the rock or were "pranks of nature" or works of the Devil placed there to confuse believers. Others, struck by the similarity of some fossils to living shells, suggested that the fossils had been left high and dry by Noah's flood. As long as the rocks were considered permanent and unchanging, the presence of fossils in them had to remain a mystery.

During the Renaissance, a number of scholars began to make observations that did not so literally adhere to the Scripture. Among the many brilliant accomplishments of Leonardo da Vinci (1452–1519) was a surprisingly modern explanation of how the seashells were found in the higher elevations of the Apennine Mountains. Da Vinci thought that the fossil seashells were the remains of organisms that had been buried by river muds carried down from the Alps. The river muds and the enclosed shells had both petrified and later uplifted to their present position in the mountains. But da Vinci was dealing with relatively recent fossils whose similarity to modern species was not difficult to perceive. Other fossils looked nothing like extant organisms and were less obviously the remains of once-living creatures.

Steno's Laws

In the mid-1600s, a young Dane by the name of Niels Steensen (1638–1686) came to the court of the Grand Duke of Tuscany to serve as court physician. Today we know him by the latinized version of his name, Nicolaus Steno. He traveled widely in the hills of Tuscany, observing the tilted, fossiliferous beds in the vicinity of Florence. In 1666, he had the opportunity as court doctor and anatomist to dissect the head of an unusually large shark that had been recently caught near Livorno. A close look at the teeth in the mouth of the shark convinced him that they were the same as the *glossopetrae* (tonguestones) so often found as fossils (Fig 1.1). Tradition held that the triangular tonguestones were the petrified tongues of snakes or dragons, but Steno could not ignore their similarity to the teeth of modern sharks.

Figure 1.1

Steno's drawing of the head of a shark, showing that the teeth are identical to the controversial tonguestones (Steno, 1669).

In 1669, Steno summarized his ideas in his revolutionary work, *De solido intra solidum naturaliter contento dissertationis prodromus* (Forerunner of a dissertation on a solid naturally contained within a solid). The title seems bizarre for a work on geology, but it reveals the central problem that Steno had to address. Because his contemporaries regarded rocks as permanent, the presence of the fossils in them was a mystery. Steno was the first to suggest in print that the enclosing rocks had not always been solid but had once been soft sediments that had subsequently hardened around the shell or shark's tooth. With the realization that rocks had not always been solid, Steno saw the tilted rocks in Tuscany in an entirely new light. Every layer had once been soft sediments deposited on a solid base. From this idea Steno derived three famous laws:

Original horizontality Sediments on a solid base must have been deposited horizontally or they would have slid to a lower point. Rocks that lie at an angle must have been tilted after the sediments were petrified.

Original continuity Sedimentary layers normally form continuous sheets that either covered the entire earth or were bounded by solid substances. Discontinuous layers that are similar on either side of a valley must have been separated by erosion from their original continuous state.

Superposition Each layer must have been deposited onto a solid layer, and therefore each layer is necessarily younger than the layer under it. In any succession of rock layers, then, the lower rocks are the oldest and the layers upward (superposed) are progressively younger.

From these basic principles, a geologist can work out the succession of strata in any region and determine the *relative ages* of beds. Steno's work clearly implies that no single noachian deluge could have deposited all the earth's layers and that earth history is longer and more complex than Genesis indicates. These implications remained unexplored for over a century. Steno converted to Roman Catholicism as the *Prodromus* was being published and shortly thereafter returned to Denmark to serve the church for the rest of his life.

Early Time Scales

In the early 1700s geologists and miners began to use the principle of superposition in their descriptions of the various visible rock layers in the mountains of Europe. They noticed that the rocks at the cores of the higher mountains tended to be the oldest and were nearly always crystalline. The younger rocks were mostly det-

rital, covering the older rocks and forming the lower mountains. Johann Gottlob Lehmann (1700–1767), a professor of mining and mineralogy at the mining academy in Berlin, formalized these rock divisions in 1756. According to Lehmann, the high mountains such as the Alps are composed of crystalline rocks and tilted strata. They contain no organic remains but do contain minerals, and so he called them the *Ganggebirge*. Lying on the Ganggebirge and forming lower mountains is a second class of rocks, which Lehmann called the *Flötzgebirge* because he thought their layers had been formed in the noachian flood. These layered rocks contain abundant fossils, but Lehmann thought they had hardened since Noah's time. A third class of rocks, mostly deposits of landslides, earthquakes, and volcanic eruptions, he thought had formed since the deluge.

One of Lehmann's colleagues, George Christian Füchsel (1722–1773), worked out the detailed sequence of the rocks of southern Germany in 1762. He divided the sequence into nine units, from the tilted basal rocks through coal bearing sequences to what we now recognize as the Permo-Triassic succession. The study of the rock record was continuing to indicate that earth history is more complicated than the Scripture portrays.

In Italy (Fig. 1.2), Giovanni Arduino (1713–1795) described the rocks of Tuscany, which had been studied

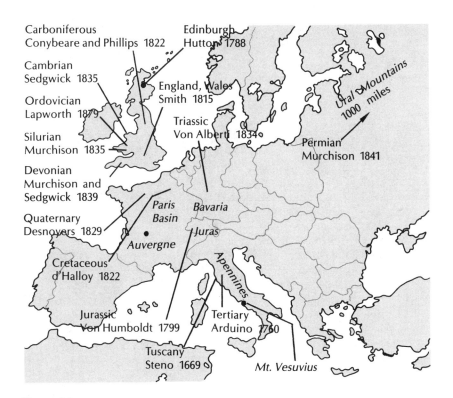

Figure 1.2

Locations of some of the important geographical features that were the basis for the construction of the geologic column. Adapted from Eicher (1976).

by Steno a century earlier. Because he was an influential professor of mineralogy at Padua, Arduino's divisions of the Apennine succession were accepted and used by most of his colleagues in Italy. In 1760, he recognized three basic divisions of rock: the *Primitive* schists, granites, and basalts that formed the cores of the high mountains; the fossiliferous, layered *Secondary* rocks (mostly limestones and shales); and a younger group of very fossiliferous sedimentary rocks, the *Tertiary* rocks found in the low hills. Arduino's threefold succession was intended as a descriptive generalization, but many later authors equated these categories with rocks formed before, during, and after the noachian deluge. Of the terms so far in this section, only Arduino's term *Tertiary* survives in the modern geologic time scale, though with a different meaning from Arduino's.

Werner and Neptunism

Perhaps the most influential figure of late eighteenth-century geology was Abraham Gottlob Werner (1750–1817). Descendant of a family that was long connected with the iron industry, Werner taught mineralogy at the Freiberg Mining Academy, which he made the center of geology in Europe. He was a charismatic figure who could spellbind an audience. Werner's disciples eagerly spread his teachings throughout Europe, and made his ideas very influential.

Werner was the most famous protagonist of *Neptunism*, the prevailing school of geological thought in Europe in the late eighteenth century (this school takes its name from the Roman god of the ocean). According to Neptunists, most rocks (including the crystalline rocks) were precipitated out of the noachian floodwaters. Contrary to popular myth, however, Werner did not explicitly tie his Neptunism to the noachian deluge. He followed Lehmann in calling the granites, schists, gneisses, slates, and basalts the Primitive rocks (*Urgebirge*), but he believed that they were the first chemical precipitates during the formation of the earth (Table 1.1). Above these were the Transition rocks (*Übergangsgebirge*), which were layered and contained a few fossils. According to Werner, these were mostly chemically precipitated but included some materials deposited by receding seas. Above the Transition series were rocks that were more stratified and fossiliferous, which were called the float beds (*Flötz-Schichten*). The *Flötz* rocks were mostly layered sandstone, limestone, shale, and coal, deposited by the last of the retreating seas. Finally, Werner recognized a fourth class of rocks, the Alluvial rocks (*Aufgeschwemmte-Gebirge*) which were loose materials deposited by running water since the retreat of the seas. Volcanic ash and cinders were also included in the Alluvial rocks.

Although Werner's fourfold division of the rocks was popular and widely used, it had its limitations. When the rock types occurred out of his sequence, Werner had to invoke additional inundations of the sea in local areas to rescue his theories. Werner was too busy in Freiburg to do much fieldwork in the later part of his life, but he encouraged his students to conduct their

---

Table 1.1

Wernerian Scheme of the Earth's Rocks

Series	Description
1. Primitive (*Urgebirge*)	Includes what are now recognized as intrusive igneous rocks and high-rank metasediments. These were considered to be the first chemical precipitates derived from the ocean before emergency of land areas.
2. Transition (*Übergangsgebirge*)	Composed of thoroughly indurated limestones, dikes and sills, and thick graywackes. These were considered to be the first orderly deposits formed from the ocean. With the Primitive rocks, they were though to be "universal" in extent, extending without interruption around the world.
3. Stratified (*Flötz*)	More commonly termed Secondary. Here were included the majority of obviously stratified fossiliferous rocks, plus certain associated "trap rocks." The "Flötzgebirge" were thought to represent the emergence of mountains from beneath the receding ocean, with products of the resulting erosion deposited on the flanks of the mountains.
4. Alluvial (*Aufgeschwemmte*)	Called "Tertiary" by some contemporaries. Poorly consolidated sands, gravels, and clays formed after the withdrawal of the ocean from the continents, and some older volcanic rocks.
5. Volcanic	Added more or less an an afterthought to include younger lava flows demonstrably associated with volcanic vents. Werner placed slight importance on these rocks and considered them to be merely the local effects of burning coal beds.

own field investigations. In particular, Werner's emphasis on methodical observations and on avoiding metaphysical speculation was a very positive influence on geology. Eventually, the work of his students forced Werner to change some of his ideas, though Neptunism in a highly modified form was influential in geology well into the nineteenth century.

One of the most controversial aspects of Werner's theories was the idea that all rocks, including those we now call igneous and metamorphic, were precipitated from water. Arduino and many other early geologists had long recognized that basalts erupted from volcanoes, but the Neptunists insisted that basalts and granites were chemical precipitates from water. This is understandable because the basalts with which they were familiar tend to be interlayered with sediments and show no obvious evidence of ever having been hot. They are extremely fine-grained and were difficult to study in the days before the petrographic microscope. Also, chemistry was in its infancy, so the idea that the crystals of granitic rock could precipitate out of water was just as plausible as crystallization out of a hot melt.

Finally, the French geologist Nicolas Desmarest (1725–1815) challenged the Neptunists to look at the volcanoes of Auvergne in France and satisfy themselves that basalts are of volcanic origin. Werner had never seen a young volcano in Saxony, but his student, the eminent geologist Leopold von Buch (1774–1853), did take up Desmarest's challenge and was eventually convinced of the igneous origin of basalt. The group who emphasized the importance of the earth's heat in the formation of crystalline rocks were known as the *Plutonists* (after the Roman name for the god of the underworld) because they advocated a fiery, volcanic origin for basalts and other igneous rocks.

Hutton, Lyell, and Uniformitarianism

During the battle on the continent between the Neptunists and the Plutonists, another development was occurring in Scotland. Among the gentlemen who met at the Oyster Club and the Royal Society of Edinburgh were some prominent thinkers: Joseph Black, the chemist who discovered carbon dioxide: John Clerk, a naval tactician; John Playfair, the mathematician; James Watt, inventor of the steam engine; Adam Smith, the economist who first analyzed capitalism; and the pioneering geologist James Hutton (1726–1797). Hutton was a gentleman farmer who spent much of his time hiking among the crags and lochs of Scotland. When he met with his colleagues, he tried out his developing ideas on them. Hutton was convinced that basalts and granites were of volcanic origin, but his thinking went beyond that of the Plutonists of the continent.

Hutton and his Scottish friends were Deists, less restricted by the church's teachings about Genesis. Hutton was strongly influenced by the scientific advances of his predecessors (such as Isaac Newton) and his contemporaries, who valued natural law, rather than supernatural intervention as God's preferred method of dealing with His Creation, and he began to see the earth as a product of natural laws. He resurrected the naturalism of the Greek and Roman philosophers, which had been forgotten or ignored by centuries of dogma. Hutton not only observed rocks that were formed by volcanic action, but also was a keen observer of how sediments formed sedimentary rocks. He began to see all rocks as products of continuing earth processes rather than products of a single supernatural creation or noachian deluge. This naturalistic idea is usually called *uniformitarianism*, because it assumes a uniformity of modern and ancient processes. Uniformitarianism has frequently been summarized by Sir Archibald Geikie's phrase: "The present is the key to the past." In contrast, the traditional viewpoint was called *catastrophism* because its proponents resorted to supernatural explanations such as a catastrophic deluge to explain the rock record. The naturalistic, uniformitarian assumption is critical to all historical geology and, indeed, to most science. Scientists must assume that natural laws continue to operate for phenomena that cannot be directly observed or they cannot theorize about them. Without uniformity of natural laws, no past event could be studied, nor could there be any understanding of the stars or planets.

Most of Hutton's predecessors and contemporaries (especially after Isaac Newton) believed that natural law had operated in the past, but the political influence of Christian dogma forced them to give supernatural processes a role as well. Although Hutton clearly did not invent uniformitarianism or even discuss it first, he followed it to conclusions that contradicted the prevailing interpretation of the Scripture. Hutton made careful observations of earth processes and noticed how weathering, erosion, and sedimentation eventually form soft sediments, which, he deduced, must be buried and transformed into rock by high temperatures and great pressures. Eventually, these rocks are uplifted, weathered, and eroded to form new sediments and new rocks, in a continuous cycle. Because there is no clear beginning or end to this slow, cyclic process, the earth must be immensely old. As Hutton (1788, p. 304) put it, "The result, therefore, of our present enquiry is that we find no vestige of a beginning, no prospect of an end."

Hutton was especially impressed by the implications of *angular unconformity*, such as the one at Siccar Point near Edinburgh (Fig. 1.3). Such a structure clearly was not formed during a single flood but required considerable time to reach its present configuration. The underlying sequence must have been deposited, consolidated,

Figure 1.3

Siccar Point, on the North Sea coast of Scotland. Late in his career, Hutton recognized the angular unconformity between the deformed Silurian slates and graywackes, and the overlying Devonian Old Red Sandstone. When John Playfair visited this site in 1805, he said, "The mind seemed to grow giddy by looking so far into the abyss of time." Photo from Institute of Geological Sciences; British Crown copyright.

uplifted and tilted, and then eroded over a long period of time. The amount of time was not recorded by the rocks, but was represented by their surfaces, which had been eroded after the rocks had been tilted and had been buried again under sediments. Then the overlying sediments had been lithified, uplifted, and again eroded into their present shape. Because there were many such angular unconformities throughout the rock sequence, Hutton realized that the earth must be much older than the 6000 years allowed by the prevailing interpretation of the Scripture. Earth history was much more complex than anything described in Genesis. Hutton was one of the first people to grasp the immensity of geologic time.

Hutton's *Theory of the Earth* was first published in 1788 and came out in two volumes in 1795. His ideas did not find immediate acceptance, partly because his prose was difficult to read but mostly because his ideas clearly contradicted Scripture. After Hutton died, his friend John Playfair presented in 1802 a clear summary of his work as *Illustrations of the Huttonion Theory of the Earth*. The geologic community found much of value in Hutton's

ideas but were still not ready to abandon their concept of the Genesis story.

Almost a generation later, Hutton's theories were finally made inescapable by the greatest geologist of the nineteenth century, Sir Charles Lyell (1797–1875). Lyell had been trained as a lawyer and enjoyed geology as a hobby, but the hobby very quickly pushed aside his legal practice. He began serious study of the rocks of Great Britain and the continent, and, between 1830 and 1833, published his three-volume *Principles of Geology*. In this work, Lyell argued Hutton's case very forcefully, using many more examples than Hutton. Indeed, his argument reads very much like an extended legal brief. He took an extreme position on uniformitarianism to combat the catastrophists. Lyell argued not only for uniformity of natural laws and processes (*actualism*), but also for uniformity of rates (*gradualism*). Lyell asserted that earth history was cyclical, with every event repeating sooner or later. However, the fossil record was just beginning to appear directional and nonrepeating, with complex life forms appearing after less complex ones.

Figure 1.4

Satirical cartoon, drawn by Henry de la Beche, one of Lyell's colleagues, poking fun at his extreme uniformitarian beliefs. Lyell believed that life had not changed significantly through time and that extinct animals such as ichthyosaurs and pterodactyls would one day reappear. Here Professor Ichthyosaurus lectures to students about the skull of a strange creature from the last creation.

In response to this, Lyell pointed out that mammals had just been found in the Mesozoic as well as the Cenozoic and predicted that they would eventually be found in the Paleozoic. Lyell felt that animals that were clearly extinct, such as iguanodonts, ichthyosaurs, and pterodactyls would return in the next cycle (Fig. 1.4). Eventually, Lyell came to realize that the fossil record was indeed directional and nonrepeating, but he never abandoned gradualism.

By mixing uniformity of processes with uniformity of rates, Lyell managed to confuse geologists for over a century. Lyell was battling with people who believed that supernatural catastrophes explained the geologic record. Because supernatural explanations are untestable and unscientific, Lyell rejected all catastrophic ideas about the earth. The gradualistic bias was so strong that geologists denied clear evidence for local natural catastrophes in geologic history. Certainly, meteorite impacts or major hurricanes do not fit into the gradualistic scenario of slow, steady, cumulative change. Yet it appears that these and many other rapid, unique events have had a major effect on the rock record.

William Smith and Faunal Succession

Despite Werner's ideas, geologists were beginning to realize that rock sequences differed regionally and that no rock type was a universal time marker. What *could* be used to tell geologic time? Ironically, the solution to this problem was worked out not by the grand metaphysical schemes of the gentlemen geologists of the time, but by a working man who spent his spare time observing the outcrops and collecting fossils.

His name was William Smith (1769–1839), and his limited schooling had prepared him for a career as an engineer and surveyor. The early phases of the industrial revolution had led to major canal-building projects in the late 1700s. Smith worked on the Somerset Coal Canal in 1793, which gave him the opportunity to view the fresh canal excavations being made across England and to collect fossils from them. Fresh outcrops are hard to find in most parts of southern England, so Smith was one of the few geologists to see a fresh slice through the stratigraphic sequence, which was then poorly known (Fig. 1.5 and Table 1.2). He made detailed notes of the rocks he encountered and the fossils they contained. Soon he discovered that similar-looking limestones could be distinguished by their fossils. Indeed, each rock unit had its own distinct fossil assemblage, and these assemblages could be recognized wherever he went in England. Smith had discovered what we now call the *principle of faunal succession*. This was sixty years before Darwin came up with an evolutionary explanation for *why* faunas change through time.

In 1799, several of Smith's geological friends accompanied him into the field and were impressed by his discovery. Smith knew the sequence of fossils so well that he was able to arrange their fossil collections in the order of the strata from which they had come. Smith's lists of fossils were widely circulated, but he was too busy to be concerned with publication and did not publish his 1799 sequence until 1813. He was working on a grander project: mapping all of the rock units he could recognize by their fossils. Finally, the *Geological Map of England and Wales* was published in 1815. It is considered by many to be the first modern geological map ever produced. The map is so accurate and meticulously researched that it can still be used (Fig. 1.6).

Smith might be considered the forerunner of the modern professional geologist because his training was specialized and because he earned his living by doing geologic work for the industry of the time. However, in his time, geology was still the domain of wealthy gentlemen, who not only were class conscious but pursued geology as a hobby, not as a "vulgar" way of making a living. Smith was ignored by the geological community for years, and other gentlemen used his work to publish their own maps and take credit for them. He was finally recognized in 1831 by the Geological Society of London as the "father of English geology." William Smith had provided geologists with the key they needed to unlock geologic history: faunal succession.

West   **East**
                                                                                                                                          18

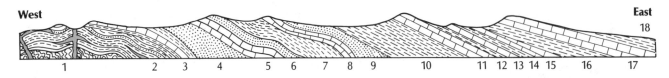

1   2   3   4   5 6   7 8 9   10   11 12 13 14 15   16   17

Figure 1.5

Simplified east–west geologic cross section from Wales to London, showing the units first recognized by Smith, Murchison, Sedgwick, and their peers. The numbers correspond to the units listed in Table 1.2. From Weller (1960).

Cuvier and Brongniart

With so many geologists working along similar lines in the late eighteenth century, it is not surprising that several people observed the same phenomena in several countries. The Abbé Jean Louis Giraud-Soulavie (1752–1813) described the sequence of fossils in the Vivarais district of France in 1779 and recognized that every rock unit contained distinctive fossils. His discovery was developed by two very prominent French scientists, the Baron Georges Cuvier (1769–1832) and Alexandre Brongniart (1770–1847). They worked out the succession of strata in the Paris Basin, with Brongniart doing most of the fieldwork. Their work was published between 1808 and 1812. They found that Werner's division of the strata did not work in the Paris Basin. Strata that contain fossils of the Transition are not hard, steeply dipping limestones as they are in the Alps but are soft horizontal chalk beds. Historians of geology have long

thought that Cuvier and Brongniart worked out their ideas of faunal succession independently of Smith. However, Eyles (1985) makes a plausible case that Brongniart may have heard Smith in London in 1802 during one of the rare episodes of peace in the midst of the Napoleonic Wars. Whether or not Brongniart developed faunal succession independently, it is clear that Cuvier and Brongniart (1808) were more concerned with documenting the "revolutions" of life on earth (Hancock, 1977). Smith was clearly the originator of the idea that faunal succession can be used to correlate localities in different places with one another, the basis of modern biostratigraphy.

Cuvier did not stop at describing the sequence of fossils in the Paris Basin. He was the most prominent man in French science, so he had good political reasons to try to explain faunal succession. Trained in geology in Germany, he was a master of politics. So great was his scientific reputation and political savvy that he managed to survive the French Revolution, Napoleon,

Table 1.2

Classic Stratigraphic Sequence of England and Wales

Unit [a]	Description	Unit [a]	Description
1	Precambrian schists	9	Triassic New Red Sandstone
2	Cambrian and Ordovician shales, limestones, and contrusive and extrusive rocks	10	Lias, impure shale and limestones that form a lowland areas (name is a corruption of *layers*)
3	Silurian shales and limestones	11	The Oolite, an upland-forming limestone
		12	Oxford Clay
	Units 2 and 3 were originally called "Transition" until broken up by Murchison and Sedgwick. Unit 1 was the "Primary" of many authors.	13	Corallian Limestone
		14	Kimmeridge Clay
		15	Portland and Purbeck Beds, mostly limestone
4	Devonian Old Red Sandstone, reddish and brownish fresh-water strata that are better developed in Scotland than in England		Units 13, 14, and 15 form another lowland zone. Units 10 through 15 are Jurassic and are the strata principally studied by William Smith.
5	Carboniferous Limestone, formerly known as Mountain Limestone	16	Gault and Greensands
6	Millstone Grit	17	The Chalk
7	Coal Measures	18	Tertiary London Clay Group
8	Permian Magnesian Limestone, grades westard into sandstone that was confused with division 9		Units 16 and 17 are Cretaceous.

Units 6 and 7 constitute the Upper Carboniferous.

[a] This succession is not complete according to modern standards, and most of these old formations are now subdivided.
Source: Weller, 1960, p. 34.

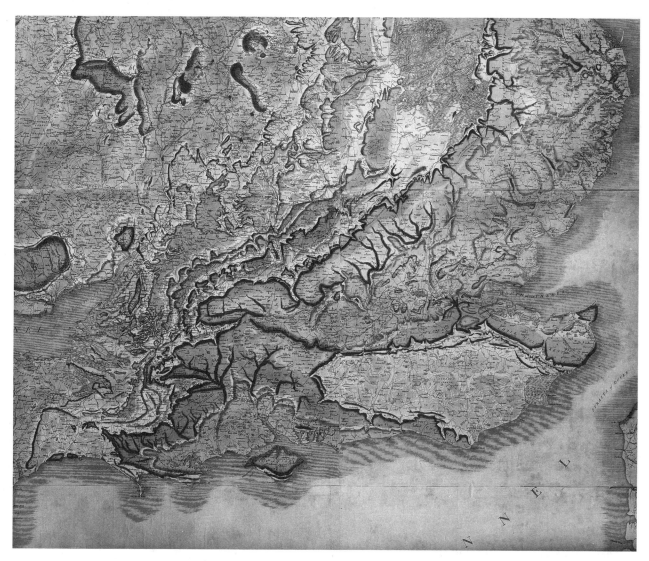

Figure 1.6

Part of William Smith's 1815 geologic map of southeast England. The map clearly shows the V-shaped eastward opening of the London Basin, the syncline drained by the Thames. The Wealden anticline is just to the south of London.

and three subsequent French kings without loss of power of prestige. Cuvier is best known as the founder of comparative anatomy and vertebrate paleontology; he was legendary for his ability to reconstruct extinct creatures from a few fragmentary bones. Because of his prominence, he had to reconcile the increasingly complex fossil record with the Genesis account of a single flood. His solution was to attribute all of the layers with extinct fossil forms to a dark, murky *antediluvian* (Latin, before the flood) period, which occurred before the creation and flood described in Genesis. During antediluvian times, he thought, there were many floods, each of which destroyed all extant life and was followed by new creations and new life forms. Cuvier explained each layer with its distinctive fossils as the flood deposits of

each successive creation. His solution was popular with the catastrophists of the time. In fact, the term *geologic time* was originally coined to describe this antediluvian period, before *historic time* as recorded in the Bible. The antediluvian period was a time of supernatural happenings and so was beyond the reach of natural law. Cuvier's compromise was ultimately abandoned because of Lyell, who doggedly argued that natural law must be extended back into the prehistoric past. To Lyell, any concession to supernaturalism was a retreat from science altogether. Two lasting contributions have survived Cuvier: the proof that life has changed progressively through time and the proof that extinction was real. The latter idea was repugnant to the theologians' concept of a caring, all-providing deity.

The Naming of the Periods and Systems

Through the early nineteenth century, many geologists were refining the knowledge of their local stratigraphic successions. Although they began by describing their rocks in Wernerian terms, they found that the old divisions of Primitive, Transition, *Flötz*, and Alluvial did not accurately describe their sections. So, independently, they coined their own terms for distinctive rock successions that had not been previously recognized. This explains the rather haphazard development of the terms used in our modern time scale (Fig. 1.7).

Jurassic rocks full of ammonites are widely exposed throughout Europe, especially in the Alps, where they have been studied by many pioneers of geology (see Fig. 1.2). The traveler and scientist, Alexandre von Humboldt (1769–1859), described the massive limestones of the Jura Mountains in Switzerland as the "Jura-Kalkstein" during a geological excursion through France, Switzerland, and northern Italy in 1795. Later geologists began to notice that the fossils of the Jura Mountain sequence are similar to the Lias and Oolite sequences described by Smith in England (see Table 1.2) and suggested that they were equivalent. In 1839, Leopold von Buch applied von Humboldt's term to the rocks of southern Germany, establishing the *Jurassic* as a system for rocks in Switzerland, Germany, and England.

The coal deposits of England had been known for a long time as the Coal Measures, and similar deposits occur in Europe. In 1808, J. J. d'Omalius d'Halloy had described the *Terrain Bituminifére* of the coal seams of Belgium. In their 1822 summary of the geology of England, William D. Conybeare (1787–1857) and William Phillips (1775–1828) coined the term *Carboniferous* (Latin *carbo*, coal) for the Coal Measures and three underlying units: the Millstone Grit, the Mountain Limestone, and the Old Red Sandstone. None of these units could be fit into Werner's Transition or *Flötz* series and therefore required a new name.

The chalk that makes up the White Cliffs of Dover can be traced into northern France, Belgium, Holland, Denmark, northern Germany, Poland, and Sweden. Smith recognized four units: two kinds of chalk, a greensand (glauconitic sandstone), and the Gault clay. While Conybeare and Phillips were summarizing the geology of England, d'Halloy was preparing a geologic map of France in 1822. He recognized a *Terrain Crétacé* (Latin *creta*, chalk) for the Secondary, or *Flötz*, rocks, which included not only the chalks but also the greensand and marl beneath them. In the same book in which they coined the term *Carboniferous*, Conybeare and Phillips modified the French term to *Cretaceous*.

The material above Arduino's Tertiary was generally loose and unconsolidated and was considered to be

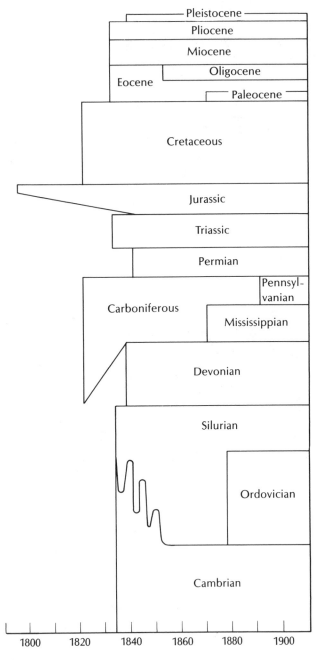

Figure 1.7

Diagram showing the dates of recognition (horizontal axis) and relative durations (vertical axis) of each of the standard divisions of the modern geologic time scale. Notice how the Paleocene and Oligocene were split off from the Eocene, the Devonian split off from the Carboniferous, and the dispute over the Cambrian-Silurian boundary finally resolved by creation of the Ordovician. Modified from Moore (1955).

deposits of recent earth processes. Werner had called these deposits Alluvium. Buckland in 1823 separated the Alluvium (deposits formed in the present) from *Diluvium* (deposits formed after the Tertiary by the noachian deluge). In 1829, Paul Desnoyers described all of the post-Tertiary deposits of the Seine Basin as *Quaternary*, in reference to Arduino's threefold terminology. Where later geologists characterized the faunal content of the Quaternary, they realized that it contained not only recent deposits but also deposits of the ice ages (first recognized as ice-deposited by Agassiz in 1837).

German geologists, such as Lehmann and Füchsel, divided the lowest part of the *Flötz* into three distinctive formations, the Bunter Sandstone, Muschelkalk Limestone, and Keuper Marls and Clays. In 1815, Friedrich August von Alberti began studying the salt deposits in southern Germany and carefully documenting the Bunter, Muschelkalk, and Keuper and their fossils. In 1834, he formally named these three units as the *Trias*, or *Triassic*, in reference to the threefold subdivision.

Sedgwick and Murchison

Most of the remaining divisions of the modern time scale were developed by two men, Adam Sedgwick (1785–1873) and Roderick Impey Murchison (1792–1871). Sedgwick was a clergyman's son who studied at Cambridge and became the professor of geology there in 1818. Geology was then an infant science, a division of "natural philosophy," so it took little time to learn all there was to know about the subject. After his appointment, Sedgwick studied geology intensively, made many field excursions, and was soon one of the leaders of the Geological Society. Many important scientists, including Charles Darwin and Roderick Murchison, were his proteges.

Murchison was a wealthy gentleman who spent his youth in the military. At the end of the Napoleonic wars, he retired and took up the gentlemanly hobby of fox hunting. At the age of thirty-two, he became interested in geology and sold his horses and hounds to learn about rocks in the field with Sedgwick, William Buckland (1784–1856), and Charles Lyell. There he would travel by carriage with his wife and maid, and frequently a guest. His wife would collect fossils and make sketches, while the men would clamber around the hills on foot. In 1827 and 1828, Murchison took Lyell with him to study geology across Europe. From this experience Lyell drew much of the evidence he marshalled in *Principles of Geology*.

In 1831, Murchison and Sedgwick decided to work together on the poorly known, structurally complex "old graywackes" beneath the Old Red Sandstone, which

were considered equivalent to Werner's Transition beds. These rocks were exposed in Wales, a region with numerous folds, faults, and unconformities. Sedgwick worked upward from the base of the section in northern Wales, piecing together a sequence based on lithologic similarities and deciphering the complex structures. He paid little attention to the fossils because they were scarce and poorly preserved. Murchison worked downward from the Old Red Sandstone in southeast Wales, carefully collecting fossils and determining the sequence of faunas with help from local collectors. Murchison and Sedgwick met in the middle of the section in 1834, and the next year they jointly presented their results. Murchison named his rocks the *Silurian System* (after the Roman name for an ancient Welsh tribe, the Silures) and described the fossils found in each of four formations. Sedgwick described the *Cambrian System* (after the Roman name for Wales) but had difficulty presenting a sequence of formations and generally ignored the fossils. Together, their work eliminated the need for the obsolete Wernerian term *Transition*.

Their different criteria for recognition of the systems soon led to problems. Murchison and others traced fossiliferous "Silurian" rocks into Sedgwick's Cambrian, so that eventually any fossiliferous rocks found in Sedgwick's area became part of the expanding "Silurian." Sedgwick still did not attempt to describe a Cambrian fauna. Instead, he concentrated on deciphering the local structural complexities. As a result, Murchison was able to recognize his "Silurian" in Europe by the similarity of fossils, whereas Sedgwick's "Cambrian" could not be recognized outside Wales. Murchison continued to rise in prestige, becoming the Director of the Geological Survey of Great Britain in 1855. As he did so, the "Silurian" expanded until it had swallowed up much of the Cambrian and was nearly equivalent to the original Transition (see Fig. 1.7). The friendship of Murchison and Sedgwick turned into a bitter feud that was carried on by their students after they both died in the early 1870s. It was resolved in 1879 by Charles Lapworth (1842–1920), who worked out the sequence of trilobite and graptolite faunas in Scotland and then in Wales. His scheme was based on the description of Czechoslovakian fossils by Joachim Barrande. Lapworth proposed a compromise unit, the *Ordovician* (after the Ordovices, the Roman name of another Welsh tribe), for Sedgwick's "Upper Cambrian" and Murchison's "Lower Silurian." He showed that all three had distinctive faunas but that the graptolites of the "Upper Cambrian" and "Lower Silurian" were very similar and distinctive from faunas above or below them. Although the term was not adopted immediately (the U.S. Geological Survey did not recognize it until 1903), the Ordovician was identifiable all over the world because it was based on a distinctive faunal aggregate. Ironically, the temporal dominance of Silurian over Cambrian has been reversed. When numerical dates became available, the Silurian turned out to be only about 35 million years in length, but the Cambrian and Ordovician are each about twice as long (see Fig. 1.7).

Before the Silurian-Cambrian feud destroyed their friendship, Sedgwick and Murchison collaborated on other stratigraphic problems. In the late 1830s, they worked on strata from Devonshire that looked like the Cambrian slates and graywackes of Wales but bore plant fossils that looked like those found in the Coal Measures. After a big controversy in the geological community, Sedgwick and Murchison realized that these beds were equivalent in age to the Old Red Sandstone but different in lithology. (The concept of facies change was still in its infancy.) In 1839, they named these rocks the *Devonian System*. They showed the validity of this new system by describing similar rocks and fossils that occurred in Germany. The Old Red Sandstone had originally been the lowest lithologic subdivision of Conybeare and Phillips' "Carboniferous", but on faunal evidence Sedgwick and Murchison showed that the Devonian was a distinct system. Once again, Smith's principle of faunal succession had provided a solution where old lithologic names and confusing facies changes had caused problems.

In 1840 and 1841, Murchison and the French paleontologist Edouard de Verneuil went to study the thick sedimentary sequences of Russia, accompanied by Russia's best geologists. Murchison not only observed rocks and fossils similar to those he had seen in Europe and Great Britain but also confirmed the value of faunal succession and the systems that had been named. Near the town of Perm in the Ural Mountains, he described a sequence of rocks above the Carboniferous which he called the *Permian*. These rocks contained fossils similar to the German *Zechstein*, the lower New Red Sandstone, and the British Magnesian Limestone. These fossils appeared to be intermediate between those found in the Coal Measures and those from the Triassic of Germany. Murchison spent three more years preparing a detailed monograph of the geology of Russia west of the Ural Mountains.

Lyell's Percentages and the Tertiary

Most of the divisions proposed so far were based on distinctive rock units, or occasionally on a rock unit and its contained fossils. During his travels in Italy in 1828, Lyell had noticed that the thick sequences of Tertiary rocks could also be subdivided by fossils. Rather than propose rock sections characterized by a distinctive suite

of fossils, he came up with a different approach. It was apparent that the fossil molluscs became progressively more modern in appearance in younger Tertiary beds. In the third volume of *Principles of Geology* (1833), Lyell formally proposed a fourfold division of the Tertiary based on the percentage of living species found at each level. The molluscan data on which he based this concept were gathered and identified by the French conchologist Paul Gérard Deshayes (1797–1875). After studying about 8000 species based on over 40,000 specimens, Deshayes had seen a similar pattern and come to similar conclusions. Lyell's four "periods" (now called epochs) of the Tertiary were as follows:

Lyell's "period"	Living species (%)
Newer Pliocene	90
Older Pliocene	33–50
Miocene	17
Eocene	3.5

Rudwick (1978) has shown that Lyell's intent was not to generate a name for a sequence of rock units and their contained fossils but to use the extinction of molluscs as a paleontological "clock." Deshayes had divided the Tertiary into three discrete "epochs," each having a catastrophist connotation. Lyell assumed that the turnover of mollusc taxa had been continuous and uniform throughout the Tertiary, so a geologist could sample the molluscan fauna anywhere and tell fine increments of time on the molluscan clock by the percentage of living species (Fig. 1.8). Theoretically, this system has great advantages. If molluscan turnover is continuous, a geologist does not have to worry about unconformities and time missing from rock sections. Molluscan turnover would be analogous to radioactive decay, a continuous process ticking away, independent of the local vagaries of preservation of rock section.

Unfortunately, the problems outweigh the possible advantages. First, molluscan turnover is not a continuous, linear process like the ticking of a clock. Changes in Tertiary molluscan faunas were as episodic as any in the rock record, with intervals of rapid turnover during mass extinctions, and also intervals of stability. Second, Lyell and Deshayes' "species" are difficult to use today because subsequent taxonomic work has raised most of them to generic rank, split off many more new species, and lumped others together. The percentage approach would be difficult to apply, even if all the taxonomy were updated. Stanley et al. (1980) have calculated Lyellian percentages for many groups, including molluscs. In gastropods, for example, they find about 50 percent of the modern species in existence at the beginning

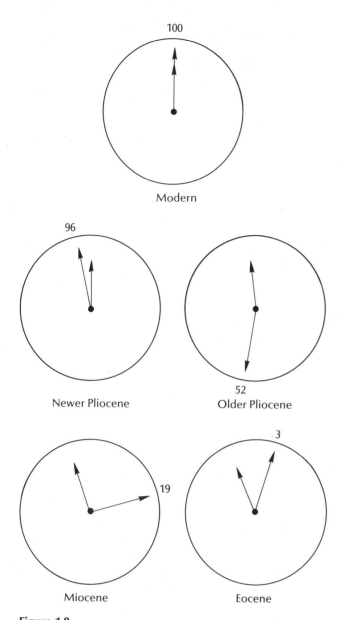

Figure 1.8

Diagram showing Lyell's original concept of the epochs of Tertiary time in terms of the faunal "clock." The numbers indicate the percent of living species. From Rudwick (1978).

of the Pliocene and less than 5 percent at the beginning of the Miocene.

Finally, the clock model does not mesh well with a system geared toward placing boundaries on discrete units based on real rocks and fossils in the field. A clock face has no real boundaries, only arbitrary subdivisions. In Lyell's concept, the Miocene was not a division of time from 5 to 23 Ma (Mega-annum, or 10^6 years before the present) but a discrete moment when approximately 17 percent of the molluscan species were of modern forms. Thus, there are no precise boundaries for his units.

This confused geologists who tried to apply traditional stratigraphic procedures to an essentially chronological concept. Lyell indicated that several areas were typical of each period, causing many subsequent quarrels among stratigraphers over which section should be "the type." Although Lyell's concepts originated in the Italian Tertiary section, Deshayes' collections were from the Paris Basin. Much of the Paris Basin fauna is restricted to that area, so the "type" fauna is difficult to recognize outside France. In recent years, the emphasis has shifted from molluscs to marine microfossils, which can be recognized worldwide and are much easier to collect and study. For practical reasons, very little of Lyell's original concept of dividing the Tertiary remains in use.

Lyell indicated that further subdivisions of Tertiary time were possible as smaller increments on the clock face. In 1839 Lyell restricted the meaning of Pliocene to his former "Older Pliocene" and coined the term *Pleistocene* for his former, "Newer Pliocene." In 1849, Edward Forbes equated Pleistocene with the ice ages, and that definition was accepted by Lyell in 1873.

In the Paris Basin, the "Upper Eocene" was poorly fossiliferous and had been labeled "Lower Miocene" by some stratigraphers. In 1854, Heinrich Ernst von Beyrich used *Oligocene* for a sequence of rocks in north Germany and Belgium that was more fossiliferous than the "Upper Eocene" of the Paris Basin. The fossils of von Beyrich's Oligocene were more advanced than those of the French Eocene, but not as modern as those of the Miocene, so this division was placed between the Eocene and Miocene of Lyell.

The base of Lyell's Eocene was split off by the paleobotanist W. P. Schimper in 1874. He recognized a series of floras in the Paris Basin that he felt were distinct from those of Lyell's Eocene and so invented the term *Paleocene* for them. Fossil plants are very difficult to recognize around the world, so the term "Paleocene" did not catch on until the invertebrates and mammals had also been studied and compared. The U.S. Geological Survey did not formally accept the "Paleocene" until 1939. As a result, many early twentieth-century publications used "Lower Eocene" for what we now call Paleocene.

The Mississippian and Pennsylvanian

After the Old Red Sandstone had been placed in the Devonian, European geologists recognized a Carboniferous consisting of a lower sequence (the Mountain Limestone and Millstone Grit) and the upper Coal Measures. Coal Measures were also recognized in North America, particularly in Pennsylvania. Below the coals was a sequence of limestones that were well exposed in the Mississippi Valley near St. Louis. These were called "Subcarboniferous" by David Dale Owen (1807–1860) in 1839. In 1870, Alexander Winchell (1824–1891) applied the term *Mississippian* to these rocks. He emphasized the fact that carbonates occurred under this coal, instead of the clastics found under the coal further east.

By 1890 the confusion over stratigraphic nomenclature reached such a point that the U.S. Geological Survey commissioned a series of correlation papers. In his 1891 report on the Carboniferous, Henry Shaler Williams recognized the Mississippian and correlated it with the Lower Carboniferous limestones of Europe. To complement this term, he proposed *Pennsylvanian* for the North American coal sequences best known from Pennsylvania. These terms were used by the most influential geology textbook in North America in the early twentieth century (Chamberlain and Salisbury, 1906) and so became accepted by most American geologists. The U.S. Geological Survey did not adopt the terms until 1953, so their older publications do not agree with the rest of the North American literature. However, detailed correlation has shown that the Mississippian-Pennsylvanian boundary is younger than the Lower-Upper Carboniferous boundary in Europe, so the terms are not equivalent.

European geologists do not use the American terms, but they are so familiar to most American geologists that few could be convinced to abandon them. Some have suggested that the Mississippian and Pennsylvanian be designated as subsystems of the Carboniferous (Moore et al., 1944). Harland et al. (1982) have made the Mississippian and Pennsylvanian international "subperiods" of the Carboniferous, which are in turn subdivided into ages based on the Russian zonation (Appendix B).

The Eras

The history of life was punctuated by several mass extinctions that dramatically changed the composition of the world's biotas. Two of the most severe were the Permo-Triassic event and the Cretaceous-Tertiary event. These events coincide with the boundaries that separate the three Phanerozoic eras, now called Paleozoic, Mesozoic, and Cenozoic. Interestingly, the names of the eras were coined before these great extinction events were recognized. In 1838, Sedgwick applied the term *Palaeozoic Series* to the less-metamorphosed rocks beneath the Old Red Sandstone. John Phillips used *Mesozoic Era* and *Kainozoic Era* as well as Sedgwick's Palaeozoic in an 1840 *Penny Cyclopaedia* article. The next year Phillips changed the spelling from *Kainozoic* (Greek *kainos*, recent) to the latinized *Cainozoic*. The Palaeozoic included all of Sedgwick's units plus the Carboniferous and the Mag-

nesian Limestone (later called Permian by Murchison). The Mesozoic included the Cretaceous, the Oolite (Jurassic), and the New Red Sandstone (part Triassic, part Permian). The Cainozoic included Lyell's Eocene, Miocene, and Pliocene "Tertiaries." Phillips based these three great divisions on the nature of the fossils found in them. Indeed, they have proven useful for indicating the major divisions of earth history.

Subdivisions of Precambrian Time

Eighty-seven percent of geologic time passed before an abundant shelly fossil record was preserved. This vast abyss of time with a poor fossil record was long ignored by most stratigraphers because little could be done with it. The rocks were labeled "Precambrian basement" and generally forgotten because stratigraphic subdivision and intercontinental correlation was impossible without a detailed fossil record. Terms such as "Huronian" and "Algonkian" were proposed for events recorded in the Canadian shield, but these terms are only applicable locally. Large-scale divisions, such as *Archean* and *Proterozoic*, have been recognized internationally and have been approved by the International Union of Geological Sciences (IUGS). In recent years, Precambrian geology has blossomed, and many more details have been deciphered.

The primary breakthrough in dating the Precambrian has been a chronology based on radiometric dates rather than on stratigraphy and fossils. Precambrian events are usually described by their radiometric age or in terms of an orogenic event, rather than as part of a relative stratigraphic sequence. Fossils have become better known in the Precambrian, but they remain too scarce through most of the interval to be of much use. Soviet geologists have developed a stratigraphy of stromatolites (Semikhatov, 1980; Bertrand-Sarfati and Walter, 1981), but these have not yet proven to be widely applicable. Stromatolites are apparently strongly controlled by environmental factors, and specimens from different continents are difficult to correlate. The study of Precambrian microfossils holds great promise but is still in its early stages (Vidal and Zoubek, 1981).

The latest Precambrian has proven to be abundantly fossiliferous as more well-preserved sections have come to light in recent years. The terms "Ediacaran," "Eocambrian," "Sinian," and "Vendian" have been proposed for this interval, and the last seems to be gaining favor (Sepkoski and Knoll, 1983; Harland et al., 1982; Cowie and Johnson, 1985). A decision by the IUGS Subcommission on Precambrian Stratigraphy is expected to resolve this dispute. Most geologists place the Cambrian at the base of the Paleozoic, but some authors have advocated including the Vendian in the Paleozoic. Because geologists frequently want to speak of the time before abundant fossils, the term "Prepaleozoic" would then replace Precambrian in importance (Dott and Batten, 1988).

RELATIVE VERSUS NUMERICAL AGE

The vast age of the earth was appreciated long before any numerical dates could be reliably calculated. Using Steno's laws, the relative time scale discussed earlier was developed, extended, and filled out in detail. Names rather than dates for the various levels of the relative time scale were a necessity, because no one could guess how old they really were. Even today, most geologists prefer to use relative time terms instead of numerical ages. There are good reasons for this. Not only are the traditional time terms more familiar, but their relative ages are stable and well known. By contrast, numerical ages are still in a state of flux for much of the Phanerozoic, and experimental error increases with earlier dates. In only a few places can numerical dates be associated directly with a stratigraphic interval. To give the

numerical age of most geologic events, the geologist must first determine its relative age by fossils and then compare this with the composite time scale, in which dates from thousands of individual localities have been compiled. This trouble is only worthwhile when speaking to a nongeologist unfamiliar with the geologic time scale or when trying to make calculations that require numerical age estimates.

Early Estimates

Most human cultures have had only a vague notion of the numerical age of the earth. Some cultures envisioned it as immensely long. Others did not have a linear sense of time, so the question "When did it begin?" would have been meaningless. The Hebrew culture was one of the first with a true linear sense of history, and their scriptures reflect this. By adding up the ages of the patriarchs listed in the Bible (the "begat" method), scholars tried to calculate the date of Creation (Fig. 1.9). There were many such attempts over the centuries of

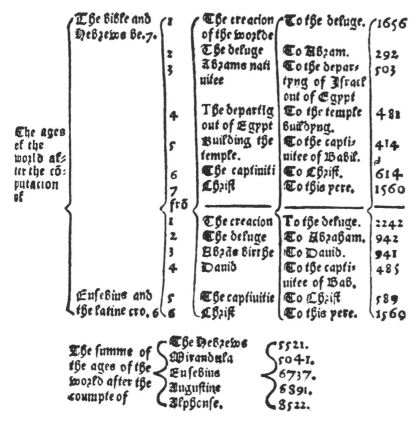

Figure 1.9

Calculation of the age of the world using the "begat" method (Cooper's *Chronicle*, 1560).

Medieval scholasticism, but the most famous was made by Archbishop James Ussher of Ireland (1581–1665) in 1650. He declared that the time of creation was in the evening of October 22, 4004 BC (Brice, 1982). This date was added as an unauthorized marginal note to the 1701 English Bible and appears as a marginal note in some Bibles even now. Until Hutton, few doubted that the earth was only 6000 years old and that man had been on the planet since the sixth day of creation.

One of Hutton's most revolutionary notions was the immense age of the earth, with "no vestige of a beginning, no prospect of an end." Lyell extended this notion and even suggested that earth history had been cyclical and nondirectional. Darwin developed his theory of evolution on the assumption that Hutton and Lyell were right about enormous amounts of time having passed and felt that this was a requirement for his concept of evolution. In 1867, Lyell himself tried to calculate the age of the earth based on rates of molluscan evolution and came up with an estimate of 240 million years since the Ordovician.

During the nineteenth century, many scientists used other methods to estimate the age of the earth. These estimates were all based on the assumption that some physical property of the earth had changed uniformly through time. By backward extrapolation, they thought it was possible to estimate the time to the beginning of the process. Of course, when rates are not uniform, extrapolation can lead to ridiculous results, as Mark Twain so clearly pointed out in *Life on the Mississippi* (1883):

> Now, if I wanted to be one of those ponderous scientific people, and "let on" to prove what had occurred in the remote past by what had occurred in a given time in the recent past, or what will occur in the far future by what has occurred in late years, what an opportunity is here! Geology never had such a chance, nor such exact data to argue from! Nor "development of species," either! Glacial epochs are great things, but they are vague—vague. Please observe:
>
> In the space of hundred and seventy-six years the Lower Mississippi has shortened itself two hundred and forty-two miles. That is an average of a trifle over one mile and a third per year. Therefore, any calm person, who is not blind or idiotic, can see that in the Old Oölitic Silurian Period, just a million years ago next November, the Lower Mississippi River was upwards of one million three hundred thousand miles long, and stuck out over the Gulf of Mexico like a fishing rod. And by the same token any person can see that seven hundred and forty-two years from now the Lower Mississippi will be only a mile and three quarters long, and Cairo and New Orleans will have joined their streets together, and be plodding comfortably along under a single mayor and a mutual board of aldermen. There is something fascinating about science. One gets such wholesale returns of conjecture out of such a trifling investment of fact.

Despite Twain's warning, many geologists tried to extrapolate backward to find the age of the earth for lack of a better method. Their assumptions seemed promising at the time, but today we recognize the problems in these methods. Three basic methods were used to make this estimate: changes in chemical composition of the ocean; rates of accumulation of sediment; and Kelvin's estimate of the rate of the cooling of the earth.

In 1715, the astronomer Edmund Halley suggested that the age of the earth could be calculated from the rate of increase in ocean salinity by estimating the total amount of salt in the ocean and the rate at which it is supplied by rivers. In 1899, John Joly made such a calculation and came up with an age of 90 million years since water first condensed on earth. His calculation is flawed by several erroneous assumptions. The main problem is that the sodium content of the ocean is not increasing but is maintained at equilibrium by a complex exchange cycle. Sodium-bearing sediments precipitate out of the water, sink to the ocean floor, and are eventually recycled to the land through tectonic processes.

The second method was to estimate the total thickness of sediments in the composite global stratigraphic column and to use an average rate of sediment accumulation to get a minimum age of the earth. This method is troubled by a host of erroneous assumptions. For example, (a) not only do rates of sedimentation vary enormously, but in retrospect it appears that most of geologic time is represented by unconformities and thus not recorded by sediment at all; (b) thicker and thicker sections were being found yearly, and (c) thicknesses of some rock types, such as shale, tend to be grossly underestimated due to postdepositional compaction. Between 1860 and 1909, there were many estimates (Table 1.3). Most of the values were under 100 million years, probably because those who calculated them were all unconsciously influenced by the estimate of Lord Kelvin.

William Thomson, Lord Kelvin (1824–1907), was one of the most eminent scientists of his time. He is most famous for his achievements in thermodynamics (the Kelvin temperature scale is named for him), but his work included engineering the transatlantic cable. He had become interested in the problem of the energy balance of the earth and sun and began to make calculations of how long it would take the earth to cool down from an initial temperature like that of the sun. Using measurements of the thermal conductivity of crustal rock and of the heat flow from the interior of the earth, Kelvin calculated that the earth had been **cooling from an initial molten state less than 100 million years ago.** This was far too short for uniformitarian geology to operate, but geologists had no better answer for a physicist of Kelvin's caliber. Kelvin's estimate of the age of the earth troubled Darwin until the day he died, because it left little room for the slow, gradual rate

Table 1.3

Estimates of the Age of the Earth

Date	Author	Maximum Thickness (feet)	Rate of Deposit (years for 1 foot)	Age[a] (millions of years)
1860	Phillips	72,000	1332	96
1869	Huxley	100,000	1000	100
1871	Haughton	177,200	8616	1526
1878	Haughton	177,200	?	200
1883	Winchell	—	—	3
1889	Croll	12,000[b]	6000[c]	72
1890	de Lapparent	150,000	600	90
1892	Wallace	177,200	158	28
1892	Geikie	100,000	730–6800	73–680
1893	McGee	264,000	6000	1584
1893	Upham	264,000	316	100
1893	Walcott	—	—	45–70
1893	Reade	31,680[b]	3000[c]	95
1895	Sollas	164,000	100	17
1897	Sederholm	—	—	35–40
1899	Geikie	—	—	100
1900	Sollas	265,000	100	26.5
1908	Joly	265,000	300	80
1909	Sollas	335,000	100	80

[a] Based on estimates of maximum thicknesses of sedimentary rocks.
[b] Spread evenly over the land areas.
[c] Rate of denudation.
Source: Holmes, 1913.

of evolutionary change he had envisioned. By 1880, the American geologist Clarence King was setting up a thermal conductivity laboratory at the U.S. Geological Survey, and in 1893, he estimated that the age of the earth was only 24 million years.

Once again, the problem with Kelvin's date was a fundamental flaw in his assumptions. Kelvin assumed that the earth was cooling from an initial molten state and that no further heat was being added to the system. In fact, the earth is being continuously supplied with new heat, but its source was discovered only a few years before Kelvin's death in 1907.

Radioactivity and the Geologic Clock

In 1896, Henri Becquerel discovered that uranium emits mysterious rays that expose a photographic plate in total darkness. In 1903, Marie and Pierre Curie showed that radioactive substances produced heat, and by 1906 R. J. Strutt had shown that the earth contained enough radioactive materials to account for all of its heat flow. In short, the earth was not cooling down as Kelvin had

calculated. Instead, all of its heat comes from radioactive decay, and nothing remains of the primordial heat that Kelvin thought he was measuring.

The discovery of radioactivity not only discredited Kelvin's estimate but more importantly, it provided the key to measuring the true age of the earth. Radioactive decay is the only long-term process we know that behaves in a clocklike, linear fashion from an initial state, unlike the rate of salinity change or sedimentation or cooling, as long as it remains a closed system. Radioactive decay was discovered by Rutherford and Soddy in 1902, when they found that radioactive elements change to other elements. By 1906, Rutherford was attempting to measure the age of minerals from their helium-uranium ratios, although leakage of helium made the results unreliable. In the United States, Bertram Boltwood was working with uranium-lead, a more stable system, and by 1905 he obtained his first dates, which were remarkably accurate, even by today's standards (Table 1.4). In 1911, Arthur Holmes, then a student in Strutt's laboratory, summarized most of the available dates and information about radioactive decay. In a little book entitled *The Age of the Earth* (1913), Holmes made a conclusive case for the radiometric method. Numerical dating methods have become more refined and widely used ever since.

The Magnitude of "Deep Time"

Geologists are so accustomed to dealing with millions of years that they often take such enormous amounts of time for granted. It never hurts to remind ourselves of the magnitude of what Playfair (1802) called "the abyss of time" or John McPhee (1980) labeled "deep time." One of the most vivid analogies takes the age

Table 1.4

Boltwood's First Radiometric Dates

Geologic Period	Lead/Uranium Ratio	Millions of Years
Carboniferous	0.041	340
Devonian	0.045	370
Precarboniferous	0.050	410
Silurian or Ordovician	0.053	430
Precambrian Sweden	0.125	1025
	0.155	1270
United States	0.160	1310
	0.175	1435
Ceylon	0.20	1640

Source: Holmes, 1911.

of the earth (4.5 billion years) and squeezes it into a calendar year. If the initial condensation of the earth takes place on January 1, then some of the major events of earth history are as follows:

February 21	Life first appears
October 25	Cambrian shelly faunas appear
November 20	Devonian fishes, first amphibians
December 7	Permian reptiles and Pangaea
December 15	Jurassic dinosaurs, **first birds**
December 25	Dinosaurs extinct
December 31, 3 PM	First hominids
December 31, 11 PM	Appearance of *Homo sapiens*
December 31, 11:58:45 PM	End of last ice age
December 31, 11:59:45 PM	Rise of Rome
December 31, 11:59:50 PM	Fall of Rome
December 31, 11:59:57 PM	Columbus discovers America

In his excellent book *Basin and Range*, John McPhee gives a perspective on "deep time" in the words of Princeton geologist Ken Deffeyes:

In geologists' own lives, the least effect of time is that they think in two languages, function on two different scales. "You care less about civilization [says Deffeyes]. Half of me gets upset with civilization. The other half does not get upset. I shrug and think, So let the cockroaches take over. Mammalian species last, typically, two million years. We've about used up ours. Every time Leakey finds something older, I say, 'Oh! We're overdue.' We will be handing the dominant-species-on-earth position to some other group. We'll have to be clever not to."

A sense of geologic time is the most important thing to suggest to the nongeologist: the slow rate of geologic processes, centimeters per year, with huge effects, if continued for enough years. A million years is a short time—the shortest worth messing with for most problems. You begin tuning your mind to a timescale that is the planet's time scale. For me, it is almost unconscious now and is a kind of companionship with the earth. It didn't take very long for those mountains to come up, to be deroofed, and to be thrust eastward. Then the motion stopped. That happened in maybe ten million years, and to a geologist that's really fast. If you free yourself from the conventional reaction to a quantity like a million years, you free yourself a bit from the boundaries of human time. And then in a way you do not live at all, but in another way you live forever. (McPhee, 1980, pp. 128–129)

FOR FURTHER READING

Albritton, Claude C., Jr., 1986. *The Abyss of Time: Unraveling the Mystery of Earth's Age.* Freeman, Cooper & Co., San Francisco. Nicely written short account of the changing perception of the scale of geologic time.

Berry, William B. N., 1987. *Growth of a Prehistoric Timescale,* 2d ed. Blackwell Scientific Publishers, Palo Alto, Calif. Best short introduction to the history of the people and ideas behind our modern timescale.

Burchfield, Joe D., 1975. *Lord Kelvin and the Age of the Earth.* Science History Publications, New York. Fascinating account of the struggle between physicists like Kelvin, who calculated a young age of the earth, and Darwin and the geologists, who needed much more geological time.

Conkin, Barbara M., and James E. Conkin, 1984. *Stratigraphy: Foundations and Concepts.* Van Nostrand Reinhold Publishing Co., New York. Anthology of the classic works of the eighteenth and nineteenth centuries that first established the ideas discussed in this chapter.

Faul, Henry, and Carol Faul, 1983. *It Began with a Stone: A History of Geology from the Stone Age to the Age of Plate Tectonics.* John Wiley & Sons, New York. Excellent readable review of the history of geologic thought.

Gould, Stephen Jay, 1983. *Hen's Teeth and Horse's Toes.* W. W. Norton & Co., New York. Delightful essays from *Natural History* magazine on topics such as Steno ("The Titular Bishop of Titiopolis"), Hutton ("Hutton's Purpose"), and Cuvier ("The Stinkstones of Oeningen").

Gould, Stephen Jay, 1984. "Toward the Vindication of Punctuational Change." In W. A. Berggren and J. A. Van Couvering (eds.), *Catastrophes and Earth History: The New Uniformitarianism.* Princeton Univ. Press, Princeton, N.J. Excellent discussion of the meaning of uniformitarianism today.

Gould, Stephen Jay, 1987. *Time's Arrow, Time's Cycle.* Harvard Univ. Press, Cambridge, Mass. Brilliant account of the ideas of Burnet, Hutton, and Lyell, which traces the influences of cyclical versus linear time in their works. It also debunks many myths about them that still appear in textbooks.

Greene, Mott T., 1982. *Geology in the Nineteenth Century: Changing Views of a Changing World.* Cornell Univ. Press, Ithaca. The major controversies in nineteenth-century geology (primarily in mountain building and tectonics) are analyzed, with a demythologized discussion of Werner, Hutton, and Lyell.

Laudan, Rachel, 1987. *From Mineralogy to Geology: The foundations of a science, 1650–1830.* Univ. Chicago

Press, Chicago. One of the most up-to-date works on the history of geology, emphasizing the ideas in the context of their times without the myths that have subsequently developed around them.

Rudwick, Martin J. S., 1972. *The Meaning of Fossils: Episodes in the History of Paleontology*. Science History Publications, New York. Includes a thorough treatment of the ideas and times of Cuvier, Lyell, and many other pioneering geologists who first described the stratigraphic record.

Rudwick, Martin J. S., 1985. *The Great Devonian Controversy: The Shaping of Scientific Knowledge among Gentlemanly Specialists*. Univ. Chicago Press, Chicago. Detailed description of the confusion and controversy that led to the recognition of the Devonian System, with a lucid analysis of the sociological structure of a scientific debate.

Secord, James A., 1986. *Controversy in Victorian Geology: the Cambrian-Silurian Dispute*. Princeton Univ. Press, Princeton, N.J. The fascinating details of the dispute that developed over the early parts of the stratigraphic record and eventually destroyed the friendship between two giants of British geology.

Close up of aggrading climbing ripples from the Colorado River, Arizona. Stratigraphic data such as these indicate rapid directional flow with excess sediment supply. Photo courtesy Paul E. Potter.

2

Stratigraphic Data

THE DATA BASE

We have seen how geologists intepreted the rock record in the past. How does our modern approach differ? In the next few chapters, we examine the process of stratigraphic interpretation. Beginning on the level of individual outcrops, we look at various models for interpreting the depositional environment responsible for the rocks. Finally, we examine the methods by which large-scale interpretations and correlations can be made.

Stratigraphic analysis begins with data collection, which consists mostly of describing, measuring, and interpreting sedimentary rock sequences in the field. This book is written with the assumption that the reader has some familiarity with the basics of sedimentary petrology; for further discussion of sedimentology, the reader is referred to any of the excellent texts listed later in this chapter. For the purposes of outcrop description, only a few of the basic concepts of sedimentology are mentioned here.

Stratigraphic data come from two basic sources: surface outcrop measurements and subsurface data, usually generated from well logs, seismic profiles, and cores. Depending on the nature of the problem and the region, the data can be entirely from the surface, entirely from the subsurface, or (ideally) a mixture of both. In this chapter we review the methods of surface outcrop description. Subsurface methods are discussed in Chapter 11.

SEDIMENTARY ROCK DESCRIPTION

When a geologist approaches an outcrop for the first time, certain things are immediately apparent. If the outcrop consists of only one uniform rock type, the description is fairly straightforward. In most cases, however, an outcrop will consist of two or more types of rocks, which may form discrete layered units that the geologist can subdivide in the field. Each unit then requires its own separate description, a measurement of its thickness, and observations on its relationship with overlying and underlying units. These observations are part of the process of measuring sections, discussed later.

The most obvious property of every clastic or detrital sedimentary rock unit is its grain size. Most names and classification schemes of clastic rocks are based on this property. The most widely used grain-size scale is the Udden-Wentworth scale (Fig. 2.1). Because clastic par-

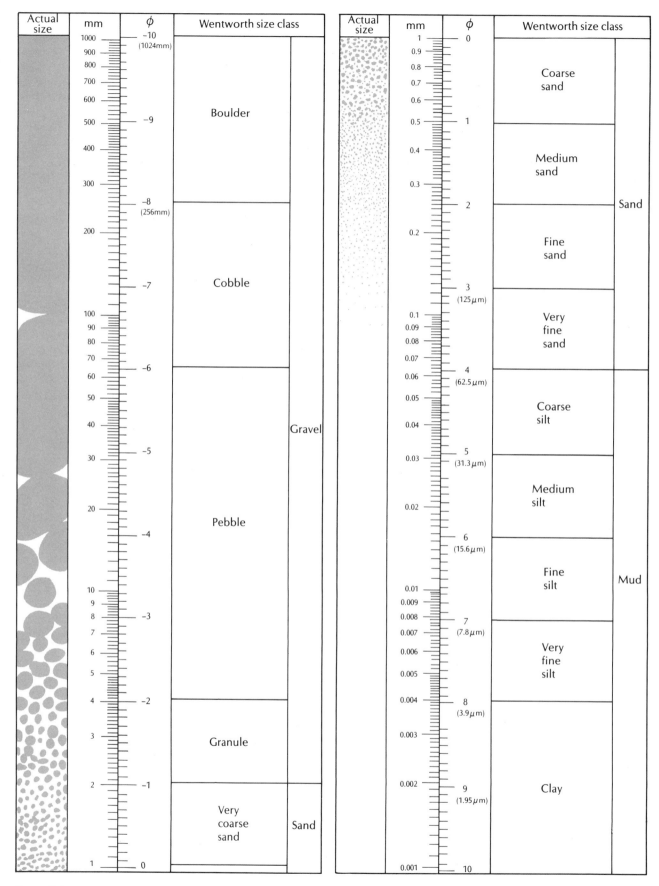

Figure 2.1

Udden-Wentworth grain-size scale and ϕ/mm conversion chart. From Lewis (1984).

ticles range enormously in grain size, the scale is based on factors of 2. Starting at 1 millimeter (mm), each larger size class is a multiple of 2 of the next smaller class; each smaller class is the next larger class divided by 2. These fractions and large numbers are often difficult to deal with, so Krumbein (1934) invented the *phi (φ) scale* based on the negative log to the base 2:

$$\phi = -\log_2 \text{ diameter} \qquad \text{(in millimeters)}$$

Phi notation converts messy fractions into positive whole numbers. Cobble and boulder sizes can then be expressed in negative whole numbers.

For unconsolidated sediments, grain size is analyzed by sieving or settling, but this is seldom possible with lithified sedimentary rocks. Even if the rocks are poorly indurated and readily break down into component grains, a sieve analysis is not usually reliable for determining the primary grain-size distribution. The size of the grains may be secondarily enlarged by authigenic overgrowths or by the suturing of two or more grains. Geologists generally estimate grain size visually in the field. Clasts larger than a few millimeters can be directly measured, and there are commercially produced grain-size comparision charts for quick visual estimates of sand-sized particles. The simplest method, however, is to take the sieved sand splits from a grain-size analysis and glue a small sample of each grain size onto a labeled card for use in comparision. These visual methods are remarkably accurate for the simple purpose of describing modal grain size.

Other properties of sedimentary particles can also be estimated visually. Roundness and sphericity can be roughly estimated and noted in the description. Sorting characteristics are also valuable information. Again, a visual image of sorting can be used. Observations should be made of the nature and percentage of material between the framework grains. In many cases, it is possible to determine what kind of cement or matrix is present in a sample with just a hand lens, a knife, and an acid bottle. All of these observations should be supplemented by observations of the hand samples in the laboratory and preferably by petrographic analysis.

Mudrocks are much more difficult to analyze in the field. Because their grain size is almost impossible to resolve with a hand lens, laboratory work is essential. The old "taste test" works well for field estimates of grain size. Because silt is gritty and clay is slimy to the taste, a mudstone with less than one-third silt or sand is creamy to chew. If the sample is gritty to the taste but no silt or sand grains are visible with the hand lens, then it is probably between one-third and two-thirds silt or sand. If it is less than one-third clay, the silt or sand is visible with a hand lens.

Conglomerates present the opposite sort of problem. Grain size and composition are visible to the naked eye in coarse-grained rocks, but they can contain a tremendous variety of rock types, with a spectrum of textures. As a result, conglomerates can be difficult to describe because there is too much "noise" and not enough "signal" in the data. Grain size is typically highly variable, so it is best to measure the maximum size of a number of clasts in a given area to get a representative sample. There is usually a greater variety of lithologies and mineralogies in conglomerate clasts than in sands. Finally, the relations between conglomerate clasts and the nature of their matrix are critical for describing and understanding conglomerates and breccias; for example, are the larger clasts in contact with one another or are they supported by a fine-grained matrix? The answer to such questions can make a big difference in interpreting how the conglomerate was deposited.

Field description of carbonate rocks is more difficult. For one thing, carbonates are susceptible to diagenetic changes, surface weathering, and dissolution. As a result, outcrop colors and appearances are often deceiving, and even a freshly broken surface may give very little information. Another problem with outcrop observations is that many critical properties can be determined only with a microscope. For example, many sedimentary and biogenic particles cannot be seen with the hand lens, and the mud content is critical to most limestone classification schemes. A bottle of 10 percent hydrochloric acid can be used for testing between calcite and dolomite, but most other properties of carbonates are better observed in the laboratory. For this reason, a good field sampling program is essential when describing and mapping carbonate rocks. An experienced observer can accomplish much more in the field after laboratory analysis.

The colors of rocks in the field can sometimes be useful, but they can also be deceptive. The color of a formation may be its most obvious feature and can be used for mapping, even from the air. Colors are diagnostic of some lithologies. For example, sandstones and conglomerates can show characteristic colors that are based on their component lithologies. Dolomites are often more buff or yellowish in the field than limestones. Other colors reflect diagenetic differences. Green and gray shales are typically colored by their abundant organic material and reduced sulfides whereas red, yellow, and brown shades in shales are often due to staining by iron oxides and hydroxides. Black shales are very rich in organic matter. Because many of these diagenetic colors are acquired long after deposition, they must be interpreted with caution. The color of a weathered surface can differ radically from the fresh outcrop, so both must be noted and described. Sometimes color differences superficially resemble depositional boundaries but are actually diagenetic boundaries. For example, the volcaniclastic sediments of the John Day Formation in central Oregon have been subdivided into a lower, green colored Turtle Cove Member and an upper, tan-colored Kimberly Member (Fisher and Rensberger, 1972). The color difference is due to a diagenetic change in zeolites and is demonstrably younger in some parts of the section than others. Parts of the Turtle Cove Member are equivalent in age, lithology, and fossil content to parts of the Kimberly Member elsewhere. In this case, the deceptive color difference has no time significance. The standard Munsell color chart (Goddard et al., 1948) is widely used for unambiguous descriptions of color. For most purposes, simple verbal descriptions are adequate.

SEDIMENTARY STRUCTURES

So far we have discussed sedimentary rocks in terms of the particles of which they are made or in terms of the bulk properties of those particles. The next level in the hierarchy concerns the properties of the rock itself, particularly the macroscopic *sedimentary structures*. Sedimentary structures are often very informative because particular types of structures result from particular depositional processes. In many cases, certain structures or combinations of structures are the best clue to the sedimentary environment.

Physical Structures

Physical (inorganic) structures are sedimentary features formed by physical processes without the influence of organisms. The most important of these are mechanical structures formed during the deposition of the sediments, *primary sedimentary structures*. They do not include features formed during diagenesis.

Plane Bedding.   The simplest sedimentary structure is plane bedding. Simple horizontal beds form in practically all sedimentary environments and under a variety of conditions, so further descriptive detail is needed to interpret them. Bedding is so common in sedimentary rocks that its origin is often overlooked because it is assumed to be inevitable. Three basic mechanisms can form plane bedding: sedimentation from suspension, horizontal accretion from a moving bed load due to a change in the competence of flow, and encroachment into the lee of an obstacle. Plane beds frequently represent rapid deposition, usually by a single hydrodynamic event. Most deposited beds, however, have been reworked, so *preservation potential* is also important. For example, submarine landslide deposits are rapidly buried, so they have a high preservation probability, but beach deposits are almost always reworked after they are deposited.

Finer-scale plane bedding [less than 1 centimeter (cm) thick] is usually called *lamination*. A number of mechanisms can form laminae. The classic example is the alter-

Figure 2.2

Seven years of Pleistocene varves near Seattle, Washington. Total thickness shown is about 12 cm. Photo courtesy J. Hoover Mackin.

nation of light and dark layers, such as glacial varves (Fig. 2.2). There also can be alternation of mineral composition, as is seen in heavy mineral lags among the normal quartz sands on some beaches. Lamination can also be due to alternation of grain sizes caused by changes in current strength during deposition. In some cases, apparent lamination is not a primary feature at all but a secondary color banding due to diagenetic effects.

Lamination in muds is usually the result of slow, steady deposition. Absence of lamination in mudstones is probably due either to *flocculation* (clumping of clays before they settle) or to secondary *bioturbation* (disturbance by organisms). As mentioned, laminated sands have usually been rapidly deposited, commonly by a single hydrodynamic event, such as the swash-backwash of surf, traction by steady flow, avalanching down a dune face, or the migration of ripples, leaving a heavy mineral lag. Truly massive deposits that show no bedding are anomalous. Frequently, such deposits show cryptic bedding when studied by X-ray or when stained (Fig. 2.3). The lack of plane bedding is usually due to bioturbation, deposition from highly concentrated sediment dispersions, or rapid deposition from suspension.

Normal graded bedding occurs when the grains in a single bed range continuously from coarse grain sizes at the base to fine grain sizes at the top (Fig. 2.4A). It is usually the result of the settling of a single, poorly sorted suspension of sand, in which the coarser fractions settle out more rapidly than the finer fractions. Graded bedding is commonly produced by *turbidity currents*, which are turbulent suspensions of water and sand that are denser than the surrounding water (Fig. 2.4B). Because of the density difference, the turbidity current flows as a discrete unit; very little mixing with the surrounding less dense water occurs. The poorly sorted

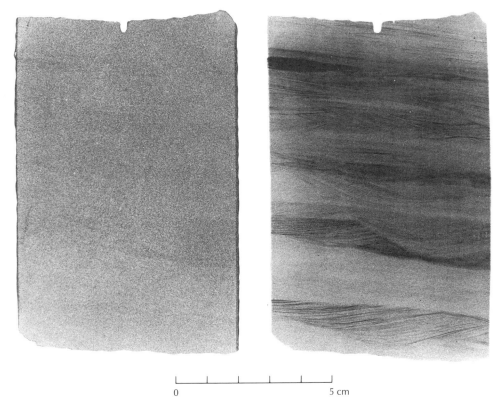

0   5 cm

Figure 2.3

A polished slice of core (left) and a positive print of an X-radiograph (right) of Berea Sandstone (Mississippian), Illinois. Only vague banding is visible on the polished slab, but X-radiation reveals an apparent dip of 10°, scour and fill, and cross bedding. From Hamblin (1965).

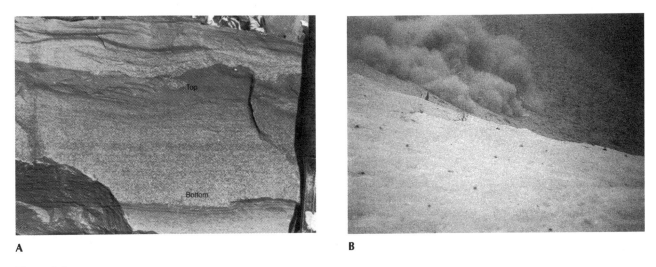

A   **B**

Figure 2.4

(A) Ordovician turbidite bed from the Martinsburg Formation, showing well-developed graded bedding. Photo courtesy E. F. McBride. (B) Side view of a turbidity current flowing down a submerged slope in Jamaica. A submarine propeller disturbed the sediment, setting up a density flow. From Moore et al. (1976).

Figure 2.5

Inversely graded debris flow deposit, Miocene Violin Breccia, Ridge Basin, California. Note the large rocks at the top of the debris flow that projected upward and were buried by the fine material at the base of the next flow. Photo by the author.

material remains suspended until the flow begins to slow down. Then the turbulence ceases, and the material settles out according to size. The depositional product of a turbidity current is called a *turbidite*.

Although they are rare, some beds exhibit *reverse* (or *inverse*) *grading*, with the coarsest material at the top and the finer material at the bottom (Fig. 2.5). The mechanism that produces inverse grading is not well understood, but it is commonly found in debris flows, probably because the dispersive pressures of a grain flow tend to push the larger particles to the top of the flow where they encounter less friction. The finer grain sizes, on the other hand, can move more easily in the mass of the flow.

Bedforms Generated by Unidirectional Currents.

As soon as flow attains a force sufficient to erode particles from the bed, sediments are transported in a set of *bedforms*, or structures on the surface of the bed. If these bedforms are later buried and preserved, they can form sedimentary structures. Flume studies have shown that there is a predictable sequence of bedforms that depend on velocity, grain size, and depth of flow (Fig. 2.6A). In sand that is finer than 0.7 millimeters

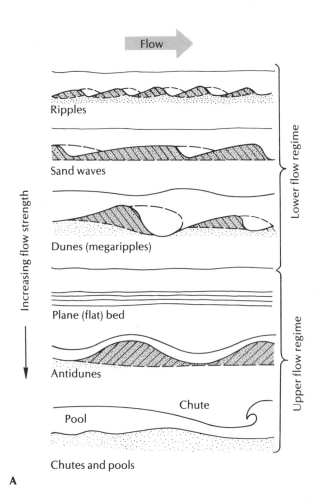

A

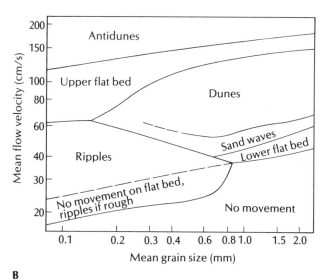

B

Figure 2.6

(A) Sequence of bedforms through increasing flow strength conditions. From Blatt et al. (1980). (B) Changes in bedforms resulting from different flow velocities (vertical axis) and grain sizes (horizontal axis). From Lewis (1984).

(coarse sand or finer), the first features to form are *ripples*. Their spacing is typically 10 to 20 millimeters or less, and their height is less than a few centimeters. As velocity increases, ripples enlarge until they form sand waves and finally dunes, which have spacings from 0.5 to 10 meters or more and heights of tens of centimeters to a meter or more (Fig. 2.6B).

In deeper currents, greater flow velocity is required to produce the larger bedforms. Changes in water temperature or clay content change the viscosity of the flow and consequently affect the settling velocity. This alters the bedforms regardless of the other variables. Ripples, sand waves, and dunes all form in much the same manner. Small irregularities on the surface of the stream bed cause a slight turbulence as the flow is diverted up and around them. Eventually, the flow over an obstacle no longer hugs the bottom but separates from it at the *point of flow separation* (Fig. 2.7), which is at the crest of the ripple or dune. The flow meets the bottom again at the *point of flow reattachment*. Beneath this zone of laminar flow is the zone of turbulence and back flow on the lee side of the ripple. This is the *zone of reverse circulation*. Sediment migrating up the ripple or dune avalanches down into this zone and is deposited by the weaker currents. This generates the inclined foreset beds that produce *cross-bedding*. Because the ripple or dune is continually eroded on the upstream side and accreted on the downstream side, these bedforms continuously migrate downstream. Meanwhile, most of the fine-grained suspended load of silt and clay is carried downstream, resulting in segregation of grain sizes.

The shape of ripples depends primarily on a balance between the bedload and the material that is settling from suspension. If there is little suspended load, the ripples are steep, with a sharp angle between the foreset and bottomset beds. If there is a large suspended load, the lee slope builds steadily, forming curved cross-strata and a tangential contact between foreset and bottomset beds. Ripples and dunes are dynamic features that change constantly. The downstream end of the zone of backflow (the point of reattachment) continuously fluctuates, so only its approximate position can be identified. Beyond the point of reattachment are turbulent eddies that scour downstream, forming troughs with their long axes parallel to flow. As the ripples or dunes migrate downstream, they fill the troughs in front of them. This natural association of troughs and ripples produces normal trough cross-stratification.

Dunes form by the same processes as ripples, only on a much larger scale (centimeters in the case of ripples, meters in the case of dunes). Whereas ripples are unaffected by changes in depth and strongly affected by changes in grain size, dunes are more strongly affected by depth and less by grain size. Dune height is limited only by depth of flow, but ripples can reach only a certain maximum height. Ripples tend to migrate in one plane (except in the case of climbing ripple drift, discussed later). Dunes, on the other hand, frequently migrate up the backs of other dunes.

With increased flow velocity, dunes are destroyed, and the turbulent flow, which was out of phase with the bedforms, changes to a sheetlike flow, which is in phase with the bedforms. Intense sediment transport takes place along a *plane bed* (see Fig. 2.6A), which can produce evidence in rocks of planar laminated sands. At even higher velocities, plane beds are replaced by

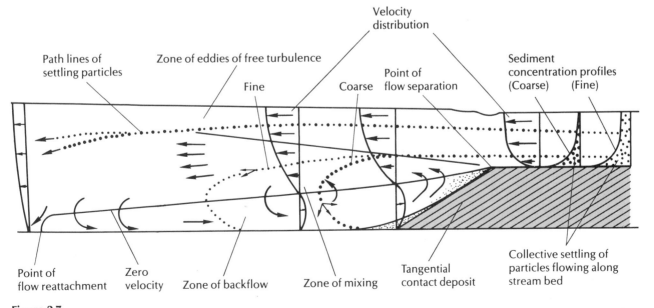

Figure 2.7

Flow pattern and sediment movement over migrating ripples or dunes. Velocity profiles are shown by the vertical bars. After Jopling (1963).

antidunes, which produce low, undulating bedforms that can reach five meters in spacing. Their fundamental feature is that their crests are in phase with the surface waves, so they migrate by accretion on the upstream side. In rocks, they are characterized by faint, poorly defined laminae. Antidunes generally show low dip angles (less than 10°) and are associated with other indicators of a high flow velocities. Because they migrate upstream, antidunes should leave evidence of a flow contrary to other current-direction indicators. It seems that antidunes are rare in the rock record, probably because they are reworked where the current slows before final burial.

Finally, at the highest flow velocities, the antidunes wash out and are replaced by chutes and pools (see Fig. 2.6A).

The three-dimensional geometry of cross-stratification is a useful indicator of flow and sediment load. Starting with stationary current ripples (Fig. 2.8A), simple trough cross-stratification, as mentioned earlier, develops from migrating ripples and dunes (Fig. 2.8B). Tabular cross-stratification (Fig. 2.8D), on the other hand, is produced by migrating sand waves. Horizontal stratification can be produced by plane bed conditions at high flow velocities. Frequently, the migration of a ripple is interrupted, eroded back, and then buried by a

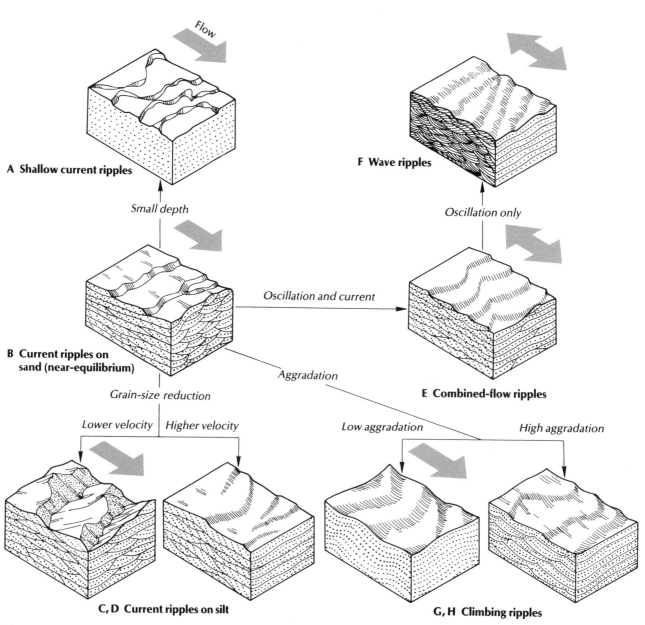

Figure 2.8

Variations in ripple forms and stratification caused by changes in velocity, grain size, depth, rate of sediment supply, and flow direction. From Harms (1979).

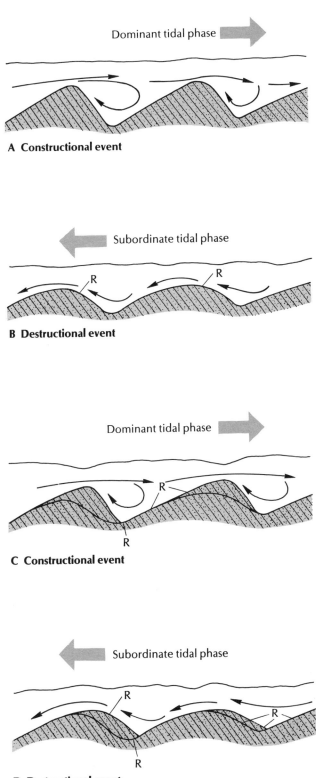

A Constructional event

B Destructional event

C Constructional event

D Destructional event

Figure 2.9

Sequence of events that forms reactivation structures. The dominant tidal phase builds cross-beds (A), which are eroded back during tidal retreat (B). The return of the constructional tide buries this erosional reactivation surface, R, with new cross-beds (C), and the process repeats (D). After Klein (1970).

new advancing bedform. Such an interruption produces a tiny erosional surface between cross-strata, known as a *reactivation surface* (Fig. 2.9). Figure 2.8 shows the natural sequence of ripple features resulting from changes in flow conditions, grain size, and sediment supply. As flow increases, incipient ripples develop into full-scale trough crossbeds at equilibrium. If the grain size then decreases, the shape of the current ripples changes, depending on flow velocity (Fig. 2.8C, D). If the current becomes less unidirectional, sinuous combined-flow ripples result (Fig. 2.8E). A fully oscillatory current (such as in waves) produces straight, symmetrical ripple marks with a distinctive lenticular cross-section (Fig. 2.8F). If the sediment supply increases, then the ripples build upward, or *aggrade*. Low aggradation produces *climbing ripples* (Fig. 2.8G). High aggradation produces sinuous ripples that are in phase (Fig. 2.8H).

Bedforms Generated by Multidirectional Flow.   Although they are formed in a different manner, wave ripples on beaches are similar to current ripples. A rotating eddy precedes a wave as it moves onshore, precipitating the sand load into troughs and ripples. As the wave crest passes, the eddy rises with the crest and disperses into the backwash. Thus the coarser grain sizes are left on the beach, and the finer sand is washed offshore, so beach sands are very well sorted. Wave ripples are not easy to distinguish from current ripples, but they do have some differences. They are usually symmetrical (or only slightly asymmetrical) with peaked crests and rounded troughs. If they are asymmetrical at all, they indicate a current direction toward the shore. Their cross-laminae also dip shoreward for the same reason.

Other waveforms are confined to tidal regions. Unlike on the beach, fine sediment in the tidal zone is moved onshore because tides flow in slowly, allowing it to settle. Retreating tides move out too slowly to scour away much of this deposition. As a result, tidal ripples are generally unidirectional, with weak backflow structures. Cross-beds are oriented in two directions, often with reactivation surfaces caused by the reversal of current direction during a tidal cycle (Fig. 2.10). This is known as *herringbone cross-bedding*. The bidirectional nature of tidal outflow currents often superimposes a weaker ripple system on the dominant sinuous ripples produced by rising tides. These two systems produce *interference ripples*, or "tadpole nests" (Fig. 2.11). The most distinctive features of tidal regions are caused by the mixing of sand- and mud-sized fractions from the asymmetrical currents. Small lenses of sand in muddy beds are called *lenticular bedding* (Fig. 2.12A), which is caused by sand being trapped in troughs in the mud as sand waves migrate across a muddy substrate. If mixing produces minor mud layers in a sandy substrate, the pattern is called *flaser bedding* (Fig. 2.12C). An equal

Figure 2.10

Herringbone cross-stratification from alternating tidal currents, Cambrian Cadiz Formation, Marble Mountains, California. Photo by the author.

mixture of sand and mud (Fig. 2.12B) characterizes *wavy bedding*.

Wind-transported sand behaves differently from water-transported sand though wind-generated ripples look superficially like water-generated ripples. Sand particles in wind move mostly by *saltation* (jumping and bouncing) and to a lesser extent by *surface creep*. Particles that are too large to move by saltation and creep accumulate as a lag, forming a desert pavement in areas of wind deflation. Because saltation is more effective in moving sand than scouring, erosion is heaviest on the exposed upwind side of a sand dune, where the impact of windblown particles is greatest. Deposition occurs on the protected lee side; because there is no zone of backflow, the lee sides do not scour. This is clearly the opposite of water ripples, which erode on the lee side.

Wind ripples migrate continuously by eroding on their upwind side and building on their downwind side until they reach an equilibrium size for the wind strength and sand supply. They are usually made of sand that is coarser than the substrate over which they migrate, and their crests are of coarser particles than their troughs. Water ripples show the opposite condition in both of these features. This is because wind ripples form by the winnowing of their crests, which leaves the coarser material behind, whereas water ripples accumulate coarser

A

B

Figure 2.11

(A) Interference pattern formed in symmetrical ripples from two coexisting wave sets in a modern tidal flat. Photo courtesy J. D. Collinson. (B) Interference ripples in a Pennsylvanian sandstone, Cape Breton Island, Nova Scotia. Photo courtesy P. E. Potter.

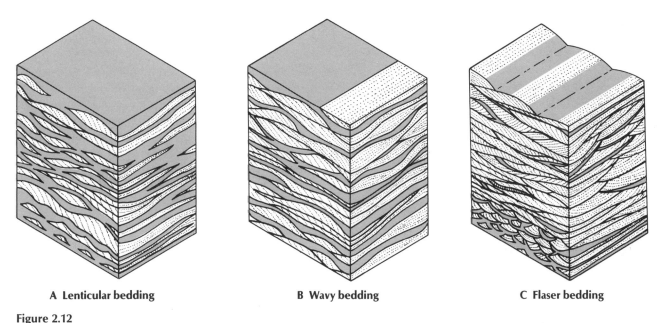

A Lenticular bedding   **B Wavy bedding**   **C Flaser bedding**

Figure 2.12

Diagrams showing (A) lenticular, (B) wavy, and (C) flaser bedding. After Reineck and Wunderlich (1968); Blatt et. al. (1980).

sediments in the troughs where the zone of backflow results in weaker currents and reduced competence. Another major difference is that wind ripples are not restricted by the shallow flow depths that restrict water ripples, so eolian dunes can be enormous (meters to tens of meters in height). Indeed, gigantic cross-strata are virtually restricted to eolian environments (see examples in Chapter 3).

Bedding Plane Structures.   The many types of sedimentary structures just discussed are formed during the deposition of the bed and are generally three-dimensional. Another class of sedimentary structures forms on the interface between beds, usually on the exposed surface of a recently deposited bed before it is finally buried. Such structures can be extremely useful because they indicate current directions and postdepositional deformation of the sediment.

Sole marks, found on the bottom surfaces of beds, are usually casts or molds of depressions that were formed in the underlying beds by currents. The filling, or sole mark, tends to have a higher potential for preservation because it is immediately buried as the depression is filled. The commonest form of sole mark is a *flute cast* (Fig. 2.13), which is shaped like an elongated teardrop that tapers upcurrent. It is formed by a slight irregularity on a mud substrate that causes flow separation and a spiral eddy. The eddy spirals around a horizontal axis parallel to the flow and scours out the rounded, deep end of the flute cast. As the spiral eddy diminishes, the scouring becomes narrower until the point of the scour is reached. Another class of sole mark is the *tool mark*,

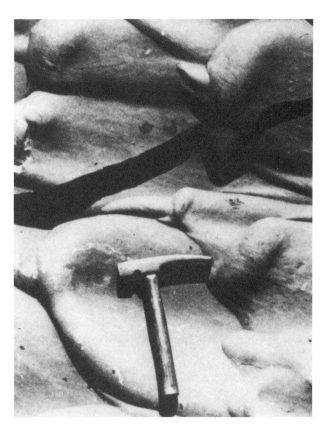

Figure 2.13

Large, closely spaced, overlapping flute casts on the base of a sandstone bed, Smithwick Formation (Pennsylvanian), Burnett County, Texas. Current flowed from left to right. Photo courtesy of F. J. Pettijohn; from Pettijohn and Potter (1964).

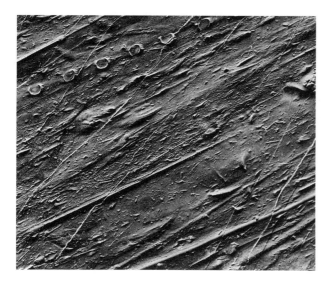

Figure 2.14

Tool marks from the base of the Carpathian flysch, Poland. The marks include circular skip casts from spool-shaped fish vertebrae, shallow brush marks, and deeper drag marks. Photo courtesy J. E. Sanders.

which is caused by the indentation of the cohesive mud bottom by any object, or "tool" (Fig. 2.14). Tool marks include *groove casts*, *brush marks*, *skip marks*, *chevron molds*, *prod marks*, and *bounce marks*. These names clearly describe the types of indentations that are left by various objects (e.g., twigs, branches, pebbles, shell fragments, and fish vertebrae) that produce them.

Subaerially exposed mud also produces sedimentary structures that can be diagnostic. Most familiar of these are *mudcracks* and *raindrop impressions*, which nearly always indicate drying of a subaerial mudflat. Because curling mudcracks always curl upward, they are also good indicators of the top side of a bed. In undeformed strata, this is not very important. However, when beds have been structurally deformed, the top is not necessarily obvious. In this case, it is critical to find *geopetal* structures, which indicate the top of the bed. Cross-beds usually have truncated tops (because the next cross-bed set scours down into the previous one) and tangential contacts between foresets and bottomsets, so they can often be used to determine top (see Fig. 3.22). Ripple crests are usually sharp whereas ripple troughs are always rounded and scooped. Normally graded beds are clear indicators of top because the coarsest material settles out first and is concentrated at the bottom (see Fig. 2.4A). Sole marks are found only on the base of the bed; the depressions that molded them are therefore on the top of the underlying bed.

Soft-sediment Deformation Structures. Sediment can be deposited so rapidly that the beds are unstable. In some cases, denser material is deposited on top of less dense material, so gravity tends to overturn it. If there is enough pore water, the whole mass can become liquified, like quicksand, and deform. Large forces placed on the sediment before lithification can deform it while it is still soft. If a mass of sediment slumps (a common occurrence on slopes in the marine environment), the sediment can become internally deformed. All of these processes produce distinctive deformation structures. The most common are *load structures*, which are irregular, bulbous features formed when a denser material has sunk into a less-dense medium (Fig. 2.15). Sometimes, balls of sand load downward into underlying mud and then are pinched off, forming *pseudonodules*, or *ball and pillow structures*. These occasionally reach enormous dimensions (Fig. 2.16A, B). Tonguelike protuberances of mud into overlying soft sediment are known as *flame structures* (Fig. 2.17). Finally, deformation of soft sediments leads to *convolute bedding* and other features that suggest intense structural deformation on a regional scale (Fig. 2.18). However, these features are formed shortly after deposition and do not imply regional structural forces. One can easily be fooled by such features into postulating spurious structural events. The best way to distinguish convolute bedding from true structural deformation is to see whether it is widespread and penetrative or restricted to a single bed. Convolute lamination should also be associated with other more diagnostic soft-sediment deformation features.

Figure 2.15

Load casts from the Smithwick Formation, Texas. Photo courtesy E. F. McBride.

A **B**

Figure 2.16

(A) Ball and pillow structure seen from below. Annot Sandstone (Oligocene), Peira-Cava, Maritime Alps, France. (B) Cross section of ball and pillow structure, showing internal lamination conforming to the boundary of the pillow. From the Ordovician Cynthiana Group, Pendleton County, Kentucky. Photos courtesy P. E. Potter.

Figure 2.17

Flame structures and graded bedding in late Pleistocene lacustrine sediments, Fraser River Valley, British Columbia. Field of view is 35 cm wide. Photo courtesy A. Rodman.

Biogenic Sedimentary Structures

Sedimentary structures formed by organisms are known as *trace fossils*, (*Lebensspuren*, German "living traces") or *ichnofossils* (Greek *ichnos*, "trace"). Besides their importance as paleontological objects, they are useful clues about depositional conditions. Frequently, they are given taxonomic names as if they were valid Linnaean genera and species, but this is not really proper. Trace fossils are fossilized behavior, not body fossils. Few "ichnogenera" can be definitely associated with a known body fossil. It is likely that one type of trace may have been produced by several types of organisms or that one organism produce several types of traces. This type of taxonomy is comparable to giving a different species name to different footprints produced by the same individual wearing different shoes. Nevertheless, the practice of giving Linnaean names to trace fossils is so well established that it persists for lack of a better system.

Figure 2.18

Convolute lamination in polished slabs of siltstone from the Martinsburg Formation (Ordovician), Pennsylvania. From McBride (1962).

Certain characteristic trace fossils have been clearly associated with specific depth and bottom conditions (Fig. 2.19). A working knowledge of the more common ichnogenera is thus a very useful skill because they are almost as diagnostic as index fossils for certain purposes.

Vertical tubelike burrows ("piperock") are commonly known as *Skolithos* and are believed to have been formed by tube-dwelling organisms that lived in rapidly moving water and shifting sands. Horizontal U-shaped burrows with many intermediate, riblike feeding traces are known

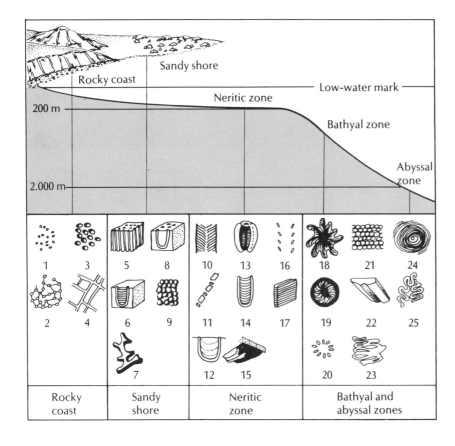

Figure 2.19

Summary diagram of the most common marine facies and depth-related trace fossils. (1–4) All borings. (5) *Skolithus*. (6) *Diplocraterion*. (7)–(8) Other deep burrows. (9) Detail of wall of *Ophiomorpha* burrow. (10) *Cruziana*. (11, 13, 16) Other trilobite traces. (12, 14, 15, 17) Feeding burrows with weblike *spreiten* between vertical parts of U tubes. (18) *Zoophycus*. (19) Another complex feeding burrow. (20) Radiating trace. (21) *Paleodictyon*. (22) Enlarged portion of one section of meandering bilobate grazing trail, preserved as sole mark. (23, 24, 25) Typical meandering or spiral surface traces. From Blatt et al. (1980); adapted from T. P. Crimes (1975).

as *Cruziana* and occur in moderate- to low-energy sands and silts of the shallow shelf. Many *Cruziana* are believed to represent the feeding traces of trilobites. Broad, branching infaunal feeding traces known as *Zoophycos* occur in low-energy muds and muddy sands in the bathyal zone. Meandering feeding traces on bedding planes are called *Nereites* and are usually found in the abyssal plains. *Diplocraterion yoyo* tells a very specific story about the sediment-water interface: *D. yoyo* is a burrow trace found between the arms of a vertical, U-shaped tube that presumably housed a burrowing, tubelike organism. When the burrow was buried by sediment, the organism moved up in its burrow; when the upper part of the burrow was eroded away, it dug in deeper. The sequence of U-shaped burrow traces thus responds like a yoyo to the rise and fall of the sediment-water interface.

The absence of trace fossils can also be informative. If there are no trace fossils in a sequence that should otherwise be heavily burrowed, there might be reason to suspect that the water was anoxic and inhospitable to organisms. In sequences that are bioturbated, individual unburrowed beds were probably deposited very rapidly, so the organisms could only rework the uppermost part. Hard substrates can also be bored by organisms that drill their way in. The presence of rock borings can indicate ancient shorelines and beachrock or an unconformity in which sediment was subaerially exposed. Organisms that bore into rock also are found in reefs.

FOR FURTHER READING

Allen, J. R. L., 1985. *Principles of Physical Sedimentology*. George Allen & Unwin, London. Most comprehensive text available on the physical processes that affect the movement of sedimentary particles.

Blatt, H., 1982. *Sedimentary Petrology*. W. H. Freeman and Co., San Francisco. Most up-to-date text on the subject.

Blatt, H., G. V. Middleton, and R. C. Murray, 1980. *Origin of Sedimentary Rocks*. Prentice-Hall, Englewood Cliffs., N. J. The most complete and rigorous book on sedimentary processes and petrology yet available; a classic.

Collinson, J. D., and D. B. Thompson, 1982. *Sedimentary Structures*. George Allen & Unwin, London. Short but thorough and well-illustrated discussion of the formation and interpretation of sedimentary structures.

Folk, R. L., 1974. *Petrology of Sedimentary Rocks*. Hemphills, Austin, Texas. Laboratory manual constructed for use at the University of Texas, which has become a classic reference.

Friedman, G. M., and J. E. Sanders, 1978. *Principles of Sedimentology*. John Wiley, New York. Slightly different approach, emphasizing sedimentary processes and environments rather than petrology.

Leeder, M. R. 1982. *Sedimentology: Process and Product*. Allen & Unwin, London. Concise and well-illustrated introduction to sedimentology. One of the best bargains on the market.

Pettijohn, F. J., 1975. *Sedimentary Rocks*, 3d ed. Harper & Row, New York. Classic textbook on sedimentary petrology, emphasizing descriptive features of sedimentary rocks.

Pettijohn, F. J., and P. E. Potter, 1964. *Atlas and Glossary of Primary Sedimentary Structures*. Springer-Verlag, New York. Beautiful photographs and descriptions of classic examples of sedimentary structures.

Selley, R. C., 1988. *Applied Sedimentology*. Academic Press, San Diego. A very thorough, readable, and up-to-date review of the basics of sedimentology, with an emphasis on the practical aspects of the discipline.

Tucker, M. E., 1982. *The Field Description of Sedimentary Rocks*. Open University Press, Milton Keynes, England. Paperback manual giving techniques for fieldwork and description in sedimentary sequences.

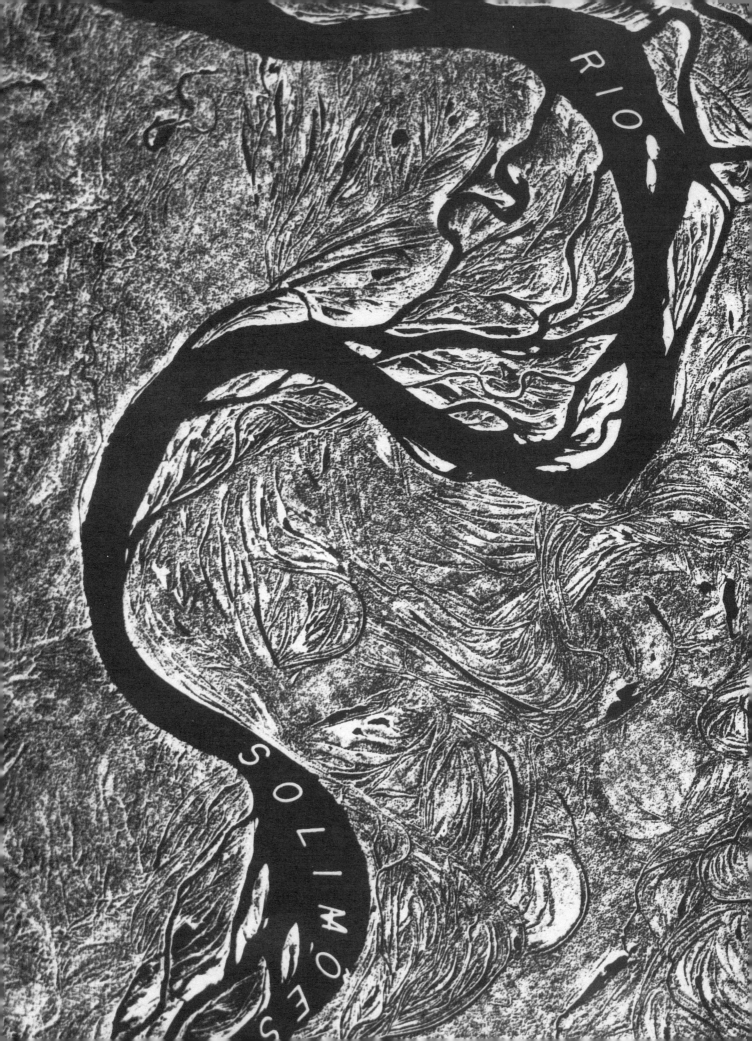

II

DEPOSITIONAL SYSTEMS

Radar image of meandering river scars, Rio Solimoes, Amazon Basin, Brazil. Photo courtesy P. E. Potter

Classical stratigraphy, as treated in the textbooks of the fifties and sixties, emphasized the *descriptive* properties of sedimentary rocks and stratigraphic units. The interpretation of these rocks received less coverage and was frequently organized along descriptive lines. For example, distinctive rock types, such as red beds, were discussed as if they had all formed under the same conditions. An approach that is more *genetic* and less descriptive has emerged in the last three decades as geologists began to study present-day depositional environments in detail to see what types of sediments and stratigraphic sequences are formed. These data are used to analyze ancient stratigraphic sequences for analogues of modern environments.

The genetic approach has become so fruitful that it is now used in virtually all interpretations of stratigraphic sequences. Take, for example, a sandstone body that is exposed in an outcrop. A good depositional model not only allows the geologist to make a reasonable reconstruction of the ancient environmental conditions but also has economic importance. A geologist who understands the depositional circumstances can make predictions about the subsurface geometry and the extent of that sandstone. If it happens to be an important reservoir rock for water or hydrocarbons, this knowledge may be crucial in determining where to drill.

This depositional systems approach has two distinct parts. By studying the modern depositional analogue, we are concentrating on the *process*, or cause, of deposition. As we shall see, there are many facets of depositional processes. These include physical processes, such as wave and current activity, gravity flows, sea-level changes, and tectonism; biological processes, such as biochemical precipitation, bioturbation, and photosynthesis; and chemical processes, such as solution, precipitation, and authigenesis. Other static elements of the environment, such as water depth and chemistry, sediment supply, and local climate and topography, are part of the depositional process as well.

The stratigraphic record, however, consists of sedimentary rocks that are the *product*, or effect, of these depositional processes. We must use the properties of these sedimentary bodies to infer the ancient processes that produced them. These properties include their overall geometry, their physical and biological sedimentary structures, as well as porosity and permeability, acoustical features, resistivity and radioactivity. After looking at a number of ancient sequences which have similar sedimentary features, we distill the local variability and construct a *facies model* that generalizes these features. This facies model is an environmental summary that will suggest what features we might find in the modern depositional system that would explain those features found in the ancient example. This model can be used not only to find a modern depositional analogue, but also as a framework for future observations and a predictor of what features should be found in other examples. The facies model, in turn, can be tested by the examination of more modern analogues or more ancient sequences with new features not seen before.

The important thing to keep in mind, however, is that a facies model is a deliberate idealization. Most of the variability of real examples has been distilled out. As a consequence, each time the geologist attempts to analyze real depositional sequences, there will be variations that are not part of the model. Some of these variations are insignificant. Others are critical to distinguishing between two very similar environments. Some are due to the fact that the case under study may be intermediate between two idealized models, and have properties of both. This is frequently the case in environments that are typically adjacent in the real world. For example, many deltaic sequences interfinger with meandering fluvial floodplain sequences, and many fluvial sequences are hybrids between braided and meandering systems. The geologist interpreting these sequences should not be too rigid about fitting their interpretation into idealized "pigeonholes." Real examples are much more complex than the idealized models we like to work with!

Nevertheless, there is a value in learning about these somewhat oversimplified models. They greatly reduce the complexity of the real world into a useful framework that can guide our thinking, and suggest what features to look for. In many cases, the models suggest what sedimentary features would discriminate between two likely models. Without these models to focus our observations, we would be back in the days of purely descriptive stratigraphy, and would not know where to begin to look for features that might be significant for interpretation.

In the following four chapters, we will examine some of the major depositional systems that are responsible for the bulk of the stratigraphic record. Some environments, such as glacial systems, are omitted because they are relatively rare. After the discussion of each of the major systems, an ancient example is provided. Many of these environments produce extremely variable outcrop patterns, however, and not every outcrop formed in a given environment will closely resemble the example. For this reason, the reader should not take the examples too literally. They are meant to demonstrate some of the major themes seen in each environment, and variability is the norm.

In the following chapters, there is only space to give a brief introduction to the more important depositional systems found in the stratigraphic record. For more detail, the reader is urged to consult some of the many excellent books listed in the end of the chapter.

Gigantic cross-beds from the eolian Navajo Sandstone, Zion National Park, Utah. Photo courtesy W. K. Hamblin.

3

Nonmarine Environments

All the depositional systems discussed in this chapter are nonmarine. These systems tend to be more familiar to us due to our personal encounters with streams, rivers, and lakes. However, most subaerial systems have a low preservation potential because they are above the *base level* of erosion, the level on the earth's surface above which sediments must eventually erode and below which they can accumulate. In most places and times, the base level is near or below sea level. Nonmarine deposits are usually preserved in the rock record only when they fill a basin that is sinking below base level, such as a graben or a subsiding downwarp. In addition, the late Cenozoic is an era of unusually low sea level relative to the rest of the Phanerozoic, so a much larger area of the globe is nonmarine at this moment. During most of the Paleozoic and Mesozoic, however, large shallow epicontinental seas dominated the continents, and areas of nonmarine deposition were few.

ALLUVIAL FANS

The sedimentary cycle begins as bedrock is weathered away from the uplifts and picked up by mountain streams. The first major sedimentary body to be formed of these newly eroded sediments is usually the alluvial fan. An *alluvial fan* is a cone-shaped deposit of coarse stream sediments, sheet flood deposits, and debris flows that forms where a narrow canyon stream suddenly disgorges into a flat valley. The sudden change of a stream from a narrow, confined channel with a steep gradient to the broad flats of the valley causes a sudden drop in the hydraulic power of the stream. The decreased competence of the stream allows the coarser material to drop and accumulate.

Alluvial fans are best known from desert environments (Fig. 3.1) though they also occur in humid environments. They are usually triangular in map view and wedge-shaped in cross section (Fig. 3.2), radiating out of the mouths of mountain canyons. A large number of fans from canyons along a mountain front can coalesce to form a pediment along the base of the mountain. In arid regions, this pediment is known as a *bajada*. The slopes of alluvial fans range from 1° to 25° but average around 5° to 10°. The larger the particle size, the steeper the slope. For an alluvial fan to be active, there must be continued elevation and erosion of the highland to supply the coarse debris. This process most often happens along rising fault scarps, which are most frequently seen in tectonically active desert areas today. Where the head-

Figure 3.1

Single alluvial fan at the mouth of a canyon in the steep east wall of Death Valley. Badwater at left; highway for scale. Note the braided streams coursing along the surface of the fan. From Shelton (1966).

land stops rising, the supply of coarse debris eventually stops, and the fan degrades and begins to merge with the surrounding valley deposits.

Sediment in alluvial fans is transported in three ways: stream flow, debris flow, and mudflow. In arid regions, the most significant stream flow occurs in flash floods, which take place when sudden desert thundershowers dump large amounts of rain in a limited area. This flow is channeled down the mountain canyons as rapidly moving waves of water with tremendous erosive power. There are many stories of enormous walls of water that suddenly sweep down on surprised hikers in canyons where no rain has been falling, and of the enormous boulders, vehicles, and even houses that these flash floods can carry. In *Basin and Range*, John McPhee relates the story of a bartender in Nevada who boarded up the doors and windows of his saloon when he heard the rumble of an approaching flash flood. So large were the boulders brought by the flood that this was the only protection he had against broken doors and windows.

As the flow of the flash flood decreases, it drops the pebbles, cobbles, and boulders wherever the strength of the flow falls below the critical competence needed to transport them. This results in a flow that is choked with more detritus than it can carry, which forms a *braided stream*. The surfaces of alluvial fans in deserts are typically covered with radiating systems of braided stream channels, most of them dry except during the rare flood (see Fig. 3.1). Each flood cuts a new channel, which causes older ones to be filled with gravel. During the high-water point of the flood, the excess water tops the channel banks and spreads out across the fan, forming a shallow sheet of sand or gravel that contains no clay or fine material. These midfan sheetflood deposits are typically well sorted, well stratified, and cross-bedded; they usually form lobes that emerge from the channel at the *intersection point* of the channel profile and the fan surface (Fig. 3.3). Because there is little silt or clay, the water can pass through the porous gravels on the fan without blocking the pores. Thus the lobe-shaped deposit, which is called a *sieve deposit* (Fig. 3.4), becomes progressively coarser toward the front of the lobe, where the porous gravel accumulates. Sieve deposits are usually formed in the proximal fan or upper midfan (see Fig. 3.2).

Freely flowing water is not the only mode of transport in an alluvial fan. When sediment becomes saturated with water, it can flow as a viscous plastic mass that behaves more like quicksand than flowing water because the grains are supported by the pressure of the water-soaked matrix trapped between them. This *debris flow* can carry enormous boulders, as well as every smaller particle, including clay, resulting in an extremely poorly sorted, chaotic jumble of boulders, cobbles, pebbles, sand, silt, and clay (Fig. 3.5). There is no stratification unless a series of flows accumulate on top of one another. Sometimes the base of a debris flow shows reverse (or

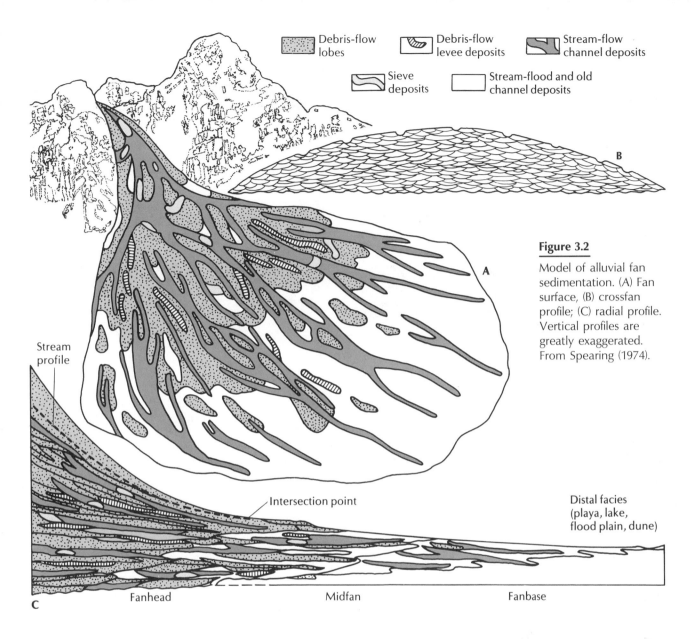

Debris-flow lobes

Debris-flow levee deposits

Stream-flow channel deposits

Sieve deposits

Stream-flood and old channel deposits

Figure 3.2

Model of alluvial fan sedimentation. (A) Fan surface, (B) crossfan profile; (C) radial profile. Vertical profiles are greatly exaggerated. From Spearing (1974).

B

A

Stream profile

Intersection point

Distal facies (playa, lake, flood plain, dune)

Fanhead

Midfan

Fanbase

C

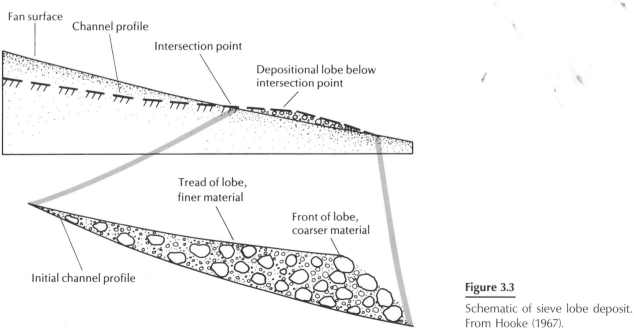

Fan surface

Channel profile

Intersection point

Depositional lobe below intersection point

Tread of lobe, finer material

Front of lobe, coarser material

Initial channel profile

Figure 3.3

Schematic of sieve lobe deposit. From Hooke (1967).

A

B

Figure 3.4

Sieve lobe deposits in a modern alluvial fan, Eureka Valley California. (A) Hummocky topography formed by sieve lobes; slope is 16° along radial profile through fan valley and sieve lobes, 20° along older and smoother parts of the fan to the right. (B) Sieve lobe deposit from the same area. Photos courtesy W. B. Bull.

inverse) grading, in which the grain size increases upward (see Fig. 2.5). Debris flows typically form lobate, tabular bodies of uniform thickness and are commonly found in the upper reaches of the fan. If the viscous fluid flow has little coarse material and is composed mostly of sand, silt, and mud, it is called a *mudflow* (Fig. 3.6). Viscous mudflows form restricted, narrow lobes like debris flows, but mudflows that are more fluid can

Figure 3.5

Debris flow on the Sparkplug Canyon fan, on the west side of the White Mountains, California. Photo courtesy W. B. Bull.

form enormous sheetflood deposits and move at velocities of 10 kilometers per hour (km/hr). There are many examples in the geological literature of catastrophic mudflows that bury villages, overrun fleeing people, and float houses and debris for miles. The mudflows emanating from Mt. St. Helens in Washington and from the Nevado del Ruiz volcano in Colombia are vivid recent examples.

Many examples of ancient alluvial fan deposits are found in the stratigraphic record (see Box 3.1). Because alluvial fans require rapid uplift, they are usually formed and buried in rapidly downdropping grabens, foreland basins, and strikeslip basins. A typical profile through an alluvial fan (see Fig. 3.2) shows a mixture of unsorted debris flows, stream-channel conglomerates (often called *fanglomerates* if they occur in a fan), cross-bedded sandstones, and sieve deposits. Grain size ranges from coarse gravels and cobbles at the top of the fan to fine sands near the base. These features are seen in the typical stratigraphic column (Fig. 3.7), which shows a coarsening-upward sequence of cross-bedded sandstones and fanglomerates. The basal part of the sequence consists of cross-bedded sandstones from the distal fan. As the uplift continues and the fan builds outward and upward, the coarser deposits of the proximal fan accumulate thicker and thicker packages. Many of these are individual debris flows that are meters in thickness. At the very end of the fan cycle, the uplift ceases, and a thin, fining-upward sequence forms during the decay of the fan.

Alluvial fans are normally limited in their lateral extent and merge abruptly with the deposits of the basin.

Figure 3.6

Mudflow on the Santiago Creek fan, from the north side of the San Emigdio Mountains, California. Photo courtesy W. B. Bull.

sandstone are most common. Debris-flow deposits are unsorted, can show reverse grading, and can contain boulders. Sieve deposits form conglomerates with no matrix of finer material, unless it has infiltrated at a later time. Lenticular bodies occur that are comprised of cross-bedded channel sand and pebble conglomerates formed by channel cut-and-fill and debris flows; they are commonest near the apex of the fan and decrease downslope. Paleocurrents radiate from the apex as well. Ripple marks and convolute lamination can occur in the finer grained sheetflood deposits. Sediments can be very immature and angular, with abundant coarse rock fragments and feldspars. Sheetflood deposits are typically oxidized, so redbeds are common.

Fossils Because alluvial fans are usually highly oxidized, fossils and organic matter are rare. High-energy flood waters and coarse conglomerates are also destructive to fossils.

However, their thicknesses can be enormous if the source uplift continues for a long time. For example, the fault graben fill of the Miocene-Pliocene Ridge Basin near the San Andreas fault in California has over 9000 meters of sandstones and fanglomerates. The Triassic graben fills of the Newark Supergroup of the Appalachians exceed 7000 meters in places (Bull, 1972).

Diagnostic Features of Alluvial Fans

Tectonic setting Alluvial fans are typically found in rifting continental grabens, foreland basins, collisional overthrust mountain belts, or other highlands undergoing rapid uplift. They are associated with meandering fluvial valleys and playa lakes.

Geometry Wedge-shaped and limited in lateral extent, alluvial fans extend from only a few tens of meters to kilometers from the source highland. Their thickness can be tremendous (7000 meters or more) if uplift is persistent.

Typical sequence Alluvial fans are comprised of coarsening-upward sequence of cross-bedded sandstone, channel lag conglomerates, and unsorted debris flow deposits. Sometimes a fining-upward sequence forms during the decay of the fan.

Sedimentology There is an extreme range in grain size, from boulders to clay; particle size decreases down-fan. Conglomerate (fanglomerate) and cross-bedded

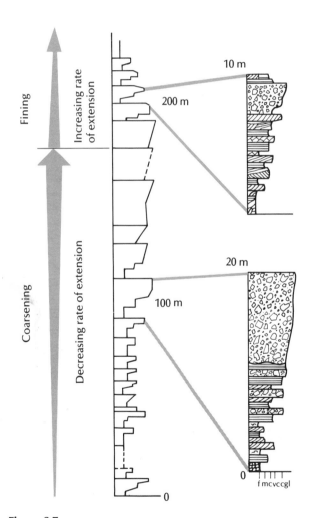

Figure 3.7

Vertical profile through cyclic alluvial fan deposits, showing the characteristic coarsening-upward profile. Based on the Devonian Hjortestegvatnet fan of Norway (Steel and Gloppen, 1980).

Box 3.1
Devonian Fanglomerates of Norway

One of the best exposed examples of ancient alluvial fan deposits occurs in the Hornelen Basin in western Norway (Steel et al., 1977; Gloppen and Steel, 1981). This is one of several fault-bounded basins formed by extension of the region during the Devonian Caledonian orogeny (Steel and Gloppen, 1980; Steel, 1988). Although the basin is about 25 kilometers wide and 70 long, it has 25 kilometers of sedimentary fill. This fill is composed of a succession of stacked coarsening-upward sequences about 100 to 200 meters thick (see Fig. 3.7). These form the spectacular sandstone and conglomeratic cliffs visible from the air (Fig. 3.8). Within each cycle are coarsening-upward subcycles, ranging in thickness from 10 to 25 meters. These cycles, as well as the thicker cycles, are attributed to varying rates of tectonic movement along the extensional fault system. There are also cycles caused by major floods or by the switching of deposition from one fan lobe to another. These units are commonly less than a few meters thick and fine upward.

The alluvial fan deposits are concentrated close to the fault-bounded margins of the basin (Fig. 3.9A) and interfinger with finer basin-fill deposits. These include sheet-flood and stream sandstones and floodbasin sandstones and shales. The fan deposits themselves consist of nearly all the typical facies. There are debris flow deposits (Fig. 3.9B) that contain boulders several meters in diameter. These conglomerates are poorly sorted, lack any pervasive stratification, and are typically clast supported. Many of these beds show inverse grading. In many cases, clasts project above the top of the bed and are buried by the finer, inversely graded base of the next bed. Some of the elongate clasts are oriented vertically, indicating high matrix strength that prevented any further settling or alignment.

Some of the conglomerates have been interpreted as sieve deposits (Fig. 3.9C). These deposits are less than a meter thick and are composed of a well-sorted, open framework of angular pebbles and cobbles. They are lenticular and sometimes banked against larger boulders or blocks. The open framework between clasts has been in-filled at a later time by fine red siltstone.

The bulk of the deposits are stream-deposited sandstones and fanglomerates (Fig. 3.9D,E), which show much better sorting and rounding. They occur as laminated sandstones (Fig. 3.9D) or show planar and trough cross-stratification (Fig. 3.9E). Individual beds vary from 15 to 100 centimeters (cm) in thickness. These are found in lenticular channels in finely laminated sands, interpreted as sheetflood deposits. Some of the debris flow deposits apparently continued below water level, forming subaqueous debris flows (Fig. 3.9F). These are characterized by inverse-to-normal grading within the bed.

Although some of these features can be found individually in other environments, the combination of the coarse conglomerates with the distinctive debris flow and sieve deposits, combined with the geometry of the deposit along the margins of the basin, clearly indicate that much of the Hornelen Basin, conglomerate and sandstone formed in Devonian alluvial fans.

Figure 3.8 Eastward view along part of the axis of the Hornelen Basin, western Norway, showing basin-wide cyclicity of alluvial fan coarsening-upward sequences. Scarps are approximately 100 meters high. In addition to the cyclicity, there is a general trend from fine-grained (foreground) to course-grained alluvial (background) fans. Photo courtesy R. J. Steel.

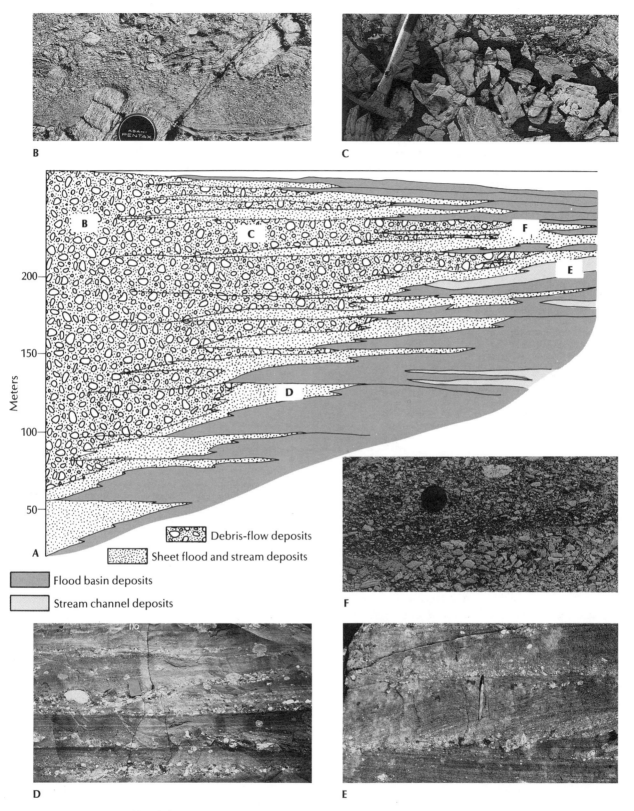

Figure 3.9 (A) Profile of the Hjortestegvatnet fan, showing the characteristic interfingering of deposits. (B) Boundary between two debris flows. The upper flow is inversely graded, and scoured into the lower flow; one large clast from the lower flow protrudes into the upper flow. (C) Possible sieve deposit, made of sorted angular quartzite pebbles with fine-grained secondary matrix filling the cavities. (D) Sheet flood deposits, composed of thin graded beds with pebble lags. (E) Stream flood fanglomerates, showing well-developed cross-bedding. (F) Inversely graded subaqueous debris flows. Photos courtesy R. J. Steel.

BRAIDED FLUVIAL SYSTEMS

Any flowing body of water that has insufficient discharge to carry its sediment load, or has easily erodible banks, forms the classic *braided* pattern (Fig. 3.10). We have already seen how most of the flow in an alluvial fan is braided. Braiding is typically found in the upper reaches of a fluvial system as it comes out of the source area. Thus braided streams usually have steep gradients, abundant coarser sediment (chiefly sand and gravel), and rapid discharge fluctuations, which lead to rapid shifting of channels over their easily eroded banks. Braided systems are also very wide and shallow during their slack-water stages. "A mile wide and a foot deep" was the way the Oregon Trail pioneers described the Platte River, which was so different from the deep, constantly flowing rivers they had known in the East and Midwest.

Large, rapid discharge fluctuations are responsible for the overloading of sediment. At flood stage, the river can carry virtually all of its load. Most of the time, however, little water is flowing, and the competence of the flow is insufficient to move most of the bedload. Some reworking occurs, but most of the changes in the morphology of the braided fluvial complex occur during flood stage. Typically, an obstacle that is difficult to erode causes a flow separation and accumulates sediment in its lee. These deposits continue to build up until they form *longitudinal bars*, one of the most characteristic features of braided systems (Fig. 3.11). Longitudinal bars continue to aggrade downstream and sometimes become islands, which can be stabilized by vegetation. Longitudinal bars are more common in the upper reaches of a braided complex and are usually full of gravel. They show high flow velocity features, including plane bedding. The abundant pebbles are normally well imbricated. On their downstream end, crossbedding may

develop; and sandier, protected channels between the longitudinal bars usually develop ripples and dunes. *Transverse* (or *linguoid*) bars (see Fig. 3.11), which are broader than longitudinal bars and more lobate in shape, are more abundant in the sandier, downstream reaches of the braided fluvial system. Such bars are in fact megaripples, formed during flood conditions and modified by slack-water conditions. Transverse bars are composed of low flow velocity features, particularly ripples, dunes, and trough cross-bedding.

There are also wedge-shaped sand bars that form at oblique angles to the flow. These, called *cross-channel bars*, form where a small channel discharges into a larger one or where the flow is forced to spread laterally or flow obliquely. These bars start with a nucleus that is high during slack-water stages. Sand, flowing around the nucleus, forms "horns" on either side that continue to grow. Eventually, as it grows and accretes other sand bars, the bar can expand into a sandflat. The cross-beds in these bars often show flow directions perpendicular to the main current.

The bars, which tend to merge and divide with the changes in flow, are eroded back along their sides and downstream ends. During flood stages, the old channels, constructed by sand bars, are swamped and obliterated. As the water recedes and the stream stabilizes again, new channels are formed that frequently cut the old. The resulting channel cross-section is thus a complex of lenticular channel sequences that cross-cut one another in great profusion. In longitudinal section (see Fig. 3.11), some bars appear lenticular, but many form long, continuous cross-bedded sand bodies. A typical stratigraphic column for a braided river (see Fig. 3.11) shows abundant trough cross-stratification in channels that are scoured into older channels. The base of the channel sequence contains the coarsest particles, commonly a lag of pebbles or cobbles. Frequently, cross-channel bar migration

Figure 3.10

Braided river channels, Muddy River, Alaska. Photo by Bradford Washburn.

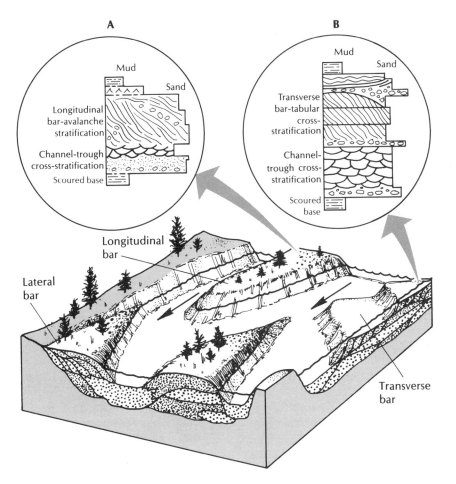

Figure 3.11

Generalized depositional model for a low-sinuousity braided channel. Sequence A is dominated by the migration of a gravelly longitudinal bar Sequence B records deposition of successive transverse bar crossbeds across a channel. From Galloway and Hobday (1983).

produces cross-beds that are perpendicular to the flow direction of the trough cross-beds. As the channel fills and becomes a sand flat, there is both lateral accretion of cross-channel bars and vertical accretion of laminated sands and, eventually, muds that fill the abandoned channel. These bar-top muds are almost the only fine sediment in such systems because most of the clays and muds remain in suspension in the fast-moving water. Frequently, these sand flats are stabilized by vegetation, causing bar-top deposits to be full of burrows and root casts. With the gravel at the base and the mud and silt at top, the typical braided system forms a fining-upward sequence (see Fig. 3.11). Unlike the meandering system described later, it fines upward abruptly at the top of a cycle; the meandering sequence fines upward gradually. An example of an ancient braided river is discussed in Box 3.2.

Diagnostic Features of Braided Systems

Tectonic setting Braided systems occur in the upper reaches of alluvial plains, relatively near source upland. Like alluvial fans, they can be associated with rapidly downdropping basins because they require an upland to provide the coarse material and high stream gradient.

Geometry Elongate, fairly straight lenticular or sheetlike sand bodies grade laterally into finer deposits of an alluvial plain.

Typical sequence There is a fining-upward sequence of channel lag gravels, abundant sandy trough cross-beds filling channels, occasional tabular cross-beds migrating across channels, topped by vertically accreted, laminated sand and mud with burrows and root casts. Unlike meandering rivers, braided systems are ephemeral and rapidly shifting, so the sequence may cross the channel and repeat several times.

Sedimentology Gravel is more common in longitudinal bars of upper reaches of the system, but sand is dominant throughout. Unlike meandering systems, there is very little silt and mud. There is abundant tabular and trough cross-stratification; vertically accreting plane beds are less common. Longitudinal bars can show high flow velocity plane beds, and ripples and dunes are common for lower flow velocities.

Fossils Braided systems are usually unfossiliferous except for root casts and burrows on vegetated sand flats.

Box 3.2
Triassic Fluvial Sandstones of Spain

In the central Iberian Range of Spain, just north of Madrid, is a sequence of Triassic sediments that demonstrate many of the features of sandy braided systems (Ramos et al., 1986). The Rillo de Gallo Sandstone (Fig. 3.12) is composed of medium-grained red sandstone with some conglomerate beds and forms the caprock in much of the Iberian Range. This unit is interbedded with conglomerates and thin-bedded sandstones and mudstones, which together represent the classic basal clastic sequence found throughout the Lower Triassic in Europe. In the type area in Germany, it is known as the Buntsandstein. As in the rest of Europe, the Spanish Buntsandstein is capped by the other two facies of the trinity that gave the Triassic its name (see page 15). Above the redbeds are carbonate tidal sediments equivalent to the German Muschelkalk, followed by mudstones and evaporites of the Keuper facies.

The Spanish Buntsandstein sequence is believed to represent the filling of grabens that were formed when the Atlantic began to rift open in the Triassic (Alvaro et al., 1979). At the base of the sequence are pebble conglomerates formed in alluvial fans (Ramos and Sopeña, 1983). Overlying and interbedded with these conglomerates are the Rillo de Gallo sandstones, which form cliffs and ledges that are several hundred meters thick and can be traced along the range for over a hundred kilometers (see Fig. 3.12). However, they pinch out rapidly as they approach the boundary faults of the ancient graben. The sandstones are medium- to coarse-grained, subrounded to subangular, with less than 15 percent matrix. In most places the sandstones are predominantly quartz (56 to 98 percent) and moderately mature, but near the border faults, the feldspar and rock fragment content can reach 30 percent.

Sketches of the outcrop pattern (Fig. 3.13A)

clearly show a cross section that would be expected from sandbar migration in a braided river channel (Fig. 3.13B). Several common facies are shown by the sedimentary structures in the sandstones. Facies TB consists of tabular cross-strata about 4 meters thick, with flat bases and tops. The foresets dip at angles of 12° to 19°, with the angle increasing downstream. This facies is attributed to the growth and migration of transverse bars (Fig. 3.13B). Some of the tabular cross-stratification (Facies TBv in Fig. 3.13C) shows clear reactivation surfaces, indicating that minor bedforms migrated over the lee side of the major bedform. This facies represents not only lateral growth of the foresets but also vertical growth of the topsets. Other tabular cross-stratified strata pass downstream into trough cross-stratification (Facies TBt). Tabular cross-stratified conglomerates (Facies TC) are interbedded with the sandstones (Fig. 3.13C). Finally, there are trough cross-stratified sandstone units (Facies T) with scoured channel (ch) bottoms. The troughs can be up to 2 meters thick and 20 meters wide. This facies is attributed to migrating channel-fill deposits, formed by the migration of megaripples during flood stage.

As the sketch of the cross section (see Fig. 3.13A) clearly shows, the Rillo de Gallo Sandstone is the product of a braided system. The abundance of gravel, the predominance of tabular cross-stratification, and the almost total lack of fine-grained deposits all strongly support this interpretation. A meandering system, on the other hand, would show almost no gravel and would consist of sandstone channels in floodplain mudstones. Meandering systems seldom show the repeated stacking of tabular cross-stratified sandstones and frequent channeling and trough cross-bedding of braided systems.

Figure 3.12 Panoramic view of Rillo de Gallo Sandstone, Guadalajara, Spain. Facies T and TB are labeled. Photo courtesy A. Ramos, A. Sopeña, and M. Perez-Arlucea.

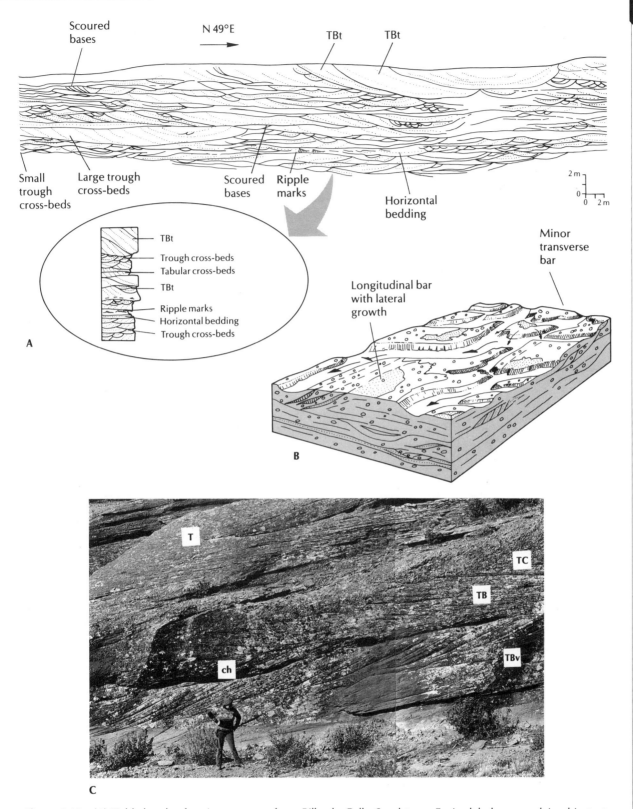

Figure 3.13 (A) Field sketch of main sequence from Rillo de Gallo Sandstone. Facies labels are explained in text. (B) Block diagram of depositional model suggested for the Rillo de Gallo Sandstone. From Ramos, Sopeña, and Perez-Arlucea (1986). (C) View of outcrop. Facies TBv in lower right show tabular cross-stratification, reactivation surfaces and vertical accretion simultaneous with forward progradation of bedforms. Paleocurrent direction is oblique to picture. Small channel (ch) on top of sandbar was filled with cross-stratified sand. Sandstone facies (TB) and tabular cross-stratified conglomerates (TC) are seen in the middle of the picture. Sandstone facies (T) is composed of large trough cross-strata. Photo courtesy A. Ramos, A. Sopeña, and M. Perez-Arlucea.

MEANDERING FLUVIAL SYSTEMS

In the lower reaches of a fluvial system, the gradient is much less steep than before, and most of the coarser material has been left behind. The relatively straight braided channels become more sinuous as they get further from the source uplands until the fully sinuous *meandering* system is established. Braided systems remain relatively straight because of their rapid, though intermittent, flow. As the stream flow becomes slower,

deeper, and steadier, secondary currents can develop. The dominant pattern is a spiral secondary flow (Fig. 3.14A), which results as water moves around a bend and is deflected toward the outer bank of the bend by centrifugal forces. The deflection is stronger near the surface, where the flow velocity is high, and less near the base, where the flow velocity is retarded by friction with the bed. This spiral current tends to erode the outside of each bend and then to transport the material laterally and downstream to the inside bend, where the

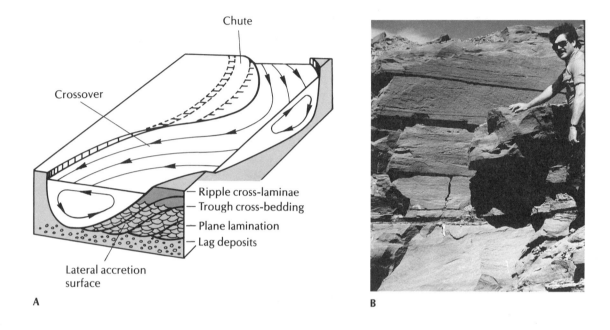

A   B

C

Figure 3.14

(A) Model of meandering river, showing secondary flow patterns and internal structure of point bar deposits. From Blatt et. al. (1980). (B) Detail of the outcrop in (C), showing sedimentary structures characteristic of decreasing flow regimes through the point bar. Trough cross-beds at base are nearly 1 meter thick; the rippled cross-beds near the top are only a few centimeters thick. (C) Cross-section of ancient point bar deposit, showing well developed lateral accretion surfaces ("epsilon cross-beds") accentuated by thin mudstone layers. From the Triassic Moenkopi Formation, north of Winslow, Arizona. Photos by the author.

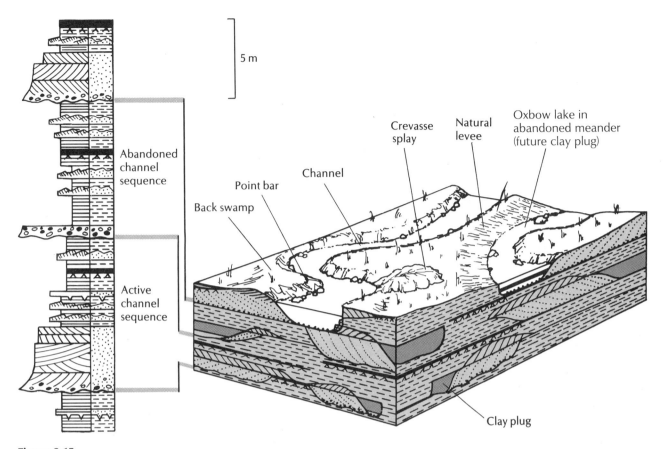

5 m

Abandoned
channel
sequence

Active
channel
sequence

Back swamp

Point bar

Channel

Crevasse
splay

Natural
levee

Oxbow lake in
abandoned meander
(future clay plug)

Clay plug

Figure 3.15

Typical three-dimensional geometry and characteristic vertical sequence of a meandering floodplain deposit, showing point bars, crevasse splays, clay plugs filling oxbow lakes, all interrupting thick sequences of vertically accreted floodplain mud-stones. After Hallam (1981) and Selley (1978).

flow is slower and less erosive. Because the flow spiral in each bend is reversed from that in the previous bend, there must be a nonspiraling flow, or *crossover* (Fig. 3.14A), in the straight stretches between bends.

Because the natural tendency of meanders is to erode their outside banks and deposit along their inside banks, they are continuously migrating laterally and becoming more sinuous. When a river becomes extremely sinuous, the broad meanders have only narrow necks separating them (Fig. 3.15). During floods, the erosive power of the water increases and eventually breaches these necks, forming *cutoffs*. Because cutoffs are straight stretches of rapid flow, most of the water is diverted from the meander. Eventually, the abandoned meander is isolated from the river and forms an *oxbow lake*, which gradually fills up with silt, mud, and vegetation. Sometimes a major load can cause the flow to breach the side of a channel and start a new sinuous course across some other part of the flood plain. This process is called *avulsion*.

The most characteristic product of the meandering river is the *point bar sequence* (Figs. 3.14, 3.15), which

forms at the inner bank of a meander. As the meander becomes more sinuous, the point bar accretes laterally by the following process. At the edge of the bank where the bed is just high enough to begin to slow the current, gravel and course sand are deposited, forming a thin, discontinuous *channel lag* (see Fig. 3.14A). As this base builds up, it diverts the strong gravel- and sand-carrying current further out into the stream where a new packet of channel lag begins to accrete. Over the old channel lag deposit, a high flow velocity plane lamination and large-scale trough cross-beds of migrating dunes are deposited (see Fig. 3.14B). As these features are built, the current that is strong enough for their deposition is in turn diverted outward (as the gravel deposition again moves farther from shore), and low flow velocity sand ripples and a climbing ripple lamination are deposited by the remaining shallow current. If this point bar is now abandoned by the stream so that it is not cut by subsequent channels, it will finally be capped by fine muds on its highest surface by the lowest flow velocity currents that manage to wash across the top of the bar.

The product of this process is the lens-shaped point bar sequence, a series of sedimentary structures that reflect, upward through the sequence, the decrease in flow velocities (Figs. 3.14A,C). If these processes continue long enough for the outer edge of the bar to migrate some distance laterally out into the stream bed, they produce a tabular, well-stratified body of sand that is progressively younger away from the axis of the meander. The top surface of this sand body is covered by arcuate swells and swales known as *scroll bars*, each of which is a product of lateral accretion of point bar sands (Fig. 3.16; see also page 40). It is important to remember that, though each level of the sequence looks continuous and almost undifferentiated from its oldest to its newest deposition all the strata of a point bar sequence are formed simultaneously. The channel lag is farthest out from the axis of the meander, the plane lamination and trough cross-beds next closest to the bank, and so on to the lowest flow velocity deposits at water's edge (this is an application of Walther's law, discussed in Chapter 7). The time planes are not the horizontal boundaries between depositional features but the sloping surfaces that parallel the inside slope of the meander (Fig. 3.14A). The deposit builds *laterally* from the axis of the meander into the stream. Occasionally, this sloping surface is visible in cross sections of the point bar. These are known as *lateral accretion surfaces* or *epsilon cross beds* (Fig. 3.14B). There is a tendency, during mapping and correlation, to treat similar-looking rock units as being of the same age, but the point bar sequence is a classic example of a time-transgressive but laterally continuous deposit. This must be kept in mind when examining ancient deposits (see Box 3.3).

Sedimentary features other than the point bar also form in the meandering system. During flood stage, the point bar is inundated and scoured by flows in shallow flood channels, or *chutes*, which may or may not cut off the meander (see Fig. 3.14A). These *chute bars* form small channels in the point bar sequence that fill with gravel and large-scale tabular crossbeds. The oxbow lakes fill with laminated clay, silt, and organic matter that settle out of suspension after the meander is abandoned. These form isolated *clay plugs* (see Fig. 3.15) in the sandy meander belt, which may block later erosion when the meander attempts to cut across them again.

The most striking difference between the meandering and braided fluvial systems is that the sandy deposits of the meandering river are narrow belts lying in a sequence of muddy floodplain deposits (see Fig. 3.15), whereas the braided system is much broader with relatively little mud or silt. The meandering channel is bordered by a *natural levee*, a low, wedge-shaped ridge that usually confines the flow within the channel (see Fig. 3.15). When floods top the banks, the levees are bypassed, only to be rebuilt as the water level drops.

The rebuilt levees form long, ribbonlike, sinuous bodies of laminated muds and small ripple cross-beds. Because they are the only high features on the floodplain, they frequently become overgrown and may preserve root casts, soil horizons, organic debris, as well as mudcracks and raindrop impressions.

During rising water conditions, the levee is frequently breached, and the water spills out onto the flood plain, forming *crevasse splay* deposits (see Fig. 3.15). These are broad tongues of sandy and muddy sediment that radiate across the floodplain away from the channel. They show small-scale cross-beds and climbing ripple drift. As discussed in Chapter 2, climbing ripple drift indicates a rapidly aggrading condition with excess sediment load relative to the water flow. This is to be expected because the water velocity is greatly reduced after it leaves the channel and spreads as a shallow sheet across the floodplain, finally coming to a stop. When the largest floods come, the entire floodplain is covered with a few centimeters of fine silt and mud. These floodplain muds are uniformly laminated as they settle out of suspension. The floodplain is frequently marshy and can build up significant humus, soil horizons, and other features of vegetation. The natural levee, crevasse splay, and floodplain deposits are all examples of vertical accretion, in which each lamination represents a separate event and is synchronous. By contrast, the lateral accretion of the channel and point bar sequence is time-transgressive.

Diagnostic Features of Meandering Fluvial Systems

Tectonic setting Meandering systems are most commonly found in low parts of the craton. They can be preserved in downdropping basins or in aggrading coastal sequences. They are associated with floodplain muds and lake deposits. They grade downstream into the deltaic system and upstream into a braided system.

Geometry Channel sequences typically form long ribbonlike bodies of sand ("shoestring sands") within a thick sequence of shales. Channel sands may be scattered randomly through the sequence, depending on where the channel shifts after avulsion (see Fig. 3.15).

Typical sequence As in the braided system, there is fining upward from a basal channel lag gravel to the sandy point bar sequence of plane beds, trough cross-beds, and ripple drift (see Fig. 3.14A). Unlike the braided system, there is a much larger fine-grained component of laminated muds formed in the oxbow lakes, natural levees, crevasse splays, and floodplain.

Sedimentology Grain sizes range from channel lag gravels to floodplain muds. Laterally accreted point bar sands show decreasing flow velocity sedimentary structures: plane beds, trough cross-beds, and ripple cross-lamination. Floodplain muds are finely laminated, vertically accreted, and may show climbing ripple drift, mudcracks, raindrop impressions, soil horizons, organic matter, and fossils.

Fossils Organic matter and fossil wood are common, particularly in the floodplain. Land vertebrates and invertebrates can occur in the floodplain muds or in the channel sands. Freshwater molluscs are particularly diagnostic.

Figure 3.16

Aerial photograph of spectacular scroll bars of an Ohio River point bar near Henderson, Kentucky, exhibiting over a kilometer of lateral accretion. Photo courtesy U.S. Geological Survey.

Box 3.3
Paleocene and Eocene Floodplain Deposits of Wyoming

One of the best exposed examples of ancient meandering fluvial deposits can be seen in the Powder River Basin of Wyoming (Ethridge et al., 1981). In this region, the coal and uranium that accumulated in these deposits has prompted massive strip mining operations. The mine walls provide clean, vertical cuts through the floodplain muds and channel sandstones, giving cross sections that look almost like idealized block diagrams. The Powder River Basin lies between the Black Hills of South Dakota on the east, the Bighorn Mountains of Wyoming on the west, and the Laramie Range on the south. These ranges were uplifted by the Laramide Orogeny in the Paleocene and early Eocene, so the basin may have had much the same configuration then as it does now. The ancient drainages converged toward the center of the basin and then drained north into Montana, as they do today.

Most of the basin was apparently covered with low, swampy floodplains, on which both mudflats and coal-producing swamps were abundant. These environments produced the thick sequences of siltstones and claystones of the Paleocene Fort Union Group and the Eocene Wasatch Formation. When the floodplain was covered by thick vegetation, low-grade lignite coal seams were produced. In some places, coal seams reach 10 meters in thickness, and a few are almost 30 meters thick! The great thickness, low sulfur content, and proximity to the surface of these coal seams make this region one of the primary coal-mining areas in North America.

Punctuating the thick coal-bearing mudstone sequence are the sandstones of the fluvial channels (Fig. 3.17A). Many of these channel sands occur as isolated lenses in the mudstones. Surrounding these channels are thin stringers of fine sandstone produced by crevasse splays (Fig. 3.17B). These consist of fine-to very fine-grained sandstone with ripples, small scale trough cross-bedding, parallel lamination, and scour and fill structures. The upper parts of the crevasse splays are commonly full of roots and burrows, and their upper portion merges into the swamp deposits.

In addition to the lenticular sandstones of the fluvial channels, there are thick tabular sand bodies that appear to represent migrating point bar sequences (Fig. 3.17C). These fine upward from medium- to very coarse-grained sandstone at the base to very fine-grained sandstones and siltstones at the top. Horizontal lamination and large-scale cross-beds are found near the bases; the cross-beds become progressively smaller in scale toward the top. The most characteristic diagnostic features of the point bar, however, are the large-scale lateral accretion surfaces, or epsilon cross-beds. In the Powder River Basin, these can be as much as five meters thick.

As the point bars and meanders migrated, they became more sinuous until the meander was cut off, leaving abandoned channels (Fig. 3.17D) that became oxbow lakes. These eventually filled with fine-grained, parallel laminated, organic-rich claystone and siltstone and eventually were buried and capped by laminated floodplain mudstones. Many of these lake deposits contain freshwater molluscs (Fig. 3.17E).

In summary, the early Tertiary rocks of the Powder River Basin show all the classic features expected of a meandering fluvial system: a predominance of floodplain muds and silts punctuated by coal seams, lenticular sand channels with crevasse splays, tabular point bar sand bodies with lateral accretion surfaces, and abandoned meander channels filled with laminated, organic-rich clays and silts.

Figure 3.17 (A) Block diagram of Paleogene floodplain deposits, Powder River Basin, Wyoming. (B) Cross-bedded crevasse-splay sand sheets alternating with floodplain muds and coals. (C) Lateral accretion surfaces in a 5-meter thick point bar channel sand, Bear Creek Uranium Mine. (D) Clay plug filling abandoned meander channel, approximately 8 meters thick. (E) Freshwater molluscs in pond deposit. Photos B, C, and D courtesy F. G. Ethridge. Photo E by author.

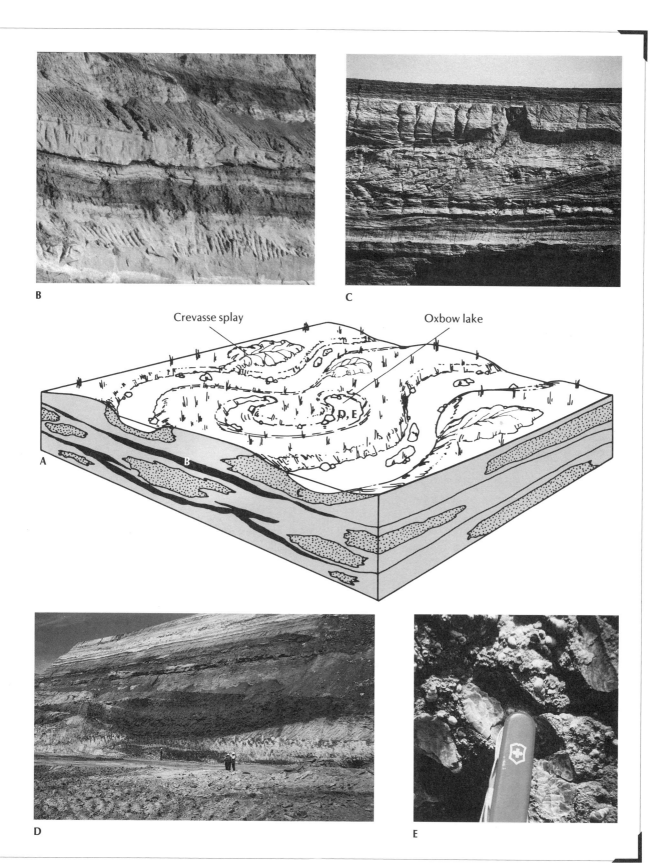

B

C

Crevasse splay   Oxbow lake

A   B   D, E

D

E

LACUSTRINE DEPOSITS

A lake is a landlocked body of standing, nonmarine water, and *lacustrine* deposits are formed by ancient lakes. In terms of geological time, most lakes are temporary features. They form in some sort of basin or impoundment and become sediment traps that eventually fill up. Lakes can vary enormously in their dimensions, from small, ephemeral ponds to bodies of water as large as the Caspian Sea (372,000 square kilometers) and the Great Lakes (245,240 square kilometers). Some lakes are deep (Lake Baikal is 1742 meters deep), but some are so shallow that their levels fluctuate dramatically with season and climate. For example, Lake Eyre in Australia (9300 square kilometers in area) is covered with water only a few times in a century and is a dry lake bed the rest of the time. About 60 percent of lakes are freshwater, but many are saltier than the ocean.

Ancient lacustrine deposits received relatively little attention in the past, but recently, they have become important source rocks for oil shale, uranium, and coal. Many ancient lakes were significantly larger than all but the largest lakes today. For example, the Triassic lakes that produced the Popo Agie Formation of Wyoming, Utah, and Colorado had a surface area over 130,000 square kilometers, and Pleistocene Lake Dieri in Australia covered over 110,000 square kilometers. Lake deposits can also vary tremendously in thickness. Lake Eyre and its Pleistocene predecessors have accumulated only 20 meters of sediment, but Devonian lacustrine sediments in Scotland are reported to be up to 4000 meters thick. Lakes are most common in regions of internal drainage, where a closed basin accumulates water. Most often they are found in regions of tectonic depression such as rift grabens. They also form in volcanic calderas (e.g., Crater Lake), glacial depressions (e.g., the Great Lakes and most of the ten thousand lakes of Minnesota), karst sinkholes, and meteorite craters, or are impounded in river valleys behind glacial moraines, lava flows, alluvium, or landslide debris. For a lacustrine deposit to be preserved, however, it must accumulate a thick sequence that is subsequently buried. This most commonly happens in grabens and broad regional downwarps. Consequently, these areas contain most of the ancient lake deposits.

Because lakes and shallow seas are both large bodies of standing water, they produce fairly similar sedimentary sequences. In both cases, the dominant sediment is low-energy silt and mud with occasional carbonates. In most other respects, however, lake shales differ from marine shales. On the average, lake basins are much smaller than epicontinental seas, so their deposits tend to be much less laterally continuous than marine shales and limestones. Along the shore of a lake, there is a rapid change in facies, interfingering with a narrow belt of fluvial deposits and even alluvial fans, which are less likely to occur along a marine coastline. Typically, lakes form a series of these facies belts that are arranged concentrically from the mudstone or marls in the center to the coarsest sandstones on the margins. Because most lakes fill with sediment over time, they tend to show a sequence that is shallowing and coarsening upward (see Fig. 3.18).

Many ancient lakes, however, are large enough to produce facies patterns similar to marine sequences. Deep lakes that are affected by pulses of sedimentation during runoff peaks can have turbidity currents, which produce turbidite sequences. The marginal fluvial sequences can prograde and fill in the lake, forming lacustrine deltaic deposits (*Gilbert-type deltas*). Usually, however, lacustrine deltas and turbidites are of a much smaller scale than their marine analogues.

Lakes also have some properties that produce sediments seldom found in marine sequences. Most lakes have negligible tides, and their waves are generally small, the shoreline belt and its associated sedimentary structures are limited, and the peculiar patterns of lake circulation have a much greater effect on the sediments than the wave-dominated flow seen in the marine environment. Typically, a lake has periods of overturn (e.g., spring

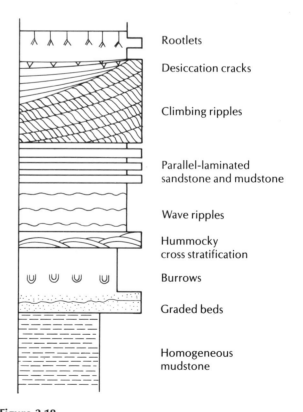

Rootlets

Desiccation cracks

Climbing ripples

Parallel-laminated sandstone and mudstone

Wave ripples

Hummocky cross stratification

Burrows

Graded beds

Homogeneous mudstone

Figure 3.18

Ideal vertical sequence developed by lake margin regression due to fluvial influx. From Galloway and Hobday (1978).

and fall) when the entire lake circulates and periods when the water is density stratified (e.g., summer and winter). This regular alternation of stagnation and overturn produces fine lamination that can be very rhythmic and laterally extensive. This lamination can reflect not only seasonal cycles but also larger-scale climatic cycles, and thus it is of great importance as a paleoclimatic tool. In proglacial lakes, the cycles of stagnant frozen winter sediments and oxygenated sediments of the summer overturn produce striking glacial varves (see Fig. 2.2). During periods of stagnation in any density-stratified lake, organic matter becomes concentrated in the deeper water. The lack of oxygen from reduced circulation and from the excess nutrients leads to stagnant, reducing conditions. Under these circumstances there will be a concentration of organic matter, producing black shales that are high in low-grade hydrocarbons, or *kerogens*. These are the "oil shales" that have received so much attention in recent years. Stagnant lake bottoms also discourage scavengers and the processes of decay and so can produce unusually complete fossils, such as the famous fossil fish of the Eocene Green River Formation (see Box 3.4).

Where the clastic input into a lake is limited, chemical sedimentation can dominate. Chemical precipitates are usually either saline or carbonate. Saline lakes are better known because they are abundant in the desert basins all over the world. Wherever evaporation exceeds inflow, the salinity can exceed 5000 parts per million (ppm) dissolved solids, producing a saline lake. (Normal marine salinity is about 3300 to 3700 parts per million). The most abundant chemicals are SiO_2 and ions such as Ca^{2+}, Mg^{2+}, Na^+, K^+, HCO_3^-, CO_3^{2-}, SO_4^{2-}, and Cl^-. As evaporation proceeds and increases the ionic concentration, dense brines sink to the bottom and precipitate evaporite minerals. Carbonates are the first to be produced, followed by gypsum. If the process continues, halite is precipitated, followed by a series of potassium and magnesium salts, many of which are unique to dry lakes. These can include a number of unusual hydrated sodium and calcium carbonates and sulfates (such as gaylussite, trona, glauberite, mirabilite, and thenardite), halides (halite and sylvite), and even borate minerals. Desiccation features, such as mudcracks, are commonly associated with the minerals found in the late stages of the drying of a lake.

Lacustrine carbonates, on the other hand, are produced where there is neither excess evaporation nor much clastic input. Unlike marine carbonates, freshwater limestones are produced mostly by inorganic precipitation. Freshwater typically contains abundant carbonate from dissolved atmospheric CO_2 and from dissolved bedrock carbonate. The carbonate ion concentration is strongly controlled by changes in pH, which fluctuates continuously in freshwater lakes. Precipitation of calcite is facilitated by two factors: plants use carbon dioxide

and raise the pH, and warmer temperature lowers the solubility of calcite. Lacustrine carbonates are usually low-magnesium calcite, precipitated in finely laminated beds with mudstone and marl. Occasionally, biogenic activity can produce freshwater limestone. Freshwater ostracods are the most common, but freshwater gastropods and bivalves are also important. Freshwater calcareous algae form a kind of stromatolite called an *oncolite*, which is a subspherical or tabular body formed when encrusting algae trap sediment. Some reach tens of centimeters in diameter and are known as *algal biscuits*.

These freshwater fossils, though rare, are important as environmental indicators. In many cases, lacustrine and marine deposits are difficult to distinguish, but the absence of normally abundant marine invertebrate fossils and/or the presence of certain freshwater invertebrates are often the best criteria. In addition, the scale and geometry, the rapid facies changes, and the tectonic setting can all yield clues to distinguish marine from lacustrine deposits.

Diagnostic Features of Lacustrine Deposits

Tectonic setting Lacustrine deposits are typically found in fault grabens or broadly downwarped basins with internal drainage or limited outflow. They are associated with other nonmarine environments, particularly fluvial sands and alluvial fans.

Geometry They are circular or elongate in plan view and lenticular in cross section. Size ranges from a few meters to 100,000 square kilometers; usually thin (less than 200 meters), but sequences can exceed 1000 meters.

Typical sequence There is a coarsening upward from laminated shales, marls, and limestones to rippled and cross-bedded sandstones and, possibly, conglomerates. The sequence frequently shows cyclic alternation of laminae.

Sedimentology There are finely laminated mudstones, commonly rich in kerogen, along with marls and freshwater limestones. Margins of the lake have fluvial muds and sands, with cross-beds, wave ripples, and climbing ripple drift. Thin turbidite sequences can form in the lake basin, along with black shales. Hypersaline lakes form a regular sequence of carbonate, gypsum, halite, and other evaporites, often associated with desiccation features such as mudcracks.

Fossils Important for distinguishing lacustrine from marine sequences, the complete absence of normal marine invertebrates is significant, as is the presence of nonmarine ostracods, diatoms, molluscs, and freshwater fish and insects. Preservation is excellent in some anoxic lake basins.

Box 3.4
The Eocene Green River Formation of the Rocky Mountains

Figure 3.19   Classic exposures of extensive, finely laminated Green River lacustrine shales. View northward across Hells Hole Canyon, Uintah Country, Utah. Photo by W. Cashion, courtesy U.S. Geological Survey.

One of the largest and best-studied ancient lacustrine deposits in the world is the Eocene Green River Formation of Wyoming, Colorado, and Utah. It is one of the world's richest deposits of oil shale, as well as having economic amounts of trona and other evaporite minerals. It extends over 100,000 square kilometers and reaches a maximum thickness of 3 kilometers in Utah (Ryder et al., 1976). The Green River Formation was deposited in four separate basins: the Green River and Washakie Basins in Wyoming, the Piceance Basin in Colorado, and the Uinta Basin of Utah. The Wyoming basins were filled by Eocene Lake Goshiute, and the southern basins contained Eocene Lake Uinta. Lake Goshiute was apparently a shallow playa during most of its history whereas Lake Uinta was a deeper, perennial lake. Although their detailed histories are different and there are different formal stratigraphic names for each basin, their overall histories are the same. Together, they spanned almost ten million years of deposition in the region.

The predominant lithology of the Green River Formation is miles and miles of finely laminated shale (Fig. 3.19). The shales include marlstones, silty organic-poor dolomicrites, dolomitic claystones, volcanic tuffs, and oil shales. The latter contain much organic matter in the form of kerogen, making them valuable sources of hydrocarbons. The kerogen is associated with calcite layers and is believed to have accumulated during algal blooms. This would indicate quiet waters and depths of 5 to 30 meters (Ryder et al., 1976). The shales show cyclic patterns of deposition that are probably caused by the effect of climatic fluctuations on lake level. These cycles

Figure 3.20   (A) Block diagram showing relationships of facies of Green River Formation. (B) Intertonguing between lacustrine shales and fluvial sands of the Uinta Formation. (C) Mudcracked shales from the desiccated lake margins. (D) Finely varved oil shale, the commonest rock type in the open lake basin. (E) Authigenic crystals of dolomitic marlstone of the saline mudflat. (F) Freshwater algal stromatolites from the lake basin. (G) Extraordinary preservation of freshwater fossils, including this frog, is common in the Green River Formation. Photo B by author. Photos C, D, E, and F by W. H. Bradley, courtesy U.S. Geological Survey. Photo G courtesy L. Grande.

B

C

D

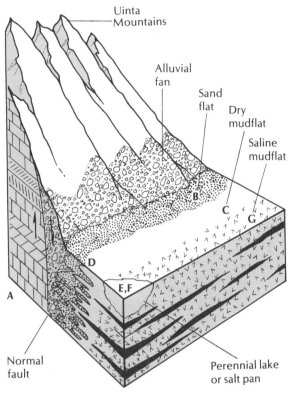

Uinta
Mountains

Alluvial
fan

Sand
flat

Dry
mudflat

Saline
mudflat

B

C

G

D

E,F

Normal
fault

Perennial lake
or salt pan

A

E

F

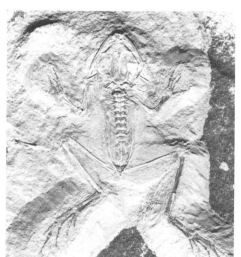

G

reach 5 meters in thickness and can be traced for 20 kilometers without change in thickness. These climatic fluctuations are believed to have operated on time scales of 20,000 to 50,000 years per cycle.

Lake Uinta was apparently a stratified, shallow carbonate- and organic-rich perennial lake (Desborough, 1978). Along its margins (Fig. 3.20A), the lake shales pass into sandstones and siltstones of the streams and deltas along the lakeshore. In some places, the lake margin is dominated by carbonate tidal flats with stromatolites, oolitic and pisolitic grainstones, and green pyritic marlstone. Seasonal stratification was so strong that the deeper parts of the lake were reducing and anoxic, allowing the deposition of oil shale. By contrast, the surface waters must have been well oxygenated because the Green River Formation is famous for its well-preserved fossils of freshwater fish. These organisms and many birds, frogs, and other unlucky victims (as well as much plant debris) sank to the stagnant bottom, where there were no scavengers to disaggregate them. They were slowly buried by the fine clay, resulting in world-famous preservation (Fig. 3.20G).

Lake Goshiute, on the other hand, appears to have been considerably shallower and more ephemeral. During the deposition of the Wilkins Peak Member of the Green River Formation, this playa lake frequently dried up (Eugster and Hardie, 1975). The best indicators of this are thin-bedded dolomitic mudstones with abundant mudcracks (Fig. 3.20C), which show complex histories of opening and filling. These suggest an exposed playa covered by occasional sheetfloods that deposited the fine laminae of silt and mud and filled older mudcracks. In some places, there are flat-pebble conglomerates made of ripped-up dolomitic mudchips. The mudstones are interbedded with wave-rippled calcarenites and oolites, suggesting shallow, agitated waters during wet episodes. Algal stromatolites occur along the lake margin (Fig. 3.20F). At the edges of Lake Goshiute, the mudflat deposits are interbedded with massive and cross-bedded sandstones of the marginal alluvial fans and streams that encroached on the lake basin (Figs. 3.20B, 3.21). Some of these sandstones are 10 meters thick and can be traced for distances of 100 kilometers. They probably represent braided streams that traversed the old lake floor during dry phases.

The center of Lake Goshiute was occupied by a lacustrine sequence that changed with lake level. During wet periods, the lake bottom was filled with stagnant, organic- and carbonate-rich water that produced oil shales and dolomitic mudstones similar to those of Lake Uinta (Fig. 3.20D). During periods

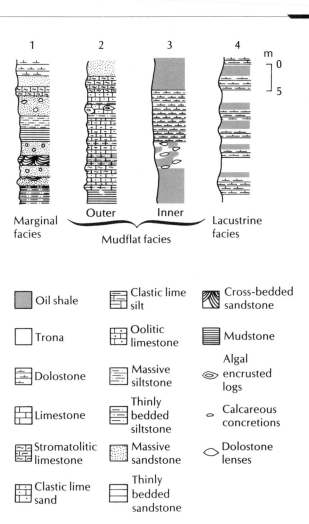

Figure 3.21 Schematic stratigraphic sections of the Green River Formation: 1, marginal facies; 2, outer mudflat facies; 3, inner mudflat facies; and 4, lacustrine facies. From Surdam and Wolfbauer (1975).

of drying, the lake bottom became hypersaline, and thick (1 to 11 meters) beds of trona and halite were precipitated (Fig. 3.20E). These conditions were similar to modern soda lakes, such as Lake Magadi in Kenya (Eugster, 1970). Abundant volcanic ash contributed to the unusual alkali chemistry of this system.

Although the thick, extensive sequence of Green River shales might at first suggest a marine environment, there are clear indications that it was lacustrine in origin. It was formed in a landlocked basin far from the ocean and contains freshwater fish and ostracods, land plants, and many other diagnostic organisms. The abundance of mudcracks, flat-pebble conglomerates, and evaporites in Lake Goshiute also suggests an ephemeral lake. The frequently stagnant water, producing organic-rich oil shale, is unique to lacustrine environments.

EOLIAN DEPOSITS

Eolian (wind-formed) sand dunes are among the best known depositional environments because of their spectacular crossbeds (see Box 3.5) and their use as paleoclimatic indicators. Sand dunes are best known from the desert environment though they do occur in coastal settings (discussed in Chapter 4). Most arid deserts occur in a narrow belt between 20° and 30° latitude in areas of persistent high pressure, or behind mountain ranges that have a *rain shadow* effect. Thus they have been used as indicators of paleolatitude, as well as clear signs of arid climate and strong prevailing winds.

Because air is only about 1/1000th the density of water, eolian processes are very different from subaqueous processes. Wind has much less competence to pick up particles than does water, so the coarsest material is left behind as a *deflation lag*, or *desert pavement*, which protects underlying fine material from being eroded. Most of the sand-sized and finer particles are carried along in a traction carpet moving just above the surface,

and the individual grains move by saltation and intergranular collision. As a result, eolian sands tend to be very well sorted and well rounded, with microscopic surface pitting and frosting. The silt and clay is carried away by the wind to be deposited where the wind diminishes. Most dune sands are nearly pure quartz although there are unusual types, such as the gypsum sands of White Sands, New Mexico.

The most characteristic feature of eolian dunes is their enormous cross-beds. Unlike water, there is no depth limitation on the "sea of air," so dune height is limited only by the strength of the wind and the supply of sand. Individual cross-bed sets can reach 30 to 35 meters in thickness, with foresets that dip 20° to 35° (Fig. 3.22). This thickness of individual sets is found throughout the entire unit; it is not just an unusual maximum. Eolian cross-beds have long, sweeping sigmoid shapes with longer, more asymptotic bottomsets than do marine dunes. Geometrically, most cross-beds are planar-tabular or trough- and wedge-shaped. Detailed study of the faces of eolian dunes (Fig. 3.23) shows that the ripples move laterally across the face of the dune. This

Figure 3.22

Large-scale wedge-planar crossbeds (middle) between tabular crossbeds in the Navajo Sandstone, Zion National Park, Utah. Photo by E. Tad Nichols, courtesy U.S. Geological Survey.

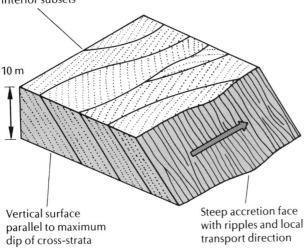

Horizontal or slightly dipping
erosional surface exposing
interior subsets

10 m

Vertical surface
parallel to maximum
dip of cross-strata

Steep accretion face
with ripples and local
transport direction

Figure 3.23

Composite sets of cross-strata deposited on an ancient
eolian dune complex. Sand drifted by wind along dune
slopes left the low crests of eolian ripples on their surfaces.
From Walker and Harms (1972).

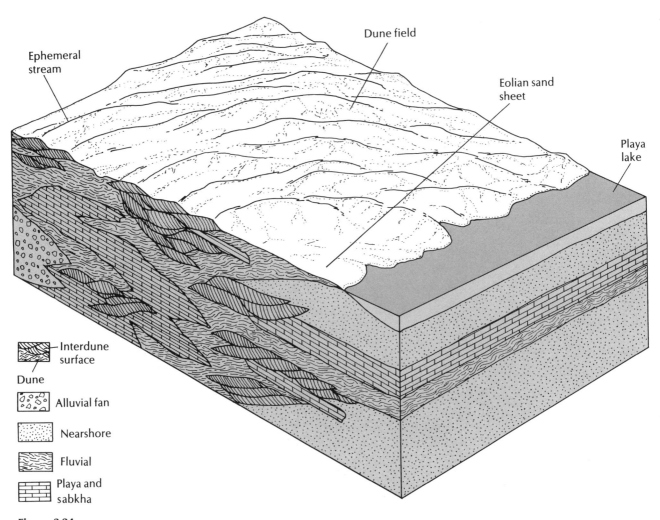

Ephemeral
stream

Dune field

Eolian sand
sheet

Playa
lake

Interdune
surface

Dune

Alluvial fan

Nearshore

Fluvial

Playa and
sabkha

Figure 3.24

Opposite page: Facies model for (A) eolian sand sheet and ephemeral stream deposits and (B) eolian dune deposits. The
block diagram (above) shows lateral facies relationships of dune and interdune systems. Diagrams A and B from Fryberger et.
al. (1979); block diagram adapted from Lupe and Alhbrandt (1979).

suggests that much of the cross-bed is formed not by sand avalanching down the face of the dune but by sand blown across the lee face of the dune.

In addition to the abundance of large-scale crossbeds, low amplitude wind ripples typically form a small-scale cross-lamination within the larger cross-bed sets. These usually migrate *up* and along the lee face of a dune, which does not occur in marine sands. Eolian ripples are typically lower in amplitude and more asymmetrical than water ripples. Other types of deposits can be found between the dune fields. Thin mud-stones with mud-cracks and raindrop impressions are the remains of ephemeral lakes between dune fields. Occasionally. coarse deflation lags from blowout areas are also found interbedded with dune deposits. A typical eolian se-

A

B

Figure 3.25

Interbedded eolian sandstones (thick units) and interdune deposits (thin-bedded sandstones and shales) of the Permian Cedar Mesa Sandstone, southeastern Utah. (A) Twenty-five-meter thick eolian cross-beds (between upper arrows), which interfinger to left with playa shales. Lower arrows indicate 12-meter thick eolian sandstones capped by playa shales. (B) Deflationary surface S covered by playa mudstones with filled mudcracks M and intruded by sand-stone dikes D. The two overlying eolian sandstones are separated by shales deposited by flooding F. From Langford and Chan (1988); photos by R. Langford.

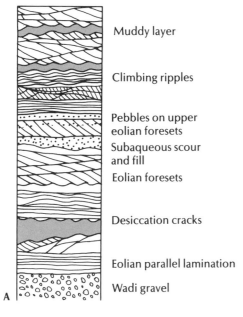

A

Muddy layer

Climbing ripples

Pebbles on upper eolian foresets

Subaqueous scour and fill

Eolian foresets

Desiccation cracks

Eolian parallel lamination

Wadi gravel

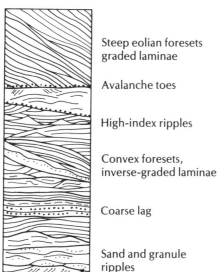

B

Steep eolian foresets graded laminae

Avalanche toes

High-index ripples

Convex foresets, inverse-graded laminae

Coarse lag

Sand and granule ripples

quence (Figs. 3.24, 3.25) is dominated by thick dune crossbeds and lesser amounts of deflation gravels, mud-cracked pond shales, and small-scale cross-lamination. Unlike most other depositional sequences, however, eolian deposits have no regular progression of sedimentary structures. Dune migration and ephemeral lake formation is a much more random process than the orderly sequence of facies belts found in most sedimentary environments.

Recently, some geologists (reviewed by Walker, 1984) have suggested that many examples of classic eolian cross-bedded sandstone are actually formed by large-scale marine sand waves and tidal ridges (discussed in

Box 3.5
Jurassic Dunes of the Navajo Sandstone, Utah and Arizona

One of the most photogenic examples of ancient cross-bedded dune sands is the famous Navajo Sandstone of the Utah-Arizona border region. This thick sandstone makes up the spectacular white cliffs of Zion National Park in Utah and the combination of the fractures and the cross-beds gives "Checkerboard Mesa" its distinctive surface pattern. In northeastern Arizona the Navajo Sandstone is about 300 meters thick, and it reaches 600 meters in southwest Utah (Harshbarger et al., 1957). Its correlative, the Aztec Sandstone in the Mojave Desert of California and Nevada, is over 900 meters thick. This sandstone also extends to southwest Wyoming, where it is known as the Nugget Sandstone and is about

150 meters thick. Along its eastern and southeastern margins in Colorado and northern Arizona, the Navajo thins to a wedge and then vanishes. This enormous sand sea must have extended 1000 kilometers north-south and 400 kilometers east-west, comparable to a small Sahara (Fig. 3.26).

The Navajo Sandstone consists of homogeneous, well-rounded, well-sorted, frosted quartz sand, with an average grain diameter of 0.2 millimeters (mm). Its most striking feature is the well-developed cross-stratification. Both wedge-planar (see Fig. 3.22) and tabular-planar (Fig. 3.27) cross-beds are well developed. Individual cross-bed foresets are typically long (15 meters) and sweeping, with concave-

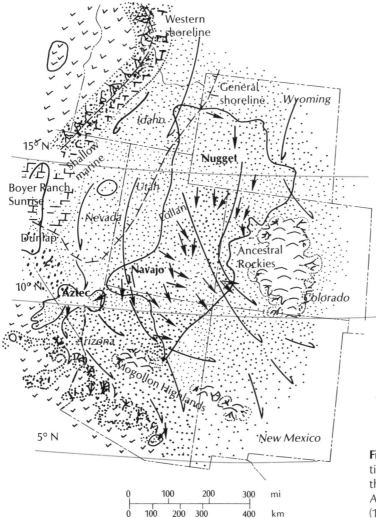

Figure 3.26 Paleogeographic reconstruction of environmental conditions during the deposition of the Navajo-Nugget-Aztec Sandstones. After Kocurek and Dott (1983).

upward surfaces. Dips are 20° or greater. Individual cross-bed sets are typically 5 to 10 meters thick and in some places reach almost 35 meters in thickness!

Ripple marks are relatively rare in the Navajo Sandstone. Where they occur, their crests are parallel to the crests of the dunes. Only ripple marks that form from crosswinds on the lee sides of the dunes have a good chance of burial and preservation. Fossils also are rare in the Navajo dune sands. Ostracods and other freshwater crustaceans have been reported from interdune areas, and a few dinosaur and pterosaur tracks (Fig. 3.28A) are known, as well as the skeletons of bipedal dinosaurs and mammal-like reptiles (Fig. 3.28B). More common are contorted beds and other slump structures. These kinds of folds are often seen in modern dunes when the sands have been wetted; they are commonly due to oversteepening on the lee side of the dune.

Compared with the spectacular cross-bedded sand dunes, the interdune deposits have received little attention. The Navajo and the underlying Kayenta Formation, which interfingers with the Navajo and contains eolian sands throughout, contain many facies that are interpreted as interdune deposits (Middleton and Blakey, 1983). In many places, the thick sets of eolian cross-bedded sandstones are separated by thin (less than 1 meter) horizontal planar-laminated sandstones. These appear to represent thin sheets of interdune sand deposited on an erosional surface between episodes of dune migration.

In the intertonguing interval are thicker sequences of planar-laminated fine-grained sandstone and siltstone. These occur as beds 2 to 6 meters in thickness, sandwiched between eolian cross-bedded sandstones. Thin laminae of very coarse-grained sandstone and granules also occur in this facies. These beds are believed to be the product of interdune ponds and mudflats. In a few places, sandfilled V-shaped desiccation cracks can be found, which indicate alternating wetting and drying of these mudflats. Occasionally, the beds are capped by a calcareous horizon that probably represents a caliche. They also include cryptalgal laminated car-

Figure 3.27 Tabular-planar cross-beds in the Navajo Sandstone, north of Kanab, Utah. Photo by E. Tad Nichols, courtesy U.S. Geological Survey.

A

B

Figure 3.28　Pterodactyl trackway (top) up the dune face in the Navajo Sandstone. Photo courtesy J. Madsen. Skeleton of a mammal-like reptile (bottom) from the Navajo Sandstone, Navajo National Monument. Photo courtesy Shuler Museum of Paleontology, Southern Methodist University.

bonates with ostracods and mudcracks that were probably formed in a small calcareous pond.

In summary, many lines of evidence in the Navajo Sandstone clearly point to an eolian origin and rule out formation by tidal sand ridges. The enormous thickness, abundance, and steepness of the cross-bedded sands, which are tabular-planar and wedge-planar cross-stratified, clearly point to a dune origin.

The high angle and the size of the dune foresets are particularly characteristic of desert sand dunes. In addition, the rapid directional fluctuations of the cross-bed sets would only be expected with rapidly fluctuating wind directions. The mudcracked inter-dune deposits indicate subaerial exposure. Finally, the presence of dinosaurs and their tracks and of freshwater ostracods rules out a marine origin.

Chapter 5). Walker and others have shown convincingly, however, that tidal sand waves have dips of no more than 6°, in contrast to the much steeper dips found in eolian dunes. In addition, the other evidence—mud-cracked layers, absence of marine faunas and presence of terrestrial trackways, tens to hundreds of meters of monotonous, thickly cross-bedded sands, and association with other terrestrial facies—argue strongly against a marine origin for these extensive, steeply cross-bedded sandstones.

Diagnostic Features of Eolian Deposits

Tectonic setting Eolian deposits are found in inland basins in latitudes between 10° and 30°, or behind mountains in rain shadows. They are commonly associated with alluvial fans, playa lakes, and other desert deposits.

Geometry Dune fields can cover hundreds of square kilometers and form thick tabular bodies with individual sets up to 35 meters in thickness.

Typical sequence The sequence is dominated by enormous cross-bed sets that are meters in thickness with foreset dips of 25° to 30°; lesser amounts of mud-cracked shales and deflation lag gravels are present. The sequence of sedimentary structures is random.

Sedimentology Extremely well sorted, well rounded, quartz-rich sand with no fine matrix and few coarse gravel lags are found. Grain surfaces are typically frosted or pitted (under the scanning electron microscope) and are highly oxidized and coated with iron oxides. Large-scale cross-beds composed of smaller-scale low-amplitude wind ripples form cross-lamination that moves along or up the lee face of the dune.

Fossils Rare vertebrate footprints, root casts, and insect burrows, are all typical.

FOR FURTHER READING

The literature on sedimentary environments has grown enormously in the past few years. However, certain key works are indispensable for further understanding and illustration of the environments discussed above.

Davis, R. A., Jr., 1983. *Depositional Systems*. Prentice-Hall, New York. The most complete and current textbook on the subject.

Ethridge, F. G., and R. M. Flores (eds.), 1981. *Recent and Ancient Nonmarine Depositional Environments: Models for Exploration*. Soc. Econ. Paleo. Min. Spec. Publication 31. An excellent collection of papers on alluvial fan, fluvial, lacustrine, and eolian deposits.

Galloway, W. A., and D. K. Hobday, 1983. *Terrigenous Clastic Depositional Systems*. Springer-Verlag, New York. A well-illustrated description of terrigenous systems, concentrating on the economic aspects of their analysis.

Reading, H. G. (ed.), 1986. *Sedimentary Environments and Facies*, 2d ed.) Blackwell Scientific Publications, Oxford. A complete, well-illustrated discussion of the major facies types.

Reineck, H. E., and I. B. Singh, 1973. *Depositional Sedimentary Environments*. New York, Springer-Verlag. Well-illustrated coverage of modern sedimentary systems, but little discussion of ancient examples.

Rigby, J. K., and W. K. Hamblin (eds.), 1972. *Recognition of Ancient Sedimentary Environments*. Soc. Econ. Paleo. Min. Spec. Publication 16. A symposium containing many classic papers on the criteria for distinguishing depositional environments.

Scholle, P. A., and D. Spearing (eds.), 1982. *Sandstone Depositional Environments*. Amer. Assoc. Petrol. Geol. Memoir 31. Gorgeously illustrated (mostly in color) treatment of most of the clastic environments, with excellent discussion.

Selley, R. C., 1978. *Ancient Sedimentary Environments*, 2d ed., Cornell Univ. Press, Ithaca, New York. Excellent short review of the major ancient environments and their formation.

Spearing, D., 1974. *Summary Sheets of Sedimentary Deposits*. Geol. Soc. America Maps and Charts Series MC-8. Seven major depositional environments are each discussed on a large chart, crammed with illustrations.

Walker, R. G. (ed.), 1984. *Facies Models*, 2d ed. Geoscience Canada Reprint Series 1, Geol. Assoc. Canada, Toronto. The best short introduction to sedimentary facies models, full of well-chosen illustrations.

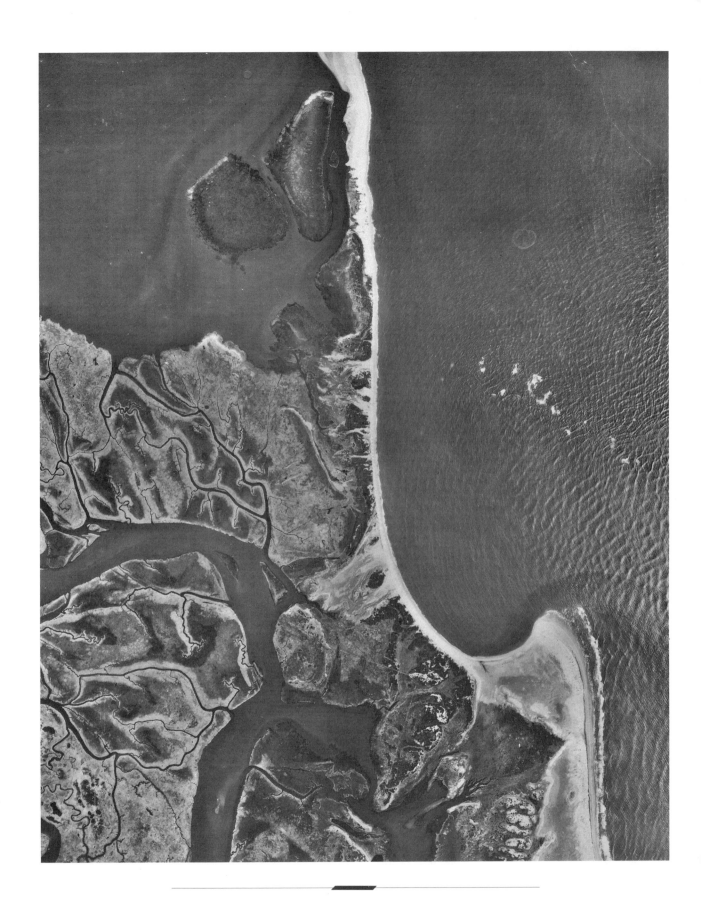

Aerial photograph of beach and tidal flat deposits, Little Egg Inlet, near
Atlantic City, New Jersey. Photo courtesy U.S. Geological Survey.

4

Coastal Environments

The boundary between marine and nonmarine rocks is usually a gradual transition. Meandering rivers begin to show the influence of the ocean in their deltas or tidal estuaries. Barrier islands, though fully marine, are formed from terrestrially derived sediments and can have subaerial deposits on them, such as sand dunes. In fact, there are a number of transitional environments that are neither fully marine nor nonmarine: The delta, the peritidal flats, lagoon, and marshes, and the barrier island complex, together form a group of closely associated facies, which we discuss in this chapter.

DELTAS

Deltas form where rivers carry more sediment into the sea than marine erosion can carry away. The Greek historian Herodotus (*c* 490 B.C.) first used this name for the deposits at the mouth of the Nile, which resemble the Greek letter *delta* (Δ) on a map. Deltas seldom form on active, subducting continental margins because there is no stable shallow shelf on which sediments can accumulate. In addition, active margins usually have coastal

mountain ranges, which limit the size of the river basin that provides the deltaic sediment.

Geologists have devoted much research and study to deltas in the last few decades because they are important host rocks for coal and petroleum. Also, because deltas *prograde* (build out) from the margins of basins, and frequently fill them during geologic time, they have great stratigraphic significance, because they make up a large part of many basin-fill sequences (see Box 4.1).

Deltas experience a combination of fluvial and marine processes and thus are very complex, with each one having more than a dozen distinct environments of deposition. These environments can be grouped into three broad divisions: the delta plain, with its meandering floodplains, swamps, and beach complex; the steeper delta front; and the broadly sloping prodelta, which grades into the open shelf (Fig. 4.1). Because deltas are constantly supplied with sediment from their river basins and so prograde outward, the time planes run parallel to the sloping depositional surface (Fig. 4.2). Deltaic sands and muds, like point bar sequences, are good examples of sedimentary bodies that build by lateral accretion and are therefore time-transgressive.

The shape of a delta is not always the triangle that suggested the name to Herodotus. It is influenced by

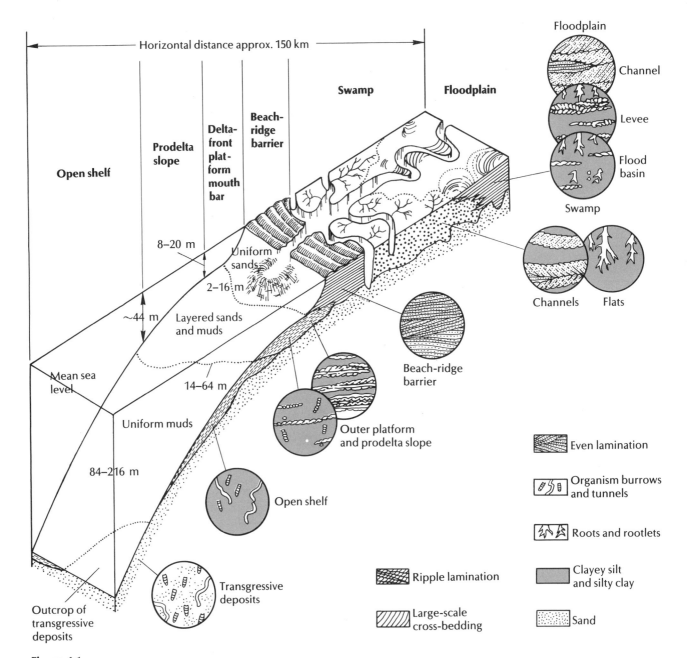

Figure 4.1

Typical cross section across the Niger Delta and adjacent marine shelf, showing sediment types, morphology, and distribution. From Allen (1970).

sediment input, wave energy, and tidal energy. *River-dominated deltas* have large volumes of sediment and tend to be lobate when there is a moderate sediment supply (Fig. 4.3A) and elongate when the sediment supply is large (Fig. 4.3B). If the sediment supply cannot keep up with the erosive powers of tides, then the delta tends to be very small. A *tide-dominated delta* (Fig. 4.3C) has many linear features parallel to tidal flow and perpendicular to the shore. A *wave-dominated delta* (Fig. 4.3D) is smoothly arcuate; the wave action reworks the sediment, making it much sandier than other types of deltas.

River-dominated deltas are lobate because of a unique hydrodynamic interaction between river water and sea. Where the river water enters the sea, there is a sharp density contrast because the river water is much less saline and therefore less dense than seawater. As a con-

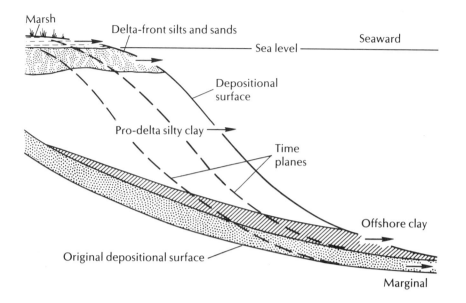

Figure 4.2

The constructional phase in a delta cycle, showing time planes paralleling the depositional surface. From Shepard et al. (1960).

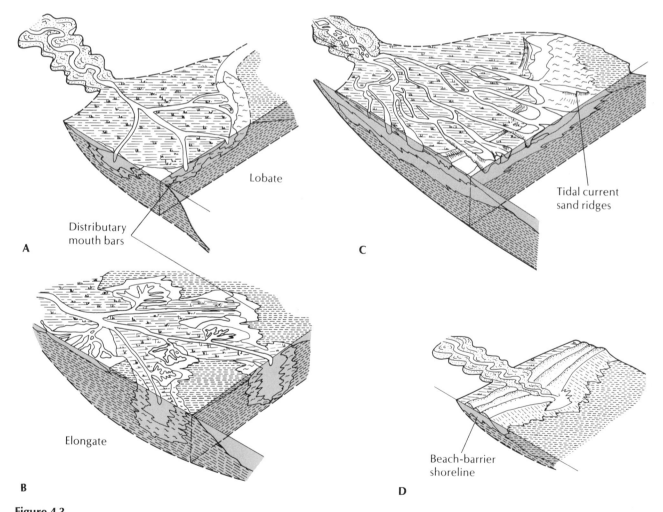

Figure 4.3

Various delta types: (A) Lobate and (B) elongate are river dominated; (C) is tide dominated, and (D) is wave dominated. From Reading (1986).

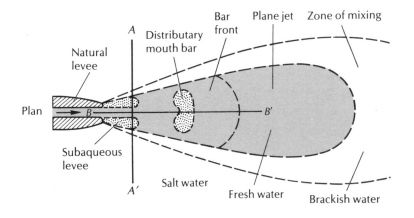

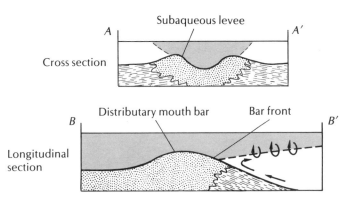

Figure 4.4

River mouth depositional patterns caused by interaction of buoyant river water mixing with marine water. From Davis (1983).

sequence, the river water forms a plane jet that spreads out and forms a layer over the seawater (Fig. 4.4). Because there is only limited mixing at the margins of the jet, there is relatively little frictional inertia. The jet often goes a long way seaward before it gradually begins to break up and mix with the seawater. The sand carried in the stream is thus deposited along the sides of the jet in *subaqueous levees*, where friction and mixing slow the flow. Further offshore, where friction and spreading begin to slow the jet, sediment is dropped in *distributary mouth bars* (Figs. 4.4, 4.5). The finest suspended matter is carried even further, eventually to settle out along the prodelta slope.

The hydrodynamics of river mouth not only influence the map view of the delta shape but also produce variability in its cross-sectional geometry. Most of the delta complex is composed of fine silts and clays that settle out of suspension in the prodelta or delta front or in the lagoon complexes. Scattered through these fine sediments are sandy wedges and tongues that are formed by the various distributaries of the river. These distributary mouth bars tend to build narrow lenses of sand and then to abandon them as the river switches to a new distributary. A typical prograding deltaic sequence is complex of discontinuous sandy lenses scattered through the silty-muddy wedge (Fig. 4.6). In cross

Figure 4.5

Aerial photo of the Pass-a-Loutre distributary of the Mississippi, looking west, on January 28, 1984. Strong offshore winds pushed water away to expose the crest of the distributary mouth bar, which is normally submerged. Photo courtesy D. Nummedal.

A Initial progradation

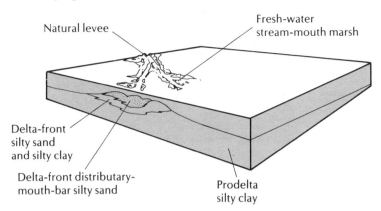

Natural levee

Fresh-water
stream-mouth marsh

Delta-front
silty sand
and silty clay

Delta-front distributary-
mouth-bar silty sand

Prodelta
silty clay

Figure 4.6

Development of the delta through time
by progradation (A, B, D) and distributary
switching (C). From Davis (1983).

B Enlargement by further progradation

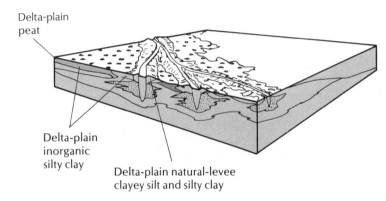

Delta-plain
peat

Delta-plain
inorganic
silty clay

Delta-plain natural-levee
clayey silt and silty clay

C Distributary abandonment and transgression

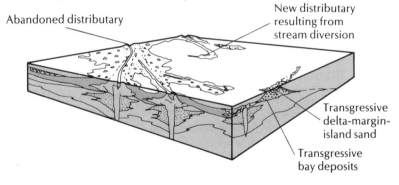

Abandoned distributary

New distributary
resulting from
stream diversion

Transgressive
delta-margin-
island sand

Transgressive
bay deposits

D Repetition of cycle

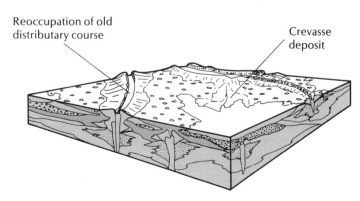

Reoccupation of old
distributary course

Crevasse
deposit

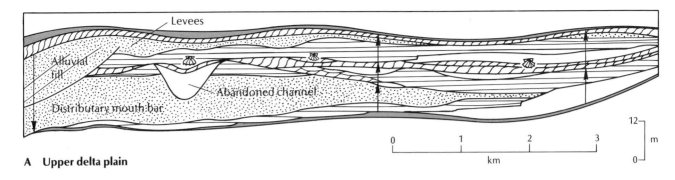

A   Upper delta plain

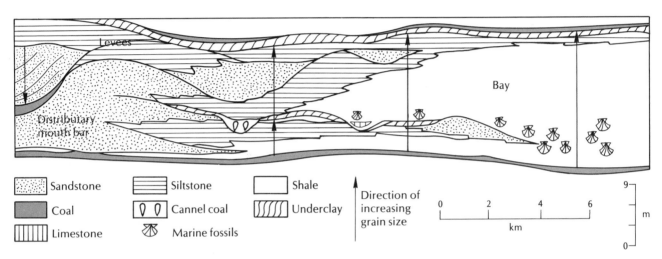

B   Lower delta plain

Figure 4.7

Cross section of (A) upper and (B) lower delta plain deposits, Allegheny Formation. Note the coarsening-upward grain size trends. From Ferm (1974).

section, examples of ancient prograding deltas have complex geometries of levee muds and silts, lenticular distributary mouth bar sands, and frequent cross-cutting channels (Fig. 4.7). The distributary channel sands are abundantly cross-bedded, with plenty of ripple cross-lamination, scour-and-fill structures, and discontinuous clay lenses. The distributary mouth bar sands are even more complexity cross-stratified because of the complex current systems that pass over them; they may show the effects of both wave and current ripples. Frequently, wood, debris, and other organic matter carried down the river during floods ends up in the distributary mouth bar. Sometimes decaying material produces gases that deform the overlying beds in gas-heave structures.

Between the distributaries on the delta plain are wide, shallow *interdistributary bays and marshes*. Like the floodplains of the meandering river, they are low, marshy mudflats cut by small narrow channels of slow-moving water (Fig. 4.8). Because there are no fast-moving currents, most of the sediment is finely laminated silt and

clay, which settles out of suspension, usually after a flood. The abundant vegetation growing there produces much organic matter, which can become peat or coal. The muds are also full of root casts and burrows, and may have shell debris in the tidal channels. Much of the interdistributary sequence is built of sand sheets from crevasse splays, deposited when floods breach the levees that protect the marsh.

At the outer limits of the distributary mouth bar is the *distal bar*, which occupies much of the upper foreset part of the delta front. Because it is at the fringes of the plane jet of river water, it shows interbedded layers of fine sand and mud (see Fig. 4.1). There are also cross-beds, cut-and-fill structures, and ripple marks from both the fluvial and the marine currents. The distal bar is within normal marine salinity but is very shallow, so it is rich in life, with shell beds, burrows, and bioturbation. The distal bar often has a "stairstep" topography due to rotated slump blocks of sand and mud; these "growth faults" are caused by the instability of dense sand lying

Figure 4.8

Interdistributary bays and marshes of the Mississippi River delta. In the center is a large crevasse splay fanning out to the right, which had breached the main distributary channel on the left during a 1976 flood, and nearly filled the interdistributary bay by 1984. Photo courtesy J.M. Coleman.

over less-dense mud. The delta front can be unstable because of oversteepening as well as overloading, so slumps, slides, and convolute bedding are very common.

The remaining slope and bottomset beds of the delta are composed of *prodelta clays and silts*. These sediments become progressively finer away from the delta and are usually finely laminated unless bioturbation is extensive. When floods enlarge the plane jet, they deposit an occasional sandy layer on the prodelta slope. Normal marine invertebrates are common in the prodelta muds, and these muds and shell beds grade outward onto the normal marine shelf (discussed in Chapter 5).

The classical stratigraphic profile of deltaic deposits (Fig. 4.9) shows a coarsening-upward sequence from the delta slope muds and silts to the distributary mouth bar sands. This is the reverse of the fining-upward sequence found in most meandering fluvial systems. The deposits on top of the distributary bar finger sands are abruptly finer, consisting of interdistributary and levee muds and silts, and frequently coals. In large prograding delta complex, there is tremendous lateral and vertical variability due to the influence of several coarsening-upward cycles from sequential distributaries.

Coal
Underclay
Sandstone, fine- to medium-grained,
    multi-directional planar and trough cross-beds
Sandstone, fine-grained, rippled
Sandstone, fine-grained, graded beds
Sandstone, soft sediment slumping
Sandstone, fine-grained, flaser-bedded,
    and siltstone

Silty shale and siltstone with calcareous
concretions, thin-bedded, burrowed
occasional fossil

Clay shale with siderite bands, burrowed,
    fossiliferous

A

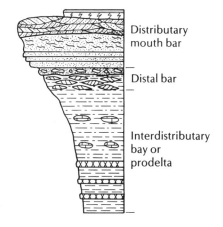

Distributary
mouth bar

Distal bar

Interdistributary
bay or
prodelta

Figure 4.9

Typical vertical sequences through lower delta plain deposits in the Pennsylvanian of eastern Kentucky. (A) Typical coarsening-upward sequence. (B) Same sequence interrupted by splay deposit. From Baganz et al. (1975).

Coal
Rooted sandstone
Sandstone, fine-grained, climbing ripples
Sandstone, fine- to medium-grained
Sandstone, medium-grained, festoon cross-beds
Conglomerate lag, siderite pebble, coal spar
Sandstone, siltstone, graded beds
Sandstone, soft sediment slumping
Sandstone, siltstone, flaser-bedded
Siltstone and silty shale thin-bedded, burrowed

Burrowed sideritic sandstone
Sandstone, fine-grained
Sandstone, fine-grained, rippled

Silty shale and siltstone with calcareous
concretions, thin-bedded, burrowed

Clay shale with siderite bands, burrowed,
fossiliferous

B

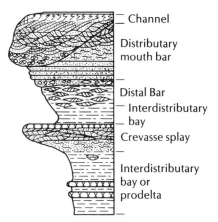

Channel

Distributary
mouth bar

Distal Bar
Interdistributary
bay
Crevasse splay

Interdistributary
bay or
prodelta

Diagnostic Features of Deltas

Tectonic setting Deltas occur along coastal plains of passive margins or in broadly downwarping cratonic basins. They are associated with meandering fluvial deposits and with shallow marine shelf deposits.

Geometry Deltas are roughly triangular in plan view and wedge-shaped in cross section. They are tens to thousands of square kilometers in area and tens to thousands of meters thick.

Typical sequence A coarsening-upward sequence of prodelta muds and clays is followed by distributary bar finger sands, and then muds and coals of interdistributary marshes and levees.

Sedimentology There is a wide range of grain sizes from coarse sand to fine mud, generally becoming finer away from land. Coal and other organic matter can be

---

Box 4.1
Pennsylvanian Deltas of the Appalachians

Coal-bearing deltaic deposits were formed in great abundance during the Pennsylvanian in many parts of the world. Some of the best exposed and most studied of these occur in the Allegheny-Cumberland Plateau region of the Appalachians, primarily in the states of Pennsylvania, Ohio, West Virginia, and Kentucky. They were formed by a series of rivers that drained northwestward out of the eroding Appalachian highlands in Maryland and Virginia (Fig. 4.10) toward the shallow seaways in western Pennsylvania and Ohio. Many of these ancient deltaic sequences are exposed in roadcuts and coal mine walls, as described by Baganz et al. (1975), Horne and Ferm (1976), Horne et al. (1978), and Howell and Ferm (1980).

The Appalachian sequences show a full spectrum of environments, from upper delta plain fluvial meander belts to barrier-island complexes (discussed later). Between these environments are well-developed deltaic plains and distributary sequences, which in cross section (see Fig. 4.7) form a series of coarsening-upward sequences of floodplain and interdistributary bay silts and muds, punctuated by channel and distributary mouth bar sands, and coal swamps. In vertical profile, these coarsening-upward sequences usually begin with interdistributary bay or prodeltaic muds at the base, which may be interrupted by sandy crevasse splays (see Fig. 4.9). Eventually, the sequence becomes sandier, with abundant ripple bedding that represents deposits produced in the distal bar. This is capped by fine-grained sandstones with some small-scale festoons and ripples that reflect deposition on a distributary mouth bar. In many places, this upper sandstone is cut by sandstone channels with multidirectional planar and festoon cross-beds and with climbing ripples from the main distributary channel. Where progradation is well advanced, the distributary channel evolves into a fluvial channel and bar deposits are eroded by meandering rivers, which leave laterally accreted point bars.

The block diagram in Fig. 4.11A shows the idealized geometry of this environment in detail. Some roadcuts near Pikeville, Kentucky, show a cross-sectional geometry that is almost identical to the model (Fig. 4.11B). In the midst of bay-fill mudstones are lenticular sandstone bodies 1.5 to 5 kilometers wide and 15 to 25 meters thick, which represent the distributary mouth bar sands. These sandstone lenses are widest at their bases, with gradational lower and lateral contacts. Grain size increases upward in the sequence and toward the center of the bar. Laterally persistent graded beds (Fig. 4.11C) are found on the flanks of the bars, as are oscillation- and current-rippled surfaces. The central part of the bar exhibits multidirectional festoon cross-beds (Fig. 4.11D). The flanks and the front of the bar show much slumping and soft-sediment deformation, associated with oversteepening along the delta front of dense sands overlying less dense saturated muds (Fig. 4.11E).

Some distributaries are rapidly abandoned, so few continuous sheet sands are formed by the lateral accretion of point bars typical of the fluvial plain. Instead, the sand bodies are discontinuous and pinch out laterally. They are capped by channels with thin levees along the edges (Fig. 4.11A). These channels are rapidly abandoned and are filled with a clay plug (Fig. 4.11G) composed of shales, siltstones, and organic debris that settle from suspension in the ponded water of an abandoned distributary. In a few examples, the abandoned channels become small estuaries, which were inhabited by marine and

locally important. There is a wide variety of sedimentary structures depending on current strength and grain size. Distributary sands tend to show small cross-beds and ripple marks. Levee muds are mudcracked. Burrows, shells, and other biogenic evidence are common in the laminated muds of the interdistributary. Prodelta muds are finely laminated (unless bioturbated), with occasional sandy layers; growth faults, slumping, and soft sediment deformation are common.

Fossils Organic material is very common in the interdistributary, where thick layers of peat or coal can form. Most of the world's great coal deposits come from ancient deltas. Shells, bioturbation, and root casts are also common in the marshes. Organic material can be trapped in the distributary sands, although this is less common. Prodelta muds have lesser amounts of organic debris but can have abundant shells and bioturbation.

brackish-water invertebrates. Some are filled with thick coals and other plant debris (Fig. 4.11F), representing swamps that grew in the abandoned channel (Hook and Ferm, 1985). The clay plug is commonly burrowed or penetrated with root casts. If coarse-grained sediments are present, they are thin-rippled, cross-bedded sands and silts deposited during floods or at sites near the distributary cutoff.

The largest levees (Fig. 4.11H) are thin (1.5 meters thick) and about 150 meters wide, but most are

smaller. They are composed of poorly sorted, irregularly bedded, partially rooted siltstones and sandstones. Thin coal beds may parallel the distributary.

Crevasse splays are another major distinctive component of the Appalachian deltas. They are miniature versions of the delta, fanning out from a breach in a levee and forming their own thin sets of channels within the interdistributary bay mudstones. They coarsen upward, but on a thinner, smaller scale, and also become finer grained away from the breached levee until they grade laterally into interdistributary bay-fill muds. The largest crevasse splays reach 12 meters in thickness and 30 meters in width. In outcrop (Fig. 4.11G), they appear as thin, sand lenses that thicken in areas where splay channels developed (shown by arrows in Fig. 4.11G). The splay channels are commonly filled with ripple drift crosslamination, indicating very rapid sedimentation in the waning flood currents that occur when the water released by the breached levee has nearly drained away. In some splay deposits, the sedimentation was so rapid that tree stumps were buried upright and fossilized in place (Fig. 4.11I).

The Pennsylvanian sequences of the Appalachians are among the best-developed and best-exposed examples of ancient deltas. Laterally discontinuous distributary bar sandstones punctuate a sequence of interdistributary muds and prodeltaic clays; within these bars are well-developed channel sequences, levees, and crevasse splays. This association, along with the coals and other plant fossils, is clearly deltaic. Fluvial sequences and barrier-island sequences would exhibit sandstone bodies of much greater lateral continuity and would not have the distinctive association of sedimentary structures that form in deltas.

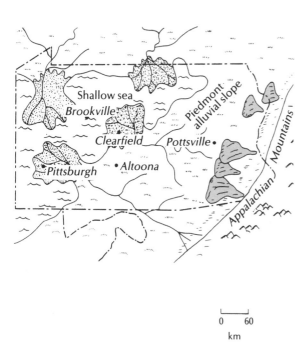

Figure 4.10 Paleogeography of Pennsylvania during deposition of the Lower Kittanning (Middle Allegheny Group). From Williams et al. (1964).

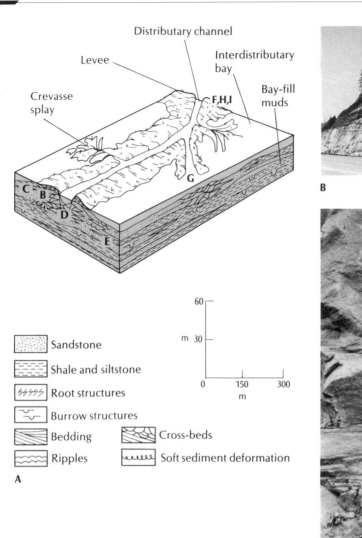

B

A

Sandstone

Shale and siltstone

Root structures

Burrow structures

Bedding   Cross-beds

Ripples   Soft sediment deformation

D

C

Figure 4.11   (A) Block diagram of distributary, showing various subfacies. (B) Lateral and vertical gradational coarsening from bay-fill shales (lower right) to distributary mouth bar sandstones. (C) Individual sets of graded beds on the flanks of the distributary mouth bars. (D) Multidirectional festoon cross-beds in the central portion of the bar, due to the dispersal of river currents as well as wave and tidal currents over the bar crest. (E) Slump structures and flow rolls on the front and flanks of the distributary mouth bar. (F) Abundant plant debris on bedding surfaces at the distal margin of the splay deposit. (G) Abandoned channel-fill with thin levees, near Ivel, Kentucky; levees dip away from the channel. (H) Levee deposits consisting of irregularly bedded, root-penetrated clay-shales and siltstones with thin local coals. (I) Tree stump surrounded by shale and siltstone, indicating very rapid deposition of the fine-grained sediment. Photos courtesy J. C. Ferm.

E

H

F

I

G

PERITIDAL ENVIRONMENTS

Coastal regions with rivers that do not carry excessive sediment do not have deltas. Instead, a system of coastal lagoons, estuaries, and tidal flats develops. The dominant influence on this system is the rise and fall of tides because these regions are typically very low and marshy at high tides but emergent during low tides. Like deltas, these environments are common on the low, marshy coastlines of passive tectonic margins, where the relief is minimal and the shoreline is often drowned by the postglacial rise in sea level. Peritidal environments have many other similarities to deltaic environments. The primary difference is the excess of river sediment that is supplied to a constructive delta, which causes it to prograde, whereas the peritidal environment shows the reworking effects of tides.

Many coastal areas have extensive wetlands, or *salt marshes*, particularly on the borders of estuaries, delta channels, and lagoons. Unlike the deltaic floodplain marshes or other freshwater swamps, these marshlands are regions where fresh and saltwater mix; thus the range of water salinity can be extreme. They are highly stressed environments, dominated by a few organisms that can tolerate the extreme ranges of salinity. They are usually clogged with vegetation, mostly with marsh grass, or with mangroves in some tropical settings. As one would expect, muds rich in organic material accumulate in the marshes, and in many parts of the geological record, these have become peats or coals. Because of the limited water circulation and the excess of decaying plant matter, the water chemistry is often very reducing, so the muds are black with unoxidized organic matter. Typically, these black muds alternate with fine silt and sand layers from tidal-channel flooding. Sometimes the bedding is slightly undulatory due to disturbance by plants, or in extreme cases, the bedding is completely destroyed by bioturbation. Interspersed with the muds are small, sandy channel-fill deposits from the tidal channels. These are typically cross-stratified, with organisms (such as oysters) that can tolerate the brackish water.

In many protected coastal regions, the strength of the tidal flow is sufficient to prevent vegetation from becoming established. In these regions, broad areas of unvegetated sediments form *tidal flats* (Fig. 4.12). In general, the absence of strong wave action and a wide tidal range are both necessary to form tidal flats. For example, the 2.5 to 4 meter tidal range of the North Sea coast of Europe produces tidal flats up to 7 kilometers wide (Fig. 4.12A), and the 5-meter tides of the Yellow Sea produce flats up to 25 kilometers wide. These flats have extremely low relief, except for the tidal channels that are carved into them. During low tide, they are completely exposed, and all the effects of the receding tidewaters can be seen in the exposed sedimentary structures (Fig. 4.12B).

Tidal flats are unique among sedimentary environments in that they experience a daily cycle of reversing flow direction and exposure. This produces *tidal bedding* (Fig. 4.13), a regular cycle of mud and sand deposition from the current fluctuation. At low water and high water, there is relatively little flow, so the finer muds settle out. During advancing and receding tides, the cur-

A

Figure 4.12

(A) Aerial view of tidal flats near Waddensee, Holland. Photo courtesy KLM-Aerocarto. (B) Patterns of ripples developed upon megaripples on tidal flats exposed at low tide. Photo courtesy R. A. Davis, Jr.

B

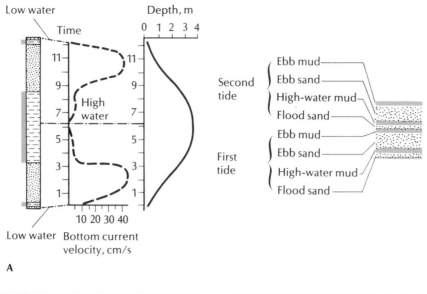

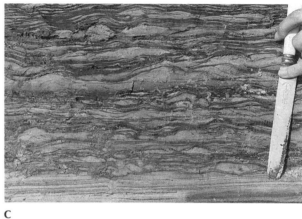

Figure 4.13

(A) Development of tidal bedding according to Reineck and Wunderlich (1968). Marker beds emplaced at high and low tide enclose bedload deposition of sand and suspension deposition of mud. Depositional phases are controlled by changes in velocity during tidal cycle, which is also shown, as is depth change. During two tidal cycles, four couplets of sand and mud comprising a tidal bed are deposited. (B) Tidal bedding, Holocene tidal flats, northwest Germany. Scale in centimeters. Photo courtesy F. Wunderlich. (C) Tidal bedding, Jurassic, Bornholm, Denmark. Photo courtesy B. W. Sellwood.

rents move sand across the mudflats, producing cross-bedded sands between the muds. In many instances, the reversing directions of cross-bedding produce herring-bone cross-beds (see Fig. 2.10), which resemble the pattern of ribs on a herring or other fish. These are considered unique to the tidal environment.

Another characteristic sedimentary feature of the tidal flats is flaser and lenticular bedding (see Fig. 2.12). The difference in current strength between the tidal cycles separates the muddy from the sandy fraction. Ripples formed by higher velocities accumulate mud in their troughs, which are not scoured out by receding tides, thus forming flaser beds. As the abundance of mud increases, wavy bedding and eventually lenticular bedding

results. Tidal cross-beds commonly show reactivation surfaces (see Fig. 2.8) because the retreating tides carve back the cross-beds that are built during the advance, only to have the next tidal surge rebuild the cross-bed over the reactivation surface. The asymmetry of the incoming and outgoing tides can have other effects. Tides going in two directions produce interference ripples (see Fig. 2.11). The difference in tidal flow results in superimposed bedforms from two flow velocities. For example, stronger incoming tides may build megaripples, on which smaller ripples are superimposed by outgoing tides (see Fig. 4.12B). The upper reaches of the tidal flat are subaerially exposed except during unusually high tides. They have many features of subaerial exposure,

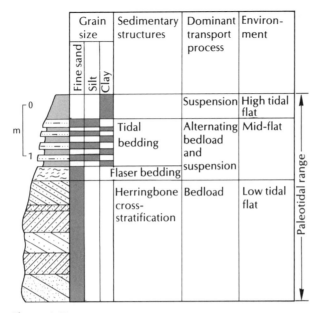

Grain size			Sedimentary structures	Dominant transport process	Environ-ment
Fine sand	Silt	Clay			
				Suspension	High tidal flat
			Tidal bedding	Alternating bedload and suspension	Mid-flat
			Flaser bedding		
			Herringbone cross-stratification	Bedload	Low tidal flat

Figure 4.14

Typical paleotidal range sequence based on the middle member of the Wood Canyon Formation, Nevada. From Klein (1972).

such as mudcracks or algal mats. These mudflats can also harbor oysters and other organisms that cannot tolerate the strong flow of the lower tidal flat. Consequently, upper tidal flat muds are often heavily bioturbated.

The typical tidal flat sequence thus shows a fining-upward column (Fig. 4.14). The lower tidal flat is coarser (mostly sand) and dominated by herringbone cross-strat-ification. The midflat region shows the rapid changes between bedload sands and suspended muds by its abundant tidal bedding, flaser bedding, and related features. The upper tidal flat is formed entirely of suspended muds, which may be mudcracked, bioturbated, and col-onized by salinity-tolerant organisms and algal mats. In three dimensions, the tidal sequence may move laterally and is punctuated by many tidal-channel deposits. The tidal sequence can also be seen to intergrade with the saltmarsh muds and the foreshore sands.

Diagnostic Features of Peritidal Environments

Tectonic setting Peritidal environments are com-mon on stable, passive tectonic margins with broad, shal-low coastal regions. They are an important component of the thick sequence that is built on passive margins.

Geometry Tabular in shape, peritidal environments usually form local sand bodies a few meters in thickness

that parallel the shoreline, because they are very sensi-tive to small changes in sea level and so do not accu-mulate very long in one place.

Typical sequence A fining-upward sequence of her-ringbone cross-bedded sands, flaser- and tidal-bedded sands and muds, and mudcracked, bioturbated, upper tidal flat muds is typical. Black, organic-rich, salt-marsh muds accumulate near and on top of the tidal flat sequence.

Sedimentology Grain size is finest in the upper tidal flat and coarsest in the tidal channels, with abrupt changes between mud and sand in the midtidal flat. Many unique sedimentary structures, such as herring-bone cross-bedding, flaser and lenticular bedding, in-terference ripples, and reactivation surfaces occur. Superimposed bedforms from two flow velocities also occur. Abundant bioturbation, organic matter, and coals and peat occur in the upper tidal flat and salt marsh.

Fossils Invertebrates that are tolerant of extreme sa-linity changes can be very abundant. Plant remains are common in the salt marsh, and algal mats can form in the upper tidal flat. Upper tidal flat muds also show extensive bioturbation and other evidence of burrowing.

BARRIER COMPLEXES

Barriers are elongate sandy islands or peninsulas that parallel the shoreline and are separated from it by la-goons or marshes. They are formed on coastlines where there is an abundant supply of sediment and where the tidal range is small enough that the longshore currents and wave action are more important than the onshore-offshore flow of tidal currents (Fig. 4.15). Long, narrow barrier islands are found on *microtidal* coasts, which have a tide range of 2 meters or less (Fig. 4.15A). Short barrier islands, broken by numerous tidal channels, are found on *mesotidal* coasts, which have a tide range of 2 to 4 meters (Fig. 4.15B). If the sediment supply is too low, then no barriers are formed; the coastline is unprotected and usu-ally erodes back. Some barrier-free coasts are influenced only by the tides, producing a *macrotidal* coast (Fig. 4.15C).

Barrier islands, like deltas and tidal flats, are known primarily from passive continental margins adjacent to coastal plains. Particularly good examples have been studied along the Atlantic and Gulf coastal plain of North America. Barrier islands migrate landward and seaward with changes in sea level and often make up an important portion of transgressive or regressive shore-line deposits. The channels and spits of barrier islands are also prone to migrating up and down the length of the barrier. Because barrier islands typically produce

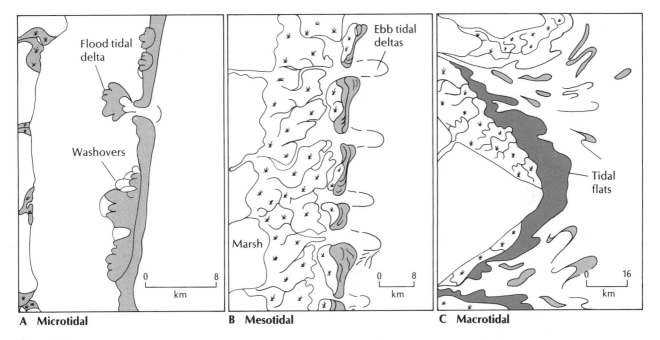

Figure 4.15

Variation in coastal sand bodies due to differences in tidal range: (A) long, narrow microtidal barriers; (B) short mesotidal barriers; (C) linear tidal-current ridges of macrotidal estuaries, perpendicular to shore. After Barwis and Hayes (1979).

long, shoestring sands of high porosity and permeability within impermeable shale sequences, they become excellent reservoir rocks for petroleum. Barrier sands have also proven to be host rocks for uranium, and beach sands can be placer mined for gold, diamonds, and heavy minerals. Marshes and lagoons behind barriers are important areas of coal accumulation.

The barrier-island complex is subdivided into several distinct subenvironments.

Shoreface

The main barrier separates the shoreface region from the backbarrier lagoons and tidal flats (Figs. 4.16, 4.17). Along the shoreface is a series of features that are caused by wave action and variation in tide levels (Fig. 4.18). The shoreface begins where the wave base first begins to feel the bottom, around depths of 10 to 20 meters, depending on the size of the waves. On the lower

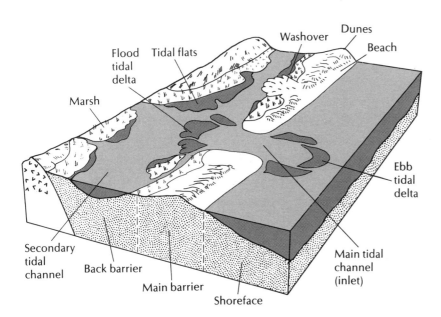

Figure 4.16

Diagram showing the various subenvironments in a barrier-island system. From Walker (1984).

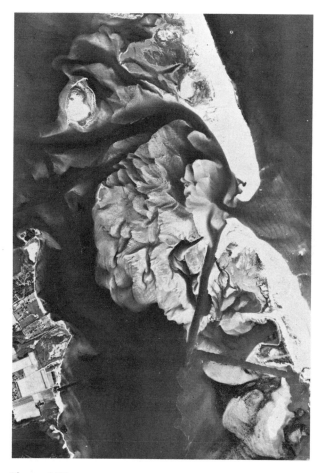

Figure 4.17

Tidal delta on the lagoon side of a tidal inlet, formed by a hurricane in September 1947. Photo courtesy M. M. Nichols, U.S. Department of Agriculture.

shoreface, the wave effect is very weak, so offshore shelf processes are also important. As a consequence, the typical deposit (Fig. 4.18) is fine sand intercalated with layers of mud. There may be planar lamination, though bioturbation and trace fossils commonly obliterate primary structures. In the middle shoreface, the shoaling and breaking of waves is strongly felt. Where the shoreward drag of the wave base is balanced by the backwash of breakers, the sand accumulates in *longshore bars*. Here the medium-fine grained sand is very well sorted, with **abundant broken shell layers. Low-angle, wedge-shaped** planar cross-beds are most common on the seaward side of these bars, but ripple laminae and trough cross-beds occur on the landward sides. The middle shoreface is less strongly bioturbated than the lower shoreface, though extensive biogenic debris and activity are common. The upper shoreface, or surf zone, feels the effect of the plunging waves in the onshore-offshore direction and also the wave-driven longshore current. Thus the sedimentary structures of the shoreface reflect varying flow directions, with multi-directional trough cross-beds interspersed with low-angle planar cross-beds, and subhorizontal plane beds.

The shoreface is particularly susceptible to erosion during the unusually strong waves generated by major storm. Most longshore bars are wiped out, and the entire sequence is often scoured away and replaced by sand dropped during the receding storm. For this reason, shoreface sequences are not common in the stratigraphic record. Often, the entire shoreface sequence is represented by storm deposits, capped by the foreshore, backshore, and dune deposits that are emplaced after the storm recedes.

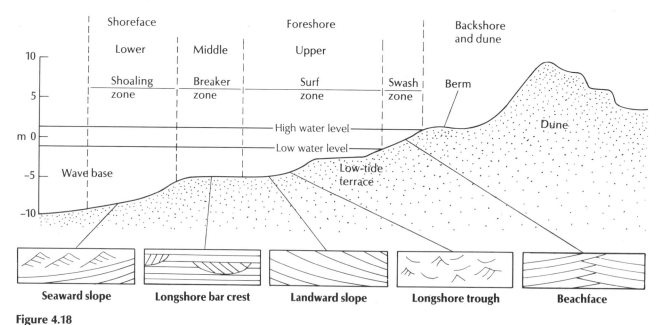

Figure 4.18

Characteristic sedimentary structures for the beach, shoreface, and nearshore bar environments. Modified from Walker (1984).

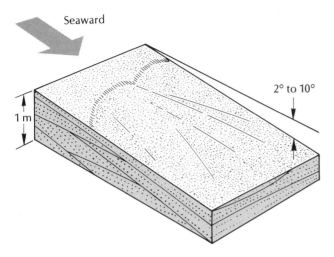

Figure 4.19
Beach stratification formed in the swash zone of shoreline waves. Fine, even laminae are deposited, as thin flows run up the beach and return as waves break. The laminae dip seaward at the angle of the beach slope, which for sand beaches ranges around 2° to 10°. Slight discordances between sets of laminae reflect changes in wave size or direction or in tide level. From Harms (1979).

Foreshore

The foreshore, or swash zone, is the region between the high and low water levels of the beach. It is influenced primarily by the swash and backwash of the breakers, so the directions of cross-bedding are mostly perpendicular to the shoreline. The most characteristic feature of the swash zone is gently laminated planar cross-beds that dip shallowly seaward (about 2° to 10°) and are broadly lenticular (Fig. 4.19). Although symmetrical ripples are commonly formed from the regular ebb and flow of the water on the beach, they are usually destroyed by the incessant reworking and thus are seldom found in the stratigraphic record. Foreshore sands tend to be coarser than those found further offshore because the wave energy is greater; they are also very well sorted due to the extensive reworking by swash and backwash. Extensive reworking often concentrates minerals by density, so beach sands may have well-developed dark laminae of heavy minerals segregated from the quartz sand.

Backshore

Above the high water level is the backshore and dune area of the barrier complex (see Fig. 4.18). The top of the foreshore is marked by a sandy terrace, called the *berm*, whose crest is just above high water level. The berm is formed during unusually high storm waves and remains there until the next big storm. Here the wind is more important than the waves, except when storm surges cause unusually high water levels. The sand brought in by these high water levels is redistributed by the wind and by storm washovers. The sand is formed into subhorizontal or landward-dipping plane beds with small cross-beds. At the crest of the barrier complex there is often a dune field, which is affected only by wind. As one would expect, there are extensive eolian trough crossbeds up to 2 meters thick, which are multidirectional due to fluctuating wind directions. They often have the curved bedding surfaces characteristic of eolian cross-beds. Barrier dune cross-beds differ from desert dunes chiefly in their smaller scale and limited, linear distribution. They also tend to be abundantly burrowed by marine crustaceans and may have well-developed roots as well, unlike desert dunes.

During hurricane-force storms, however, the dune crest is breached in many places, and waves that overtop the barrier deposit lobes or sheets of sand into the lagoon. These are called *washover deposits* (Figs. 4.16, 4.20, 4.21). They are dominated by landward-dipping horizontal strata from storm washover episodes (see Fig. 4.20). Each layer is usually thin (a few centimeters to 2 meters in thickness) and limited in lateral extent, forming a lobate sheet. Between washover episodes, the sand is reworked by the wind into dune trough cross-beds. Where the washover sand enters the lagoon, small-scale delta foresets form. Washover sands are generally fine- to medium-grained and not as well sorted as beach or dune sands.

Lagoons and Tidal Flats

Behind the barrier is a region of lagoons and tidal flats, in which the quiet waters allow fine silt and clays to settle out of suspension (like in the tidal flats described previously), forming a sequence of mudstone and shale. Usually there is an abrupt shift from the coarse, clean, well-sorted sands of the backshore dunes to the finely laminated clays of the lagoon. The lagoonal waters become stagnant and anaerobic, forming black organic-rich muds. In other instances, the lagoons are overgrown with vegetation, forming salt marshes, coal and peat swamps, or algal mats. In some parts of the world, the lagoons become hypersaline, and evaporite minerals are formed. Large parts of the area around the lagoon are within the tidal range, and a tidal-flat sequence of herringbone crossbeds, flaser beds, and reactivation surfaces accumulate. Unlike ordinary tidal flats and salt marshes, however, the area around the backbarrier lagoon is punc-

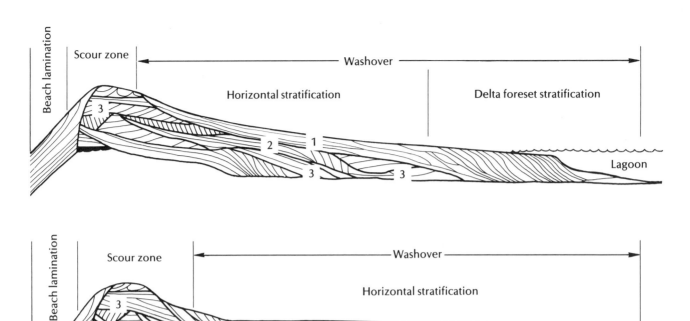

Figure 4.20

Sedimentary structures that occur in washover-fan sands: 1, newly deposited washover; 2, old washover; 3, eolian deposits. After Schwartz (1975).

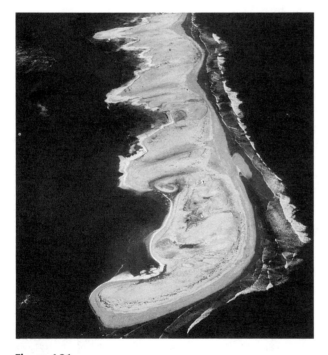

Figure 4.21

Washover fans created on the lagoonal side of Nauset Spit, Massachusetts, during the 1978 blizzard. Photo courtesy S. Leatherman.

tuated by sands that show the effects of the nearby barrier and the storms that breach it. Storm washover layers with their delta foresets (see Fig. 4.20) are frequently found. Another characteristic feature is the flood tidal delta (see Figs. 4.16, 4.17) formed by sand washed in through the tidal inlet. The sands of such deltas are typically well sorted, with planar cross-beds that are laterally discontinuous because they form a lobate sheet at the mouth of the tidal inlet. Lagoonal shale sequences are easily distinguished from other shale sequences by their proximity to backshore dune deposits and by frequent intrusions of washover and flood tidal inlet sands.

Barrier Island Dynamics

Barrier island complexes are dynamic systems, capable of great lateral migration and variability. For example, studies of modern tidal inlet channels have documented rapid migration by lateral growth and accretion (Figs. 4.22, 4.23B). As the channel and spit sequence migrates laterally, it leaves behind a sequence of deposits that are very different from the classic barrier bar sands (Fig. 4.23A). The sequence begins with a channel floor lag gravel of shells and pebbles. It is then covered by a thick sequence of planar and trough cross-beds of the deep channel, which show many reactivation surfaces. The shallow channel deposits can be plane-parallel laminae or shallow, bimodal trough cross-beds. Eventually, the channel sequence is capped by normal beach spit deposits, mostly made of steeply seaward-dipping planar cross-beds or landward-dipping cross-beds behind the spit. The lateral migration of the inlet (Fig. 4.23B) forms a sequence of channel deposits with time planes that cut obliquely through the sequence of sedimentary structures. The inlet sequence in the barrier complex is by definition time-transgressive.

Even more important than lateral longshore migration is the migration of the barrier complex onshore and offshore. Barrier island sands that have accumulated vertically in a single place are seldom found in the stratigraphic record. Instead, because of changes in relative sea level and sediment supply, the sands migrate onshore and offshore rapidly, forming a large part of the stratigraphic sequence in many basins. When there is excess sediment supply relative to sea level change, a prograding sequence is formed (Fig. 4.23C). Although progradation usually takes place during falling sea levels, it can take place during rising sea level if the sediment supply is sufficient. In either case, the coarse sands of the beach, the finer sands of the upper shoreface, and the silts and muds of the lower shoreface build outward across the offshore muds, and in turn are followed by the lagoonal muds of the backbarrier. Although horizontal layers of differing grain sizes are formed, it is clear that the time planes cut through the lithologic units and that the entire sequence is time-transgressive.

A prograding (regressive) sequence thus produces a coarsening-upward profile (Fig. 4.24A), from offshore muds to shoreface silts and sands and to the coarser sands of the beach and dune complex. Many examples of these sequences exist in the rock record (see Box 4.2). Because prograding barriers migrate laterally, they are capable of producing extensive sand sheets and shoestring sands that interfinger with the underlying and overlying shales.

Transgressive barrier sequences are much less well known in the stratigraphic record, for reasons discussed in Chapter 7. The rise in sea level must be very rapid relative to the supply of sediment or else shoreface processes erode the sediments back rapidly. If the sediment supply and sea level rise are "ideally" balanced, then the barrier island sands migrate shoreward over the lagoonal muds, producing a different kind of coarsening-upward sequence (Fig. 4.24B). The lagoonal muds are capped by flood tidal delta sands and washover sands, interspersed with tidal flat and marsh muds. Eventually, the washover sands are topped by the crossbedded eolian dunes. Adding the lateral migration of the barrier inlet sequence (Fig. 4.23C) to either the prograding or the transgressive pattern results in extremely complex barrier island sequences.

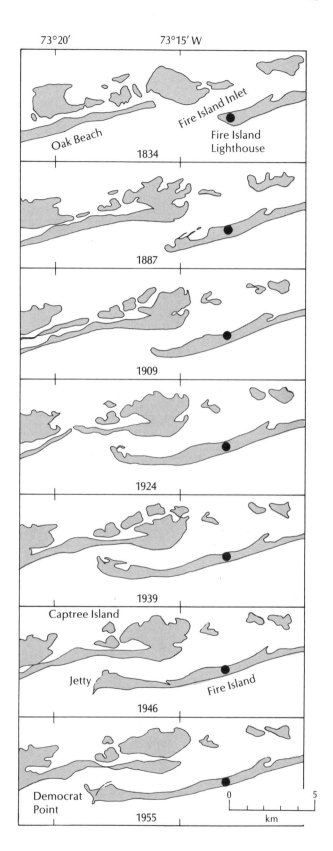

Figure 4.22

Migration of the tidal inlet along Fire Island, south shore of Long Island, New York, from 1834 to 1955. After Friedman and Sanders (1978).

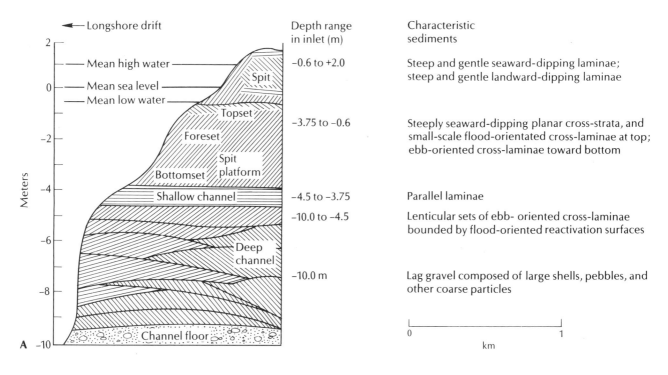

← Longshore drift

2

Mean high water ———————————— −0.6 to +2.0

0   Mean sea level ————————————
Mean low water ————————————

−2

−4

Shallow channel   −4.5 to −3.75

−6

Deep channel

−8

Channel floor

A −10

Depth range in inlet (m)

−0.6 to +2.0

−3.75 to −0.6

−4.5 to −3.75

−10.0 to −4.5

−10.0 m

Characteristic sediments

Steep and gentle seaward-dipping laminae; steep and gentle landward-dipping laminae

Steeply seaward-dipping planar cross-strata, and small-scale flood-orientated cross-laminae at top; ebb-oriented cross-laminae toward bottom

Parallel laminae

Lenticular sets of ebb- oriented cross-laminae bounded by flood-oriented reactivation surfaces

Lag gravel composed of large shells, pebbles, and other coarse particles

Labels within figure A: Spit; Topset; Foreset; Bottomset; Spit platform

Scale: 0 ——————— 1 km

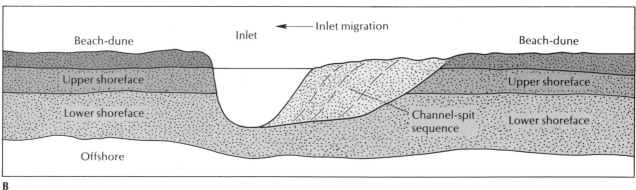

← Inlet migration

Beach-dune

Inlet

Beach-dune

Upper shoreface

Upper shoreface

Lower shoreface

Channel-spit sequence

Lower shoreface

Offshore

B

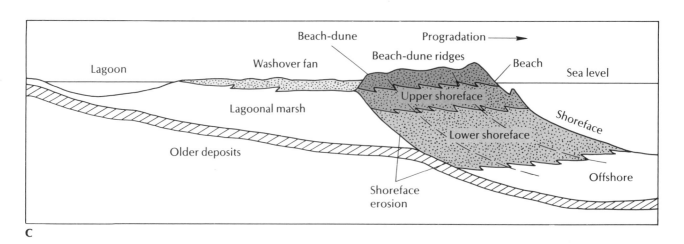

Beach-dune

Progradation →

Beach-dune ridges

Lagoon

Washover fan

Beach

Sea level

Lagoonal marsh

Upper shoreface

Shoreface

Lower shoreface

Older deposits

Offshore

Shoreface erosion

C

Figure 4.23

(A) Vertical sequence of sedimentary structures formed by the migration of Fire Island inlet, New York. From Walker (1984). (B) Section parallel to shore showing lateral migration of tidal inlet sequence. (C) Section perpendicular to shore showing onshore-offshore migration of barrier island sequence, here responding to regression. (B) and (C) from Scholle and Spearing (1982).

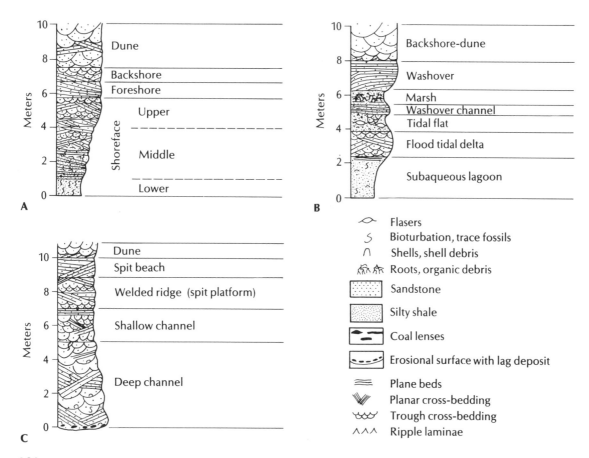

Figure 4.24

Stratigraphic sequences of three end-member facies models of barrier islands: (A) prograding model, (B) transgressive model, (C) barrier-inlet model. From Walker (1984).

Diagnostic Features of Barrier Complexes

Tectonic setting Barrier complexes occur in stable, sloping coastal plain settings, particularly along passive tectonic margin sequences or in broad cratonic seaways. They are closely associated with peritidal and deltaic sequences and with shallow marine shelf sands and muds.

Geometry A barrier complex typically forms elongate, shoestring sands within marine shale sequences that are tens of meters in thickness, a few kilometers in width, and tens to hundreds of kilometers long. If the barrier complex progrades, it can produce a tabular sandstone body extending for tens to hundreds of kilometers.

Typical sequence A coarsening-upward is typical. For prograding barriers, offshore muds are capped by shoreface silts and sands, and finally by medium- and fine-grained beach and dune sands. For transgressive barriers, the lagoonal muds interfinger with washover and flood tidal inlet sands and are capped by dune sands.

Sedimentology Because of extensive reworking, beach sands can have mineralogically mature quartz mixed with occasional heavy-mineral lags. Lagoonal muds can be organically rich and can accumulate peat and coal; in a few cases, evaporites are known. A variety of sedimentary structures occur, but shallow-dipping tabular cross-beds and plane beds are most common on the shoreface and washover fan. Beach sands have a unique type of cross-stratification, formed by swash and backwash. Eolian dunes produce larger-scale tabular and trough cross-beds but are abundantly burrowed and impregnated by root casts (unlike desert dunes).

Fossils The shoreface has a diverse and abundant shelly invertebrate fauna, and extensive infauna, with many types of burrows and trace fossils. The high-energy wave environment, however, typically breaks up and abrades most of the shell debris. Dunes have both burrows and root casts. Lagoons are dominated by plants and mudflat-dwelling animals, depending on the salinity. If there is a wide fluctuation in salinities, then only organisms that can tolerate those conditions are found there.

Box 4.2
Ordovician Shoreline Sequences of South Africa

One of the world's best exposures of ancient shore-line sequences occurs in the cliffs of Table Mountain, which towers over Cape Town, South Africa (Fig. 4.25). As described by Tankard and Hobday (1977), Rust (1977), and Hobday and Tankard (1978), the Lower Paleozoic Table Mountain Group includes the Ordovician Graafwater Formation (Rust, 1973), a tidal flat deposit capped by the Peninsula Forma-tion, which is a barrier and shelf deposit. To the north, the Graafwater integrades at its base with the Piekenierskloof Formation, a braided alluvial deposit. To the south, the Peninsula Formation is capped by a series of younger units representing offshore sands and muds and also glacial deposits. The Cedarberg Formation contains late Ordovician brachiopods, which suggests an age for the under-lying units. The Table Mountain Group is one of the few good examples in the stratigraphic record of a nearly continuous transgressive barrier sequence. As discussed in Chapter 7, conditions conspire to wipe out most transgressive sequences unless sea level rise is rapid and balanced by continuous sediment influx. Most coastal sequences in the stratigraphic record appear to be regressive, caused by prograding shorelines and/or a relative drop in sea level.

The 70-meter thick Graafwater Formation dem-onstrates many features that are characteristic of a tidal flat, backbarrier deposit. Three basic facies are recognized by Tankard and Hobday (1977). The quartz arenite facies (Fig. 4.26A) is composed of supermature fine- to medium-grained quartz sand-stones, with a few pebbles and cobbles up to 12 centimeters in diameter in basal channel lags. Bed thickness and grain size both decrease upward within each unit, which ranges in thickness from 5 to 120 centimeters. The individual sandstone units are separated by erosional surfaces draped with green siltstone (Fig. 4.26E). Trough cross-stratifica-tion is common in the sandstones, with minor planar cross-bedding. Even more characteristic are well-developed herringbone cross-beds (Fig. 4.26E) and reactivation surfaces (Fig. 4.26F). Straight-crested, slightly asymmetrical ripple marks about 4 centime-ters in wavelength are common on the sandstones that are draped by mud layers. Wave-generated symmetrical ripples, double-crested ripples, and in-terference ripples also occur. These indicate both unidirectional and oscillatory flow, both of which occur in tidal flats. The ripple marks show a distinc-tive interference pattern known as *ladderbacks* (Fig.

4.26D), and some have mudcracks, indicating subae-rial exposure (Fig. 4.26C). The combination of sand-stone channels, herringbone cross-beds, reactivation structures, and ladder-back ripplemarks led Tankard and Hobday (1977) to attribute this facies to shal-low subtidal and low-tide terrace environments.

Covering the quartz arenite faces is a heterolithic facies with rapidly alternating mudstones and quartz sandstones (hence the name). The sandstone beds range from 2 to 15 centimeters in thickness and are cross-bedded in solitary sets. The intercalated mud-stones exhibit both flaser and lenticular bedding (Fig. 4.26B) and abundant mudcracks (Fig. 4.26C). These features seem clearly to indicate a midtidal flat origin for the heterolithic facies. It grades up-ward into the mudstone facies (Fig. 4.26A), which consists of horizontally and wavy laminated silt-stones and red-brown or maroon-brown mudstones. Trace fossils, such as *Skolithos*, indicate extremely shallow conditions. Tankard and Hobday (1977) at-tribute this facies to high-tide mudflats and supra-tidal flats.

The three facies of the idealized sequence (Fig. 4.26A) bear a striking resemblance to the tidal flat model sequence (see Fig. 4.14). The facies alternate through the entire section to produce a semicyclic, fining-upward sequence (Fig. 4.27). This cyclicity is attributed to frequent shore zone regressions. The sedimentary structures indicate that the flow was asymmetrical, with dominant ebbcurrents and also a strong longshore movement of water. From the thickness of the units, Tankard and Hobday (1977) postulate a tidal range of 2 to 3 meters, comparable to that of mesotidal flats now found in the Nether-lands. The alternation of the quartz arenite facies with the heterolithic and mudstone facies produces a repetitive, laterally continuous sequence of sands and thin mudstones that clearly show their cyclicity (Fig. 4.28).

The tidal flat mudstones pass upward into and locally intergrade with landward-dipping sandstone plane beds of the Peninsula Formation (Fig. 4.29B). These plane beds dip at about 4° toward the ancient land direction and have small-scale cross-bedding in them. Hobday and Tankard (1978) attribute these beds to washover fans behind the barrier (see Fig. 4.21). This is indicated not only by their similarity in structure but also by their interfingering with tidal flat muds. In many places, the plane beds of the washover fans are interrupted by small-scale (30

Figure 4.25   Aerial view of Capetown, South Africa, showing the 750-meter thick nearshore deposits of the Peninsula Formation making up Table Mountain in the background. Photo courtesy M. P. A. Jackson, University of Texas Bureau of Economic Geology.

to 180 centimeters thick) channel sandstones, which may represent small tidal channels cut through the back barrier flats. The presumed tidal-channel sandstones have unidirectional foresets in individual channels, but adjacent or superimposed channels can have directions that are nearly opposite. Herringbone cross-beds and reactivation structures are also present, both indicating periodic current reversals commonly produced by tides.

Another plane-bedded sandstone facies occurs in the Peninsula Formation, but these dip 3° to 8° in the ancient *seaward* direction (Fig. 4.29D). The thinly laminated plane beds are very evenly spaced and include heavy-mineral layers characteristic of modern beaches. Hobday and Tankard (1978) interpret this facies as produced by the beachface of the barrier-island sand spits.

Much of the Peninsula Formation is composed of a large-scale channel facies (see Fig. 4.28A, top; Fig. 4.29C). These channels are up to 40 meters thick and are composed of trough cross-stratification with highly variable directions. At the base are thick (80 to 250 centimeters) trough cross-beds composed of

coarse sand and pebbles with a seaward inclination. In some places, they include reactivation surfaces and herringbone cross-stratification. The crossbed directions suggest a very high-energy ebb flow, probably due to storm-surge ebbs and rip currents. The thick basal cross-beds are overlain by progressively smaller-scale (15 centimeters thick) sets that become thinner upward. The entire sequence of the large-scale channels is reminiscent of the sequence formed by the lateral migration of tidal inlets in a barrier-island complex. These produce the bulk of the barrier-island sand body (see Fig. 4.24B). The Peninsula Formation shows all the classic facies of the barrier island: washover fans, small back barrier tidal channels, large-scale tidal inlet deposits, and laminated beachface sandstones. Combined with the obvious tidal flat deposits of the Graafwater Formation, they constitute one of the most clearcut and best exposed ancient examples of tidal flat and barrier-island sedimentation. A block diagram clearly shows how the facies geometry of the Graafwater and Peninsula Formations was produced (see Fig. 4.29A).

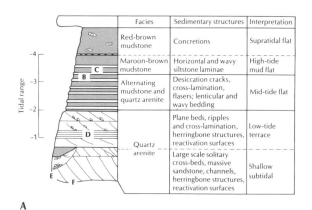

Facies	Sedimentary structures	Interpretation
Red-brown mudstone	Concretions	Supratidal flat
Maroon-brown mudstone	Horizontal and wavy siltstone laminae	High-tide mud flat
Alternating mudstone and quartz arenite	Desiccation cracks, cross-lamination, flasers; lenticular and wavy bedding	Mid-tide flat
Quartz arenite	Plane beds, ripples and cross-lamination, herringbone structures, reactivation surfaces	Low-tide terrace
	Large scale solitary cross-beds, massive sandstone, channels, herringbone structures, reactivation surfaces	Shallow subtidal

A

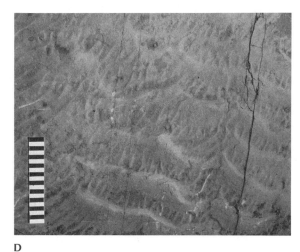

D

B

E

C

Figure 4.26 Characteristic sedimentary structures in the intertidal deposits of the Graafwater Formation. (A) Idealized model of sequence in the Graafwater, with depositional interpretation. (B) Flaser and lenticular bedding, showing some soft sediment deformation due to dewatering. (C) Mudcrack polygons enhanced by sand filling. (D) Interference ripples ("ladderback ripples"). (E) Herringbone crossbedding in the Loop Sandstone near Graafwater. (F) Reactivation structures dipping to right (arrows). A and E courtesy I. C. Rust; remaining photos courtesy A. B. Tankard.

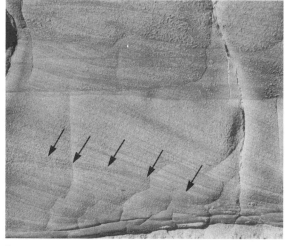

F

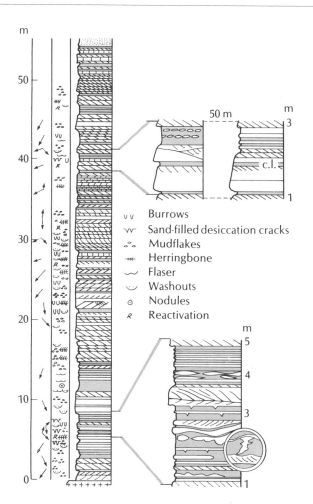

Burrows
Sand-filled desiccation cracks
Mudflakes
Herringbone
Flaser
Washouts
Nodules
Reactivation

Figure 4.27 Vertical section through the Graafwater Formation, Chapman's Peak Drive, South Africa, showing repetitive intertidal shales and sandstones. From Rust (1977).

Figure 4.28 Exposure of the vertically repetitive, cyclical intertidal deposits at Chapman's Peak Drive, shown in Fig. 4.27. Photo courtesy A. B. Tankard.

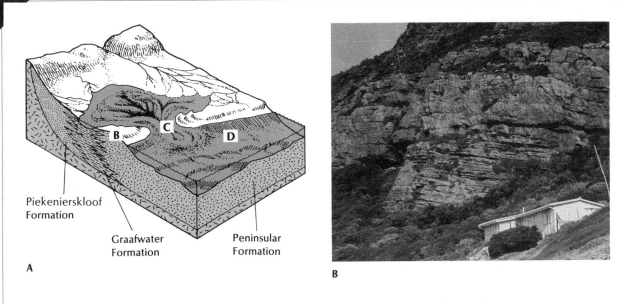

Figure 4.29 (A) Generalized model for the deposition of the Lower Table Mountain Group, including braided alluvial plain, fan delta, tidal flat, tidal estuary, barrier island, washover fan, tidal channel, inlet, and offshore tidal sand bar. (B) Landward-dipping plane beds of barrier overwash (lower sequence) overlain by tidal channel deposits (upper sequence). (C) Rapidly alternating cross-bed sets from the tidal channel. (D) Seaward-dipping spit beach-face plane beds, overlying ebb-dominated tidal inlet sequence. Photos courtesy A. B. Tankard.

FOR FURTHER READING

The works by Davis, Galloway and Hobday, Reading, Rigby and Hamblin, Scholle and Spearing, Selley, and Walker that are listed in Chapter 3 are the primary sources for information in this chapter as well. In addition, the following sources may prove useful:

Coleman, J. M., 1976. *Deltas: Processes of Deposition and Models for Exploration.* Continuing Education Publishing, Champaign, Ill. Excellent synthesis of the deltaic system by one of the leading authorities.

Davis, R. A., Jr. (ed.), 1978. *Coastal Sedimentary Environments.* Springer-Verlag, New York. Excellent review chapters of most of the coastal systems.

Davis, R. A., Jr., and R. L. Ethington (eds.), 1976. *Beach and Nearshore Sedimentation.* Soc. Econ. Paleo. Min. Spec. Publication 24. Many classic papers on beach sedimentation.

Ginsburg, R. N. (ed.), 1975. *Tidal Deposits.* Springer-Verlag, New York. The best collection of papers (45 in all) on tidal sedimentation.

Klein, G. DeV., 1977. *Clastic Tidal Facies.* Continuing Education Publication, Champaign, Ill. Excellent short review of tidal sedimentation.

Morgan, J. P., and R. H. Shaver (eds.), 1970. *Deltaic Sedimentation. Modern and Ancient.* Soc. Econ. Paleo. Min. Spec. Publication 15. A collection of papers describing most of the major delta systems of the world.

Thick sequence of Eocene turbidites and hemipelagic mudstones of the basin plain facies near Zumaya, northern Spain. Photo courtesy G. Shanmugam.

5

Clastic Marine and Pelagic Environments

CLASTIC SHELF DEPOSITS

About 5 percent of the earth's surface is covered by seawater that is less than 200 meters deep. Such areas, known as *continental shelves*, are continuous with the coastal plain sequences of the continents and have slopes of only a tenth of a degree. In tropical regions, the shelves are areas of carbonate sedimentation (discussed primarily in the next chapter). However, where the water temperatures are too cold for carbonate or where siliciclastic sediment overwhelms carbonate sediment, the shelves are covered by detrital particles from the continents, mostly fine sands, silts, and muds.

In the geologic past, shallow marine siliciclastic sedimentation was much more widespread. During episodes of high sea level, all the continents were partially covered by vast areas of epicontinental, or *epeiric*, seas. Modern examples of epeiric seaways, such as Hudson Bay, the Yellow Sea, and the Caspian and Arafura Seas, do not begin to approach the scale of ancient epicontinental seas. Epeiric marine sandstones and shales make up a major portion of the geologic record on the continents and can be very thick. For example, the Proterozoic Transvaal Group of South Africa is 12,000

meters in thickness, and marine quartz sandstones reach thicknesses of 2000 meters in the Cambro-Ordovician rocks of southern Africa (see Box 4.2).

Despite their importance in the stratigraphic record, modern continental shelf sediments are still not well understood. Naturally, it is difficult to study sediments under tens to hundreds of meters of water, and only recently have suitable techniques (diving craft and underwater photography, box coring, high resolution seismic profiling, and side-scanning sonar) been developed to undertake these studies. More importantly, present-day shelf sediments may not be the best analogues for ancient epeiric seas. The modern continental shelves were completely emergent during the last glacial maximum 18,000 years ago, and their sedimentary cover was undoubtedly affected by shoreline and fluvial processes. Between 17,000 and 7000 years ago, sea level rose rapidly (nearly a centimeter per year) due to melting of the glaciers and covered the outer shelf very rapidly. Emery (1968) suggested that most of the sand and gravel on the outer continental shelves was *relict* from the last low-sea-level episode and has been little reworked since then. Swift, Stanley, and Curray (1971) have since argued that most shelf sediments seem to be in dynamic equilibrium with modern marine processes. In many instances the sediment may have first reached the shelf by fluvial or deltaic processes but has since been reworked by marine processes. They referred to this as the *palimpsest* effect after the ancient parchments, reused by medieval authors, that bear traces of older writing under the new. Thus, modern continental shelf sediments may be only partial analogues of ancient epeiric sediments.

Continental shelf sedimentation is important for economic reasons. Shelf sands have high porosity and permeability and usually occur as isolated lenses or sheets in predominantly impermeable shales, so they are good stratigraphic traps for hydrocarbons. Some of the surrounding marine shales also are believed to be good source rocks for oil and some shallow marine sands have produced enormous amounts of oil and gas.

The depositional processes in the continental shelf system are most strongly affected by position on the shelf. The inner shelf is shallow enough to experience wind-driven, storm-driven, and tidal currents. The outer shelf is well below wave base of even the strongest storms or tidal currents, so most current flow is due to oceanic circulation, upwelling, and the effects of density-stratified water columns. It is believed that about two-thirds of the outer shelf sediments are relict or palimpsest because this region was covered during the most rapid phase of sea-level rise and has had relatively little time to reach equilibrium. The inner shelf, on the other hand, has accumulated a considerable blanket of sediment in the last 7000 years. It is probably the best analogue for ancient epeiric seaways.

On shelves that experience mesotidal and macrotidal ranges (tides greater than 2 meters), there are tidal currents of 50 to 100 centimeters per second (cm/s). These currents flow back and forth across shallow shelves, such as the North Sea, producing a strong reversing flow, that is intensified as it passes through narrow straits. These currents strongly affect *sand ribbons*, or *tidal ridges*, which run parallel to the main direction of tidal flow; these features are long stringers of sand up to 40 meters high, 200 meters wide and 15 kilometers long (Fig. 5.1). Sand ribbons are asymmetrical in cross section (Fig. 5.2), with erratic cross-stratification on the gentle slope and well-developed foreset stratification with a slope of about 5° on the steeper slope. Apparently, sediment is moved *along* the steep face of the ridge by the stronger tidal currents. Tidal reversals generate currents over the gentle slope (indicated by megaripples), and eventually this sediment migrates up to the crest and slides down the steep side. At very high tidal-current velocities (greater than 125 centimeters per second), there is active scour, which produces gravel lags (Fig. 5.3). If the tidal current is less than 100 centimeters per second, tidal sand waves form. At tidal currents slower than about 50 centimeters per second, sheets of smooth sand

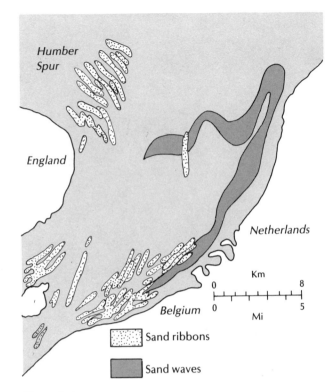

Figure 5.1

Numerous shore-parallel sand ribbons of the North Sea, and a belt where sand waves are found. From Galloway and Hobday (1983).

form, and at even slower velocities, sand forms in patches on the muddy bottom.

Tidal sand waves have crests of 3 to 15 meters and wavelengths of 150 to 500 meters. In cross section they are composed of a set of low-angle surfaces (dipping at 5° to 6°) with smaller cross-bed sets dipping down these low-angle master bedding surfaces (Fig. 5.4A). The length and the height of the cross-bed sets resemble eolian cross-stratification in some ways, and for this reason some authors have argued that classic examples of eolian strata might be formed in tidal sand waves (discussed in Chapter 3). However, a closer look at the marine sand wave cross-beds clearly shows some

Figure 5.2

Cross section of sand ribbons shows superposed smaller-scale bedforms that migrate obliquely up the gentle slope and at right angles across the steeper side. Complex internal stratification is inferred between the major sand-ribbon foresets. From Galloway and Hobday (1983).

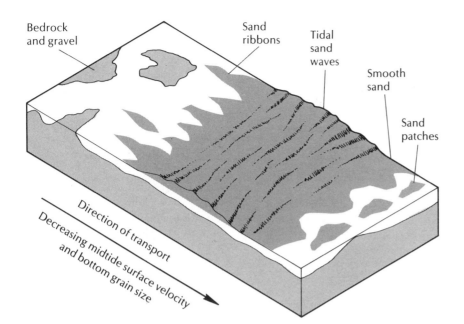

Bedrock and gravel

Sand ribbons

Tidal sand waves

Smooth sand

Sand patches

Direction of transport

Decreasing midtide surface velocity and bottom grain size

Figure 5.3

Succession of facies zones across a tidal transport path. From Blatt et al. (1980).

significant differences: (a) Tidal sand wave cross-beds seldom dip more than 6°, whereas eolian foresets typically dip 20° to 30°. (b) Tidal sand wave cross-bed sets are seldom more than a few meters thick, but eolian cross-beds are usually 10 to 35 meters thick, and there can be hundreds of meters of these cross-beds in succession. (c) The tidal sand wave foresets are actually master bedding surfaces covered by dipping sets of small cross-beds (Fig. 5.4B, upper right) whereas eolian cross-beds are remarkably uniform in their thickness and continuity (see Fig. 3.22).

Most shallow inner shelves, which have weak tidal currents (velocities less than 25 centimeters per second), are usually quiet during normal conditions, but during storms, waves with very deep wave bases are formed. These waves produce strong currents that move obliquely onto the shelf and shoreline and accentuate any bottom irregularities that are oriented obliquely to the shoreline. During the Holocene transgression, shallow shoreface ridges were apparently exposed to deeper water and stronger storm currents, which formed *linear sand ridges* up to 10 meters high, 1 to 2 kilometers wide, and tens of kilometers long. Periods of intense storm activity produced variable cross-bedding within these bars, greatly modifying them. During quiet periods, the bars were covered with wave ripples, and fine sediment accumulated in their troughs as their tops were reworked. These processes produced a coarsening-upward sequence. A gravel lag at the bottom seems to be a relict of an older erosional surface that was produced during lower sea levels in the Pleistocene.

Linear sand ridges are best known from storm-dominated coasts such as the northwest Atlantic shelf off North America (Fig. 5.5). It appears that they began

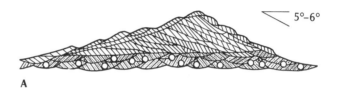

5°–6°

A

B

Figure 5.4

(A) Walker's (1984) model of the internal geometry of marine sand waves. (B) Cross section of a large sand wave in the Lower Greensand (early Cretaceous) near Leighton Buzzard, England. Main cross-bed set is about 5 meters thick but is truncated by thinner, horizontally bounded sets near the top. The compound internal cross-bedding within the larger sets distinguishes these sand waves from eolian deposits. Photo courtesy R. G. Walker.

as abandoned barrier complexes built over the shallow marine mud because many other relict features are still visible on the shelf (for example, shoal retreat massifs produced by submergence of ancient peninsulas). On shelves where there is no evidence of relict barrier islands, the sand supply for these ridges may come from turbidity currents that are set in motion by major storms.

Another feature of storm-dominated sandy shelves is *hummocky cross-stratification* (Fig. 5.6). It is well known from ancient examples but has never been produced in the laboratory. This bedform is composed of low mounds and hollows of very fine sand and silt that have sharp bounding surfaces and no apparent directionality. It is formed at water depths of 5 to 15 meters, where

strong storm waves produce water displacements of several meters and velocities of more than 1 meter per second. A storm surge picks up the finer material into suspension but immediately drops it as the flow reverses. This sediment is deposited in irregular, hummocky sheets that are partially eroded by the next surge, producing the cross-bedding. Hummocky cross-stratification apparently forms in the zone below the fairweather wave base (since it is not reworked between storms) but above the storm wave base (Fig. 5.7).

Another feature known from ancient marine sands that displays storm and wave influence and no tidal effects is *wave-ripple cross-bedding*. Wave ripples differ from current ripples not only in their symmetry but also

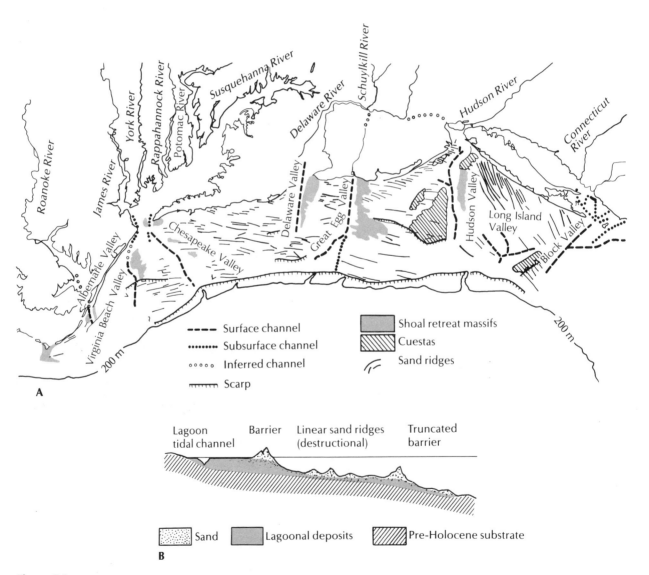

Figure 5.5

(A) Major features of the Middle Atlantic Bight and (B) schematic profile across this shelf. The shelf is composed of a transgressing barrier and progressively abandoned linear sand ridges that overlie lagoonal and pre-Holocene deposits. From Swift et. al. (1973) and Swift (1974).

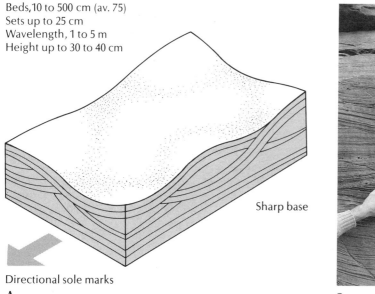

Beds,10 to 500 cm (av. 75)
Sets up to 25 cm
Wavelength, 1 to 5 m
Height up to 30 to 40 cm

Sharp base

Directional sole marks

A

B

Figure 5.6

(A) Diagram of hummocky cross-stratification, showing its low-angle, curved intersections and upward-domed laminae (Walker 1984). (B) Hummocky cross-stratification from the Cape Sebastian Sandstone, Oregon. Photo courtesy J. Bourgeois.

in having an irregular, undulatory lower bounding surface, a less troughlike shape, bundled upbuilding of foreset laminae, swollen lenticular sets, and offshooting and draping foresets (Fig. 5.8). These bedforms are unusual in that they clearly display the effects of rapidly reversing wave flows. Yet they are made of poorly sorted sand and mud and often show lenticular and flaser bedding. If they are formed during high-energy storm waves, however, the finer material can be carried in suspension and deposited where the waves shoal and lose their energy. Rapid reworking by the oscillatory currents might then explain not only the wavy, truncated ripples but also the drapes of mud that fill the hollows during shoaling.

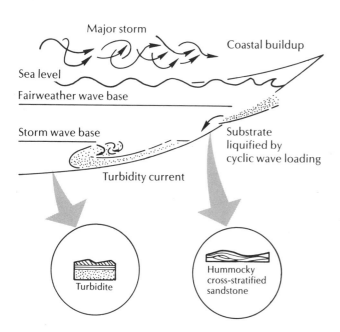

Major storm

Coastal buildup

Sea level

Fairweather wave base

Storm wave base

Substrate liquified by cyclic wave loading

Turbidity current

Turbidite

Hummocky cross-stratified sandstone

Figure 5.7

Storm winds create coastal buildup, and cyclic loading of substrate by storm waves may liquefy the substrate. The liquefied sediment may flow and accelerate basinward, becoming a turbidity current with all the sediment in suspension. Deposition from this flow below the storm wave base would result in turbidites with Bouma sequences. Above the storm wave base, waves feeling the bottom would rework the turbidity current deposits into hummocky cross-stratification. Hummocky cross-stratification could also form above the fairweather wave base but would probably be reworked into other structures by fairweather processes. From Walker (1984).

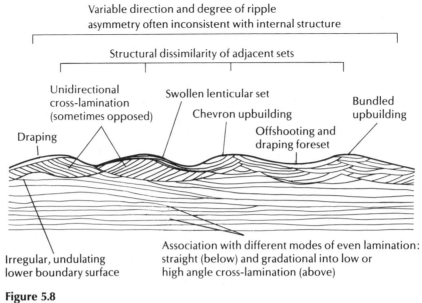

Figure 5.8

Diagnostic features of wave-ripple cross-bedding. From de Raaf et al. (1977).

The typical sequence of a shallow siliciclastic shelf is clearly affected by storm and tidal processes, but *changes in relative sea level* are the primary source of the sediments and structures on which these forces act. Many of the structures we have discussed began as shallow-water features during episodes of lower sea level and were modified during later rises in sea level. Because sea level has fluctuated many times in the geologic past, it may also be responsible for the form of many ancient shelf deposits. Storm-dominated shelf sequences, for example, would be expected to vary upward from burrowed shelf muds (well below storm wave base) to storm-graded beds (just below storm wave base), then to hummocky cross-stratification (between normal and storm wave base) and to trough cross-bedding (reworking by helical storm currents). This coarsening-upward sequence would be expected on a prograding or regressive shelf whereas the reverse would be found on a transgressive shelf (Fig. 5.9A, B). A transgressive tide-dominated shelf (Fig. 5.9C) would proceed from the high-energy gravel lags to the large-scale cross-beds of sand waves and sand ribbons to the bioturbated silts and muds found between the sand bodies; a regressive tidal shelf would produce the reverse sequence, coarsening upward. If both storm and tidal processes occur, then the sedimentary structures would be mixed. In a shelf on which accumulation is balanced by relative sea level change, there would be little net change in grain size (Fig. 5.9D), and the combined storm and tidal influence would produce a mixture of gravel lags, large-scale cross-beds, hummocky cross-stratification, trough cross-beds, and occasional burrowed mud and silt layers.

Diagnostic Features of Clastic Shelf Deposits

Tectonic setting Clastic shelf deposits are extensive in passive continental margins but more restricted in convergent margins. They are also found in epicontinental seaways and gently downwarping cratonic basins. They are associated with deeper marine shale sequences or with shallow marine limestones, and with deltas and coastline deposits.

Geometry Tabular sequences of shale with long, sheetlike or lenticular sandstone bodies are found. The total shelf package may cover thousands of square kilometers and be hundreds of meters in thickness.

Typical sequence Fining- or coarsening-upward sequences occur, depending whether the sequence is transgressive or prograding. Burrowed glauconitic muds are typically followed by storm-graded beds, hummocky cross-stratification, and trough cross-beds (storm-dominated) or by large-scale cross-beds and lag gravels (tide-dominated).

Sedimentology Quartz and clay minerals are dominant, though carbonate minerals and shell fragments can also occur. Glauconite is particularly characteristic of shallow marine muds and sands. Hummocky cross-stratification, wave-generated cross-beds, large-scale cross-stratification, and storm-graded beds are all characteristic of this environment. Burrowed silt and mud with abundant shell debris and trace fossils form the bulk of most shallow marine sequences.

Fossils The most diagnostic feature of a shallow marine deposit is its fossils. Most marine invertebrates

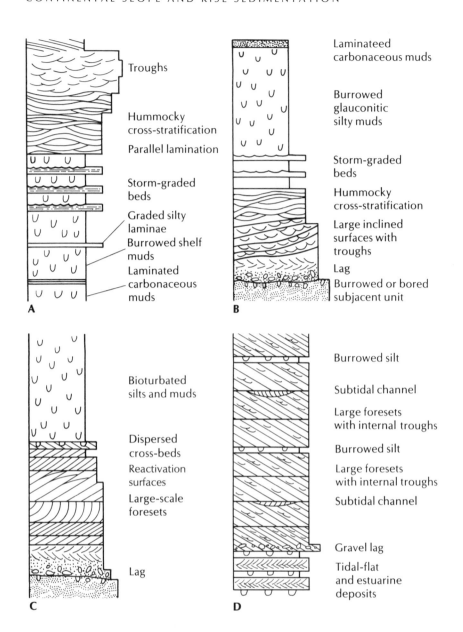

A

Troughs

Hummocky cross-stratification

Parallel lamination

Storm-graded beds

Graded silty laminae

Burrowed shelf muds

Laminated carbonaceous muds

B

Laminateed carbonaceous muds

Burrowed glauconitic silty muds

Storm-graded beds

Hummocky cross-stratification

Large inclined surfaces with troughs

Lag

Burrowed or bored subjacent unit

C

Bioturbated silts and muds

Dispersed cross-beds

Reactivation surfaces

Large-scale foresets

Lag

D

Burrowed silt

Subtidal channel

Large foresets with internal troughs

Burrowed silt

Large foresets with internal troughs

Subtidal channel

Gravel lag

Tidal-flat and estuarine deposits

Figure 5.9

Four characteristic stratigraphic profiles of marine shelf sequences due to variations in relative sea level change, and storm or tide activity. (A) Prograding (regressive) storm-dominated shelf; (B) transgressive storm-dominated shelf; (C) transgressive tide-dominated shelf; (D) balanced accumulation, storm and tide dominated. From Galloway and Hobday (1983).

require a shallow, well aerated bottom with a narrow range of salinities and thus are found in greatest abundance on the shallow shelf. If such restricted organisms are found in a sedimentary sequence, it is almost certainly shallow marine.

CONTINENTAL SLOPE AND RISE SEDIMENTATION

Between the nearly horizontal shelf and the deeper ocean floor is an abrupt boundary, called the *shelf-slope break*; below the shelf-slope break is the steepest part of the ocean floor, the *continental slope* (Fig. 5.10), which is narrow (10 to 100 kilometers) and slopes downward at

an average angle of 4° to 6°. Sediments found on the slope are under the influence of gravitational downslope movement and seldom remain on the slope. Much of the sediment on the continental slope has slumped or washed off the shelf-slope break during storms or earthquakes.

The characteristic sedimentary features of the slope are gravity-transported features: large exotic slide blocks (*olistoliths*), slumped and deformed shales, debris flows full of a chaotic assemblage of exotic brecciated blocks (*olistostromes*), and turbidites (Fig. 5.11). Some of these features can be striking—and easily misinterpreted as well. For example, spectacular syndepositional folding can be developed by slumping blocks of mud (Figs. 5.12A,B). To some geologists, these features might suggest that the sequence has been tectonically deformed,

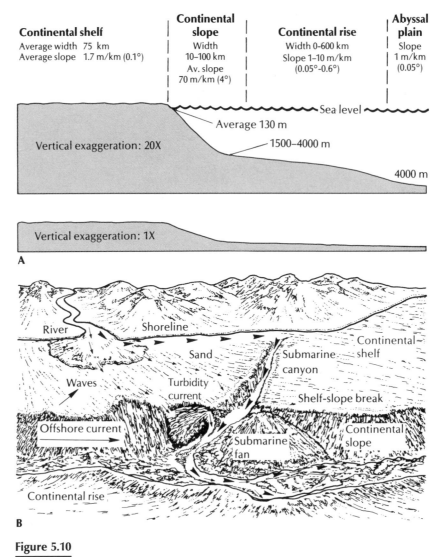

Figure 5.10

(A) Typical dimensions of a modern continental shelf. (B) Diagram of continental shelf, showing route of transport of sand (solid arrow) and mud (dashed arrow) from river mouth to deep-sea basin floor. From Reineck and Singh (1973).

but the folding is entirely due to soft-sediment deformation. Likewise, olistostromes (Fig. 5.12C) are not easily distinguished from other types of sedimentary breccias or tectonic melanges. Tectonic melanges tend to be more highly sheared and deformed than olistostromes (Hsü, 1974) though some olistostromes are secondarily sheared. Olistostromes tend to be much more limited in extent and are bounded by other deep-marine sediments because they are sedimentary bodies and not formed by tectonic accretion. Olistostrome clasts are commonly sedimentary boulders and may show some sign of rounding and transport; melange blocks are formed by fracture and may be highly deformed. Further discussion of melanges is found in Chapter 14.

In addition to its gravity-displaced deposits, the continental slope is also overlain by deposits of hemipelagic mud that settle out of suspension from the oceanic water column. These muds are commonly reworked by *contour currents*, which flow parallel to the slope. Contour currents are the product of normal oceanic circulation as water masses of different densities move relative to one another. They usually have speeds of 5 to 30 centimeters per second, though speeds of 70 centimeters per second have been recorded. This is sufficient to keep most of the clay in suspension and to transport the fine sand and silt. The resulting deposit, called a *contourite* (Fig. 5.13), is finely laminated or faintly cross-bedded silt and sand, interbedded with mud, and forms thin, imbricated

Processes

Rock fall

Sliding

Slumping

Mass flow (e.g., debris flow)

Turbidity current

Reduction of shear stress (mainly) along discrete shear planes

"Liquefaction"

Reduction of shear stress within the mass as a whole

Suspension

Deposits

Olistholiths

Slide

Slump

Mass flow deposit

Turbidite

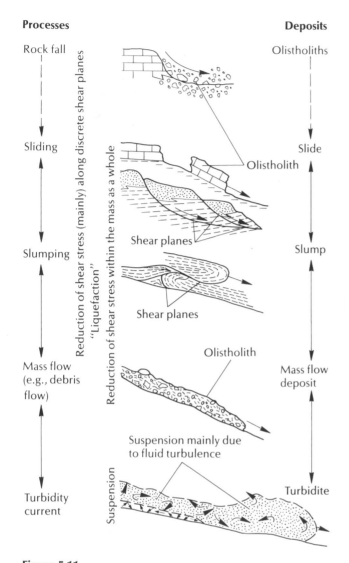

Olistholith

Olistholith

Shear planes

Shear planes

Olistholith

Suspension mainly due to fluid turbulence

Figure 5.11

Processes of mass-gravity transport and their deposits. From Reading (1986).

lenses that fill bottom scours. Contourites supposedly show neither the lateral continuity nor the graded bedding of turbidites, but in practice they can be hard to distinguish.

At the base of the continental slope is a broad, gently sloping region that grades into the sea floor and is known as the *continental rise* (see Fig. 5.10A). The slope-rise break occurs at a depth of about 1500 meters and the slope can be up to 600 kilometers in width. The gradient of the continental rise is very gentle, from 0.05° to 1.0°, or about 1 to 10 meters per kilometer. Continental rises occur primarily on passive margins; on active margins, the slope usually drops directly into the sub-

A

B

C

Figure 5.12

(A) Convolute bedding caused by submarine landsliding in the Miocene Castaic Formation, Ridge Basin, California. (B) Giant submarine slump fold within the Castaic Formation. (C) Submarine landslide conglomerate, showing normal grading with a rippled Bouma C division capping the bed. The clasts are imbricated in an upslope direction at the top of the 1.5 meter thick channel sequence. Photos A and B by the author; photo C courtesy H. E. Cook.

A

B

Figure 5.13

(A) Core of contourite sediment from the North American Atlantic continental rise off New England in 4745 meters of water. Centimeter-wide beds of hemipelagic claystones alternate with coarser, cross-laminated contourite siltstone. Photo courtesy C. D. Hollister. (B) Possible Ordovician contourite showing ripple forms with wavelengths of about 9 centimeters and amplitudes of about 1 centimeter. Both base and top have sharp contact with enclosing hemipelagic mudstone. Outcrop is from the Hales Limestone, Nevada. Photo courtesy H. E. Cook.

ducting trench. The bulk of the continental rise sequence is made of material that has slid down the continental slope or down through the submarine canyons cut into the continental shelf (see Fig. 5.10B). Between episodes of sliding, hemipelagic muds settle out of suspension and eventually form thick sequences of shale.

The most important process of sediment transport in the continental rise is the turbidity current, a turbulent suspension of particles that is denser than the surrounding water. Like other density currents, a turbidity current flows as a discrete surge, separate from the medium through which it travels (see Fig. 2.4B). Its high velocity and turbulent motion enable it to carry much larger particles than would normally be found far out on the rise. This is the key to a mystery that long puzzled geologists: Why were fairly coarse sandstone beds of great thickness and lateral extent found in sequences that had deep marine shales and all the other earmarks of very deep water? In 1950, Kuenen and Migliorini first proposed the turbility current as a mechanism for

bringing sand and fine gravel to the deep seafloor. The velocity and destructive power of the turbidity current were accidentally recorded in 1929, when the Grand Banks earthquake triggered a submarine landslide that sent a turbidity current down the side of the Newfoundland shelf and slope. As it flowed, it broke the various segments of the transatlantic telegraph cables that were stationed along the face of the slope. By recording the time at which each cable went dead and plotting its position on the slope, scientists calculated velocities of 100 kilometers per hour at the beginning of the flow and 25 kilometers per hour at the end. The turbidity current covered over 280,000 square kilometers of the North Atlantic seafloor with well over a hundred cubic kilometers of sediment.

The lithologic product of a turbidity current is called a *turbidite*. Every turbidite represents a single catastrophic flow and has a regular and predictable sequence of sedimentary features known as the *Bouma sequence*, after the man who first analyzed turbidite sequences in detail.

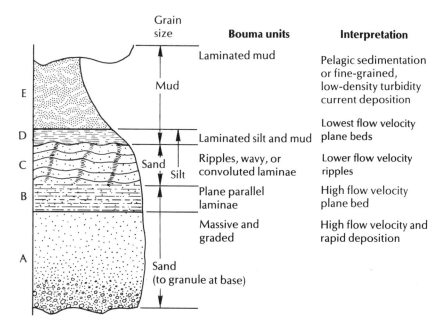

Figure 5.14

The classic Bouma turbidite sequence, showing trends of sedimentary structures, grain sizes, and depositional conditions. After Bouma (1962).

The classic Bouma turbidite sequence (Fig. 5.14) begins at the scoured base with a massive graded bed (unit A), which represents the coarsest material to settle out of suspension as the turbidity current slowed. Above this comes plane lamination (unit B), which is believed to represent high flow velocity plane beds. Unit C displays ripples and wavy lamination, indicative of lower flow velocities. Unit D is laminated silt, and E is laminated mud that settled out of suspension during the waning of the turbidity current. In some cases, unit E is topped by laminated hemipelagic mud that settled from suspension in the episodes between turbidity currents. In practice, these muds are often very difficult to tell from the laminated muds of unit E.

Not all turbidity currents have produced a complete Bouma sequence. Typically, turbidite sequences accumulated as broad sheets and lobes of sediment that spread out from the submarine canyon and built an extensive *submarine fan* complex (see Fig. 5.10B). Deposits near the top (proximal part) of the rise or within the submarine canyons are the most coarse, with debris flows and coarsely graded beds predominating (Fig. 5.15). "Classic" turbidites are found in the middle part of the fan. At the lower (distal) end of the fan, only the finer material is left, producing thin-bedded turbidites dominated by the fine sand, silt, and clay. The fan channels

are braided, much like the channels on alluvial fans, and they have distinctive levee and overbank deposits, like those seen in fluvial systems. As the fan sequences accumulated, they built the characteristic wedge shape of the continental rise. This pile of turbidites and hemipelagic muds can accumulate into enormous thicknesses of rhythmically alternating graded sandstones and shales. Many examples of rapidly sinking deepwater tectonic basins are known that have accumulated thousands of meters of turbidites and shales (Box 5.1).

A number of other sedimentary structures are also characteristic of the slope and rise. The strong currents that scour the cohesive, muddy bottom were the cause of many types of sole marks, such as flute casts and groove casts, and many of the tool marks described in Chapter 2. Because sand and mud layers of different densities are often interbedded, load casts and other types of soft-sediment deformation are common. In addition there is a characteristic suite of trace fossils, particularly those bottom-dwelling organisms that systematically feed through the muddy bottom (see Fig. 2.19, abyssal zone).

After understanding the slope and rise processes, one can predict the characteristic stratigraphic sequence (see Fig. 5.15A). The slope is dominated by fine hemipelagic muds, interrupted by submarine channel sandstones, and

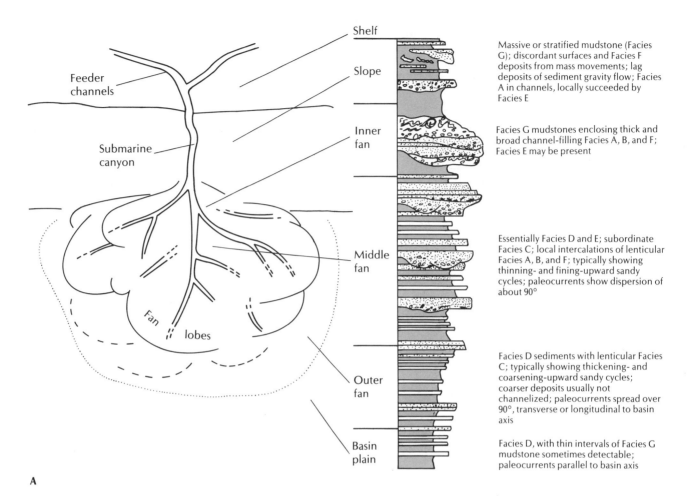

Shelf

Slope

Inner
fan

Middle
fan

Outer
fan

Basin
plain

Feeder
channels

Submarine
canyon

Fan   lobes

Massive or stratified mudstone (Facies
G); discordant surfaces and Facies F
deposits from mass movements; lag
deposits of sediment gravity flow; Facies
A in channels, locally succeeded by
Facies E

Facies G mudstones enclosing thick and
broad channel-filling Facies A, B, and F;
Facies E may be present

Essentially Facies D and E; subordinate
Facies C; local intercalations of lenticular
Facies A, B, and F; typically showing
thinning- and fining-upward sandy
cycles; paleocurrents show dispersion of
about 90°

Facies D sediments with lenticular Facies
C; typically showing thickening- and
coarsening-upward sandy cycles;
coarser deposits usually not
channelized; paleocurrents spread over
90°, transverse or longitudinal to basin
axis

Facies D, with thin intervals of Facies G
mudstone sometimes detectable;
paleocurrents parallel to basin axis

A

Figure 5.15

(A) Geometry of submarine fans and the associations of facies, according to Mutti and Ricci Lucchi (1972). From Lewis (1984). (B) Distribution of submarine fan facies. From Shanmugam et al. (1985).

chaotic olistostrome breccias and slump deposits. Frequently, slope sequences show slump scars or soft-sediment deformation from slumping. The upper part of the rise is composed of turbidite fans, which prograde as they build outward. This implies that the sequence *coarsens upward* from the thin, outer-fan sandstones to the complete turbidites and channel sandstones of the midfan, to the large upper-fan channels with occasional olistostromes. If the lobes changed course, many such sequence may have accumulated vertically. The lower part of the rise and some of the abyssal plain are covered by pelagic muds, with occasional thin, very-fine sandstones from the distal edge of each turbidite.

Because submarine fans represent pulses of sedimentation filling a deep marine trough, it has long been assumed that the accumulation of thick turbidite sequences was a reflection of tectonism that caused the trough to sink rapidly. Recently, however, Shanmugam and Moiola (1982), Mutti (1985), and Shanmugam et al.

(1985) have shown that most episodes of submarine fan deposition are highly correlated with low sea levels. For example, during the Quaternary, turbidite deposition took place only during the episodes of glaciation and low sea level. During interglacials, there was only pelagic deposition on the fan surface. Because the lower sea level exposed and eroded the continental shelf, more sediment accumulated at the heads of submarine canyons. Eventually, this sediment became destabilized and avalanched down the canyons in turbidity currents. As we saw in the case of the shelf sediments, sea-level changes appear to be the major controlling influence of deposition.

Deep marine sandstones are important because they clearly indicate paleoceanographic conditions. Also, they sometimes are the main sedimentary fill of deep, rapidly dropping, tectonic basins. Because turbidite sands are fairly coarse and are interbedded with impermeable shales, they have been important hydrocarbon reservoir rocks.

Facies A Thick to massive, channeled and amalgamated, poorly sorted coarse Ss and Cgl, with thin or no mud intervals; all gradations to Facies E

Facies B Thick to massive, lenticular sorted Ss with parallel to undulating laminae, common mud clasts, and erosional bases; thin mud intervals

Facies C Couplets of even, parallel bedded M-F Ss and minor homogeneous muds; Ss may show complete Bouma succession, some broad, shallow channels; common sole marks

Facies D Couplets of parallel-bedded, laterally continuous F-VF SS/siltst. and thicker muds. Ss with regular to convolute to ripple-drift laminations. Bouma base cut-out sequences common

Facies E Thinner, irregular and discontinuous beds of slightly coarser Ss and Silst. than D; also thinner muds. Ss with basal graded and structureless intervals; sharp upper contacts with mud

Facies F Thick intervals of mildly deformed chaotic deposits derived from sliding or mass flow

Facies G Thick muds with often obscure continuous parallel bedding

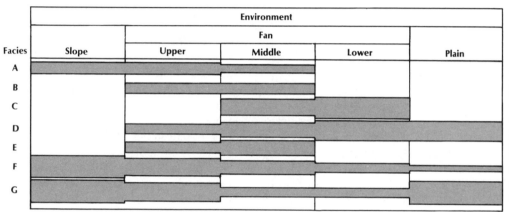

Diagnostic Features of Continental Slope and Rise Sediments

Tectonic setting Slope and rise sediments are found along all continental margins, but thick continental-rise wedges occur primarily in rifted passive margins. Tectonically down-dropped or downwarped basins may also reach sufficient depths to accumulate thick sequences of shales and turbidites. They are typically associated with deep marine pelagic sediments.

Geometry A wedge or thick lens builds against the continental margin. It may be thousands of meters thick, hundreds of kilometers wide, and extend thousands of kilometers along the base of the slope.

Typical sequence Slope sequences are mainly hemipelagic muds, with contourites that are interrupted by submarine channel sands and by olistostromes and slump deposits. The main part of the fan is built of prograding turbidites, which coarsen upward from finer, thin, distal sands at the base to coarser, complete Bouma sequences, to channel sands at the top. The lower fan and abyssal plain are covered with pelagic muds and episodic fine sand layers from distal turbidites.

Sedimentology The coarser fraction of a graded bed can have many other minerals besides quartz, but the bulk of slope and rise sediment is fine sand, silt, and clay. Bouma sequences display a series of sedimentary structures reflecting decreasing flow velocities, from graded beds to upper flow velocity plane beds, to lower flow ripples and to finely laminated silts and muds. Olistostromes and slump deposits are also characteristic of this system, as are sole marks, soft-sediment deformation, and certain types of ichnofossils.

Fossils Pelagic organisms, particularly benthic foraminifers, are the most diagnostic, but they are never very abundant. Clastic particles in turbidites can be made of broken shell debris from the continental shelf.

Box 5.1
Tertiary Turbidites of the Northern Apennines

The most famous and best exposed of the many well-studied ancient turbidite sequences around the world are those of the northern Apennine Mountains of Italy. The classic paper on these deposits by Mutti and Ricci Lucchi (1972; English translation, 1978) became the standard on which most later studies were based, and their facies terminology has influenced many sedimentologists. These thick flysch sequences were deposited in deep basins formed by the collision of Africa with the southern Mediterranean and the consequent rotation of major crustal blocks that make up the Italian Peninsula.

Mutti and Ricci Lucchi (1972) described the Apennine turbidites in terms of seven lithofacies, which they labeled A through G. *Facies A* (see Fig. 5.15) is represented by poorly sorted coarse-grained sandstones and conglomerates. These occur in massive beds (1 to 2 meters thick) and are found in ancient channels. Thin lenses of silt or fine sand may be present, but they are rare. These conglomerates, believed to have moved by debris flows, are apparently confined to the feeder channels in the slope and upper fan. *Facies B* (Fig. 5.16A) consists of medium-fine- to coarse-grained sandstones in thick beds that are lenticular but more laterally continuous than Facies A. Their most diagnostic structure is thick, parallel or broadly undulating current laminae. Facies B is closely associated with and intergrades with Facies A, and probably represents debris flows with some current reworking. The nature of the sedimentary structures suggests that they may be high flow velocity features, such as antidunes and plane beds. Both Facies A and B appear to be produced by rapidly moving turbidity currents that fill the feeder channels along the slope and the inner fan.

Facies C (Fig. 5.16B) consists of the classic turbidites of most authors and is the dominant facies on the middle and outer fan (see Fig. 5.15). This facies is mostly medium- to fine-grained sandstones with well-developed Bouma sequences. The thick sandstone beds are generally planar and laterally continuous, though there may be broad, low-relief channels in places. The beds are typically 50 to 150 centimeters in thickness and may be separated by thin shale partings. These sandstones were deposited by the classical turbidity-current mechanism previously described.

Facies D (Fig. 5.16C) consists of fine- to very fine-grained sandstones that have pronounced current laminae with parallel or undulating surfaces or climbing ripple drift. The most diagnostic feature is the planar nature of the beds, which are typically 3 to 40 centimeters in thickness and show tremendous lateral continuity. Some beds can be traced for tens of kilometers. Facies D sands typically alternate with Facies G hemipelagic muds. Facies D appears to represent the lower velocity portion of the Bouma cycle (Bouma intervals B—E) and results from low-energy turbidity currents that have left their coarse material behind. Facies D is found on the inner and middle fan but is more common on the outer fan; it is one of the dominant facies on the deep-sea plain.

Facies E (Fig 5.16D) sandstones differ from those in Facies D in having higher sand—shale ratios (typically 1:1) and thinner, more irregular, less continuous sandstone beds with frequent wedging and lensing. Typically, Facies E sandstones beds are coarser-grained than Facies D, are more cross-bedded, and have ripples or dunes on the top. This facies was probably produced by liquefied grain flows and small currents and appears to have as overbank deposits from the channels in the upper and middle fan (see Fig. 5.15).

Facies A through E are products of turbidity currents in various guises; Facies F and G are both associated with turbidites. *Facies F* is Mutti and Ricci Lucchi's (1972) label for any type of slump, slide, olistostrome, mudflow, or other type of chaotic deposit. These are found primarily on the slope and upper fan, where gravity sliding predominates. *Facies G* is pelagic and hemipelagic mudstones and shales. These are usually massive but can be vaguely stratified, especially when they are interbedded with one of the sandy facies. Of course, these muds occur in all parts of the deep sea, but they are dominant on the slope, inner fan, and deep-sea plain. The occurrence of these facies and their depositional interpretation are summarized in Fig. 5.15.

The cliffs of the Apennines show spectacular examples of associations of facies that represent various parts of the marine slope-rise complex. The slope and inner fan are usually composed of thick sequences of Facies G pelagic muds that have been cut by slumps, olistostromes, and thick fan channels (Fig. 5.17A) filled with Facies A and B conglomerates and sandstones. Some of these channels reach 1 to 2 kilometers in width. The middle fan is characterized by smaller, lenticular channels filled with

A

B

C

D

Figure 5.16 Outcrops of submarine fan deposits from the classic examples of Mutti and Ricci Lucchi (1972). (A) Facies B, showing fine- to medium-grained sandstone with almost no mud intervals. The dominant structures are thin laminae that gently undulate, like those formed by antidunes. The units are punctuated by numerous erosional surfaces. (B) Facies C, alternating thick, graded sandstones and thin mudstone interbeds. (C) Facies D, couplets of fine-grained, finely laminated sandstones with convolute bedding and climbing ripple drift, alternating with thicker mudstones. (D) Facies E, thinly stratified, irregular, discontinuous sandstones, slightly coarser than Facies D, interbedded with thinner muds. The sandstones are not laminated, but may be graded, and frequently have dunelike top surfaces, which are in sharp contact with the overlying mudstone. Outcrops shown are from the Marnoso-arenacea Formation, Senio Valley, Italy; photos courtesy E. Mutti.

Facies A and B conglomerates and sandstones that represent more distal (outer) fan channels. These channels are carved into planar laminated turbidites of Facies C, D, and E. In the outer fan (Fig. 5.17B), the fan channels have become very thin or have disappeared, so the sequence is planar, composed mostly of Facies C and D turbidites with tremendous lateral continuity. This continuity is even more striking in the deep-sea plain association (Fig. 5.17C), but the sequence is much thinner and finer-grained than the outer fan. Typically, the deep-sea plain association is dominated by thin, fine-grained Facies D turbidites alternating with Facies G hemipelagites.

Although there are some superficial similarities between deep-marine sandstones and those from other environments, their many unique or characteristic features make them easy to recognize. Well-developed graded bedding in Bouma sequences, coupled with distinctive sedimentary structures, are the most characteristic features. Only deep-sea fans and plains produce such thick and laterally continuous sequences of alternating sandstones and mudstones. Finally, there are many diagnostic trace fossils, and often foraminifers and other marine fossils as well, to distinguish these deposits from those of shallower water.

A

B

C

Figure 5.17 Large-scale outcrops display various parts of the submarine fan. (A) Proximal (or inner) fan association, showing a well-developed submarine channel about 2 kilometers wide, filled with Facies A and B beds. (B) Distal (or outer) fan association, composed of thick, continuous, repetitive Facies C and D turbidites. (C) Deep-sea plain association, composed of Facies D turbidites and Facies G hemipelagic shales, showing extreme parallelism and lateral continuity of the beds. Photos courtesy E. Mutti.

PELAGIC SEDIMENTATION

Deep ocean basins cover more than 50 percent of the earth's surface. This vast area of oceanic crust, which lies at depths of 4 to 6 kilometers or more, is mostly a flat, featureless abyssal plain with occasional seamounts and mid-ocean ridges. In the past, geologists paid little attention to deep-sea sediments, partly because they were not likely to contain oil but more importantly because so little was known about the deep sea until the advent of modern marine geology in the 1950s. According to Davis (1983), "it is safe to say that more has been learned about the geology of deep ocean since 1965 than in all previous time."

Marine geology essentially began with the voyages of the British naval vessel *H.M.S. Challenger* from 1872 to 1876. Before these voyages, it was widely believed that the waters of the deep sea were stratified by age, with trilobites still living at abyssal depths. The work of the *Challenger* scientists forever demolished this notion. John Murray (Murray and Renard, 1891) described the deep-sea sediments that were recovered by the *Challenger*, and his general outline of deep-sea sedimentation is still valid. The modern era of marine geology began in 1947 with the Swedish Deep Sea Expedition, which used the first piston-coring device for taking a continuous sample of deep-sea sediments. In the 1950s and 1960s, marine geology continued to develop at institutions such as Lamont-Doherty Geological Observatory in New York, Woods Hole Oceanographic Institution in Massachusetts, and the Scripps Institution of Oceanography in California. In 1968, the maiden voyage of the

Glomar Challenger (Fig. 5.18), the primary vessel of the Deep Sea Drilling Project (DSDP), began a revolution in the earth sciences. The DSDP cores soon provided confirmation of seafloor spreading and the best and most detailed data ever gathered about the evolution of the earth's oceans and climates. Many other important discoveries resulted from the study of deep-sea cores, such as the confirmation of the orbital variation theory of the ice ages and the proof that the Mediterranean dried up 6 million years ago. One of the most surprising discoveries was that the oldest oceanic crust is Jurassic in age and that all older seafloor has been subducted. The cycling of crust is so rapid that only the oceanic crust of the last 150 million years still survives though the earth has had oceans for almost 4 billion years.

The major conclusion reached by Murray, confirmed by later work, is that *most of the terrigenous sediment from the continents (except for suspended clays) is trapped in the continental shelves and rises and never reaches the deep-sea floor.* Instead, deep-sea sediments are *pelagic*, settling out of the overlying water column. The main constituents of deep-sea sediments are *terrigenous clays* from the continents; *biogenic skeletal material* of calcareous, siliceous, or phosphatic marine organisms; and a minor amounts of *authigenic* components, which develop diagenetically in deep-sea muds and oozes. In some places, deep-sea sediments also include occasional volcanic ash layers, eolian dust, or tektites from extraterrestrial impacts.

These constituents are not randomly distributed across the seafloor but show a definite pattern (Fig. 5.19). *Siliceous oozes* are dominant in equatorial and polar regions where oceanic upwelling occurs due to boundary

A

B

Figure 5.18

(A) The *Glomar Challenger*, the vessel that was the primary workhorse of the Deep Sea Drilling Project in the 1970s. It is now retired and managed by Texas A&M University. The vessel is 122 meters long and handles 7320 meters of drill pipe. Photo courtesy Scripps Institution of Oceanography. (B) Scale of deep-sea drilling operation at a depth of 5500 meters. If the drawing were accurate, drill string width could not be detected at this scale, yet by sonar beacons and steering devices on the drill, the vessel can relocate the same hole many times and recover continuous core. From Kennett (1982).

currents between large water masses. *Calcareous oozes*, on the other hand, are dominant in tropical, subtropical, and tenperate waters where the conditions are warm enough for calcite secretion and where the oceanic floor is not so deep that they are dissolved. Terrigenous clays are found in the centers of oceanic water masses where there is limited productivity and the sea floor is so deep that calcite dissolves. Glacial sediments are common in the polar regions, particularly around the Antarctic ice cap.

More than 47 percent of the deep-sea floor is covered by calcareous oozes, which are made of the carbonate tests of planktonic microfossils. The dominant carbonate microfossils are foraminifers, particularly the family Globigerinidae, which are single-celled protozoans with a porous calcitic skeleton (Fig. 5.20A). Coccolithophorids, submicroscopic photosynthetic organisms covered with tiny porous calcareous plates, are also very important in many regions (Fig. 5.20B). Pteropods, tiny pelagic

snails with cone-shaped or coiled shells made of aragonite, are common in certain shallow tropical waters, such as the Mediterranean or the Gulf of California.

All of these organisms live in great abundance in surface waters wherever the temperature is high enough, primarily in temperate and subtropical latitudes. In shallow waters their skeletons can form great thicknesses of calcareous ooze. Near the surface, seawater is supersaturated with calcium carbonate ($CaCO_3$), so these organisms have plenty of material with which to build their skeletons. At a depth of about 500 meters, however, the high pressure and cold temperature change the water chemistry so that calcite becomes undersaturated; this chemical change is related to the production of CO_2 by the respiration of organisms and also to the decay of dead organisms. When these organisms die, their calcareous skeletons sink toward the seafloor, dissolving as they get deeper. At a certain depth, called the *lysocline*, the rate of carbonate dissolution reaches a maximum

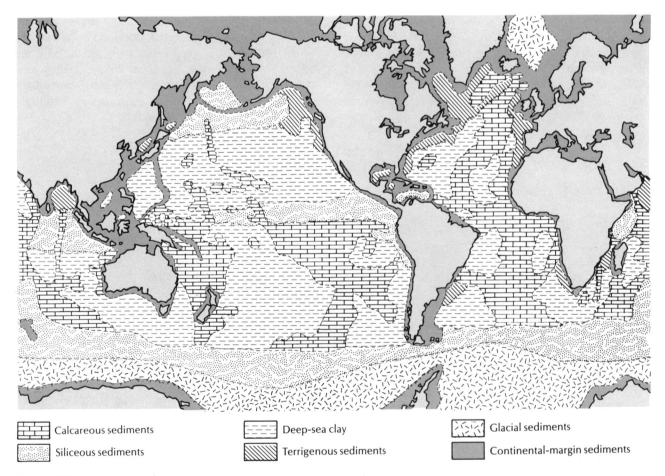

| Calcareous sediments | Deep-sea clay | Glacial sediments |
| Siliceous sediments | Terrigenous sediments | Continental-margin sediments |

Figure 5.19

The global pattern of deep-sea sediments. Calcareous oozes are restricted to low latitudes. Most siliceous oozes lie close to the poles, although some occur in the equatorial Pacific and Indian Oceans. The areas of low productivity in the center of oceanic gyres have mostly pelagic clays. From Riley and Chester (1976).

A

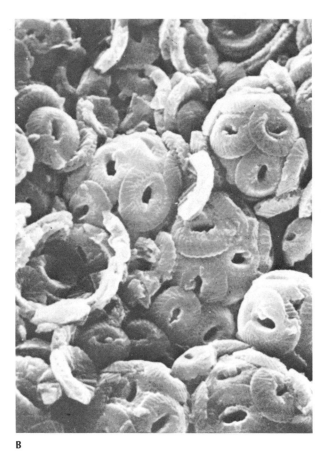

B

Figure 5.20

(A) Foraminifers (large objects) and radiolarians (smaller, coarse-meshed objects) from the Pacific Ocean floor. Diameter of radiolarians is about 100 to 200 micrometers. Photo courtesy Scripps Institution of Oceanography. (B) Button-shaped calcareous plates (coccoliths) of microscopic algae known as coccolithophorids. Most are less than a few tens of micrometers in size. Photo courtesy D. Noël.

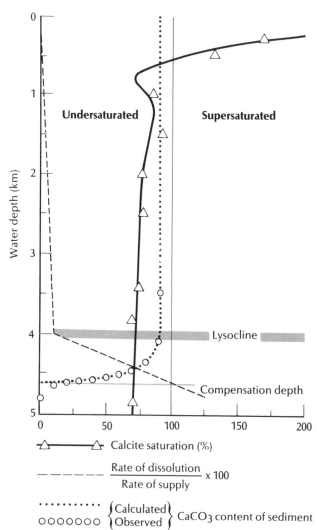

Figure 5.21

Variations in carbonate saturation with increasing depth in the equatorial Pacific Ocean. Modified from van Andel, Heath, and Moore (1975).

(Fig. 5.21). Dissolution continues with depth until a level is reached at which the rate of supply of calcite is balanced, or compensated, by the rate of dissolution; little or no calcareous material persists below this depth. This critical level is called the *carbonate compensation depth (CCD)*. The depths of the lysocline and CCD vary, depending on the depth of the ocean basin and the kind of oceanic circulation patterns. Most dissolution occurs at depths of 4 to 5 kilometers, though in some places calcite survives to water depths of 6 kilometers or more. Because the depth of most ocean basins is 4 to 5 kilometers, it is clear that only the shallower ocean basins, the floors of which are above the CCD, can retain calcareous ooze.

It seems surprising that these tiny calcareous skeletons can survive dissolution from the first zone of under-

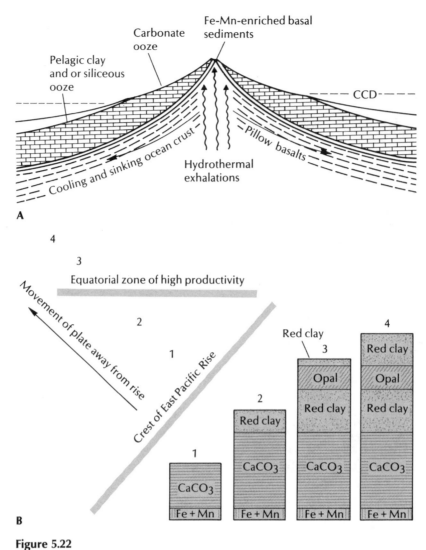

Figure 5.22

(A) Sedimentation across a mid-ocean ridge. (B) Modification of sediments as a plate migrates across belts of different biotic productivity in the ocean. From Broecker (1972).

saturation at 500 meters to the CCD at 4000 meters or more. This delay of dissolution is apparently caused by the protective organic coating on some shells (particularly coccoliths) or by the relatively large sizes of some foraminifers. In some regions, the sedimentation rate is so high that calcareous tests are buried rapidly enough to protect them from corrosive seawater. This is particularly true of calcareous oozes deposited on the mid-ocean ridges, which have been called the "snow-covered mountains of the deep ocean." The crest of the ridge (Fig. 5.22A) is typically covered by a thick layer of calcareous ooze. As spreading continues and the oceanic crust begins to sink away from the ridge, a bathymetric zonation begins to develop. First the ptero-pods dissolve, at about 1.5 to 2 kilometers, because they are made of highly soluble aragonite. At the CCD

the foraminifers and coccoliths dissolve, unless they are buried. Below the CCD only a carbonate-poor resid-ual clay remains. Because oceanic crust is mobile, it can migrate under water masses and areas of varying sedi-mentation. The result might be a stratigraphic sequence in the crust that begins with ridge-crest calcareous oozes, then ridge-flank red clays, and then (if it migrates un-der a belt of equatorial upwelling) a siliceous ooze (Fig. 5.22B).

Calcareous oozes form only in the deep sea because calcareous materials are swamped by terrigenous clastics on the continental margins. For example, the sediments of the Gulf of California are rich with microfossils, but they are so diluted by terrigenous clays that an ooze cannot form. In the deep sea, calcareous oozes are domi-nant where the waters are warm and relatively shallow.

They are not greatly restricted by currents, oceanic up-welling, fertility, or nutrients because the surface of the ocean is supersaturated with calcium carbonate (the most important mineral to calcareous plankton). In some cases, high fertility due to upwelling raises the CCD because fertility increases biogenic activity and therefore increases CO_2 concentration, which pushes the equilibrium toward dissolution. However, at the equator, high fertility lowers the CCD because calcareous shell building outstrips the supply of organic matter. In general, however, the primary limiting factors of calcareous oozes are depth and temperature.

Siliceous oozes are composed of the skeletons of planktonic organisms that build their skeletons of opaline silica, $SiO_2 \cdot nH_2O$. The most important of these are radiolarians, protozoans with radially symmetrical internal tests (see Fig. 5.20A), and diatoms, microscopic single-celled algae with porous, platelike or rodlike tests. Radiolarians dominate siliceous oozes in tropical waters, and diatoms are enormously abundant in polar siliceous oozes. Other minor components of siliceous oozes include sponge spicules (skeletal elements of siliceous sponges) and silicoflagellates (microfossils with simple tubular or ringlike tests).

Although radiolarians and diatoms are abundant in the same surface waters that teem with calcareous microplankton, their distribution on the seafloor is very different (see Fig. 5.19). Unlike calcite, *silica is undersaturated in all seawater.* Silica is transported from the land by rivers, but the amount is very small. Because it is a rare nutrient, recycling is important. Siliceous oozes can form only where oceanic upwelling brings silica and other nutrients up from the bottom. This happens primarily in areas of upwelling, particularly where oceanic water masses meet in zones of convergence or where cold bottom currents flow upward along a continental rise. Undersaturation is most severe in surface waters, where the abundant biotic activity depletes silica as rapidly as it becomes available. Silica has no compensation depth; its chance of dissolution is highest as the tests begin to sink and decreases with depth, the opposite of calcite (Fig. 5.23). For this reason, siliceous oozes can be found on any ocean bottom including the deepest, but are scarce except where upwelling allows siliceous microplankton to flourish in the surface waters.

Murray and Renard (1891) described the sediments that are found over much of the ocean floor as "the red clay" and thought it was a product of submarine eruption and weathering. However, when deep-sea clays were first studied by X-ray diffraction in the 1930s, it turned out that Murray was at least partly wrong. Most deep-sea clays are derived from the land and settle out of suspension in the deep sea. Their distribution on the ocean floor (Fig. 5.24) clearly indicates this. Kaolinite, the dominant clay in tropical soils, is found offshore

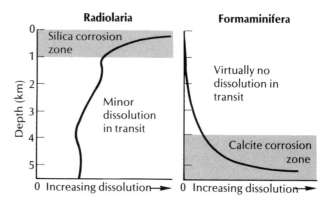

Figure 5.23

Comparison of dissolution profiles of radiolarians and planktonic foraminifers based on field experiments. Most radiolarians and diatoms dissolve in shallow waters, where silica is undersaturated due to high biotic productivity. By contrast, most dissolution of carbonate occurs at depths below about 3.5 kilometers as seawater concentration of carbonate decreases. From Berger, Bé, and Sliter (1975).

from the mouths of tropical rivers. Illite, the stable clay in most temperate soils, is most abundant in middle latitudes. Chlorite, found mainly in polar soils, is restricted to high latitudes. Montmorillonite, the only clay that is produced by weathering of submarine volcanics, is most common around mid-ocean ridges (especially the East Pacific Rise and the Mid-Atlantic Ridge) and near island arcs.

The ocean floor sediments contain not only clay minerals but also authigenic minerals (zeolites, manganese oxides and hydroxides), volcanigenic debris (mostly plagioclases and pyroxenes), eolian quartz and feldspar, manganese nodules, phosphatic fish teeth and earbones (otoliths), foraminifers which build skeletons from sand, and a few sponge spicules and radiolarians. In short, the "red clay" is everything that reaches the overlying water column and does not dissolve. As was shown in Fig. 5.19, this happens primarily in the centers of oceanic gyres, where there is only a single water mass and little chance for mixing or upwelling.

In summary, deep-sea sediments are entirely pelagic, mainly derived from terrigenous clays that settle out of suspension or from biogenic skeletal material. Their distribution follows a predictable facies pattern (Fig. 5.25). Calcareous oozes (which become limestones and chalks) are found wherever the water is sufficiently warm and shallow. Siliceous oozes (which become cherts, diatomites, and radiolarites) are found in abundance only in regions of upwelling. Where the ocean is not shallow enough, warm enough, or rich enough in nutrients from upwelling, terrigenous clays dominate. An ancient example of this pattern is discussed in Box 5.2.

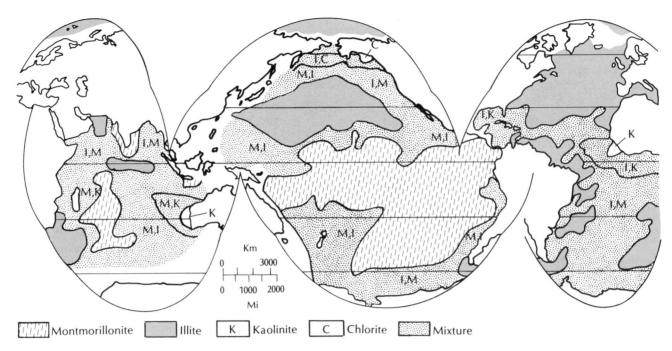

Figure 5.24

Dominant clay minerals on the ocean floor. Modified from Berger (1974).

Diagnostic Features of Pelagic Sedimentation

Tectonic setting Pelagic sediments form only in deep ocean basins. These are uplifted and become part of subaerially exposed sequences in only a few places, such as subduction zones, where deep marine pelagic sequences (chalks and cherts) are scraped up into the accretionary wedge over a trench, along with ophiolites and other ocean floor rocks. Some epicratonic basins have reached sufficient depth, or are so starved of clastic debris, that they have accumulated thick sequences of chalk, cherts, or black shales.

Geometry Deep-sea deposits form vast, thin, tabular sheets that are limited in area only by the size of the ocean basin.

Typical sequence Homogeneous, thinly bedded, finely laminated chalks, cherts, or shales with little variation in grain size are typical.

Sedimentology Calcite (chalks), opaline silica (chert), and clay minerals (illite, kaolinite, chlorite, and montmorillonite) are the dominant components. Minor authigenic minerals, volcanic and eolian dust, phosphatic fish fragments, and manganese nodules also occur. Few sedimentary structures are known other than fine lamination and extensive bioturbation and burrowing.

Fossils Biogenic oozes are composed almost entirely of planktonic microfossils of types that clearly indicate their deep-sea origin. Deep-sea clays have minor amounts of planktonic microfossils (mostly siliceous or phosphatic).

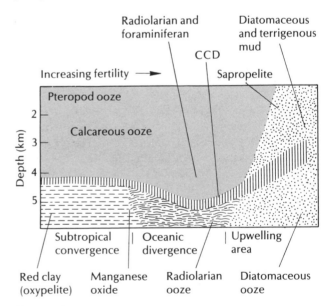

Figure 5.25

Distribution of major facies on a depth versus fertility graph. From Berger (1974).

Box 5.2

Cretaceous Shelf and Pelagic Deposits of the Western Interior of North America

Cretaceous exposures in the Western Interior of North America show most of the environments discussed in Chapters 4 and 5. During the Cretaceous, the central part of North America was covered by a vast epicontinental seaway that was 5000 kilometers long, from the Arctic to the Gulf of Mexico, and 1400 kilometers wide, from Utah to Iowa, during its maximum transgression. In some places, more than 2200 meters of marine sediments accumulated. The seaway was deepest over the modern-day Great

Plains and shoaled westward over a region that would become the Rocky Mountains (Fig. 5.26). Consequently, a distinct set of facies was deposited around the margin of this great Cretaceous ocean. The center of the seaway had shallow (not much more than 300 meters deep, according to Hancock, 1975) but open marine conditions, producing pelagic shales and chalks. A series of shales and sandstones were deposited on the shallower shelf that rimmed the seaway; in some places, the transi-

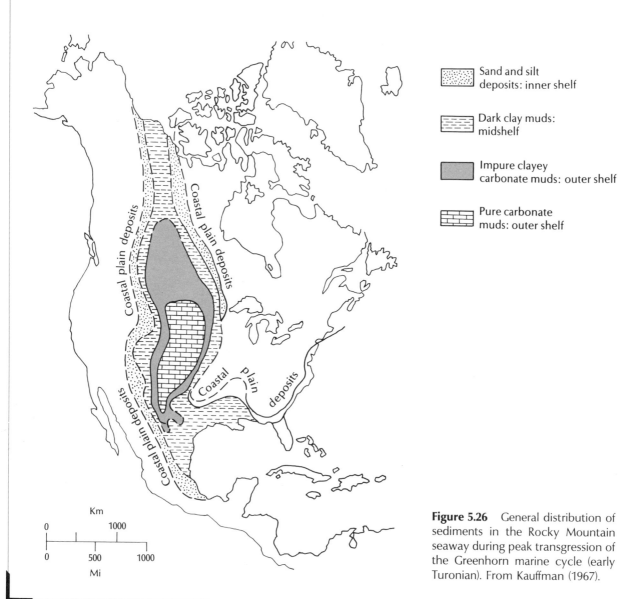

Figure 5.26 General distribution of sediments in the Rocky Mountain seaway during peak transgression of the Greenhorn marine cycle (early Turonian). From Kauffman (1967).

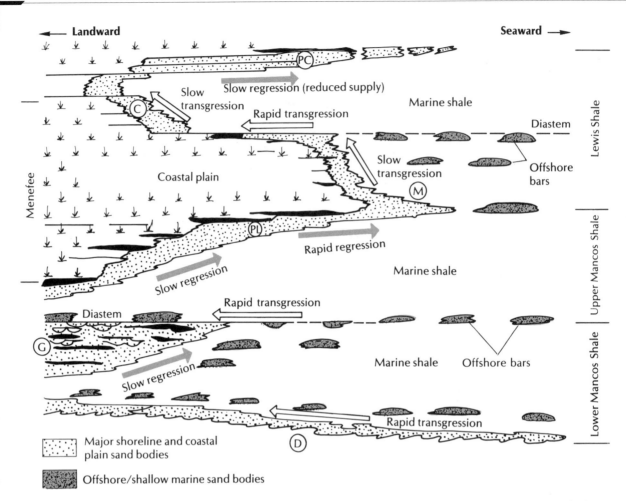

Figure 5.27 Typical stratigraphic pattern for Cretaceous rocks of the Western Interior, showing coastal plain, shoreline, and offshore deposits and indicating periods of transgression and regression. Major sandstone deposits are symbolized as follows: D, Dakota; G, Gallup; PL, Point Lookout; M, Mesaverde; C, Cliff; PC, Pictured Cliffs. From Reading (1986).

tion to nonmarine environments is preserved as well.

Sea level fluctuated continuously, so these facies belts did not accumulate in one place. Instead, they moved onshore and offshore with every change in sea level. In addition, extensive mountain building took place in the western United States throughout the Cretaceous, so orogenic episodes contributed much clastic sediment and caused local regressions. This produced a very distinctive facies pattern for the Western Interior Cretaceous (Fig. 5.27). In the center of the seaway to the east (the Plains region), deep-water shales and chalks prevailed. In the west (mostly Utah, Idaho, and Montana), orogenic activity produced nonmarine shales, sandstones, and conglomerates. In between, the marine and nonmarine shales were separated by a zigzagging series

of sandstones that represented various coastal and shallow shelf sands that were responding to the fluctuating shoreline. This shoreline area is concentrated in what is now New Mexico, Colorado, Wyoming, eastern Montana, Saskatchewan, and Alberta. When these areas were uplifted and deformed in the latest Cretaceous and Cenozoic, many of these coastal and shallow marine deposits were exposed and are now found in the cliffs of the Rocky Mountains and Colorado Plateau.

The shallow marine shales of the Cretaceous seaway are punctuated by a number of long, narrow, shallow marine sand bodies (see Fig. 5.27). Because they are porous reservoir rocks that are enveloped by impermeable shales, they have been intensively studied and exploited for oil. Walker (1984) has reviewed some of the many studies of these enigmatic sand bodies. Dimensions vary greatly, but typically they are about 15 meters thick, about 10 kilometers wide, and tens of kilometers in length (Fig. 5.28). A particularly well-exposed example is the Semilla Sandstone Member of the Mancos Shale, in the San Juan Basin of New Mexico (Fig. 5.29A). As described by La Fon (1981), this example shows a sequence (Fig. 5.29C) typical of most of these linear sand bars (Fig. 5.29B). The sequence coarsens upward from marine shales to mudstones and siltstones with thin cross-laminated or graded sandstones. This silty sandstone facies contains small clay chips (1 millimeter thick and 2 centimeters across) and is extensively bioturbated and homogenized (Fig. 5.29D). As a consequence, the bedding is very poor, though there are shaly zones within this facies that are laminated. The silty sandstone facies reaches a maximum thickness of about 20 meters but thins laterally until it pinches out beyond the margins of the bar (Fig. 5.29B). This facies contains a fauna of oysters, clams, gastropods, ammonites, crabs, and callianassid shrimp burrows that suggest well-agitated, shallow, normal marine waters.

The silty sandstone facies grades upward into a cross-bedded sandstone facies that caps the deposit (Figs. 5.28, 5.29). It is composed of thin-bedded (less

than 15 centimeters), coarsening-upward, fine- to medium-grained sandstone. It is only slightly burrowed and contains abundant small-scale trough cross-beds (Fig. 5.29E). In some places, the beds show hummocky cross-stratification. The tops of beds are ripple-marked and are separated from overlying sandstone beds by clay layers about 1 centimeter thick. The base of each bed is usually paved with rounded, flat, oval clay chips up to 9 centimeters across. The base may also contain broken oyster shells, snail-shell fragments, shark teeth, and bleached phosphate pebbles. The top of the sand bar is capped with shale, phosphatic bone debris, and shark teeth, which suggests slow deposition; there is no evidence of emergence or of beach deposition.

The origin of Cretaceous offshore sand bars such as the Semilla Sandstone is still under discussion. They do not fit any of the modern analogues discussed earlier in this chapter. Most of the authors who have described these Cretaceous sand bars emphasized storms as the major transporting agent. Storms are one of the few forces that can transport sand this far offshore into the muddy shelf. In addition, the presence of hummocky cross-stratification and of storm lags of mudchips and broken oyster shells indicates high transport energy. Tidal currents

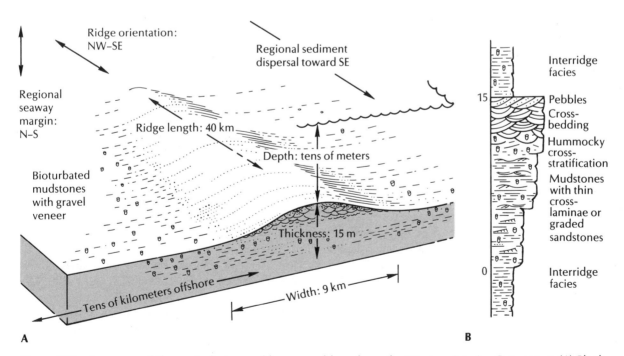

Figure 5.28 Summary of the main features of linear sand bars from the Western Interior Cretaceous. (A) Block diagram. (B) Typical stratigraphic sequence. From Walker (1984).

are thought to be of lesser importance. Modern tidal sand waves and ridges rest unconformably on transgressive surfaces with gravel lags, and these do not occur in most of the Cretaceous examples. Both tidal currents and storms are capable of producing the cross-bedding that is found at the top of the sequence. A number of models have been proposed concerning the flow patterns of storm waves necessary to produce these offshore sand ridges, but there is no clear consensus yet.

Recently, Bergman and Walker (1987) have shown that many of these offshore sand ridges have erosional surfaces capped by gravels. This suggests that the sand ridges were subaerially exposed during periods of low sea level. If this is so, then the presence of sand and gravel ridges on the Cretaceous shelf may be related to episodes of canyon cutting during periods of low sea level. When sea level rose again, the gravel source was cut off by the transgression. The sands and gravels were reworked and spread out, concentrating in the erosional channels and hollows. Long sand ridges surrounded by marine shales were the eventual result. As we saw with turbidites, sealevel changes were apparently the premier controlling factor in the Cretaceous as well.

Further offshore from the coastal sands and offshore sand bars, marine conditions existed and pelagic sedimentation prevailed. The major sedimentary product of these conditions is thick sequences of finely laminated shale and limestones, which underlie most of the Great Plains: the Pierre Shale, the Mancos Shale, the Greenhorn and Niobrara Limestones, and many other named units. Although the shales may appear monotonous at first, there is actually a distinct sequence of shale and chalk facies that formed in progressively deeper water. The facies that originated nearer to shore are predominantly pelagic shales, but the facies from the center of the seaway, which was deeper and relatively starved of mud, are clean limestones and chalks (Fig. 5.30A). However, unlike some of the modern examples of pelagic sedimentation discussed earlier, the Western Interior Cretaceous seaway was never as deep as the modern ocean. Because water depths were seldom more than 300 meters, there was no significant carbonate dissolution. Nevertheless, most of the features of the Cretaceous seaway deposits clearly indicate open, pelagic conditions with relatively little influence from shoreline clastics upon the center of the seaway.

Kauffman (1969) recognizes at least 12 distinctive facies in succession, each resulting from part of the

A

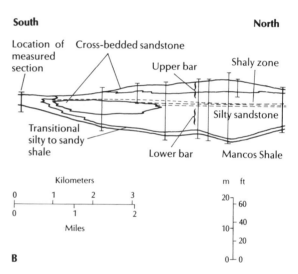

B

overall transgressive-regressive cycle (Fig. 5.30B). No more than 10 of these can be found in a single local section, however. Facies 1 and 2 are coastal-fluvial and shallow marine sandstones, respectively. Facies 3 is slabby, platy, shaly sandstone and siltstone, and Facies 4 is a slightly more calcareous version of the same (Fig. 5.30C). These beds have thin sandstone laminae and starved ripples, and are low in organic carbon. They are moderately bioturbated and contain scattered small oysters, fish teeth, fragments of inoceramid clams, and abundant benthic foraminifers. According to Kauffman (1969, 1985) and Hancock (1975), this facies represents a proximal offshore, or very shallow shelf environment that was current-swept, warm, well aerated, and generally normal marine.

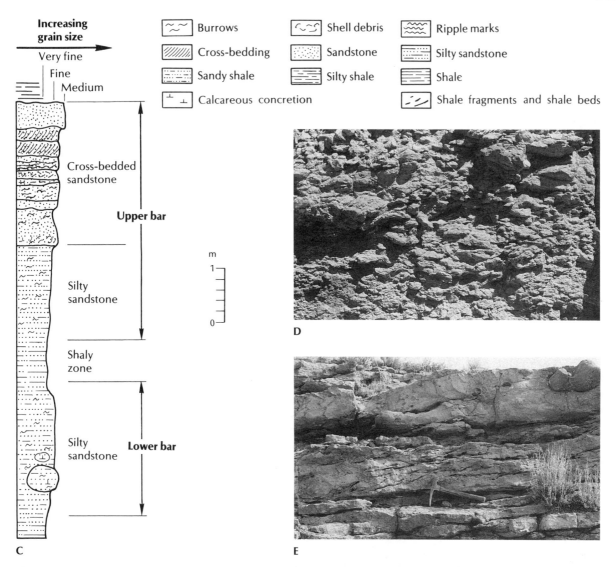

Figure 5.29 An ancient example of a Cretaceous marine sand body, the Semilla Sandstone of the San Juan Basin, New Mexico. (A) Outcrop, showing silty sandstone facies (lower two-thirds of cliff) capped by cross-bedded sandstone facies (top of cliff). (B) Cross section through Holy Ghost bar of the Semilla Sandstone. Note that the bar gradually thins and pinches out. (C) Stratigraphic section through the bar. (D) Lower silty sandstone facies, showing extensive bioturbation. (E) Cross-bedded sandstone facies, showing erosional contacts and pockets created from erosion of clay pebble clasts. Drawings based on La Fon (1981); photos courtesy N. La Fon.

Facies 5, on the other hand, is a dark organic-rich clay shale that is very finely laminated and contains only a few silt laminae, probably reworked from distal storm deposits (Fig. 5.30C). It is very poorly bioturbated, and well preserved and undisturbed volcanic ash layers are present. There are calcareous and sideritic nodules and many other geochemical features that suggest calcite saturation but reducing (low-oxygen) conditions. Only rare oysters and a few benthic foraminifers that are able to survive with little oxygen are found in this facies. According to Kauffman (1969, 1985), this facies probably represents middle offshore, quiet water and highly reducing and oxygen-depleted conditions. The stagnant bottom waters were hostile to benthic organisms and burrowing infauna, thus explaining the lack of fossils and bioturbation.

Facies 6 is a dark clay shale that is noncalcareous and completely lacking in silt or sand (Fig. 5.30D). There is very little organic carbon and only moder-

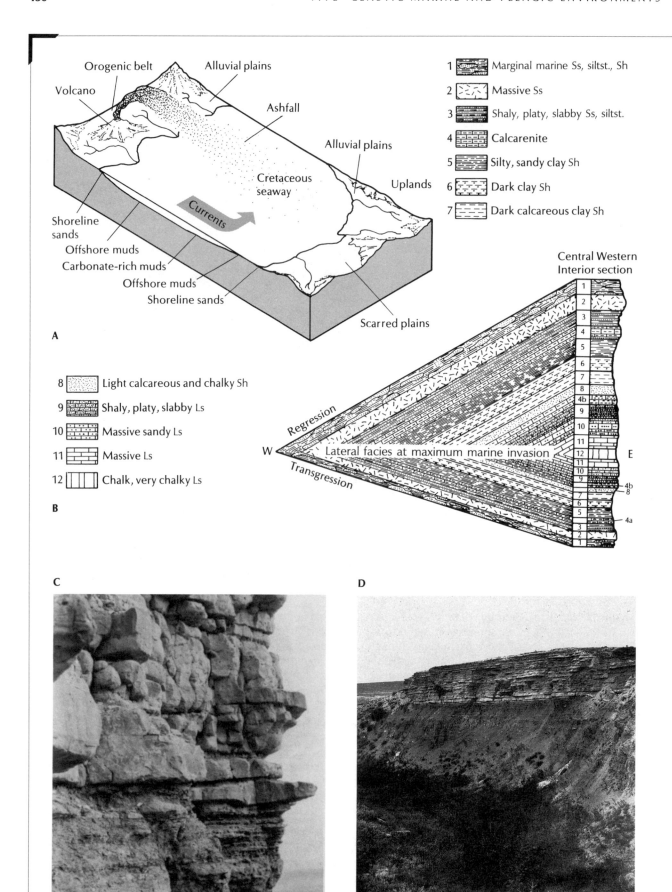

A

1 ▭ Marginal marine Ss, siltst., Sh
2 ▭ Massive Ss
3 ▭ Shaly, platy, slabby Ss, siltst.
4 ▭ Calcarenite
5 ▭ Silty, sandy clay Sh
6 ▭ Dark clay Sh
7 ▭ Dark calcareous clay Sh

8 ▭ Light calcareous and chalky Sh
9 ▭ Shaly, platy, slabby Ls
10 ▭ Massive sandy Ls
11 ▭ Massive Ls
12 ▭ Chalk, very chalky Ls

B

Orogenic belt · Alluvial plains · Volcano · Ashfall · Alluvial plains · Uplands · Cretaceous seaway · Currents · Shoreline sands · Offshore muds · Carbonate-rich muds · Offshore muds · Shoreline sands · Scarred plains

Central Western Interior section

Regression · Transgression · W · Lateral facies at maximum marine invasion · E

C

D

E

F

G

Figure 5.30 (A) Block diagram showing the Western Interior seaway during the peak of the Greenhorn cycle. (B) Kauffman's (1967) model of Cretaceous marine sedimentation in Western Interior, patterned after the Greenhorn cycle with complex interfingering removed. The western edge of the diagram would be in east-central Utah, and the eastern margin represents western Kansas and Oklahoma. The numbers in the key correspond to Kauffman's 12 facies. (C) Contact between silty shales (Facies 5) of the Blue Hill Shale and the overlying shoreface sandstones (Facies 4) of the Codell Sandstone, exposed at Liberty Point, west of Pueblo, Colorado. (D) Contact between finely laminated black shales (Facies 6) of the Graneros Shale and the overlying marly limestone (Facies 7) of the Greenhorn Limestone in southwestern South Dakota. (E) Shaly, platy, slabby limestone with fragments of *Inoceramus* shells (Facies 9) of the Lincoln Member, Greenhorn Limestone, Pueblo County, Colorado. (F) Exposure of massive, rhythmically bedded limestone (Facies 11) of the Bridge Creek Member, Greenhorn Limestone, Russell County, Kansas. (G) Finely bedded chalk (Facies 12) from open marine pelagic deposits, Niobrara Limestone, Gove County, Kansas. Photos C and F from Pratt et al. (1985); photo D by N. H. Darton and photo E by G. R. Scott, courtesy U.S. Geological Survey; photo G courtesy Kansas Geological Survey.

ate bioturbation. The sparse fauna include a few low-oxygen tolerant oysters, and most of the foraminifers are low-oxygen tolerant agglutinated forms. This facies probably represents middle offshore (125 to 200 meters deep), quiet water conditions that were occasionally slightly reducing but usually oxygenated. The lack of silt suggests that the facies was deep enough and far enough offshore that it was beyond the reach of distal deposits.

Facies 7 and 8 are calcareous clay shales that grade into Facies 9, a shaly, platy, slabby limestone (Figs. 5.30D, E). All of these units are thin bedded; facies 8 and 9 contain abundant fecal pellets and a diverse warm-water fauna of snails, clams, ammonites, both planktonic and benthic foraminifers, and even corals. These suggest quiet, outer-shelf conditions, with warm, well-oxygenated, and well-circulated normal marine water.

Facies 10 is sandy, calcarenitic massive limestone. It contains some fine-grained quartz sand and is commonly mottled and bioturbated. Some of these beds are made almost entirely of inoceramid bivalve shell fragments. The quartz sand and the shell fragments suggest that this deposit formed in the open sea on a high wave energy bottom. Facies 11 is a massive, pure limestone that is very fine grained and well cemented. It commonly forms ledges. Its most striking feature is rhythmic cycles of marl passing up into limestone, in units 0.6 to 1 meter thick (Fig. 5.30F). The alternation of marl and limestone, along with details of the petrography, suggest that these deposits reflect rapid changes in water depth (20 to 100 meters or more) in response to changes in sea level. There appear to be many minor breaks or near breaks in sedimentation.

Facies 12 is the open-sea, pure-chalk facies found only in the deepest (150 to 300 meters) water (Fig. 5.30G). It is finely bedded, with thin chalk beds that can be traced for hundreds of miles (Frey, 1972; Hattin, 1982). This incredible lateral persistence would be found only in open pelagic conditions. Most of this chalk is composed of calcareous microfossils, primarily coccoliths and foraminifers. It is comparable to the thick Cretaceous chalk deposits found all over Europe, best known from the White Cliffs of Dover. In the Western Interior of North America, however, it is seldom white, but olive-gray to olive-black because it is rich in fecal pellets and organic matter. Common fossils include oysters, inoceramids, and fish bones and scales. Ammonites, belemnites, and rudistid clams are rare. The bottom was apparently very soupy because many types of oysters, bryozoans, barnacles, serpulid worms, and sponges attached themselves to any abandoned shell that could serve as a substrate. This facies is typical of the classic exposures of the Niobrara Chalk in Kansas (Fig. 5.30G) and elsewhere in the Plains states.

The Western Interior Cretaceous seaway provides excellent examples of the offshore sand bar (barrier) environment and the open pelagic environment. Although these examples are not typical of every comparable deposit in the geologic record, most of the salient features are present. We shall examine the Western Interior Cretaceous seaway again in another context in Chapter 13.

FOR FURTHER READING

In addition to the references listed at the end of Chapter 4, the following sources are worth consulting:

Bouma, A. H., 1962. *Sedimentology of Some Flysch Deposits—A Graphic Approach to Facies Interpretation*. Elsevier, New York. The classic work on turbidites, which first proposed the concept of a Bouma sequence.

Bouma, A. H., W. R. Normark, and N. E. Barnes (eds.), 1985. *Submarine Fans and Related Turbidite Systems*. Springer Verlag, New York. Excellent recent book giving many examples of modern and ancient submarine fans, grouped by active and passive margins

Burk, C. A., and C. L. Drake (eds.), 1974. *The Geology of Continental Margins*. Springer-Verlag, New York. A massive compendium of papers covering nearly every continental margin in detail, including a classic paper by Berger on deep-sea sediments.

Doyle, L. J., and O. H. Pilkey (eds.), 1979. *Geology of Continental Slopes*. Soc. Econ. Paleo. Mineral. Spec. Publ. 27. An excellent compilation of papers on modern and ancient slope deposits.

Dzulynski, S., and E. K. Walton, 1965. *Sedimentary Features of Flysch and Greywackes*. Elsevier, New York.

A classic on turbidites, with stunning photos of sole marks from the Carpathian flysch.

Heezen, B. C., and C. D. Hollister, 1971. *The Face of the Deep*. Oxford Univ. Press, New York. An eye-catching collection of underwater photographs of the seafloor, and of the processes that affect it.

Kennett, J. P., 1982. *Marine Geology*. Prentice-Hall, Englewood Cliffs, N.J. The best single text on marine geology, with excellent chapters on marine sedimentation.

Lisitzin, A. P., 1972. *Sedimentation in the World Ocean*. Soc. Econ. Paleo. Mineral. Spec. Publ. 17. A summary and synthesis of the state of the field in 1972, emphasizing the international (especially Soviet) literature. Slightly dated by the work of the Deep Sea Drilling Project.

Middleton, G. V., and A. H. Bouma (eds.), 1973. *Turbidites and Deep Water Sedimentation*. Soc. Econ. Paleo. Mineral., Pacific Section Short Course, Anaheim, Calif. An excellent short review of the main processes of marine sedimentation.

Mutti, E., and F. Ricci Lucchi, 1972. *Turbidites of the Northern Apennines: Introduction to Facies Analysis*. [Translated 1978 by T. H. Nilsen from *International Geology Review*, 20(2): 125–166; reprinted by the American Geological Institute, Falls Church, Va.]

The classic study of ancient turbidites, which set the standard for all later studies.

Stanley, D. J., and D. J. P. Swift (eds.), 1976. *Marine Sediment Transport and Environmental Management*. John Wiley, New York. Despite the title, this is an excellent collection of papers on shelf sedimentation.

Swift, D. J. P., D. B. Duane, and O. H. Pilkey (eds.), 1972. *Shelf Sediment Transport: Process and Product*. Dowden, Hutchinson, and Ross, Stroudsburg, Penn. A classic work on shelf sediments.

Tillman, R. W., and C. W. Siemers (eds.), 1984. *Siliciclastic Shelf Sediments*. Soc. Econ. Paleo. Mineral. Spec. Publ. 34. Details of many recently studied ancient examples of shelf sediments.

Tillman, R. W., D. J. P. Swift, and R. G. Walker, 1985. *Shelf Sands and Sandstone Reservoirs*. Soc. Econ. Paleo. Mineral., Short Course 13. The most up-to-date reference on shelf sedimentation.

Warme, J. E., R. G. Douglas, and E. L. Winterer (eds.), 1981. *The Deep Sea Drilling Project: A Decade of Progress*. Soc. Econ. Paleo. Mineral Spec. Publ. 32. An excellent collection of papers summarizing the revolution in the earth sciences brought about by the DSDP.

Aerial view of ooid shoals beneath the clear waters of the Great Bahama Bank near Cat Cay. The strong currents that rolled and produced the ooids have also produced large channels and ripples. Photo courtesy Shell Development Company.

6

Carbonate Environments

THE "CARBONATE FACTORY"

Although limestones can form in nonmarine lacustrine and pelagic environments (Chapters 3 and 5, respectively), most are formed in shallow marine environments. Limestones form a major portion of the stratigraphic record on all continents, particularly in ancient epicontinental seaways and shallow shelves. Because of their great abundance, limestones have been studied for many practical reasons. Limestone bedrock produces karst topography and extensive cave formations, and its solubility makes it important for hydrologic purposes. Economically, limestones are quarried for cement and building stone and mined for the ore deposits formed by hydrothermal fluids that moved through fissures in them. Moreover, the origin, diagenesis, and porosity of limestones are critical to the petroleum industry because limestones are important reservoir rocks.

Carbonate sediments are fundamentally different from clastic sediments in many ways (Table 6.1). As James (1984, p. 209) put it, "carbonate sediments are born, not made." While terrigenous clastic sediments have been eroded from bedrock and *transported to* the basin of deposition (*allochthonous*), carbonate sediments form chemically or biochemically *within* the basin of deposi-

tion (*autochthonous*). Many carbonate particles undergo little or no hydraulic transport, so the physical sedimentary processes that make cobbles, pebbles, sand, and mud are less important than the composition of the carbonate particle itself, whether precipitated inorganically or formed biogenically. Grain size is chiefly a function of the type of skeletal particle, not hydraulics.

Another striking difference between carbonate and clastic sediments is the extremely limited conditions under which carbonates form the dominant sediment. Clastic sediments are dominant in virtually every sedimentary environment, both marine and nonmarine, whereas carbonates are the dominant sediments only in clear, warm, shallow, tropical to subtropical seas—the "carbonate factory" (Fig. 6.1). Most carbonate that does not remain on the shallow shelf is transported only short distances down the slope (basinward) or up the slope (shoreward) into peritidal regions. Freshwater limestones and pelagic oozes are almost the only exceptions to this rule, and they are much less abundant than shelf carbonates.

For carbonates to form, the shelf environment must meet a number of restrictive conditions, met only on shallow shelves between 30° north and south latitudes, because carbonate is produced by environmentally sen-

Table 6.1

Differences between Siliciclastic and Carbonate Sediments

Carbonate Sediments	Siliciclastic Sediments
Most sediments occur in shallow, tropical environments.	Climate is no constraint, so sediments occur worldwide and at all depths.
Most sediments are marine.	Sediments are both terrestrial and marine.
The grain size of sediments generally reflects the size of organism skeletons and calcified hard parts.	The grain size of sediments reflects the hydraulic energy in the environment
The presence of lime mud often indicates the prolific growth of organisms whose calcified portions are mud-sized crystallites.	The presence of mud indicates settling out from suspension.
Shallow-water lime sand bodies result primarily from localized physico-chemical or biological fixation of carbonate.	Shallow-water sand bodies result from the inter-action of currents and waves.
Localized buildups of sediments without accompanying change in hydraulic regimen alter the character of surrounding sedimentary environments.	Changes in the sedimentary environments are generally brought about by widespread changes in the hydraulic regimen.
Sediments are commonly cemented on the seafloor.	Sediments remain unconsolidated in the environment of deposition and on the seafloor.
Periodic exposure of sediments during deposition results in intensive diagenesis, especially cementation and recrystallization.	Periodic exposure of sediments during deposition leaves deposits relatively unaffected.
The signature of various sedimentary facies is obliterated during low-grade metamorphism.	The signature of sedimentary facies survives low-grade metamorphism.

From James (1984).

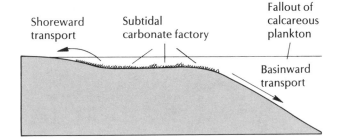

Figure 6.1

Sketch of carbonate shelf, showing main regions of accumulation of carbonate sediments. From James (1984).

a few red algae, ahermatypic corals, molluscs, and echinoderms can live. The water chemistry must be normal marine because these organisms are tolerant of a very narrow range of salinity.

The water must also be clear, free of terrigenous clastics and only slightly turbulent, because terrigenous quartz sand can abrade the softer calcite and terrigenous mud can foul gills or digestive tracts, thus hampering the biological activity of most carbonate-secreting organisms. If the water is too turbulent, lime mud is stirred up, cutting down the light and choking the marine invertebrates.

With such restrictions, it is no surprise that there are so few active carbonate environments today. Why, then, were they so abundant in the geologic past? In this case, it appears that the present is an imperfect key to the past. Throughout most of geological history, epeiric seas were widespread and provided enormous areas of the shallow, tropical marine conditions that are ideal for carbonate production. The modern analogues that have been the most studied are the Persian Gulf and the Bahama platform, but these are not nearly so large as the ancient epeiric carbonate seas. The difference in size is because Pleistocene and Holocene sea levels have been much lower than was typical of most of earth history (particularly the Paleozoic and Mesozoic) and because earth temperatures during the present interglacial are much lower than was typical of most of the geologic past. Our understanding of ancient carbonate environments is hampered by the scarcity of good modern analogues.

PERITIDAL ENVIRONMENTS

Because carbonates are so depth-restricted, the position of sea level is a natural boundary between environments. Peritidal carbonate environments undergo very different processes from shelf carbonates, which seldom suffer subaerial exposure. The occasional drying means that most organisms cannot survive in the intertidal envi-

sitive organisms. The shelf waters must be warm because these organisms are tolerant of only a narrow range of temperatures, and they must be shallow because many of them (primarily reef-forming corals and lime-mud-secreting algae) need light for photosynthesis. Most carbonate production takes place in the upper 10 meters of seawater and drops off rapidly below this; below 80 to 100 meters, the light is so dim that only

ronment. Salinity and water temperature also undergo extreme fluctuations, further limiting the types of organisms that can live there. These chemical changes have proven to be important for economic reasons as well. Peritidal carbonate flats are one of the major environments of dolomitization and apparently have been responsible for many of the thousands of meters of dolomite thickness in the geologic record. Peritidal carbonate flats form laterally persistent, evenly bedded limestones and dolostones, which are host rocks for lead and zinc ores and for petroleum. For this reason, there has been intensive study of modern carbonate tidal flats,

beginning with Ginsburg's work in the Bahamas in the 1950s (summarized in Ginsburg, 1975) and then continuing with studies of tidal flats in Bermuda, Bonaire, Florida, the Persian Gulf, and Shark Bay, Australia.

The shallow intertidal environment is close to optimum for the growth of many carbonate-forming organisms, so tidal flats are capable of building up at rates much faster than the normal rates of sea-level rise or shelf subsidence. The classical tidal flat model, then, is a *shallowing-upward* sequence from surf zone to subtidal to intertidal to supratidal deposits to eolian dunes (Fig. 6.2). This sequence can be repeated many times in the

Figure 6.2

(A) The Three Creeks area along the west coast of Andros Island, Bahamas; tidal channels cut an intertidal marsh composed largely of algae, with some ponds in the intertidal areas. (B) Diagram showing major features of the peritidal environment. From Stanley (1989).

A

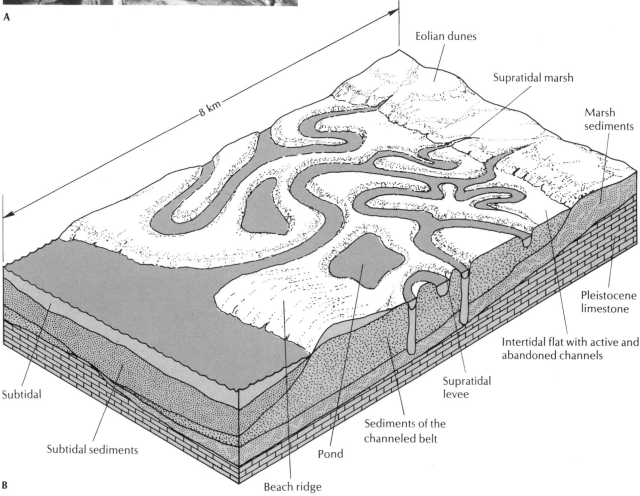

B

rocks because minor fluctuations in relative sea level make big differences in the position of the intertidal zone. Ideally, such repetitive shallowing-upward sequences form on a transgressive carbonate bank, where repeated changes in relative sea level cause the repetition of the sequences. In the past, many parts of the continents were covered with extensive epeiric carbonate seas having hundreds of square kilometers of peritidal zone, so thick, tidal-flat, shallowing-upward sequences are very common in the cratonic record.

Carbonate tidal flats occur on shores that are shielded from daily wave action but are immersed by daily tidal fluctuations and by occasional storms. As on clastic coastlines, the rare storm can have an enormous effect on intertidal environments, producing considerable local erosion. Many tidal flat deposits are formed above mean high tide by storm action. All of the subenvironments in the vertical succession of tidal flats form simultaneously in adjacent belts, so they are another classic example of units that build laterally and are time-transgressive.

The two modern examples of carbonate-forming areas—the arid, evaporitic tidal flats of the Persian Gulf (Fig. 6.3) and the humid, normal marine tidal flats of the Bahamas (see Fig. 6.2)—encompass most of the features seen in ancient carbonate rocks. If the tidal flat is on an extensive shallow shelf, the waves may be slowed at a distant shelf break. In many cases, however, the outermost environment is the surf zone, which absorbs the wave energy. On the *lower foreshore*, the deposits are below the wave swash, so they are poorly sorted, coarse-grained shell deposits with a micritic matrix. Longshore currents produce large-scale festoon cross-bedding. The *upper foreshore*, on the other hand, is within the swash and backwash zone of the waves, which results in laminated, thick-bedded, well-sorted lime sands and gravels, with planar cross-beds dipping about 15° seaward. This environment is subject to frequent sub-

aerial exposure and rapid changes in water chemistry. As a result, the calcite frequently dissolves slightly and then reprecipitates to cement the pore spaces, forming *beachrock*.

Inside the protective barrier is an extensive area of marine subtidal *lagoon*, which is never subaerially exposed. Naturally, it is very sheltered, so the main deposit is extensive lime sand and mud banks, with abundant pellets and skeletal fragments. These lagoonal sands and muds are areas of great biotic productivity, so they are frequently homogenized by burrowing organisms. Subtidal lagoon lithology is so similar to that of normal marine carbonate shelves that they are difficult to distinguish except for the close association of the lagoon with distinctive intertidal deposits.

The most diagnostic environment is the *intertidal zone* itself (see Figs. 6.2, 6.3). Here, the most important factors are the daily alternation of tidal immersion and subaerial exposure. Most of the zone is covered by rippled, fine carbonate mudflats, which are often covered by algal mats and stromatolites (Fig. 6.4) and by mud cracks (Fig. 6.5) from subaerial exposure. The flats are cut by channels for the outgoing tide waters. These channels are eventually abandoned and filled by linear or sinuous stringers of carbonate sand. Sedimentary structures, such as tidal bedding and reactivation surfaces, clearly indicate the reversing currents in the channels. Another characteristic feature of the tidal channels are breccias made of shell debris and rip-up clasts, particularly fragments of laminated mud cracks (Fig. 6.6).

Because the intertidal environment is protected from normal waves, the rare storms tend to have a disproportionate effect. Storms flood the tidal flat with waters carrying much suspended sediment and leave a storm lag of debris and mud, which is eroded and reworked during normal conditions. Shinn, Lloyd, and Ginsburg (1969) have suggested that the tidal flat is like a river delta turned inside out: that is, the sea is the "river"

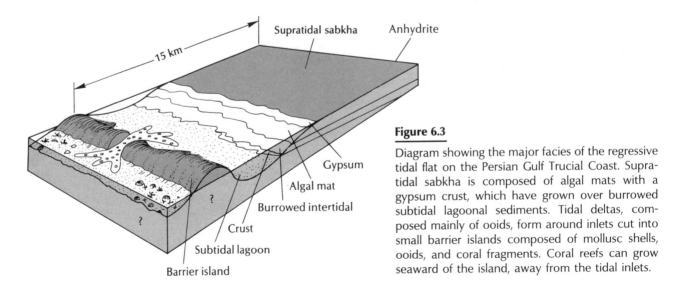

Figure 6.3

Diagram showing the major facies of the regressive tidal flat on the Persian Gulf Trucial Coast. Supratidal sabkha is composed of algal mats with a gypsum crust, which have grown over burrowed subtidal lagoonal sediments. Tidal deltas, composed mainly of ooids, form around inlets cut into small barrier islands composed of mollusc shells, ooids, and coral fragments. Coral reefs can grow seaward of the island, away from the tidal inlets.

A   B

Figure 6.4

(A) Hamelin Pool, western Australia, showing rippled intertidal carbonate sands between algal mats. (B) Modern subtidal to intertidal columnar stromatolites at Hamelin Pool. Photos courtesy R. N. Ginsburg.

Figure 6.5

Large mud cracks in thick algal mats on the edge of a brine-filled pond, Inagua Island, Bahamas. Gypsum is precipitating in the cracks and on the upturned edges. Photo courtesy E. A. Shinn.

A   B

Figure 6.6

(A) Flat, lime mud-chip pebbles accumulating on the beach ridge in Fig. 6.2B. (B) Ancient flat-pebble limestone conglomerate from the Paleozoic. From Stanley (1989).

supplying sediment *onshore* to the channeled flats, which are the "delta."

The intertidal environment is distinctive and can be recognized by a number of features. Algal mats and dome- and pillar-shaped stromatolites (see Fig. 6.4) are characteristic, and they produce a rock with even lamination that shows *fenestral*, or sheet-shaped, porosity due to gas-filled voids and the shrinking of the desiccated algal mat. Other desiccation features—such as mud cracks, mud chips, channel breccias filled with rip-up clasts and shells, and evaporite minerals—are also characteristic.

The *supratidal* environment is above the high tide level, so it is immersed only during unusually high spring tides or during storm surges. As a result, desiccation features, including thick layers of mud cracks and algal lamination, are predominant. The algal stromatolites are much more sheetlike or dome- like than pillarlike, forming laterally linked hemispheroids. In arid climates (such as the tidal salt flats, or *sabkhas*, of the Persian Gulf; see Fig. 6.3), these intertidal zones are areas of maximum deposition of evaporites, which often form laminae of gypsum or anhydrite and halite. In many cases, the mud cracks fill with evaporites. Sometimes, the evaporite laminae are distorted by soft-sediment deformation or form diapir structures. A particularly characteristic texture, called *bird's-eyes*, results when small, ellipsoidal pores form in the algal mat by shrinkage, gas bubbles, or air escape during flooding (Fig. 6.7). Above the supratidal salt flat may be eolian dunes, similar to those found in clastic barrier-island complexes except that they are made of carbonate sand.

One of the most striking features of shallowing-upward carbonate sequences is that they often build up into thick sequences of repeated shoaling cycles that can be hundreds of meters thick (Fig. 6.8). Because the rate of carbonate deposition is faster than the rate of platform subsidence or sea-level rise, carbonate sediment rapidly accumulates until it reaches sea level. This explains a

single shallowing-upward sequence but not the repeated cycles spreading out over vast platforms.

The mechanism for this cyclicity has been widely discussed, and two end-member models have been proposed (Wilkinson, 1982; James, 1984; Hardie, 1986). The *eustatic model* proposes that rapid fluctuations in sea level are responsible for the rapid change in depth of the facies. During periods of stability or slowly rising sea level, the sequence progrades outward in laterally accreted tidal flat sequences (Fig. 6.9, steps 1 and 2). If there is a rapid sea level rise, the platform is flooded, temporarily shutting off tidal flat deposition (step 3). Eventually, new tidal flat sequences prograde out from the shallow end, building another cycle on top of the last drowned sequence (step 4).

The *autocyclic model* proposed by R. N. Ginsburg (1971; Hardie, 1986) suggests that the control of the cycles is not sea-level fluctuation but the rate of carbonate sedimentation as controlled by source area. As in the eustatic model, peritidal sequences prograde outward during periods of stable or slowly rising sea level (steps 1 and 2). But the source area for the carbonate sediments is the subtidal area, which is reduced by the progradation and eventually becomes too small to furnish carbonate for further growth of the prograding wedge. At this point, the lateral growth of the shallowing-upward sequences stops (step 3). Once sea level has risen far enough, however, a new area of subtidal shelf is provided for renewed carbonate production, and the prograding peritidal sequences can build out again across the previous cycle (step 4).

Whichever model prevails for the small-scale cycles, there seems to be no doubt that the larger-scale cycles are controlled by sea-level changes and possibly also by climatic changes (Fischer, 1964; Aitken, 1978; Hardie, 1986). A component of these cycles may also be caused by tectonic changes in the platform as well. For example, Fischer (1964) was the first to propose that the various frequencies of cyclic carbonate deposition in the Triassic

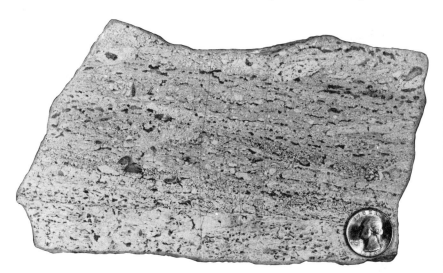

Figure 6.7

Bird's eye limestones form in intertidal and supratidal areas where algal layers produce flat mats. Gas bubbles and worm burrows have disrupted the bedding, producing holes that give a "bird's eye" appearance to the rock. An example from the Ordovician of Oklahoma; some of the algal mats have broken into flat-pebble conglomerates. From Stanley (1989).

Figure 6.8

Bedded Cambrian carbonates from Fortress Lake, British Columbia, showing dozens of repetitive small-scale shallowing-upward sequences between subtidal-intertidal limestones (dark) and supratidal dolomites (light). Photo courtesy J. D. Aitken, Geological Survey of Canada.

Lofer cyclothems of the Alps might have been caused by climatic changes due to variations in the earth's orbit. New information on the frequency of sea-level changes in geologic history (discussed later) indicate that large-scale eustatic changes could have significant effects.

The main problem with these models is dating. To prove correlation between facies and orbital variations or particular parts of the sea-level curve, precise dating is needed. So far, this has not been possible in many critical cases because peritidal limestones contain few fossils that are also found in the open ocean and can be correlated worldwide. Nevertheless, this is an area of active interest and research (see Hardie, 1986).

Diagnostic Features of Peritidal Environments

Tectonic setting Like other carbonate shelves, peritidal environments are restricted to clear, shallow, subtropical to tropical marine shelves, which occur in a

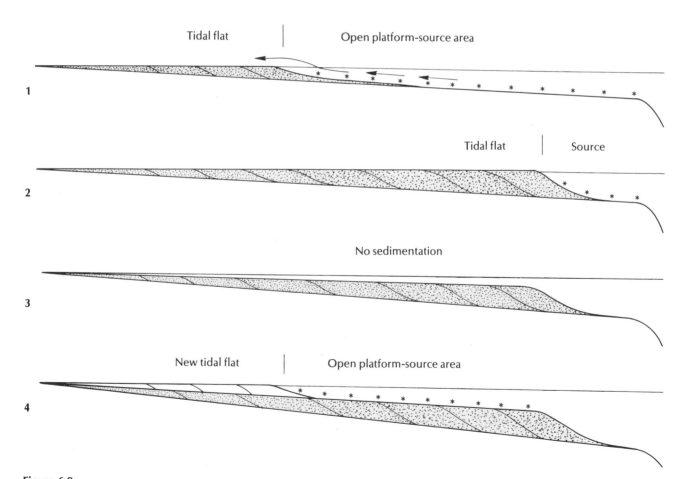

Figure 6.9

A sketch illustrating how two shallowing-upward sequences can be produced by progradation of a tidal flat wedge. These general conditions apply to both eustatic and autocyclic models. The asterisks show areas of carbonate production. From James (1984).

stable tectonic setting with little relief. These conditions occur primarily in subsiding passive margins or in epeiric seaways.

Geometry Peritidal environments typically form thin (a few meters thick) but laterally continuous beds representing the various shoreline facies belts, which may build thousands of square kilometers of shelf and form extensive tabular bodies.

Typical sequence A shallowing-upward sequence from subtidal lagoon muds or beach sands to intertidal rippled, burrowed, and mud-cracked flats to supratidal algal flats (with or without evaporites) and eolian dunes is typical.

Sedimentology Features of the intertidal and supratidal environment are very distinctive: algal lamination and stromatolites, mud cracks, tidal channel breccias with shells and mud-chip rip-up clasts, bird's-eye texture, dolomitization, evaporite minerals (especially gypsum, anhydrite, and halite), which may rise diapirically.

Fossils Stromatolites made by cyanobacteria ("blue-green algae") and other resistant types are most common, due to the desiccation and the extreme ranges of temperature and salinity. Some molluscs and burrowing worms are capable of surviving on the tidal flat, but most shell debris is washed in by storms.

SUBTIDAL SHELF CARBONATES

Shallow carbonate banks are rare phenomena today, but judging from the great abundance of shallow marine limestones, they were numerous in ancient epeiric seas. Modern shallow carbonate banks occur on continental shelves with little or no clastic input from rivers (e.g., Florida, the Persian Gulf, the coast of Yucatan) or on isolated oceanic platforms (e.g., Bermuda, the Bahamas). During major epeiric transgressions, however, the continents must have been covered with a few meters of shallow, warm, clear water, which produced enormous areas of carbonate banks, reefs, and shoals. Box 6.1 describes an ancient example of this. As long as subsidence is relatively slow, the carbonate-secreting community of organisms can keep up with it by building up substrate fast enough to remain within the photic zone. If the carbonate bank progrades across an epeiric basin, it can accumulate thicknesses approaching 2000 to 3000 meters of limestone (e.g., the Cretaceous of central Mexico, the Triassic Dolomites of the Italian Alps).

The deposits of the shelf have many similarities with peritidal and reef carbonates, and some major differences. The water is always normal marine, so the organisms found on the shelf tolerate only a narrow range of salinities. The normal marine fauna is much more diverse than the peritidal fauna, however, and a diverse fossil fauna and extensive bioturbation are typical. Because most of the organisms are directly or indirectly dependent on light, the depths are shallow, ranging from low-tide level down to 100 or 200 meters. Like other shallow carbonate environments, carbonate banks form in a narrow range of temperatures (10° to 30°C), so they are found in tropical or subtropical latitudes. Unlike peritidal lagoons, however, the water is always well oxygenated. Finally, the carbonate bank is below normal wave base but above storm wave base. Most carbonate shelf sediments are very muddy, but in some areas there is evidence of strong currents or storm activity.

Because rates of sediment production depend on the growth of organisms and not on hydraulic processes, bedding in carbonates is typically irregular and discontinuous, with variable thicknesses over a few meters. A local spurt of growth or concentration of carbonate particles can build a bed that thickens rapidly over an area of a few meters. Frequently, there are small hiatuses in the growth pattern, which may result in shaly partings between limestone beds. Some geologists have suggested that these represent episodes of terrigenous sedimentation that temporarily hamper the growth of organisms. Wavy bedding, nodular bedding, flaser bedding, and ball-and-pillow structures are also common but are produced by compaction and not by the hydraulic processes found in terrigenous systems. Stromatolites may also be responsible for wavy bedding. Cross-bedding and ripple marks occur only in areas of moderate current flow, such as exposed banks covered by skeletal sand or ooids. Of course, there are examples of evenly laminated, laterally extensive limestones, but these appear to be the exception. Most lamination is destroyed by the active bioturbation of this environment. Where it survives, even rhythmic lamination is due to some unusual fluctuation in climate and water chemistry that is produced by local conditions.

The sheltered central and inner part of most carbonate banks is an area of low-energy currents and waves; it may be protected by reefs or shoals or lie in coastal lagoons. The bottom is covered with intensely bioturbated pelletal muds. The lime mud is made of aragonite needles precipitated as the skeletons of certain types of calcareous algae. There are sand-sized particles of benthic invertebrates that live and die in the mud. In modern shelves, these are chiefly echinoderms, benthic foraminifers, and molluscs, but in ancient seas, brachiopods, bryozoans, and trilobites were also numerous.

Surrounding the low-energy muds along the outer edge of the platform is a higher-energy environment that feels the effects of waves and storms. Naturally, the current action results in coarser carbonate particles, particularly rippled skeletal sands and ooids. Ooids are a particularly good indicator of current energy. They are formed by the rolling and winnowing of small carbonate particles, which build up concentric rings of aragonitic needles as they roll back and forth through the lime mud. The faunas of the exposed banks are less diverse and abundant compared with those of the quiet lagoons, though an algal mat and related fauna usually form between storms. Along the edges of the banks are lower-energy environments where grasses and burrowing organisms predominate. Skeletal sand and ooids often spill over the banks during storms and become cemented into lumps and grapestone. Wherever plants have locally formed stabilized mounds, the environment is relatively sheltered, and a much greater faunal diversity is possible.

One of the best modern analogues of a shallow carbonate bank is the Great Bahama Bank, which covers about 100,000 square kilometers with an average water depth of 5 meters or less (Fig. 6.10A). It is surrounded by deep channels and straits of depths of 650 meters or more, and is dish-shaped in profile, with a low rim (Fig. 6.10B). Tidal ranges are 0.5 meters or less, so wind is the main influence on current flow. The central basin, which has very low energy currents, is covered with pelletal muds. Because the storms and winds come from the north and east, coralgal reefs and skeletal sands rim the higher energy northern and eastern margins of the bank; behind them are broad banks of grapestone. The

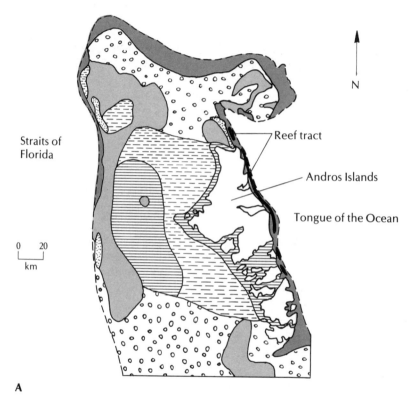

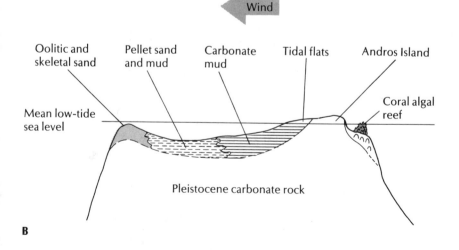

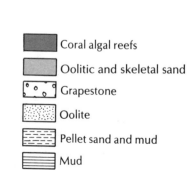

Figure 6.10

(A) Generalized map of the carbonate facies of the Bahama Banks near Andros Island. From Sellwood (1978). (B) Cross section of the Bahamas Banks, showing principal facies. From Blatt et al. (1980).

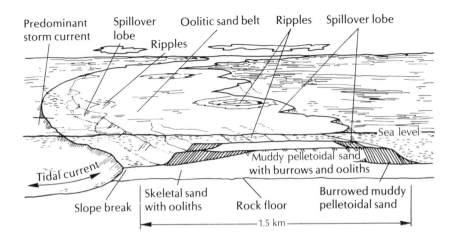

Figure 6.11

Diagram showing aerial and cross-sectional view of oolitic shoals with tidally generated, bidirectional cross-stratification. From Davis (1983).

quieter northwest margin is rimmed by oolitic shoals, in response to the relatively high tidal currents in this area (Fig. 6.11; see also the illustration on page 134). The edges of the shoals grow very rapidly, despite the higher wave energy, because nutrient-rich waters rise from the deeper marginal channels, making possible the abundant life from which skeletal sands eventually form after they are reworked by waves. The channels themselves (some of which are as deep as 4 kilometers) are floored by pelagic ooze, with some skeletal sand that has spilled over from the shelf.

The Persian Gulf (Fig. 6.12; see also Figure 6.3) is probably the largest modern example of an inland carbonate sea (as opposed to the offshore bank exemplified by the Bahamas). An area of 350,000 square kilometers is covered with water 20 to 80 meters deep. It is almost totally landlocked in a region of desert climate, so evaporitic tidal flats are very common. There is relatively little storm or wave action, and circulation is very restricted, so most of the current activity is due to tides, with a range of several meters. Because the current and wave activity is much weaker than at the Great Bahama Bank, skeletal sands are less important than lime mud. The shallow margins of the Persian Gulf are rimmed by muds rich in molluscs. In the deeper waters below wave base in the central gulf, the dominant sediment is dark, argillaceous lime muds. Only the shoals along the southwestern shore are agitated by the winds coming out of the northeast, producing a belt of skeletal sand. Reefs grow just inland of the skeletal sand belt on the southern shore.

There is tremendous lateral variation of facies in carbonate shelves and thus no standard vertical sequence. Under normal conditions, the dominant lithology is pelletal mudstones with abundant fossils. This produces a widespread, thick sheet of limestone with normal marine fossils and remarkable consistency over a wide area. If the shelf is becoming more shallow, then a shoaling-upward sequence of normal marine and restricted intertidal deposits, and possibly evaporites, is produced (Fig.

6.13). In some shelves, regular episodes of influx of terrigenous mud have produced shaly interbeds within the limestones. Shallow intracratonic basins typically fill with homogeneous limestones that thicken toward the center of the basin. If, however, there is rapid fluctuation of sea level in a shallow basin, then a rhythmic alternation of limestones and shales may be the result.

Diagnostic Features of Subtidal Shelf Carbonates

Tectonic setting All shelf carbonates require warm, clear, shallow, well-oxygenated, normal marine conditions, which typically exist only on continental shelves or epeiric seas in low latitudes with no significant clastic input.

Geometry Subtidal shelf carbonates form extensive, thick, homogeneous sheets that can cover thousands of square kilometers and reach hundreds of meters in thickness.

Typical sequence Under normal conditions, a uniform skeletal pelletal mudstone with remarkable vertical homogeneity is produced. If the sequence is shoaling upward, then the subtidal limestones are capped by intertidal sequences and possibly evaporites.

Sedimentology Although the mineralogy is almost exclusively aragonite, calcite, and dolomite (with rare shales and evaporites), textures are highly variable. Pelletal muds rich in biogenic debris are common, as are ooids, skeletal sands, and bioturbated muds. Bedding of variable thickness, with wedge- and lens-shaped units, is particularly characteristic, as are nodular bedding and flaser bedding caused by compaction.

Fossils The most diagnostic feature is the abundance of fossils of the normal marine fauna, which are tolerant of only a limited range of salinities, light conditions, turbulence, and oxygen content. The shallow carbonate shelf supports by far the greatest diversity of marine faunas of any environment on earth.

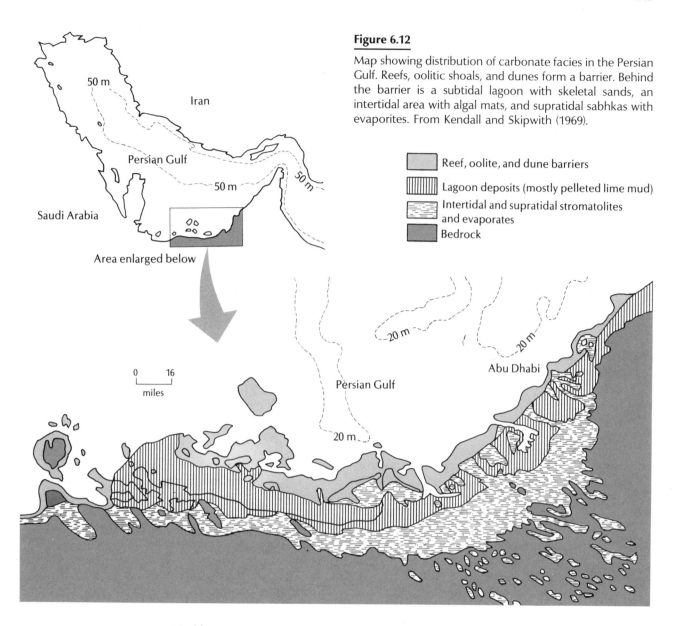

Figure 6.12

Map showing distribution of carbonate facies in the Persian Gulf. Reefs, oolitic shoals, and dunes form a barrier. Behind the barrier is a subtidal lagoon with skeletal sands, an intertidal area with algal mats, and supratidal sabhkas with evaporites. From Kendall and Skipwith (1969).

Reef, oolite, and dune barriers

Lagoon deposits (mostly pelleted lime mud)

Intertidal and supratidal stromatolites and evaporates

Bedrock

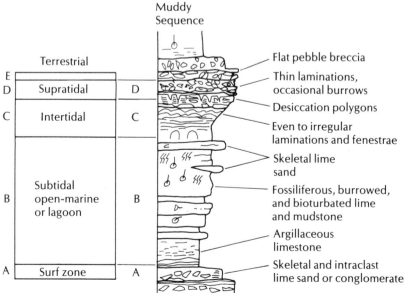

Figure 6.13

Hypothetical shallowing-upward sequence on a low-energy carbonate shelf. From James (1984).

Box 6.1
Devonian Shallow Marine Carbonates of the Helderberg Group, New York

Shallowing-upward carbonate sequences are common in the stratigraphic record, as reviewed by Hardie (1986). One of the best exposed and most completely described sequences occurs in the Devonian deposits of the Mohawk and Hudson valleys of New York. The Catskill Mountains are rimmed by a resistant limestone cliff extending for over 400 km, known as the Helderberg Escarpment. The Lower Devonian Helderberg Group (Fig. 6.14) is made up of a number of carbonate units (described by Rickard, 1962), which represent various shallow, subtidal carbonate environments. Some of the most detailed study of the Helderberg Group has concentrated on the basal Manlius Formation, which Laporte (1967, 1969, 1975) has interpreted as a peritidal deposit.

Detailed examination of the Manlius outcrop (Fig. 6.15) shows that it is composed of a series of shallowing-upward cycles, averaging 1 to 2 meters in thickness (Goodwin and Anderson, 1980). Within these cycles, three facies can usually be recognized: supratidal, intertidal, and subtidal (Fig. 6.16A), each characterized by a distinctive set of sedimentary structures. Not every cycle produces all three facies, but two out of three are usually present. All three facies occurred at the same time in belts parallel to the ancient shoreline but have been superimposed by the gradual shallowing during each cycle. In the seaward direction, the peritidal deposits of the Manlius intertongue with, and pass into, the shallow marine carbonate bank, represented by the Co-

eymans Formation, a brachiopod-rich crinoidal calcarenite.

The *supratidal* facies is represented by unfossiliferous, laminated, mud-cracked, dolomitic, pelletal carbonate mudstone. The most diagnostic features are the well-preserved desiccation cracks (Fig. 6.16B). Internally, desiccation resulted in spar-filled, irregular voids in the algal mat, which is known as bird's-eye texture (Fig. 6.16C). In detailed structure, the internal lamination is typically wavy and curled upward, due to crinkling and cracking of the algal mat. Between algal laminae are carbonaceous films from the algae, and abundant pellets. These so-called bituminous limestones once led geologists to think that the Manlius had been formed in deep, stagnant water.

The supratidal limestones are extensively dolomitized. Dolomite rhombs have replaced the original limestone in many places, and dolomite has also filled the cracks. Laporte (1967) interpreted this dolomite in light of studies of modern supratidal dolomitization (Shinn and Ginsburg, 1964; Shinn et al., 1965). Evaporation at the algal mat brought magnesium-rich seawater upward by means of capillary action. Surface evaporation then precipitated dolomite at the mat surface in a fine-grained layer and spurred replacement of the limestone by dolomite throughout the mat.

The supratidal model is supported by the almost total lack of marine fossils, except for a few skeletal fragments that washed in by high tides or storms.

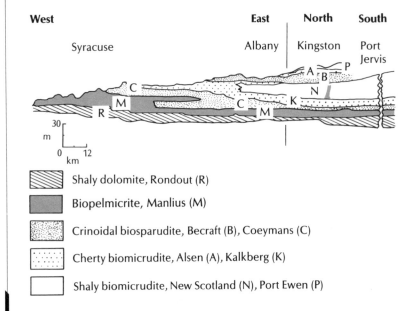

West   **East**   **North**   **South**

Syracuse   Albany   Kingston   Port Jervis

Figure 6.14 Restored section of the Helderberg Group in New York, showing lithology of individual formations. Note the interfingering of lower and middle Helderberg units and the time equivalence of parts of the Manlius, Coeymans, Kalkberg, and New Scotland Formations. The Helderberg grades downward into the late Silurian-early Devonian Roundout Formation. From Rickard (1962).

Shaly dolomite, Rondout (R)

Biopelmicrite, Manlius (M)

Crinoidal biosparudite, Becraft (B), Coeymans (C)

Cherty biomicrudite, Alsen (A), Kalkberg (K)

Shaly biomicrudite, New Scotland (N), Port Ewen (P)

Figure 6.15 Shoaling-upward sequence within the Manlius Limestone, Helderberg Group, Perryville Quarry, New York. The cycle begins with an erosive base (lower arrow), overlain by reworked stromatoporoid heads. This grades up into a fossiliferous lime wackestone and faintly laminated lime mudstone. Brown dolomitic algal laminites form the top of the sequence just beneath the upper arrow. Scale is about 2 meters long. Photo courtesy P. W. Goodwin and E. A. Anderson.

Only a few disarticulated ostracod valves, some burrow mottles, and the persistent algal lamination remain as evidence of life on what must have been a hot, hypersaline, frequently desiccated environment that was hostile to all but certain cyanobacteria.

The *intertidal* facies is represented by sparsely fossiliferous, pelletal carbonate mudstone interbedded with skeletal calcarenite (Fig. 6.16D). These 2 to 6 centimeter-thick beds were once described as "ribbon limestones." They contain abundant mud cracks and limestone mudchip conglomerates, indicating episodes of desiccation and erosion of the mud-cracked surface. There are also abundant algal stromatolites and oncolites, which are known to prefer intertidal environments. The pelletal carbonate mudstone—with its mud cracks, mud-chip conglomerates, stromatolites, and pellets—was frequently scoured and then covered by a skeletal calcarenite representing submergence of the intertidal flat. The fossil types are few but occur in large numbers. Such a low-diversity, high-abundance pattern is common in highly stressed environments, where only a few tolerant species can survive, but

their numbers are virtually unchecked by competitors or predators. The fossils include ostracods, tentaculitids (Fig. 6.16E), small spiriferid brachiopods, trepostome bryozoans, and spirorbid worms. The more elongate fossils show a strong alignment, indicating that many of the fossils were reworked by strong tidal currents. The rapid alternation between the mud-cracked, stromatolitic mudstones (indicating relative emergence) and the skeletal calcarenite (indicating submergence) are particularly diagnostic of an environment with rapid alternation of tidal cycles of desiccation and submergence.

The *subtidal* facies of the Manlius Formation is characterized by fossiliferous, pelletal carbonate mudstones interbedded with small biostromes of stromatoporoids. The carbonate mudstones are massive- to medium-bedded and contain a diverse and abundant fauna of rugose corals, brachiopods, ostracods, snails, and codiacean algae, which could survive only under continual marine submergence. Biotic activity is also shown by extensive burrow mottling, which disrupts lamination and gives the mudstones their massive appearance. These mudstones interfinger (Figs. 6.16F, 6.16G) with tabular masses and globular individual heads of stromatoporoids, which apparently grew in tidal creeks and channels in the subtidal lagoon. Although the stromatoporoids did not grow on a scale to qualify as a reef, they probably did form wave-resistant structures that sheltered the subtidal lagoon.

When all three of the facies are associated, they clearly indicate peritidal conditions. The repetition of these facies in numerous, thick cycles indicates that the Manlius is typical of the type of shallowing-upward limestone sequences found in many parts of the geologic record.

Overlying and laterally interfingering with the Manlius is the next unit in the Helderberg Group, the Coeymans Formation (see Fig. 6.14). It consists of coarse-grained, gray or blue limestone with thin, irregular bedding. There is very little clay material. It has a normal Devonian marine fauna of brachiopods, high-spired gastropods, straight-shelled cephalopods, bivalves, crinoids, trilobites, and tabulate corals. Fifty to eighty species are represented, compared with only 5 to 30 species in the more extreme intertidal and shallow subtidal conditions represented by the Manlius. The most striking feature is the abundant cross-beds (Fig. 6.17), which indicates strong current activity. All of these features suggest considerable wave or current activity in a normal marine shelf environment, well below the tide level and just below or at wave base. In some places there

Dolomite

B C

Mean high water

Mean low water

D

E

F

G

Mg-rich water

Supratidal

Laminated, dolomitic,
mud-cracked pelmicrites

Intertidal

Thin-bedded pelmicrites
and biopelsparites

Subtidal

Laminar stromatoporoids
in biopelmicrudites

Hemispherical stromatolites
in biopelmicrudites

∪∪∪ Mud cracks   Laminar stromatoporoids

Stromatolites   Hemispherical
                       stromatoporoids

Oncolites

A

B

C

D

E

F

G

are coral bioherms composed of tabulate and rugose corals, crinoid columnals, bryozoans, and stromatoporoids.

Finally, the overlying Kalkberg and New Scotland Formations (see Fig. 6.14) are composed of fine-grained, irregularly bedded silty limestones, which contain an even more diverse fauna (over 300 species) of brachiopods, trilobites, scattered tabulate and rugose corals, and especially bryozoans (Fig. 6.18). The dominance of fine silt and the irregularity of the bedding suggest very little winnowing by current activity. The abundance of delicate bryozoans is another indicator of relatively quiet conditions. These features, taken together, suggest a shallow, marine shelf environment, well below normal wave base, for the Kalkberg and New Scotland Formations. Thus the Helderberg facies belts, taken in sequence, represent tidal flat and lagoonal deposits (Manlius), well-agitated subtidal waters near wave base (Coeymans), and finally open shelf marine waters below wave base (Kalkberg and New Scotland). The overall trend (see Fig. 6.14) suggests a sequence that was deepening through continued transgression to the west. These patterns are typical of many carbonate peritidal and shelf sequences throughout the geological record.

Figure 6.17 Cross-bedded crinoidal layer of Coeymans-type lithology and stratified beds of Manlius type in the upper Olney Member, near Paris, New York. Subtidal carbonate shelf facies. Photo courtesy L. V. Rickard and the New York State Geological Survey.

Figure 6.16 (A) Generalized facies model for the Manlius Formation, showing representative rocks of each facies. (B) Supratidal mud cracks, Elmwood Member. (C) Negative print of thin section showing bird's-eye texture due to desiccation; larger vertical vug may be a gas trackway. Supratidal facies, middle Thacher Member, Catskill, New York. (D) Negative print of thin section showing skeletal calcarenite lying unconformably on calcareous mud; sand grades up into mud that is truncated above by a second skeletal calcarenite and mudstone. Fossils are abundant and include conical tentaculitids, bryozoans, and ostracods. Intertidal facies, lower Thacher Member. (E) Weathered bedding surfaces strewn with tentaculitids. Intertidal facies, lower Thacher Member, Wiltse, New York. (F) Weathered outcrop showing lateral interfingering of encrusting stromatoporoid masses with skeletal carbonate mudstones. Truncated stratification of non-stromatoporoid beds suggests lateral erosion of sediments followed by accumulation of stromatoporoid masses. Subtidal facies, upper Thacher Member, Clarksville, New York. (G) Outcrop of stratified biopelmicrites (Elmwood and Clark Reservation Members) overlain by massive bed of tightly bound stromatoporoids (Jamesville Member). Subtidal facies. Photos B–E and G courtesy L. F. Laporte; photo F courtesy L. V. Rickard and the New York State Geological Survey.

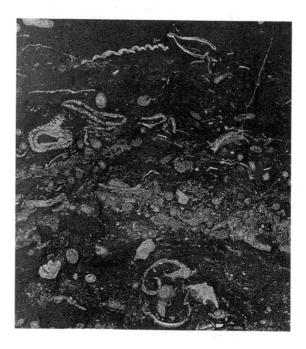

Figure 6.18 Print made from acetate peel showing subtidal, burrowed New Scotland biomicrudite with encrusting bryozoans, trilobites, and brachiopods, Bronck's Lake, New York. Photo courtesy L. F. Laporte.

REEFS AND BUILDUPS

Unlike any of the systems discussed previously, carbonate reefs and buildups are entirely self-generated. They are sediment systems that are built entirely by the organisms growing in them. The term *reef* has been used so loosely over the years (reviewed by Dunham, 1970, and Heckel, 1974) that the term *buildup* is preferably applied to any body of carbonate rock that has built up topographic relief above the surrounding environment. A reef is a buildup that has grown in the wave zone and has a wave-resistant framework. The term *bioherm*, or *biostrome*, is generally applied to any *in situ* accumulation of benthic organisms, whether or not they are topographically high or wave-resistant.

Reefs and buildups are interesting for many reasons. For the paleontologist, they are the best examples of a paleoecological community, preserved intact in its own detritus. The presence of a reef implies an adequate supply of nutrients, as well as the usual restrictive conditions of carbonate sedimentation, and so provides a valuable indicator of paleoceanographic conditions. Reefs also give clues to understanding the paleo-environment at the times in the past when building up was especially common. For petroleum geologist, buildups serve as important stratigraphic traps for oil.

A reef community is very complex. The most essential part is the framework builder, and through geologic history, a number of colonial organisms have been responsible for this process (Fig. 6.19). Today's reef-

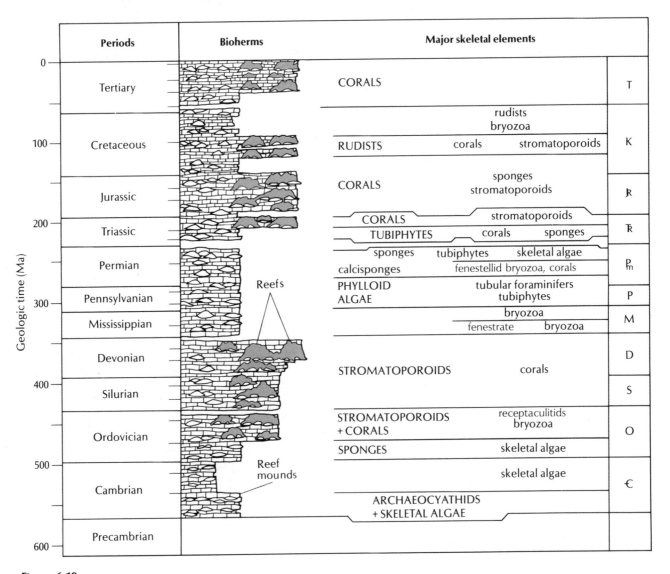

Figure 6.19

An idealized stratigraphic column showing the Phanerozoic history of bioherms. Gaps indicate periods when there were no reefs or bioherms; narrow columns indicate only reef mounds, and wide columns indicate times of both reefs and reef mounds. Major reef-builders are given in capitals; lesser, in lower case. From James (1984).

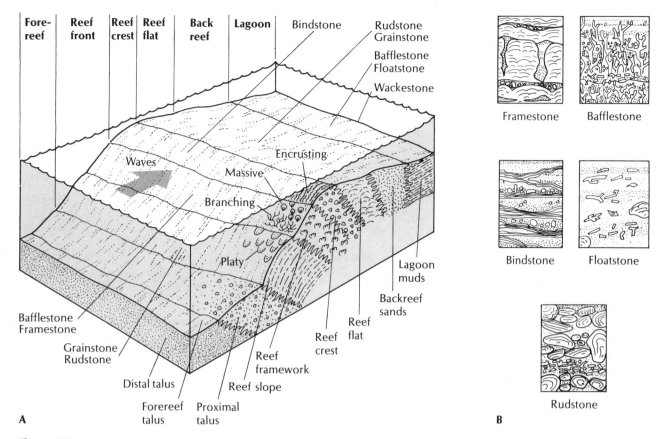

Figure 6.20

(A) Idealized cross section of platform margin reef facies, indicating reef margin zones and the nature of accumulating sediment. After Scoffin (1987). (B) Different types of reef limestone found in these facies, as recognized by Embry and Klovan (1971).

builders may be hermatypic scleractinian corals, coralline algae, bryozoans, or sponges, but in the geologic past they included cyanobacterial stromatolites, archaeocyathids, crinoids and blastoids, stromatoporoids, receptaculitids, tabulate and rugosid corals, rudistid bivalves and even richthofenid brachiopods (Fig. 6.19). The framework builders are essential to the structure of the reef, making it wave-resistant and forming baffles against strong currents and sediment. As these organisms die, new framework builders grow on top of them, which enables a reef to build upward in response to sea level. The framework organisms are kept in check by a community of bioeroders, including boring algae, worms, sponges, molluscs, and coral-eating fish and echinoids. These organisms continually weaken the framework until storms topple it, thus allowing new growth when sea level is not rising. The framework builders typically make up only 10 percent of the total volume of the reef. In the many cavities and crevices and protected interreef patches, lime mud that contains broken skeletal debris and reef fragments accumulates. As this interstitial material fills cavities, the reef becomes even more massive and resistant. Eventually, cementa-

tion locks both the framework and interstitial material into a solid, massive limestone unit.

The classification terms for reef limestones reflect the various relationships between framework and interstitial material (Fig. 6.20). Embry and Klovan (1971) divided reef carbonates into allochthonous interstitial mud and autochthonous reef rock that grows *in situ*. If more than 10 percent of the allochthonous particles are larger than 2 millimeters in diameter, and they are matrix-supported, the rock is a *floatstone*. If they are clast supported, it is a *rudstone*. *Framestones* are made of autochthonous massive framework builders that grew *in situ* whereas *bindstones* are composed of tabular or lamellar fossils that were bound together during deposition. A *bafflestone* is made of *in situ* stalked organisms that trapped sediment by baffling currents.

The reef organisms themselves give a number of clues about the environment. First, their diversity is sensitive to environmental conditions. A high diversity of both growth forms and taxa indicates nearly optimum conditions, such as a plentiful supply of nutrients and a minimum of physical stresses. During stable, optimum conditions, reef organisms can diversify and specialize, subdividing the niches so that many more organisms can live on the reef. A low diversity, on the other hand, indicates that the environment is unpredictable or stressful, or is just being colonized. Rapid and extreme fluctuations in temperature and salinity, low light levels, or intense wave activity all hamper the growth of reef organisms, so only the hardiest can continue to grow.

Second, the growth form of the framework organisms tells of the wave energy and sedimentation rate (Table 6.2). For example, delicate branching corals can survive only in low-energy wave environments, yet they grow rapidly because they are being buried by the high sedimentation rate. Encrusting bryozoans or corals, on the other hand, are capable of resisting intense wave activity.

Third, the reef is divided into various regions (sometimes called facies) because of environmental conditions. For example, a modern reef system shows a spectrum of hydrodynamic conditions, from calm water with periodical subaerial exposure in the backreef, to weakly moving water in the core of the reef, to intense wave and current activity along the wave front. The reef organisms and their growth forms are appropriately zoned to reflect this difference in current activity and strength.

In general, most reefs can be divided in three distinct facies: the reef core, the reef flank, and the interreef facies (Fig. 6.21). The *reef core* is massive, unbedded, and composed of the framework builders filled in by an interstitial matrix of lime mud and skeletal sand. The crest of the reef core is subject to the greatest wave energy, so the growth forms are low and encrusting. If the crest is relatively protected, then hemispherical, massive forms with a few interstitial branching forms can occur.

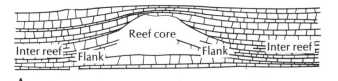

A

B

Figure 6.21

(A) A sketch illustrating the three major reef facies in cross section. From Walker (1984). (B) Reef and reef-flank deposits over 100 meters thick, from the Peechee Formation, Upper Devonian, Flathead Range, southern Rocky Mountains, Alberta. R denotes reef core. Photo courtesy B. Pratt.

Table 6.2

Growth Forms and Environments of Reef-building Metazoa

	Environment	
Growth Form	**Wave Energy**	**Sedimentation**
Delicate, branching	Low	High
Thin, delicate, platelike	Low	Low
Globular, bulbous columnar	Moderate	High
Robust, dendroid branching	Moderate to high	Moderate
Hemispherical, domal, irregular, massive	Moderate to high	Low
Encrusting[a]	Intense	Low
Tabular[a]	Moderate	Low

[a] Encrusting and tabular forms are very difficult, if not impossible, to differentiate in the rock record, yet they indicate very different reef environments.

From James (1984).

The upper part of the reef front is also made of encrusting and massive framework organisms (Box 6.2). Below 30 meters or so, the reef flank is below wave base, so the framework builders become much more branching and platelike in response to the quiet water conditions and the reduced light levels. Hermatypic corals depend on light for their symbiotic algae, as do coralline algae. As a result, modern reefs reach depths of no more than 70 meters because below this point the light level is too low.

Behind the reef crest is the reef flat, which feels considerable current and wave activity, but is usually sheltered from the worst pounding of the waves. In this region, pavements of skeletal debris and coral rubble accumulate, along with shoals of well-washed lime sand. Because the water is shallow, the framework builders grow rapidly and form irregular clumps and patches separated by current-worked channels of sand and rubble. The most sheltered area is the backreef, which feels waves only during storms. It is covered by carbonate sand caused by the biological breakup of carbonate organisms and moved around by storm waves. In the lagoons, the lime mud settles out of suspension, forming a dominantly muddy matrix. This sheltered environment provides optimum living conditions, limited only by rapid burial in the mud. Among the framework builders, stubby, dendroid forms, and large globular forms are common. A great diversity of interstitial organisms live in the mud, especially crinoids, delicate bryozoans, brachiopods, molluscs, ostracods, and calcareous green algae. The diversity of organisms alone is often sufficient to allow one to recognize the sheltered backreef.

Below the growing part of the reef core is the *reef flank*. Debris from the reef washes or slumps down the reef front and into the flank beds, which dip gently away from the core. As a result, the flank beds are usually composed of bedded limestone conglomerates made of reef debris and of lime sands that have washed in during storms. Reef flank beds are another example of deposits that accumulate at an original depositional angle and thus are exceptions to Steno's law of original horizontality. These beds grade into the interreef facies as they get further from the reef and their dips decrease.

The *interreef facies* is very similar to other shallow-water subtidal carbonate bank deposits. They are usually composed of thin-bedded, pelletal lime muds that are rich in skeletal sands. Often, the only indication of an interreef location is the rare presence of debris from the reef itself, which has been brought there by extreme storm waves. In basins with restricted circulation, evaporite minerals are common in the interreef facies.

One of the most fascinating features of reefs is *reef succession*. Reef communities do not appear on the seafloor fully developed but evolve and change as they colonize and stabilize the substrate (Table 6.3). The *pioneer (stabilization) community* is composed of organisms that encrust or send down roots and stabilize the shoal sands. Calcareous green algae, sea grasses, and crinoids are the commonest pioneers, but only a low diversity of them can be maintained in such a turbulent, unstable environment. Once the sediment is stabilized, however, other algae, corals, and sponges take root and grow. On this stabilized framework, the *colonizing community* of the main framework builders takes hold and soon dominates. Typically, a few species of branching or lamellar corals grow very rapidly, outstripping their burial by sediment. After they are well established, the *diversification stage* takes place. At this stage, the bulk of the reef mass forms as the reef grows up to sea level. A great diversity of reef-building taxa and growth forms occur, as well as of interstitial organisms, in response to the diversity of habitats. As the reef community reaches sea level and more turbulent water, the *domination stage* occurs, in which laminated, encrusting forms cover a reef core that is largely filled in.

Diagnostic Features of Reefs and Buildups

Tectonic setting At the edges of carbonate banks where upwelling from deeper waters brings up nutrients, reefs and buildups form. They are extremely limited by depth, temperature, salinity, and nutrients. Like other carbonate bank deposits, buildups are found in shallow, low-latitude, passive margins or epeiric seas free of clastic input.

Geometry Small, local moundlike or banklike accumulations that show rapid lateral changes in facies and thickness form. Patch reefs may only be a few meters high and wide, but some large reef complexes are hundreds of meters thick and kilometers wide.

Typical sequence Other than reef succession, there is no typical stratigraphic sequence of reefs and buildups. The lateral relationships of the various reef facies are more important than the vertical pattern, though both may work in concert to form a complex of massive reefs and bedded interreef limestones.

Sedimentology Framework builders are dominant, so entire deposit grows and is bound together *in situ*. Interstitial lime mud, skeletal fragments, and reef-rock breccias are formed in crevices between the framework organisms. Reefs are formed exclusively of calcite or aragonite, though the reef core is highly susceptible to dolomitization. The backreef flats may become evaporitic in the appropriate environment.

Fossils Reefs are formed almost entirely of characteristic fossils whose ecology determines the growth and shape of the reef. Reef organisms are extremely sensitive to temperature, salinity, light, and terrigenous mud, so they are excellent indicators of environmental conditions.

Table 6.3

Stages of Reef Growth

Stage	Type of Limestone	Species Diversity	Shape of Reef Builders
Domination	Bindstone to framestone	Low to moderate	Laminate, encrusting
Diversification	Framestone (bindstone), mudstone to wackestone matrix	High	Domal, massive, lamellar, branching, encrusting
Colonization	Bafflestone to floatstone (bindstone) with a mudstone to wackestone matrix	Low	Branching, lamellar, encrusting
Stabilization	Grainstone to rudstone (packstone to wackestone)	Low	Skeletal debris

From James (1984).

Box 6.2
Devonian Reefs of the Canning Basin, Australia

One of the world's best exposed and most spectacular ancient barrier reef complexes is exhumed along the northern margin of the Canning Basin in northwest Australia. The Canning Basin is the largest sedimentary basin in Western Australia, covering an area of 530,000 square kilometers and containing 13,000 meters of Ordovician through Cretaceous sediments. The northern margin consists of a belt of Middle and Late Devonian reef complexes about 350 kilometers long and 50 kilometers wide. These reefs grew on the edge of a block-faulted basin that resulted in a relief of several hundred meters above the basin to the south during the Late Devonian. The barrier reef belt may have once continued another 1000 kilometers to the west and north to connect with similar reefs in the Bonaparte Gulf Basin. Several river gorges expose spectacular sections through these reefs, most notably at Windjana Gorge (Fig. 6.22).

As described by Playford (1980, 1981, 1984; Playford and Lowry, 1966), the reef complexes can be divided into three distinct facies and several subfacies (Fig. 6.23A): the platform, the marginal slope, the basin facies. The *reef subfacies* occurs as a narrow rim around the platform and is composed of massive limestone and dolomite built up by colonial cyanobacteria and stromatoporoids. These formed a resistant framework that sheltered the backreef lagoons. The reef framework is typically an interlocking mass of stromatoporoids and the ?alga *Renalcis* (Fig. 6.23B), which encrusted all available surfaces. The voids between the framework builders were later filled with fibrous and sparry calcite (Fig. 6.23B). Other reef-building organisms included corals, brachiopods, and sponges.

The *backreef subfacies* (Pillara and Nullara Limestones) is composed of well-bedded limestone and dolomite, which was deposited in the shelf lagoon behind the reef rim (Fig. 6.23C) and extends horizontally for tens of kilometers. In some places, these rocks are interbedded with terrigenous sediments. The commonest rock type of the backreef is dominated by stromatoporoids, which built small biostromes behind the main reef front. The cavities between the stromatoporoids (Fig. 6.23D) are encrusted with the ?alga *Renalcis* and filled with laminated pelletal lime mud. Mud-cracked limestones

with a bird's-eye texture are also common, indicating that some parts of the lagoon were emergent at times (Fig. 6.23E). In other places, corals such as *Hexagonaria*, rather than stromatoporoids, form small biostromes. Large areas are covered with oolitic limestones, indicating an environment of strong wave and current action. Consequently, few organisms lived on such a mobile substrate, except for algal oncolites (Fig. 6.23F) and the bivalve *Megalodon* (Fig. 6.23G). In section, these oolitic limestones exhibit not only the oncolites and ooids but also pellets and a pronounced bird's-eye texture (Fig. 6.23F).

The *forereef subfacies* is composed primarily of talus deposits that rolled down the 30° to 35° reef-front slope. These deposits consist mainly of skeletal sands, conglomerates, and megabreccias. In some places blocks of reef limestone several meters across have rolled into the forereef deposits and distorted them (Fig. 6.23H). Some of these blocks were later colonized on their upper surfaces and eventually buried (Fig. 6.23I). Between the blocks of the megabreccia, the reef front was colonized by crinoids and brachiopods that lived in the sheltered part of the forereef. Playford (1984) has also described submarine debris flows made of reef talus that cascaded down channels cut into the reef front, producing inversely graded breccia beds similar to those found in clastic alluvial fans (Chapter 3).

The *basin facies*, found in the basins between reef fronts, is composed of horizontal terrigenous shales, sandstones, and conglomerates interbedded with thin beds of limestone. There is little or no material derived from the limestone platforms, which apparently was trapped in the forereef slope.

The reef complex shows pronounced ecological zonation (Fig. 6.24). The diversity of organisms in the backreef is considerably greater than in the stromatoporoid-dominated community of the reef itself. In addition to stromatoporoids, there are gastropods, bivalves, corals, oncolites, stromatolites, and a number of other algal structures in the backreef. The basin and marginal slopes, on the other hand, were favored by more open-ocean organisms such as fish, crustaceans, tentaculitids, nautiloids, ammonoids, and conodonts. The reef slope was colonized by crinoids, sponges, and brachiopods.

In summary, the Devonian reefs of the Canning Basin demonstrate most of the classic features seen in modern reefs: framework-building reef organisms, a sheltered backreef lagoon, a forereef talus, and the open-sea deposits of the basin. Each of these environments had its own range of environmental conditions, resulting in a pronounced ecological zonation of organisms in and around the reef.

A

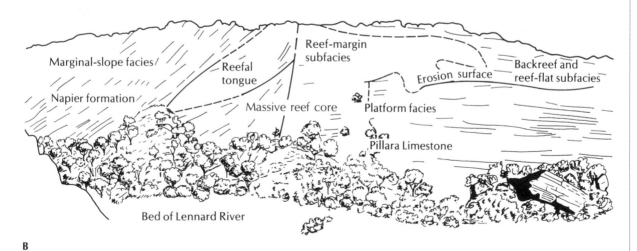

B

Figure 6.22 Panoramic photograph (A) and equivalent sketch (B) of the southeast wall of Windjana Gorge, showing facies relations in the Frasnian reef platform margin. Photo courtesy P. E. Playford.

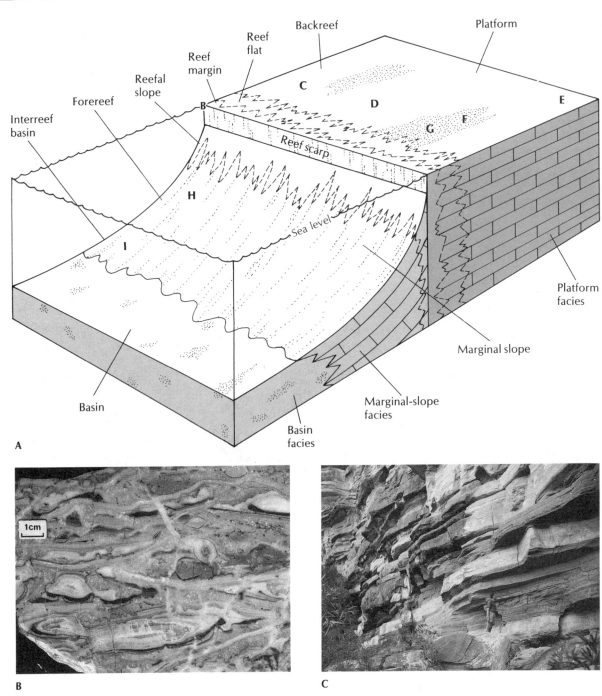

Figure 6.23 (A) Diagram showing location of representative facies in Devonian reefs of the Canning Basin, Australia. (B) A polished slab from the reef crest, showing laminar stromatoporoids and the ?alga *Renalcis* (speckled) surrounded by cavities filled with fibrous calcite. (C) Well-bedded limestone and dolomite in the Pillara Limestone, from the reef flat subfacies. (D) Cut and polished slab of a toppled stromatoporoid colony encrusted with the ?alga *Renalcis*. The cavity system has been filled with laminated pelletal mud of the backreef lagoon. (E) Mud cracks in bird's-eye limestone from the supratidal backreef area. (F) Algal oncolites in fine-grained oolitic pelletoid bird's-eye calcarenite, from the intertidal-supratidal backreef subfacies. (G) Cross sections of the bivalve *Megalodon* in life position from the backreef immediately behind the reef front. (H) Rolled blocks of reef limestone in the fore-reef facies. The forereef silty limestones and calcareous siltstones have been distorted by the rolled blocks of limestone from the edge of the reef. (I) Contact between blocks of crinoidal forereef calcarenite in the forereef megabreccia. All photos courtesy P. E. Playford.

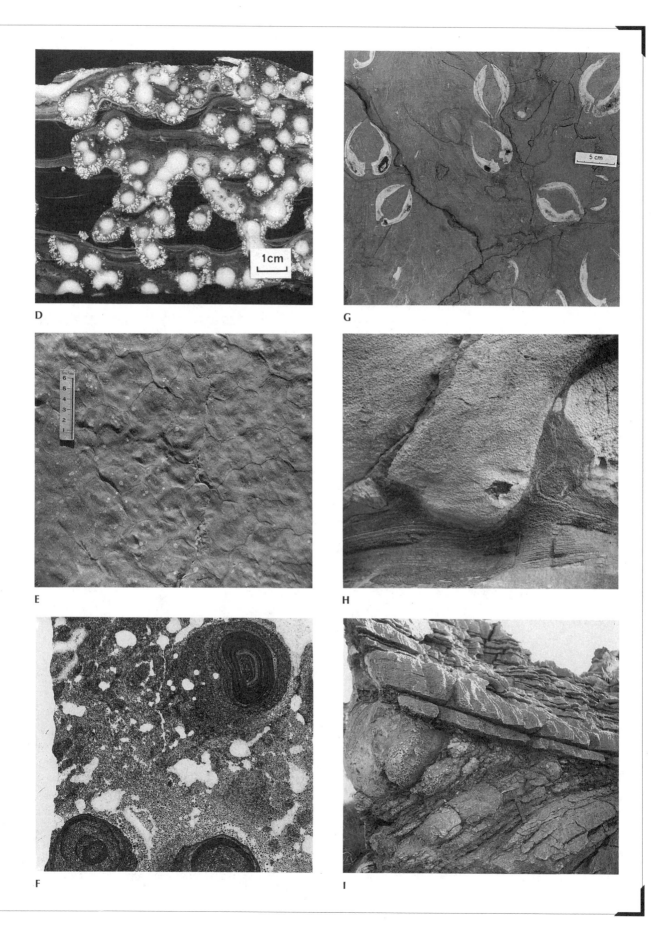

D

G

E

H

F

I

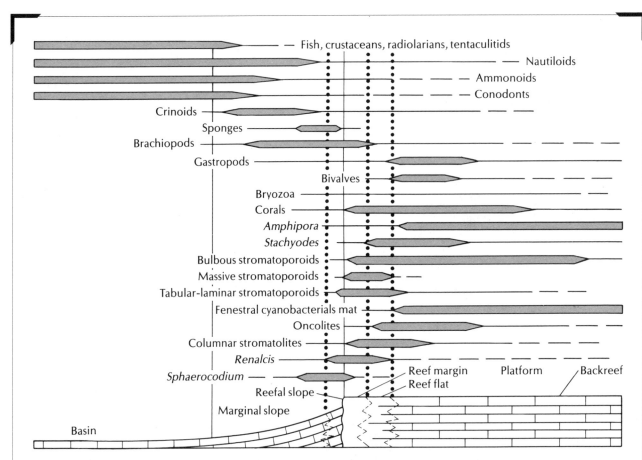

Figure 6.24 Ecological zonation of Frasnian (late Devonian) organisms in the Canning Basin reef complex. The basinal fauna is predominantly free-swimming organisms, such as fish, cephalopods, conodonts, crustaceans, radiolarians, and tentaculitids. The marginal slope is dominated by animals with the ability to anchor to the substrate, such as crinoids, sponges, and brachiopods. The reef crest is made of wave-resistant colonial organisms which build the reef framework, such as corals, massive and laminar stromatoporoids, and algae. The backreef contains organisms that prefer the sheltered conditions and can tolerate elevated temperatures, salinities, and occasional desiccation, such as gastropods, bivalves, corals, bulbous stromatoporoids, and stromatolites.

FOR FURTHER READING

From the references in Chapter 3, the books by Davis, Reading, and Walker contain excellent chapters on carbonate environments. In addition to these, the reader may wish to consult the following.

Bathurst, R. G. C., 1975. *Carbonate Sediments and Their Diagenesis*. Elsevier, New York. Although primarily oriented toward carbonate sedimentology, it contains some excellent reviews of modern carbonate environments.

Friedman, G. C. (ed.), 1969. *Depositional Environments in Carbonate Rocks*. Soc. Econ. Paleo. Mineral. Spec. Publ. 14. An excellent collection of papers on some ancient carbonate environments and their interpretation.

Frost, S. H., M. P. Weiss, and J. B. Saunders (eds.), 1977. *Reefs and Related Carbonates—Ecology and Sedimentology*. Amer. Assoc. Petrol. Geol. Studies in Geology 4. Collected papers on many modern and ancient reefs, emphasizing the Caribbean.

Laporte, L. F. (ed.), 1974. *Reefs in Time and Space*. Soc. Econ. Paleo. Mineral. Spec. Publ. 18. Another collection of papers on modern and ancient reefs, with several classic papers.

Quarterly Journal of the Colorado School of Mines, 1985–6. *Carbonate Depositional Environments, Modern and Ancient*. Part 1, "Reefs," by N. P. James and I. G. MacIntyre, 80(3): 1–70; Part 2, "Carbonate platforms," by P. M. Harris, C. H. Moore, and J. L. Wilson, 80(4): 1–60; Part 3, "Tidal Flats," by L. A. Hardie and E. A. Shinn, 81(1): 1–74; Part 4, "Periplatform Carbonates," by H. T. Mullins, 81(2): 1–63. Four excellent and very up-to-date reviews of carbonate environments in four successive issues of this journal.

Scholle, P. A., D. G. Bebout, and C. H. Moore (eds.), 1983. *Carbonate Depositional Environments*. Amer. Assoc. Petrol Geol. Memoir 33. A stunning, massive compendium of all carbonate environments, beautifully illustrated with abundant color photo-graphs and artwork. The next best thing to being there!

Scoffin, T. P., 1987. *An Introduction to Carbonate Sediments and Rocks*. Blackie and Son, Ltd., London. Excellent, well-illustrated, short introduction to carbonate rocks and their depositional environments.

Wilson, J. L., 1975. *Carbonate Facies in Geologic History*. Springer-Verlag, New York. The most complete and well-illustrated account of ancient carbonate environments.

III

THE ROCK
RECORD

View east from Toroweap Point, taken by J. Hillers from
the same spot, but almost 100 years earlier, as the
photograph on the cover. Courtesy U.S. Geological Survey.

In Part I, we surveyed the data base and the assumptions on which stratigraphic interpretation is built. In Part II, we inferred ancient depositional environments from local outcrops. Now we shall try to integrate information and inferences from outcrop evidence into the larger context.

As we first begin to examine the rock record, we discover that it is preserved and exposed only in limited places on earth, and is full of gaps. This leads us to ask the questions: What conditions are ideal for the formation and preservation of stratigraphic sequences? How completely do these sequences record geologic history? The limited exposure of outcrops on the modern surface of the earth means that many deposits of the same age are no longer connected. To reconstruct the earth history that these outcrops represent, we must ask the questions: How do these outcrops and environments relate to one another in space and time? How can they be correlated, and what is the time significance of these correlations? In addition to correlating outcrops, we can learn much by mapping and plotting the geometry of the rock units under study. In Chapter 9, we examine some of the many techniques that can be used to analyze the depositional history of strata, and ultimately to analyze the sedimentary history of the basin in which they formed.

Correlating rock units by their sedimentary properties is only one of a number of methods that can be used. One of the most important methods of correlation is biostratigraphy, or correlation by the fossil content of the rocks. Biostratigraphy requires another level of understanding, since the biological properties of the fossils are as important as the physical properties of the sedimentary rocks that contain them. In recent years, the geophysical properties of strata have become very important in correlation, particulary in the subsurface. Well-logging uses a number of different physical properties of the rock that can be detected by a device pulled through the drillhole, and is the primary tool of modern subsurface correlation. Seismic stratigraphy provides a wealth of information about subsurface units that could not otherwise be detected. The changes in the earth's magnetic field over geologic time has made it possible to correlate rocks by the way in which they record these field changes, giving us the discipline of magnetostratigraphy.

The geochemical properties of strata have become equally important. The regular pattern of change of certain stable isotopes through geological time has allowed correlation by this geochemical pattern. However, it is the unstable isotopes, decaying at a predictable rate, that gives us numerical age estimates, rather than mere relative correlation. Integrating these numerical ages with the relative sequence of events recorded in strata is often the most difficult and challenging part of stratigraphy. The geologist not only has to try to develop a framework of correlation and put dates into it, but also has to be aware of the relative strengths and pitfalls of each method. Integrated chronostratigraphy requires the skillful and well-informed use of many different disciplines, a training that all geologists should have.

Finally, from the depositional framework of local basins discussed in the beginning of this section, we conclude with the larger framework of plate tectonics. What factors ultimately control the crustal movements that form basins in the first place? In the area of basin formation and tectonics, stratigraphy has been completely revolution-alized since the 1960s. For the first time in over two centuries, we now have a global model that explains and predicts *why*, *when* and *where* sedimentary basins form. The ramifications of this understanding are enormous, and just beginning to change stratigraphy. For this reason, the stratigraphy of the twenty-first century is likely to be very different from stratigraphy as it was understood through most of this century. Indeed, it is a fortunate time to be a stratigrapher interpreting the rock record!

The "Pulpit," composed of *Protoceras* sandstone channels, Poleside Member of the Brule Formation, Badlands National Park, South Dakota. Taken by N. H. Darton in 1898. Courtesy U.S. Geological Survey.

7

Lithostratigraphy

FACIES

"Layer cake" Geology and Facies Change

You will remember that many catastrophist geologists thought that the rock record had been laid down in uniform sheets over the whole world during the noachian deluge. If this had been the case, then the same rock layers would éverywhere be evenly distributed over the earth with little difference in lithology or thickness anywhere. This concept of the rock record is often called "layer cake" geology, and it is surprising how many geologists are still subject to this kind of thinking. The impressive sequences of rock layers in places such as the Grand Canyon do tempt us to imagine that such sequences might extend indefinitely without change in thickness or lithology, each layer representing the same period of time everywhere (Fig. 7.1).

One of the first applications of uniformitarianism to geologic thinking was a comparison of modern depositional conditions with their ancient rock equivalents. Any examination of modern depositional environments (except for pelagic marine deposits) shows that most do not extend for great distances laterally but eventually change into other depositional environments. Many

sedimentary rock types are being deposited simultaneously in different areas, and no single rock type is deposited over a very wide area at one time. Rock types are diagnostic of local environments, not of ages. These ideas first appear in the geologic literature in the late 1700s. For example, the famous French chemist Antoine Lavoisier (1743–1794) produced a diagram (Fig. 7.2) in 1789 that shows that gravels should occur near the shore but that fine clays could be expected far from shore. Other geologists had noticed that the lithology of formations changes over distance. In 1839, Sedgwick and Murchison named the Devonian because the rocks of Devonshire, though they were the same age as the Old Red Sandstone, were different from it in lithology.

Steno (1669) coined the term *facies* (Latin, "aspect" or "appearance") to mean the entire aspect of a part of the earth's surface during a certain interval of geologic time (Teichert, 1958). The Swiss geologist Amanz Gressly (1814–1865) first expounded the modern concept of facies in 1838. In his study of the Mesozoic sequences of the Jura Mountains of Switzerland (type area of the Jurassic), Gressly went beyond previous stratigraphers who named rock units based on local isolated sections. He traced rock units laterally and found that they changed in appearance. Gressly's use of the term *facies* designated lateral changes in appearance of

Figure 7.1

The panorama from Point Sublime in the Grand Canyon, looking east. Drawn by W. H. Holmes. From Dutton (1882). Courtesy U. S. Geological Survey.

a rock unit and emphasized the fact that rock units were not uniform in lithology but changed as their depositional environment had changed. Continental geologists subsequently standardized this meaning in their literature. According to Haug (1907), "a facies is the sum of lithologic and paleontologic characteristics of a deposit in a given place," and this definition prevails among many stratigraphers today. It implies that facies change over distance, as Gressly originally noted.

The word *facies* was so useful, and its use became so widespread, that its meaning broadened beyond Gressly's original intention. Outside Europe, it came to denote major stratigraphic sequences whereas Gressly's concept of facies was restricted to lateral changes within a single rock unit. For some geologists, the term referred to just about any similar association of rocks occurring in certain geographic, oceanographic, or inferred tectonic settings. It has even been used by metamorphic

Figure 7.2

Antoine Lavoisier's diagram of the relationships of coarse littoral (*Bancs Littoraux*) and finer pelagic (*Bancs Pelagiens*) sediments to the northern French coastline. Lavoisier recognized that gravel could be moved only by waves near the shore whereas fine sediments can be carried into deeper water. He also saw that distinctive organisms inhabited each environment. If sea level rose, flooding of the land (*la Mer montante*), both littoral and pelagic sediments would migrate landward. If sea level fell (*la Mer descendante*), they would shift seaward. From Lavoisier (1789).

petrologists to denote metamorphic rocks of similar assemblages of minerals (Eskola, 1915). At one time, there was much discussion about whether this broad usage of *facies* had made the term so vague as to be useless (Moore, 1949; Teichert, 1958; Weller, 1958; Markevich, 1960; Krumbein and Sloss, 1963).

In recent years, however, a broad definition along the lines of "a body of rock with specified characteristics" (Reading, 1986, p. 4) or "associations of sedimentary rock that share some aspect of appearance" (Blatt et al., 1980, p. 618) has prevailed and seems to be understood by most geologists. More specifically, a *lithofacies* denotes a consistent lithologic character within a formation (e.g., "shale lithofacies" or "evaporite lithofacies") and excludes its fossil character (which was included in Gressly's concept). Similarly, the term *biofacies* applies to a consistent biological character of a formation (e.g., "oyster bank facies"). In this sense, a biofacies is also an ecological association.

Because of geologists' emphasis on depositional environments during the last two decades, the word *facies* has become closely associated with genetic interpretations. Currently, it is most often used in the environmental sense, as the lithologic product of a depositional system such as "fluvial facies" or "shallow marine facies" (see, for example, Reading, 1986; Walker, 1984). This is the way we have used the word in the first two parts of this book. Most modern geologists prefer to keep this long established and useful word and to indicate

its meaning by context, rather than coin a new terminology to separate Gressly's meaning from broadly descriptive facies terms (e.g., "sandstone facies," strictly speaking a lithofacies) and from more interpretive terms (e.g., various environmental or tectonic facies). "Sandstone facies" is clearly a descriptive term, but "fluvial facies" is just as clearly an interpretive term indicating the product of a depositional environment.

Sedimentary environments that produce changes of facies occur in natural associations. The facies that are formed next to one another in a vertical section of rock will be the same as those found adjacent to one another in a modern depositional system. This fact was described by Johannes Walther (1894), who stated that "only those facies and facies-areas can be superimposed, primarily, which can be observed beside each other at the present time." The principle that *facies that occur in conformable vertical successions of strata also occurred in laterally adjacent environments* is known as *Walther's law of correlation of facies.*

Walther's law is a useful tool for stratigraphers. In many instances, the interpretation of the environments that deposited a few units will be clear, but a conformable unit between two of them, or one that lies above or below them, cannot be so easily interpreted. Walther's law can be used to limit the possible interpretations. For example, in the traditional model of the coal cyclothem sequence (Fig. 7.3), there are shales and limestones both above and below the coal. Environ-

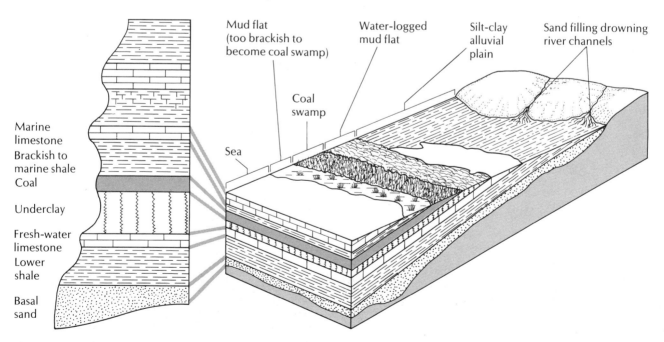

Figure 7.3

Diagram showing the traditional model for the ideal Pennsylvanian coal-bearing cyclothem of the Illinois Basin. The facies that occur laterally in adjacent sedimentary environments also occur in vertical succession, according to Walther's law. (Compare this model with the more recent interpretation of the cyclothem sequence in Fig. 7.7). From Shaw (1964).

mental associations and Walther's law suggest that the lower shale and limestone were formed in freshwater lakes or fluvial settings, but the upper shales and limestones were marine. Marine and nonmarine fossils in the limestones confirm this interpretation, but there are fewer fossils in the shale. Here Walther's law helps rule out unlikely interpretations.

Nevertheless, Walther's law must be used with caution. Environmental changes can be extremely rapid and leave no record, so there might be a gap in the sequence of facies. A classic example of such a gap is found in the Upper Devonian/Lower Mississippian Chattanooga-New Albany Shales found widely in the Appalachians and Midwest (Lineback, 1970). Because these shales overlie shallow marine carbonates, they had long been interpreted as shallow lagoonal deposits, despite many indicators of deepwater deposition. Closer examination showed that there is a subtle unconformity present between the shales and the carbonates (Cluff et al., 1981; Conkin et al., 1980), which is indicated only by pyrite

and phosphate mineralization. The Chattanooga-New Albany Shales were in fact deepwater, basinal shales, deposited unconformably on shallow marine limestones. The basin had dropped before the shales were deposited and had left a record of this rapid deepening only a few centimeters thick. The apparent "sequence" of these depositional environments is not a true conformable sequence after all because it is interrupted by a major unconformity. Walther's law does not apply across unconformities!

Transgression and Regression

Certain facies associations are common in the rock record. For example, most clastic shorelines show a series of depositional environments that are progressively finer-grained in the offshore direction (Fig. 7.4). This leads to facies belts of coarse sands and gravels (mostly fluvial in origin), finer sands (beach and shallow offshore

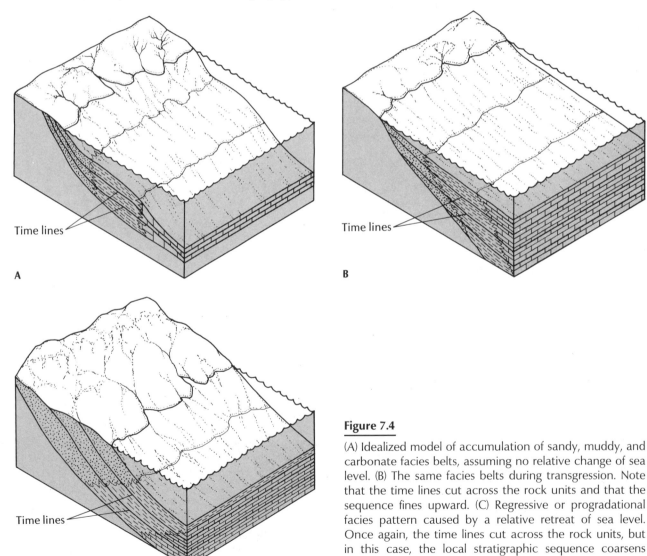

Figure 7.4

(A) Idealized model of accumulation of sandy, muddy, and carbonate facies belts, assuming no relative change of sea level. (B) The same facies belts during transgression. Note that the time lines cut across the rock units and that the sequence fines upward. (C) Regressive or progradational facies pattern caused by a relative retreat of sea level. Once again, the time lines cut across the rock units, but in this case, the local stratigraphic sequence coarsens upward.

environments), and silts and clays (shallow shelf muds, prodeltaic clays). If relative sea-level changes (either by eustatic sea-level change or by local uplift or downwarping), deposits of these facies belts accumulate. Three facies patterns are possible.

Facies belts pile up vertically (Fig. 7.4A) when the relative rate of sea-level rise is exactly balanced by the sedimentary output of the land, which causes the shoreline to remain in approximately the same place. This condition is so unusual that it is almost never encountered in the stratigraphic record.

Facies show a *transgressive* pattern (Fig. 7.4B) when the sediment supply is overwhelmed by a relative rise in sea level, which causes the shoreline to move landward.

Facies show the pattern in Fig. 7.4C when the shoreline moves seaward due either to excess sediment supply from land (*progradation*) or to a relative lowering of sea level (*regression*). Because it is difficult to tell whether sediment supply or sea-level change is primarily responsible, many geologists use the term *regression* to mean *any* movement of the shoreline away from the land.

Transgressive-regressive facies patterns form most of the marine stratigraphic record, so recognition and understanding of them are critical. Each process leaves a distinctive pattern of facies, both vertically and laterally. Because transgression brings deeper-water deposits landward with their progressively finer grain sizes, the stratigraphic column becomes *finer upward* at every point. Conversely, during regression or progradation, coarser, shallow-water deposits can build out (prograde) over the finer deposits of the formerly marine shelf; regressive sequences thus become *coarser upward*. There are many exceptions to this generalization, of course. We have seen that offshore sand bars do occur in the continental shelf (Chapter 5) in the midst of fine-grained shales. In many cases, however, transgression and regression can be inferred from a single vertical section by this simple rule of thumb.

Let us look at one more aspect of transgressive-regressive sedimentary packages. An ancient depositional surface of a given age existed in several different environments simultaneously. Thus, in the stratigraphic record, the nearshore sandstones are equivalent in age to some of the offshore shales and limestones. As the various transgressions and regressions moved in and out, they deposited a continuous spectrum of facies across these depositional surfaces. Thus, the *time lines* (which represent depositional surfaces of equal age) *cut across the facies boundaries*. The rock units or facies, such as the shale or sandstone, are of *different ages in different places*; that is, they are *time-transgressive* (Fig. 7.4).

Other lateral relationships between rock bodies are of a more descriptive nature. A rock body that gets thinner and thinner as it is traced laterally until it van-

ishes is said to *pinch out*. Pinchouts are important in oil geology because they are often natural *stratigraphic traps*. If a porous sandstone reservoir rock pinches out between two impermeable shales, the oil will be trapped at the thinnest part of the pinchout. Typically, the boundary between two laterally adjacent rock bodies is very jagged due to rapid changes in the shoreline; the result is a series of pinchouts that form an *intertonguing* pattern (Fig. 7.5). Rock bodies that do not have distinct boundaries between them are said to *grade* into one another.

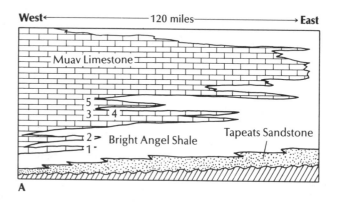

A

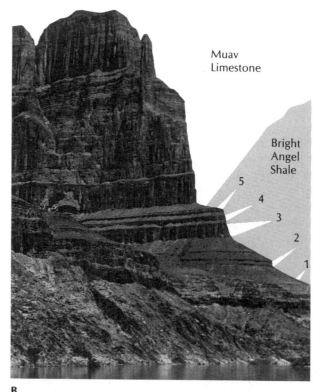

B

Figure 7.5

(A) Diagram summarizing the intertonguing relationships between Cambrian Bright Angel Shale and Muav Limestone in the western Grand Canyon. After McKee and Resser (1945). (B) Outcrop showing intertonguing; the shale members of the Bright Angel shale are shaded. Photo courtesy P. W. Huntoon.

Lateral gradation from porous sandstone to impermeable shale, if overlain by another impermeable layer, can also provide a stratigraphic trap. Oil accumulates where the zone of "shale-out" is relatively narrow.

Asymmetry of Transgressive and Regressive Cycles

The symmetrical facies record of marine transgression and regression (Fig. 7.6A) so commonly encountered in textbooks is now being reconsidered. For years, stratigraphers have noticed that the classic fining-upward sequence of transgressive facies is extremely rare and that coarsening-upward patterns are very common, even when sea level is rising. Fischer (1961) and Swift (1968) pointed out that during transgression, the reworking effect of the rising sea level tends to obliterate facies that had previously been deposited. The preservation of a transgressing shallow marine or coastal sequence is much less likely than was thought, because these deposits are reworked by shoreface and shallow marine processes before they are buried. As a result, unless subsidence is very rapid, transgression leaves a very thin sequence, making the sea level appear to have risen very rapidly.

Most of the deposits formed during a sea-level rise are regressive sequences that have built out and coarsened upward during the progradation of nearshore facies (Fig. 7.6B). These regressive sequences are less likely to be reworked before burial by later depositional processes and so are preserved more often. Therefore, frequent changes in relative sea level have an asymmetric effect. Minor episodes of regression during a large-scale transgression produce almost all the stratigraphic record. The result is an asymmetric cycle of progradational wedges during the transgressive phase and a similar sawtoothed pattern during the regressive phase (Fig. 7.6C).

This revised concept of how transgression and regression are recorded in the rock record has far-reaching implications. For example, the classic model of the fining-upward transgressive coal cyclothem of Illinois (see Fig. 7.3) may be obsolete and should be reconsidered. The traditional model placed the base of each cycle at the base of the channel sandstone, which represents a fluvial channel sand (Fig. 7.7 right-hand side); transgressive deposits covered the fluvial sands and muds with progressively deeper-water facies. If the revised concepts are correct, then the natural break in the sequence is not the channel sand but the first marine shale or limestone above the coal (Fig. 7.7 left-hand side). This represents the rapid transgression of the marine facies across the basin. The entire sequence of marine limestones and shales, deltaic distributary channel sands, and interdistributary subaerial shales and coals then follows.

This is exactly what one would expect of a prograding delta sequence. The cyclothem sequence, contrary to popular wisdom, is not a series of transgressive episodes interrupted by rapid sea-level falls, but of frequent thin, rapid transgressions followed by prograding deltaic basin fills (Fig. 7.8).

Although this asymmetrical model has gained widespread acceptance, it must be tested further because it does not explain all examples of apparently symmetrical transgressive-regressive cycles. Bourgeois (1980) has pointed out that transgressive shelf sequences are well documented on tectonically active margins, where rapid

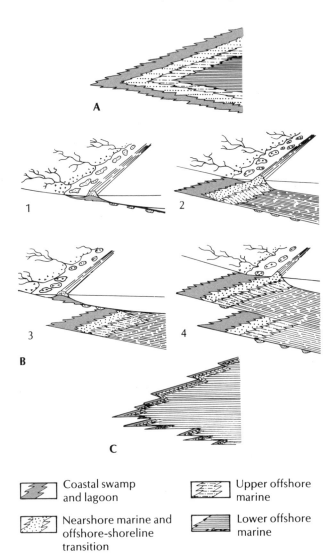

	Coastal swamp and lagoon		Upper offshore marine
	Nearshore marine and offshore-shoreline transition		Lower offshore marine

Figure 7.6

(A) Traditional symmetrical model of marine transgression and regression. (B) Stages of deposition of shoreline deposits, showing reworking and apparent rapid transgression followed by thick progradational sequences. (C) The pattern that results from this process. Even the basal transgressive sequence is made out of numerous small regressive wedges interrupted by thin, apparently rapid transgressions. From Ryer (1977).

subsidence of the basin produced transgression despite high rates of sedimentation. Most of the examples of thin transgressive sequences just discussed were from passive margins or from seaways, which have less subsidence and sediment supply.

Asymmetrical cycles also seem to have been produced in areas of relatively high sediment supply. For example, Brett and Baird (1986) found that the Devonian Hamilton Group of upstate New York showed both symmetrical and asymmetrical patterns in the same cy-

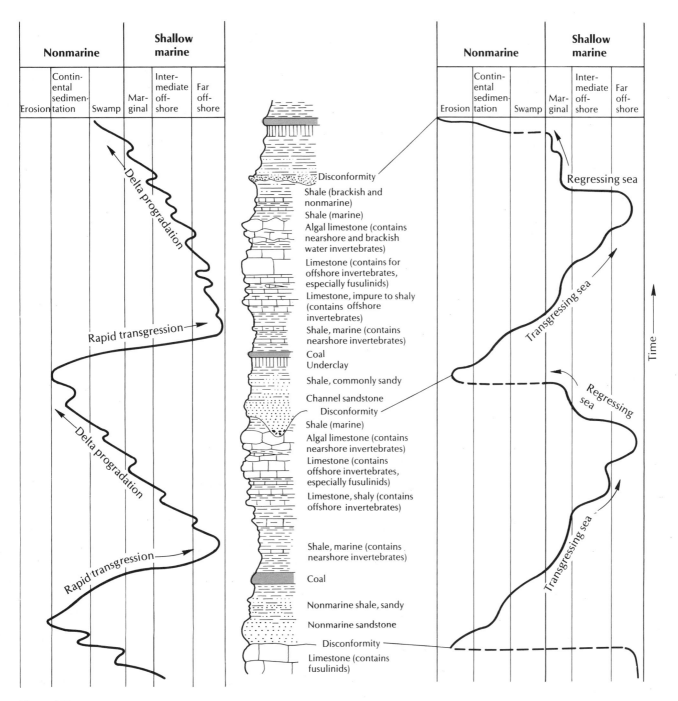

Figure 7.7

Classic Carboniferous cyclothem sequence with two facies interpretations that depend on whether one adopts the symmetrical or the asymmetrical transgression-regression model. On the right is the traditional model (after Crowell, 1978), which treats the disconformity beneath the channel sandstone as a rapid regression followed by slow transgression. On the left is a more modern interpretation (from Friedman and Sanders, 1978), which places a rapid transgression at the base of the first marine limestone and considers the rest of the sequence to be regressive, progradational delta front. Correlation of these coal-bearing cyclothems with the more open marine cyclothems support the latter model (see Fig. 7.8).

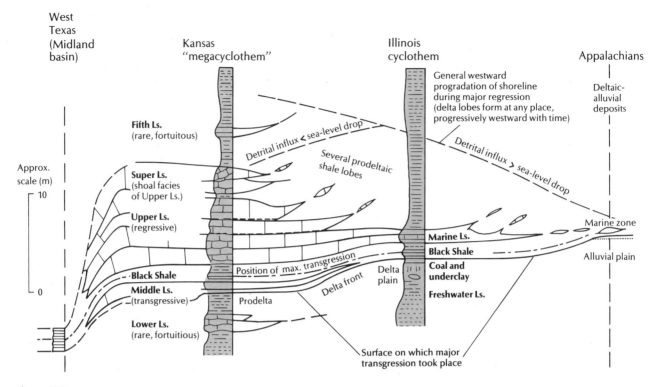

Figure 7.8
Generalized restored cross section of typical Upper Pennsylvanian eustatic cyclothem along the axis of the midcontinent seaway, showing relations between members and facies between Kansas and Illinois cyclothems and with marine deposits in the Appalachians and west Texas. The correlation clearly shows that transgression is relatively abrupt and that most of the sequence is regressive and progradational after the apparently rapid transgression (see Fig. 7.6). After Heckel (1977, 1980).

cles. The more onshore deposits are asymmetrical and shallow upward as sediment aggraded upward to the limit imposed by wave energy, in a region of abundant sediment supply. When sea level rose higher, the initial sediment starvation was followed by inundation and deep-water sedimentation. On the other hand, the offshore shales show symmetrical cycles and thinner deposits because their sediment supply was always relatively low. The geologist must keep all of these possibilities in mind when examining the local outcrops for evidence of transgression or regression.

A FRAMEWORK FOR ACCUMULATION

We have seen that sediments can accumulate due to relative changes in sediment supply or sea level, but what conditions are necessary for their accumulation and preservation in the rock record? How complete a record do they preserve? Standing on the rim of the Grand Canyon, you see below you a sequence of sediments almost a mile thick (Fig. 7.1). The casual tourist might

get the impression that all of geologic time is represented by some sort of sediment in the walls of the canyon. It seems natural to assume that such sediments are continuously accumulating in the permanent rock record, because sedimentation goes on all around us.

Yet a more critical look around the canyon suggests that the vast sedimentary sequence is visible only because the Colorado River has carved the canyon and carried away all of the sediments that once filled it. During a raft trip down the Grand Canyon, you pass rockfalls and landslides, shoot rapids over and around boulders that have fallen into the river and gulp mouthfuls of Colorado River water, muddy and gritty with sediments that are being actively transported to the Gulf of California. Indeed, the entire Colorado Plateau is a region of rapid uplift and erosion, which is the reason for the depth of the gorge in the first place. Very little permanent sedimentary record is accumulating there to represent the Holocene.

Most of the terrestrial environments discussed in Chapters 3 and 4 are subject to a similar fate. Any feature above sea level is doomed to be eroded eventually, though this fate may be forestalled temporarily if the feature is uplifted rapidly enough. Any sediments that accumulate above sea level are fated to be reworked and redeposited somewhere downhill or downstream unless the proper circumstances entomb them. The term *base level of erosion* has long been used (Powell, 1875; Gilbert, 1914) to describe the level below which erosion (chiefly stream erosion) cannot occur. In most cases, base level is sea level, though mountain lakes and other ephemeral features can temporarily establish local base levels that are above sea level. Joseph Barrell (1917) applied this concept in a broader sense than his predecessors. To him, base level was the level above which sediments cannot accumulate permanently. Dunbar and Rodgers (1957) called this the *base level of aggradation*. It is an imaginary surface of equilibrium between the forces of erosion and deposition. Barrell's definition emphasizes the fact that most terrestrial sedimentary environments are above base level and will ultimately be reworked or eroded. Base level was developed as a concept to explain why terrestrial deposits add relatively little to the sedimentary record and have low preservation potential. The main exception to this generalization occurs in nonmarine basins that drop down so far that they accumulate a thick enough pile of sediment and establish their own local base level.

Because sedimentary environments below base level are much more likely to be preserved, we would expect marine sedimentary deposits to dominate the stratigraphic record, and they do (Fig. 7.9). Still, the shallow marine environment is subject to fluctuations in sea level and to strong coastal currents, so erosion also can prevail there. The lowest point below base level is, of course,

the deep marine environment. Here the sedimentary record is the most continuous and has the best chance of preservation. Yet the deep-sea record has limitations, too. Rates of sedimentation tend to be extremely slow, so the strata are very thin; deep-sea currents occur that can locally scour or dissolve the sea floor sediment; and most of the deep-sea record has not been uplifted onto the land where it could be studied by geologists. Only the growth of marine geology in the last two decades has enabled us to study this superlative sedimentary sequence for the first time.

In summary, preservation of the sedimentary record is the rare exception rather than the rule. Indeed, special circumstances are needed for sediments to become part of the permanent rock record. Every stratigraphic sequence we examine is composed of small packages of sediment that have been preserved only by luck; the other sediments formed at the same time were destroyed. A rise in sea level helps preserve many packages of sediment, but ultimately, preservation depends on one thing: *net subsidence*. The rate of basin subsidence must exceed the rate of sea-level drop or of erosion so that the subsiding basin remains below base level until— again by luck—it is exposed for the geologist to examine.

Many of the best examples of nonmarine depositional sequences discussed in Chapter 3 were trapped in rapidly down-dropped basins, especially fault grabens. Rising sea level can build up a series of deltaic, barrier-island, and shallow-shelf sediments, which have a good chance of preservation if they sink into some sort of basin before the next sea-level drop can expose them to erosion.

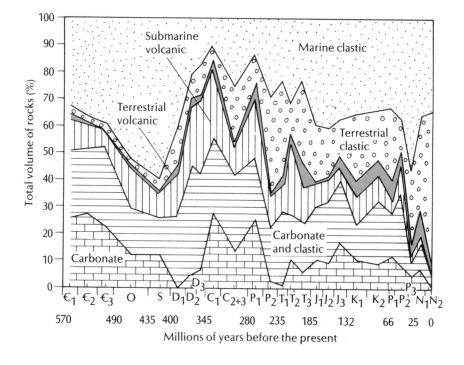

Figure 7.9

Relative percentages of various lithofacies through the Phanerozoic. Notice that marine clastic and carbonate rocks typically make up 60–80 percent of the rock record through geologic history. After Ronov et al. (1980).

GAPS IN THE RECORD
Diastems and the Completeness of the Rock Record

In the early 1900s, the discovery of radiometric dating forced stratigraphers to examine the discordance between the radiometric age of the earth and the estimates based on the thickness of the sedimentary record. If radiometric dates were correct, then the average rate of accumulation of sediments was of the order of meters per thousands of years. Yet studies of modern rates of sedimentation indicated that typical sedimentation rates are on the order of meters per year to meters per hundreds of years, implying that the stratigraphic record is very incomplete. To use Ager's phrase (1981) it is "more gaps than record." There must be far more unrecorded time than is inferred from the major unconformities.

In 1917, Joseph Barrell first articulated the idea that every bedding plane could account for some of this unrepresented time. He came to the conclusion that there are many small-scale, obscure unconformities in the stratigraphic record, which he called *diastems*. Barrell used the concept of base level to explain the intermittent

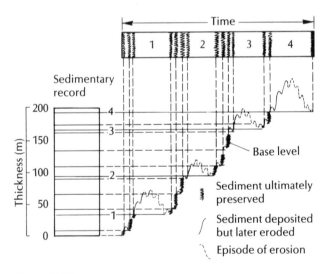

Figure 7.10

Barrell diagram (after Barrell, 1917) showing the relationship between the sedimentary record and the actual time represented by it. Because base level fluctuates up and down, most time is not represented by the sedimentary record. Some sediments are never deposited because the basins have a low base level (dashed lines), and others are destroyed by erosion after being deposited (solid lines). Only during rises in base level that are not immediately followed by erosion does sedimentary record accumulate (wiggly lines), forming a temporal record (top) made up of only these relatively short intervals. Barrell concluded that there must be many cryptic unconformities between bedding planes, which he called *diastems*. Major unconformities are numbered 1–4.

nature of sedimentation. As sediments are washed toward the sea, they move constantly into deeper water until they reach depths (according to grain size) at which water energy is no longer sufficient to transport them. Below base level, sediment will accumulate (and be preserved) whenever it is available; but above this level, it cannot accumulate permanently and will eventually be transported further until it reaches a site below base level. However, base level is not always fixed but fluctuates due to grain size, scouring bottom currents, changes in wave base, and marine slumping.

Base level also fluctuates with changes in the environment and the basin. The history of an aggrading sedimentary basin might at first glance appear to be a continuous rise in base level as the basin is filled (Fig. 7.10). Superimposed on this upward trend, however, are oscillations of eustatic sea level as well as even smaller-scale local trends that temporarily lower base level and cause erosion. The net result is that sediment can accumulate and be preserved only during part of the net aggradational phase of the major oscillations. At all other times, there is either erosion of the recently accumulated record or nondeposition. The effect of this oscillation of base level is that time is represented sporadically. Out of the total span of time, only short intervals had a chance to be preserved. Thus, the sedimentary record is not a continuous record of the earth's history but is a series of thick packages representing very small portions of the total time elapsed.

The concept of base shifting level has also been applied to nonmarine environments. For example, sediments can be trapped in a terrestrial basin and can remain there until the basin is filled and buried. In this case, the basin floor is a local base level. Although such basins are above the ultimate base level near sea level, a local disequilibrium can be maintained over the short term. These terrestrial deposits show the same episodic accumulation as those formed in the marine environment. Loope (1985) has drawn a Barrell diagram (Fig. 7.11) for the Permian dune deposits of the Cedar Mesa Sandstone of the Canyonlands of Utah. What appears to be a continuous record of eolian deposition is actually the record of a few episodes of dune migration, triggered by short regressive events in the adjacent marine environment. The bulk of geologic time is represented by the bedding planes between dune sequences, when episodes of deflation and soil formation eroded and stabilized the dune sequence.

Unconformities

Gaps in the rock record come in many different sizes, from undetectable bedding-plane diastems to huge unconformities. Dunbar and Rodgers (1957) defined an

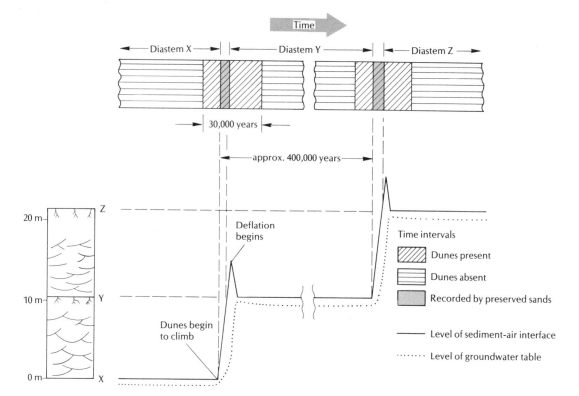

Figure 7.11

Modern application of the Barrell diagram to the Cedar Mesa Sandstone of Utah, an eolian deposit. Preservation of sediments occurs only while the sand dunes climb; after the dune stabilizes with a high groundwater table, no further accumulation occurs, and deflation begins. From Loope (1985).

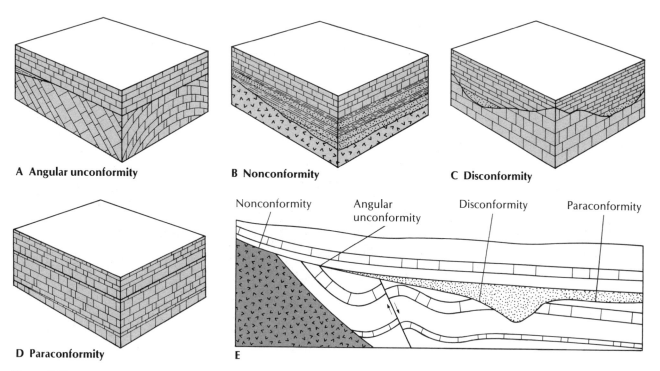

A Angular unconformity   **B Nonconformity**   **C Disconformity**

D Paraconformity   **E**

Figure 7.12

Four major types of unconformity. (A) Angular unconformity between tilted and undeformed sediments. (B) Nonconformity between crystalline (igneous or metamorphic) and sedimentary rocks. (C) Disconformity between two parallel bodies of sediment, but with some evidence gap of an erosional between them, such as a channel or other erosional surface. (D) Paraconformity, in which the sediments are parallel but there is no direct physical evidence of erosion; the unconformity is detected by determining the ages of the existing units (usually by biostratigraphy). (E) Changing character of unconformities over distance if the younger unit laps across different types of bedrock. After Dunbar and Rodgers (1957).

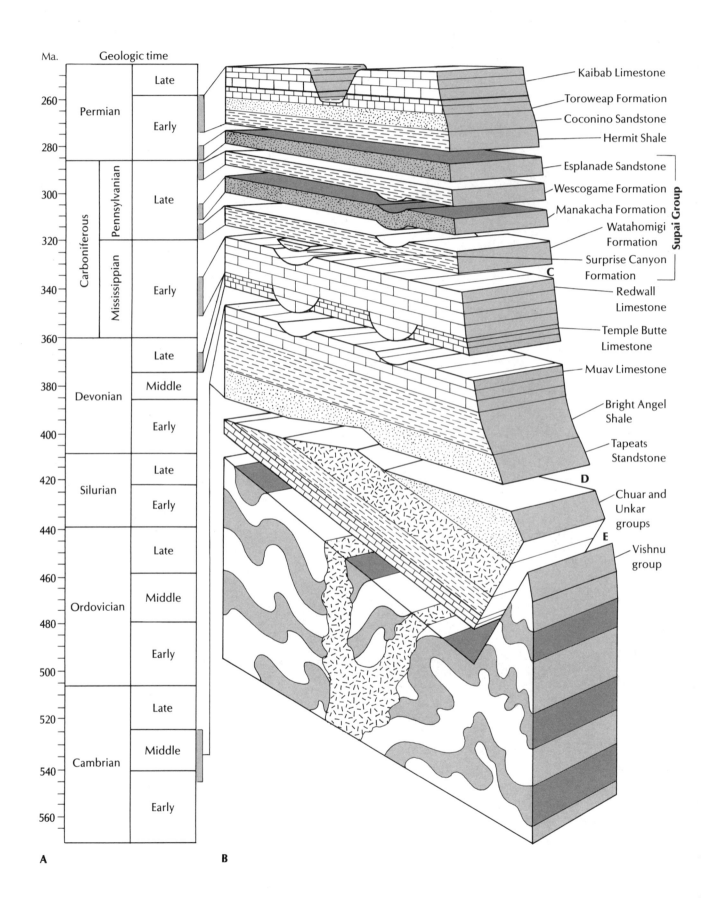

Ma.   Geologic time

260   Permian   Late
                Early
280
              Pennsylvanian   Late
300   Carboniferous
320   Mississippian   Early
340
360   Devonian   Late
                  Middle
380
              Early
400
420   Silurian   Late
                  Early
440
              Late
460   Ordovician   Middle
480
              Early
500
520   Cambrian   Late
                  Middle
540
              Early
560

A   B

Kaibab Limestone
Toroweap Formation
Coconino Sandstone
Hermit Shale
Esplanade Sandstone
Wescogame Formation   Supai Group
Manakacha Formation
Watahomigi Formation
Surprise Canyon Formation
Redwall Limestone
Temple Butte Limestone
Muav Limestone
Bright Angel Shale
Tapeats Standstone
Chuar and Unkar groups
Vishnu group

C
D
E

C

D

E

Figure 7.13

Although the Grand Canyon section may appear to the casual viewer to be a complete record of geologic time, it is remarkably incomplete. (A) The actual temporal span of the Paleozoic units of the Grand Canyon that are shown in B; less than 10 percent of Paleozoic time is actually represented, including part of the Middle Cambrian, a bit of Late Devonian, some Early Mississippian, and three short slices of the Pennsylvanian but no Ordovician or Silurian. Most of the Early Permian, but no Late Permian, is represented. (C) The disconformity between the Redwall Limestone and the overlying Pennsylvanian rocks shows a clear paleovalley fill. (D) The angular unconformity between the tilted Precambrian sediments and the overlying Cambrian deposits is shown. Some of the Precambrian units formed resistant paleotopography in the Cambrian, so that the Cambrian deposits lap over them irregularly. (E) Elsewhere in the Grand Canyon, the Cambrian sediments lap nonconformably over Precambrian schists and granites (rocks with vertical foliation). Photo C by E. D. McKee, photo D by N. W. Carkhuff; courtesy U.S. Geological Survey. Photo E by the author.

unconformity as "a temporal break in a stratigraphic sequence resulting from a change in regimen that caused deposition to cease for a considerable span of time. It normally implies uplift and erosion with the loss of some of the previously formed record." Unconformities are of four types (Fig. 7.12):

Angular unconformity This familiar type of unconformity describes the situation in which the sequence beneath the unconformity is tilted at some angle to the strata above it.

Nonconformity The erosional surface truncates massive plutonic igneous or metamorphic rocks and is covered by sediment.

Disconformity Sedimentary sequences above and below the unconformity are parallel, but there is an erosional surface between them (e.g., a river channel base, an ancient karst surface, or a soil horizon).

Paraconformity (obscure unconformity) The sequences above and below the unconformity are parallel, but there is a normal bedding plane contact and no obvious erosional surface between them. In this case, the unconformity is recognized by other evidence; such as fossils, which indicate that the beds are of significantly different ages. Some authors doubt the need for the paraconformity category, arguing that most reputed paraconformities, if traced laterally, eventually exhibit some physical evidence of erosion and thus prove actually to be disconformities (Davis, 1983).

The Grand Canyon section illustrates these categories (Fig. 7.13). For example, all of the late Proterozoic and Paleozoic sedimentary rocks lie nonconformably on Precambrian schists and granites (Fig. 7.13E). The late Proterozoic Unkar and Chuar Groups are tilted and overlain by the Cambrian sequence in an angular unconformity (Fig. 7.13D). The thick Cambrian succession of Tapeats Sandstone, Bright Angel Shale, and Muav Limestone is overlain by a disconformity that spans the entire Ordovician, Silurian, and most of the Devonian (Fig. 7.13A). Channeled into the Muav Limestone are thin sequences of the Devonian Temple Butte Limestone (Fig. 7.13B), which in turn were eroded and filled in by the Mississippian Redwall Limestone. The recently named Pennsylvanian Surprise Canyon Formation (Fig. 7.13C) is disconformably eroded into the Redwall Limestone. There are many other gaps in the remaining Pennsylvanian and Permian parts of the sequence (Figs. 7.13A,B), some of which have been considered to be paraconformities. Most, however, show some evidence of erosional breaks and are disconformities. This apparently complete record, as impressive as it appears, is indeed "more gaps than record." As the geologic time column (Figs. 7.13A,B) shows, only small parts of the late Cambrian, late Devonian, mid-Mississippian, and Permo-Pennsylvanian are represented. Most of the Proterozoic,

all of the Ordovician and Silurian, and all of the Mesozoic and Cenozoic have been eroded away or were never
deposited.

The most important criteria for recognizing unconformities in the field are usually sedimentary in nature.
These include a basal conglomerate above the unconformity; a deeply weathered erosional surface, sometimes with a clear soil horizon; truncation of bedding
planes or of clasts in the lower sequence; surfaces that
have been postdepositionally altered by burrowing,
boring, or cementation into hardgrounds; and surfaces
that show relief due to an erosional episode.

Unconformities can also be recognized from paleontological clues. For example, if one or more faunal zones
is missing, an unconformity is present. Many authors
have suggested that gaps in the evolution of a lineage
are evidence for an unconformity. Recently, however,
paleontologists (Eldredge and Gould, 1977) have argued
that the gaps may be due to rapid evolution of lineages,
not missing sections. If several lineages show an abrupt
change at the same level, however, it is likely that the
gap is due to an unconformity rather than to simultaneous rapid evolution.

Structural criteria can also help in recognizing an
unconformity. Strata are not the only geologic phenomena that are truncated by an erosional surface; occasionally, dikes or faults are truncated as well. If the dips
above and below a surface are greatly discordant, the
surface is an angular unconformity.

All of these criteria must be used with caution. Large-
scale dune cross-beds, for example, may give the appearance of an angular unconformity in a local outcrop (Fig.
7.14). If the dipping flank beds of a reef or the dipping
foreset beds of a delta are truncated by a horizontal
erosional surface, the disconformity can appear to be an
angular unconformity. Compaction of saturated sediments can cause soft sediment deformation between
layers, which gives the appearance of structural deformation. In a limited outcrop, the discordance between
the deformed and undeformed layers might be erro-
neously interpreted as an unconformity.

Obviously then, regional studies are necessary for
deciding whether an apparent erosional surface is an
unconformity and, if so, how significant it is. In some
cases, regional studies reveal that a disconformity
changes into a nonconformity or angular unconformity
over distance (see Fig. 7.12E). In the Grand Canyon, the
Cambrian Tapeats Sandstone laps nonconformably over
the basement schists (see Fig. 7.13E) and forms an angular unconformity with the tilted Proterozoic metasedi-
ments (see Fig. 7.13D). In other cases, a conformable
sequence changes into an unconformity over distance,
especially as it thins out toward the margin of a basin
(Fig. 7.15A). Although relatively continuous deposition
takes place in sequences A and B in the center of the
basin through times T_1 to T_{10}, the margin of the basin

is subject to rapid fluctuations in base level (Fig. 7.15B).
At about time T_5, base level begins to move downward
and seaward (i.e., to the right), so that deposition of
sequence A ceases on the basin margin. When base level
reaches point P at time T_7, it begins to move upward
and landward again, preventing strata B_1 and B_2 from
being deposited toward the margin of the basin. Also
at time T_7, there is slight uplift of the landward part of
sequence A, so that strata E_6 through E_9 are progres-

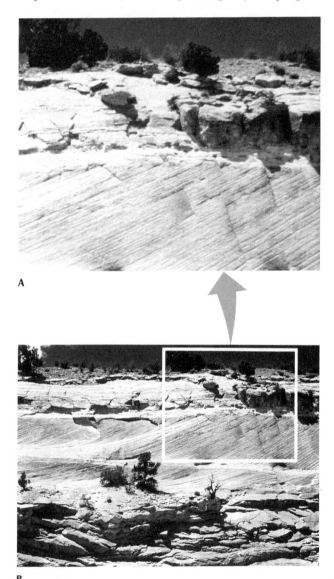

A

B

Figure 7.14

What appears to be an angular unconformity in the
close-up view of a limited outcrop (top) turns out to be a
disconformity between steeply cross-bedded eolian deposits of the Navajo Sandstone and the overlying horizontally bedded Carmel Formation (bottom), 3 kilometers west of
Boulder, Utah. If the outcrop were very limited, it would
be easy to postulate episodes of uplift and tilting that never
happened. Photos by the author, based on an idea from
Dunbar and Rodgers (1957).

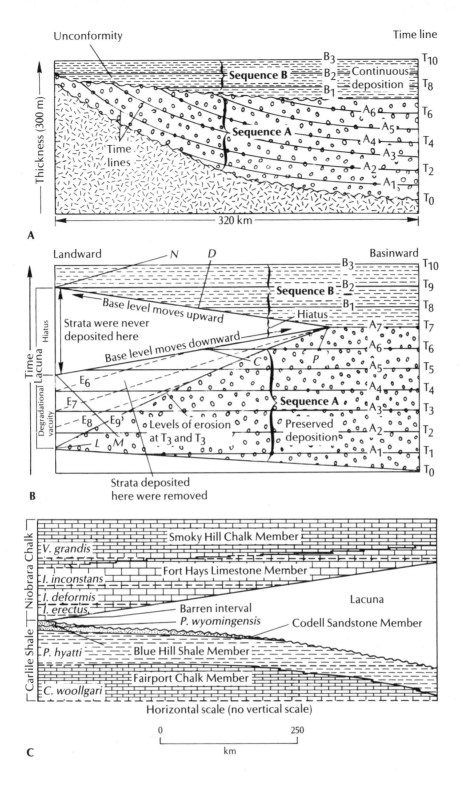

Figure 7.15

The origin and history of an unconformity is revealed in cross sections at the edge of a sedimentary basin. (A) Physical relationships between hypothetical rock units, in which sequence A rests nonconformably on eroded igneous rocks and is in turn overlain by sequence B. Note that there is continuous deposition between A and B in the basin center (right), but an unconformity develops toward the basin margin. (B) The thickness axis is replaced by a time axis, so the curved time lines above become straight. As deposition of sequence A continues without interruption from time T_0 to T_7 in the center of the basin, uplift on the basin margin moves the base level downward and erodes strata along erosional surfaces E_6 through E_9. This produces a degradational vacuity. At point P, the base level reverses direction and starts to rise again. However, little of sequence B is preserved because its base level starts out so low. This produces a hiatus (Wheeler, 1964). The combination of a degradational vacuity and hiatus produces a lacuna (Wheeler, 1964). (C) Example of a stratigraphic lacuna from the Cretaceous of Kansas. Time planes are provided by biostratigraphic zone boundaries. The stratigraphic gap becomes progressively larger toward the right. After Hattin (1975).

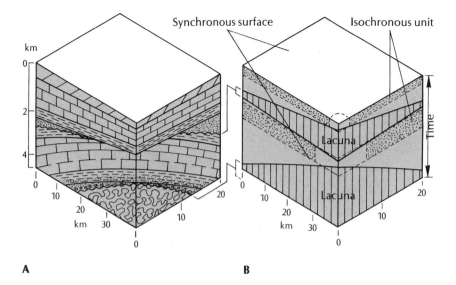

Figure 7.16

Diagram showing two unconformity-bounded successions of strata overlying crystalline basement rock. All dimensions of the block are real thicknesses. (B) The same relationships in an area-time diagram, in which the vertical axis is time rather than stratigraphic thickness. Vertically ruled areas represent lacunae of erosion and nondeposition at the position of the unconformities in A. The preserved succession can be divided into isochronous units bounded by time-parallel synchronous surfaces. From Krumbein and Sloss (1963).

sively removed by erosion. Eventually, depositional surface NP comes to lie directly on erosional surface LP. Between them is a large time gap in the rock record (a *lacuna*), composed of time intervals when strata were never deposited (a *hiatus* in the sense of Wheeler, 1964) and when strata were removed after deposition (the *degradational vacuity* of Wheeler, 1964). Figure 7.15C shows an example of a lacuna composed of both a hiatus in the Niobrara Chalk and a degradational vacuity in the underlying Carlile Shale.

The preceding idea can be visualized by imagining a hypothetical three-dimensional package of rock with unconformities (Fig. 7.16A). If the vertical axis is converted from rock thickness to time, then most of this "area-time continuum" is represented by lacunae rather than by rock (Fig. 7.16B). The rock packages are bounded by unconformities, which are *diachronous* (transgress time) in this example. Within the rock packages, however, are sequences of rock that have time significance. Rock units (or anything else) that occupy the same interval of time are *isochronous*; events that are short and simultaneous are *synchronous*. Isochronous units are bounded by hypothetical synchronous surfaces. We explore the significance of time and rock units in Chapter 8.

Catastrophic Uniformitarianism

In his provocative book *The Nature of the Stratigraphical Record* (1981), Derek Ager debunks many working assumptions about the rock record that are common among twentieth-century geologists, especially the "myth of continuous sedimentation." Areas that have apparently continuous records also have very low rates of accumulation (millimeters per year). At these rates, it would take 200 years to bury a typical *Micraster* sea urchin found in the Cretaceous chalks. The continuity in such stratigraphic sections is clearly an illusion be-

cause sections that span the same time interval in other basins are often orders of magnitude thicker. One centimeter in a so-called continuous section can be represented by tens of meters of section in some other basin. Each interval of time is represented by a very thin section or, more likely, by no rock at all. Ager shows that even "classic" sections have obvious and obscure unconformities, and he presents an unconventional picture of what even the "best" sections represent (Fig. 7.17). His metaphor for the accumulation of the sedimentary record is not the conventional "gentle, continuous rain from heaven" but is more like the child's definition of a net: "a lot of holes tied together with string. The stratigraphical record is a lot of holes tied together with sediment" (Ager, 1981, p. 35). It is like the life of a soldier, "long periods of boredom and brief periods of terror" (Ager, 1981, pp. 106–107).

Part of the reason for the discontinuities in the record is the nature of sedimentation itself. The extreme gradualist bias that geologists inherited from Lyell caused them to over-emphasize the gradual accumulation of sediment from daily processes and to ignore the large-scale catastrophic events. Perhaps the only place where sedimentation bears any resemblance to the conventional concept of "gentle, continuous rain from heaven" is the deep-sea floor, where planktonic microfossils and clay particles rain down continuously from the surface waters. Even here, though, scouring by bottom currents and dissolution due to oceanographic changes are taking place. In terrestrial and shallow marine realms, permanent preservation occurs only when a local basin drops below the base level. Most sedimentary environments do not show permanent net accumulation of sediment except those that have a high chance of preservation, such as deltas, downdropping basins and grabens, and growing reef complexes. The importance of gradual accumulation is probably overemphasized because rates of accumulation are so low. Ager states that gradualistic biases have

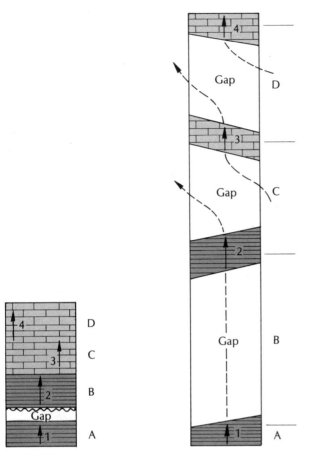

Figure 7.17

Comparison of the conventional picture of a part of the stratigraphic record (left) with the probable true picture (right). These gaps also conceal the immigration and emigration of fossil taxa (1, 2, 3, 4), giving the impression that they range continuously through time (left). From Ager (1981).

led geologists to neglect the importance of rare catastrophic events. For example, during normal conditions, the Gulf of Mexico coastal region accumulates sediment at an average rate of 10 centimeters per thousand years. Yet any given segment of coast has a 95 percent probability of being hit every 3000 years by a major hurricane, which scours much deeper than the 30 centimeters accumulated since the last hurricane. Much of the stratigraphic record in this region may be accumulated storm deposits. The daily, uniformitarian processes of the marine environment only serve to rework the top of the last storm deposit, and these efforts are usually wiped out by the next storm. Ager cites many other examples of catastrophic events that have disproportionate effects on the stratigraphic record. Among the best-known examples are the turbidites and olistostromes (discussed in Chapter 5), which slump or flow down the continental slope in a matter of minutes or hours, but make up most of the stratigraphic record of those marine environments. As Ager (1980, p. 50) points out, "the periodic catastrophic event may have more effect than vast periods of gradual evolution." He calls this the Phenomenon of Quantum Sedimentation, or *catastrophic uniformitarianism*. It is uniformitarian in that natural laws and processes are invoked (not the supernatural processes of biblical catastrophists), but it denies Lyell's gradualism of rates.

Dott (1983) suggested that the term *episodic sedimentation* be used to avoid confusion with the supernatural catastrophism. Ager (1981) emphasizes the importance of rare, high-energy events, but Dott points out that even normal fair-weather processes are inherently discontinuous and episodic. For example, a prograding shoreline (Fig. 7.18A) deposits a relatively continuous

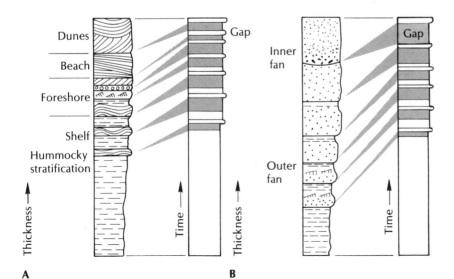

Figure 7.18

Facies models of (A) a prograding shoreline and (B) a subsea fan, replotted to emphasize the importance of gaps in deposition, as shown by the time plots to the right of each column. Notice that the gaps increase toward the top of each section. From Dott (1983).

sequence offshore; but as the shelf, fore-shore, beach, and dunes make their way across a given area, they are subject to more energetic and variable conditions. Hence, their record is more episodic. Similarly, the distal parts of the submarine fan (Fig. 7.18B) are influenced by the almost steady rain of pelagic sedimentation. However, channels of the inner fan are products of episodic turbidity-current activity, so there are significant gaps between them. The geologist must be alert to possible gaps wherever there is a small break in the bedding of an outcrop. Continuity may be an illusion.

Stratigraphic Completeness and Resolution

The implications of the incompleteness of the stratigraphic record have been discussed and quantified by Sadler (1981; Sadler and Dingus, 1982; Anders, Krueger, and Sadler, 1987). Sadler compiled nearly 25,000 recorded rates of sediment accumulation representing nearly every type of depositional environment. He found an inverse relationship between rate of accumulation and the time span represented by the section (Fig. 7.19). In

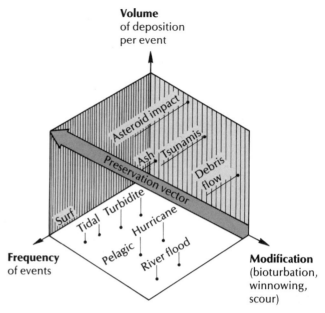

Figure 7.20

Three-dimensional conceptual graph of relationships among volume of deposition, frequency of events, and postdepositional modification, with 10 examples of processes plotted. Preservation potential increases along the thick vector shown. For example, river floods are frequent, but their deposits are low in volume and highly likely to be reworked, so their preservation potential is low. Debris flows are high in volume but are very rare and also likely to be reworked, so they are rarely preserved. Turbidites, on the other hand, are frequent, have moderately high volumes, and are seldom reworked, so they have high preservation potential. From Dott (1983).

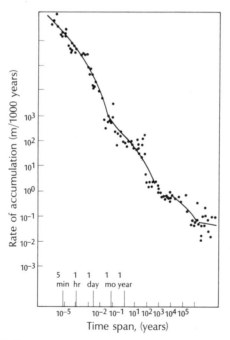

Figure 7.19

Typical accumulation rates for fluvial sediments. Note that the more rapid the rate of accumulation, the shorter the time span that sediment represents. This tradeoff means that sections that have high rates of sedimentation (and high resolution) have a limited temporal span and/or large unconformities. Sections with low sedimentation rates may have low resolution, but they span longer intervals of time and are thus more "complete" in this sense. From Sadler and Dingus (1982).

other words, the higher the rate of accumulation, the less total time can be represented by the section. This has important implications for any study that attempts to resolve time very finely. For example, in nearly complete sections of deep-sea sediments, spanning millions of years, each centimeter of sediment represents several thousand years. Events 5000 years apart can be resolved easily, but the odds of resolving events 1000 years apart are only 2 in 3, and the odds for 100-year resolution is only 1 in 30. Terrestrial sections usually span less total time and have much higher, more episodic sediment-accumulation rates; the resolution of events is therefore much coarser. For example, in the Eocene Willwood Formation of Wyoming, events 500,000 years apart can be resolved easily, but events 10,000 years apart have only a 1 in 2 chance of being resolved. For resolving 1000-year intervals, the odds drop to 1 in 9 and for 10-year intervals to only 1 in 256. Clearly, studies that depend on resolving small intervals of time may be completely impossible in many sections that were once thought to be fairly complete.

This tradeoff between thickness of section, completeness, and total time represented is a consequence of the interaction of three factors (Dott, 1983): frequency of event, volume of deposition per event, and modification by bioturbation, winnowing, or scouring (Fig. 7.20). Some events have tremendous energy and depositional volume but are infrequent (e.g., asteroid impacts). Most terrestrial events have small depositional volumes and energies, but some happen frequently (e.g., tides and floods), and some are exposed to environments in which they are likely to be reworked and modified (e.g., floodplain muds and subaerial debris flows). Those events that are infrequent, have low volumes of sediment, and are likely to be modified have the least chance of preservation in the rock record. They are shown at the low end of the "preservation vector." Events that occur frequently, have large volumes of sediment, and are unlikely to be modified have the best chance of becoming significant parts of the stratigraphic record. Few natural events occur at the optimum point of the preservation vector, but some, such as tidal and turbidite deposits, come close. From this, it is not surprising that clastic marine deposits (see Fig. 7.8) make up 30 to 50 percent of the volume of the stratigraphic record throughout geologic time.

FOR FURTHER READING

Ager, D. V., 1981. *The Nature of the Stratigraphical Record*, 2d ed. John Wiley & Sons, New York. A provocative, amusing, and highly readable "ideas" book, which has profoundly influenced modern stratigraphic thinking. "Must" reading.

Dott, R. H., Jr., 1983. "Episodic Sedimentation—How Normal Is Average? How Rare is Rare? Does It Matter?" *Journal of Sedimentary Petrology*, 53:5–23. Humorous, lively, but thought-provoking essay on the nature of episodic sedimentation and its implications for our treatment of the stratigraphic record.

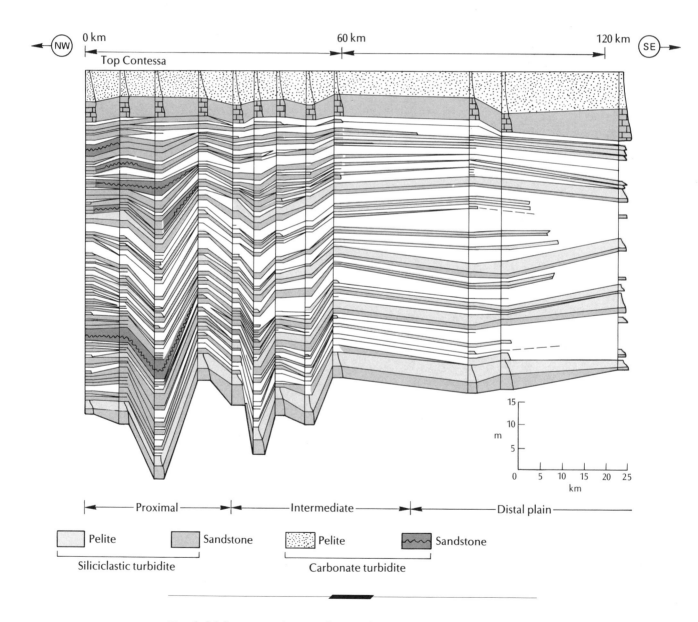

Detailed lithostratigraphic correlation of basin plain turbidites, from the Cenozoic of Italy. From Ricci Lucchi and Valmori (1980).

8

Lithologic Correlation

CORRELATION

As we saw in Chapter 7, there is no place on the surface of the earth where the entire geologic record is represented by a single section. Even classic exposures such as the Grand Canyon lack all but small parts of the Precambrian and Paleozoic (Fig. 7.13). The classic stratigraphic column of England and Wales (see Fig. 1.6) is slightly more complete, but many important intervals are missing. In most areas, the stratigraphic record must be patched together from many short, local sections. The process of demonstrating the equivalence or correspondence of geographically separated parts of a geologic unit is called *correlation*.

The word *correlation*, like the word *facies*, has had a number of contradictory and confusing meanings. Before the 1960s, geologists frequently used the term *correlation* to mean *time-equivalency*. As we shall see, geologists today think that very few rock units are time-equivalent over significant distances. To avoid the problems of proving time-equivalency, most modern geologists define correlation as equivalency of lithologic units, without time implication. Usually it is clear from context

whether lithologic correlation or time correlation is meant. In this chapter, we concentrate on lithologic correlation.

Establishing lithostratigraphic correlation can be simple but is usually very complex. The simplest method is to establish physical continuity between exposures, either by "walking it out" in the field or by tracing it on maps or cross sections. However, outcrops rarely enable us to trace beds continuously, so other procedures are necessary. Some rock units have distinctive diagnostic features that make them easy to recognize in different outcrops. If there is a strong lithologic similarity between two discontinuous exposures that lie in the same position in the sequence, they can be tentatively correlated. Sometimes the sequence itself is distinctive enough that clear correlations can be made, even though some units may no longer show lithologic similarity because of facies change. In this case, their position in the sequence is adequate for correlation (Fig. 8.1).

Other criteria have been used in the absence of better means of correlation. Unconformities can be obvious markers that span long distances, and there is evidence (discussed later) that many regional unconformities are

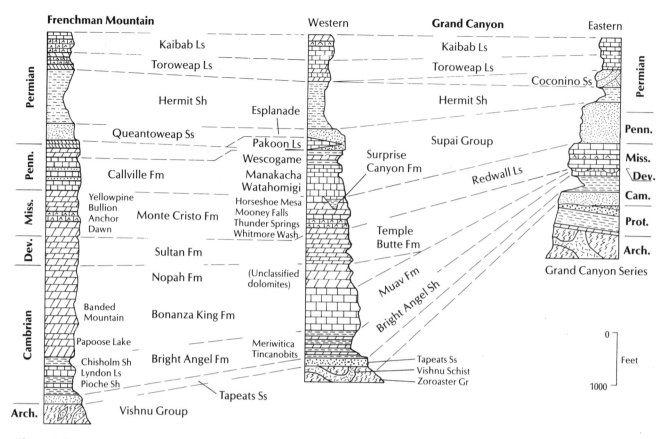

Figure 8.1

Correlation between the eastern and western Grand Canyon Paleozoic sequences (right) and the comparable sequence at Frenchman Mountain, just east of Las Vegas, Nevada. Notice that most units can be correlated by similarity in lithology, even if they change names and thicknesses over distance. Other units can be correlated by similarity in position between units that do not change from one exposure to the other. For example, the Pennsylvanian rocks are limestones in the west and sandy shales of the Supai Group in the Grand Canyon, but they are overlain by the Hermit Shale and underlain by the Mississippian limestones in both places, which establishes their approximate correlation. From Bachhuber et al. (1987).

worldwide and synchronous. Some unconformities are excellent tie points between sections, but unfortunately, the rocks above and below an unconformity are probably not correlative. As we saw in Chapter 7, the lacuna above and below an unconformity can increase in size over distance (Fig. 7.15B). If the erosional surface cuts through a sequence of units, then the units below the unconformity are of different ages in different places. In some regions, structural deformation and metamorphism occurred at restricted times in geologic history. In such cases, the degree of structural development or metamorphism can be used for tentative correlation in the absence of better indicators.

Correlation is seldom simple or straightforward. Some of the problems are purely artificial. For example, when geologists have different concepts of what constitutes good marker beds on which to base their boundaries

between units, the boundaries can change arbitrarily. This can lead to absurdities. Suppose that two state geological surveys define the boundary of a rock unit differently. The boundaries are mapped consistently within both states, but at the state line, the boundary is "offset" by the difference in definition. Regional maps that attempt to compile information from several states frequently show these "state-line faults." An unwary map user might be misled into thinking that real faulting had occurred precisely on the state line.

Other problems are due to real changes in the rocks. A facies change can cause one lithology to change gradually into another, and the boundary between them, the *stratigraphic cutoff*, must be arbitrary. Sometimes facies changes and intertonguing cause groups of units to merge or split; then the scheme of stratigraphic cutoffs must be even more arbitrary and complex (Fig. 8.2).

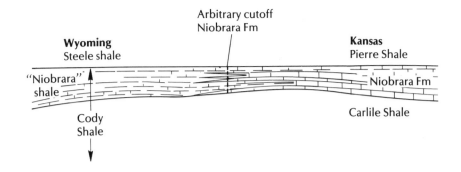

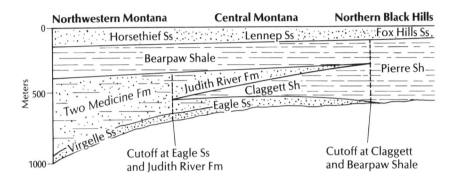

Figure 8.2

Facies change and stratigraphic cutoff. In the top diagram, the chalk and limestone of the Niobrara Formation of Kansas are shown to pass, through intertonguing and gradation, into a shale in Wyoming. Because the two lithologies are not given the same name, there must be an arbitrary cutoff along the line at which the unit loses its dominant carbonate lithology. In the lower diagram, the intertonguing interval between the Pierre and Bearpaw Shales is subdivided into distinct formations. This requires an arbitrary cutoff between the Claggett, Bearpaw and Pierre Shales, and between the Judith River, Eagle, and Two Medicine Sandstones. From Krumbein and Sloss (1963).

Time Correlation

Catastrophist geologists saw each rock layer as a phase of the receding Noachian floodwaters and therefore considered a layer to be isochronous, or time-equivalent, wherever it was recognized. The concept of facies change showed that rock types are not good time markers (see p. 169), but "layer-cake" notions still persisted. When facies analysis and biostratigraphy were applied to some classic sections, surprising conclusions often emerged. For example, the famous Catskill sequence of upstate New York (Fig. 8.3) was once divided into units based on gross lithology. The "Catskill Group" was coarse-grained sands, gravels, and redbeds, the "Chemung Group" was all the sandy shales, and so on. These rock units were placed in vertical succession, and each was thought to represent a different period of time. Detailed tracing of the rock units and their fossils, however, later demonstrated that the time lines cut across the lithologic units. Each of the old "groups" was markedly time-transgressive, spanning much of the Devonian.

In North America, layer-cake thought persisted well into the 1950s, primarily due to the influence of E. O. Ulrich (1857–1944). As described by Dunbar and Rodgers (1957, pp. 284–288), Ulrich considered the European ideas of facies changes to be relatively unimportant in his work on the Paleozoic of North America. He viewed each rock unit and its distinctive fauna as the product of one isolated advance of the shallow seas over the continental interior. Thus, he interpreted the units in the Devonian Catskill sequence as the products of several separate advances rather than as one continuous facies sequence. For him, any difference in fauna or lithology, no matter how small, was proof of difference in age and evidence of an unconformity.

Naturally, this reasoning produced correlations and paleogeographic reconstructions that seem bizarre today. In every example in his 1916 paper, for example, he was exactly wrong. It took decades for North American geologists to undo the damage his erroneous correlations had done. His influence was so great that his layer-cake ideas still lurk behind the work of North American geologists, almost fifty years after his work was discredited.

Contrary to Ulrich's ideas, rock units can be noticeably diachronous, even over short distances. Gazing into the Grand Canyon, one sees cliffs of Tapeats Sandstone, slopes of Bright Angel Shale, and more cliffs of Muav Limestone above the Precambrian basement (see Fig. 7.13). From the canyon rim, these look continuous and uniform in thickness, so it seems reasonable to think in terms of "the time when the Tapeats Sandstone was deposited." The stratigraphic distribution of trilobites, however, shows that even over a distance of only 180 kilometers all of these formations are markedly time-transgressive (Fig. 8.4). The Muav Limestone in the west is equivalent in age to the Tapeats Sandstone in the east.

The Grand Canyon furnishes another striking example of this phenomenon. The resistant reddish cliffs of

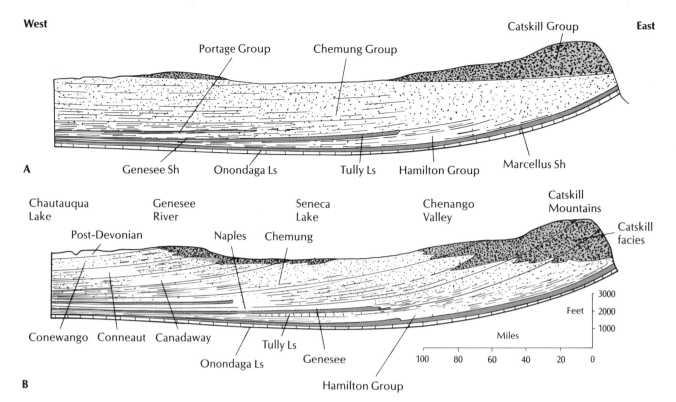

Figure 8.3

East-west cross sections across the Devonian Catskill sequences of southern New York. (A) Traditional interpretation (pre-1930), which treated the units in a layer-cake sequence of Hamilton and Portage shales, Chemung sandstones, and Catskill redbeds. (B) Present interpretation, following Chadwick and Cooper, which incorporates modern concepts of facies change. Time planes are shown by the curved lines; each unit consists of Catskill redbeds on the east, sandy facies in the center, and shales on the west. After Dunbar and Rodgers (1957).

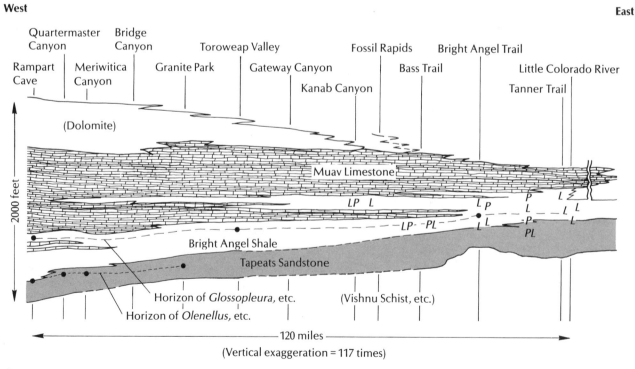

Figure 8.4

Time relations of transgressing Cambrian marine sediments in the Grand Canyon. The dotted lines indicate biostratigraphic time planes that show that the Tapeats Sandstone in the east is equivalent in age to the Muav Limestone in west; *L* and *P* denote *Lingulella* and *Paterina* fossils. After McKee and Resser (1945).

the middle part of the canyon wall are composed of the Mississippian Redwall Limestone (see Fig. 7.13). It, too, seems constant in thickness and extends up and down the canyon as far as one can see. A detailed analysis of this unit from southern Nevada to the center of the Grand Canyon at Bright Angel presents a different picture (Fig. 8.5). The Redwall records a series of almost a dozen episodes of east-west transgression and regression which span most of the Mississippian. The transgressive base of the unit spans at least one biostratigraphic zone, which represents about 2 to 3 million years. In the Grand Canyon itself, the Redwall is composed of two transgressive-regressive packages, with a big (but scarcely visible) unconformity in the middle that spans about 6 to 7 million years. Even the major subdivisions of the Redwall Limestone are several million years younger in the east than they are in the west!

In his influential book, *Time in Stratigraphy* (1964), Alan Shaw attacked the problem of the time significance of rock units. He developed a model for the nature of sedimentation in the broad, epeiric seas that produced most of the stratigraphic record of continental cratons.

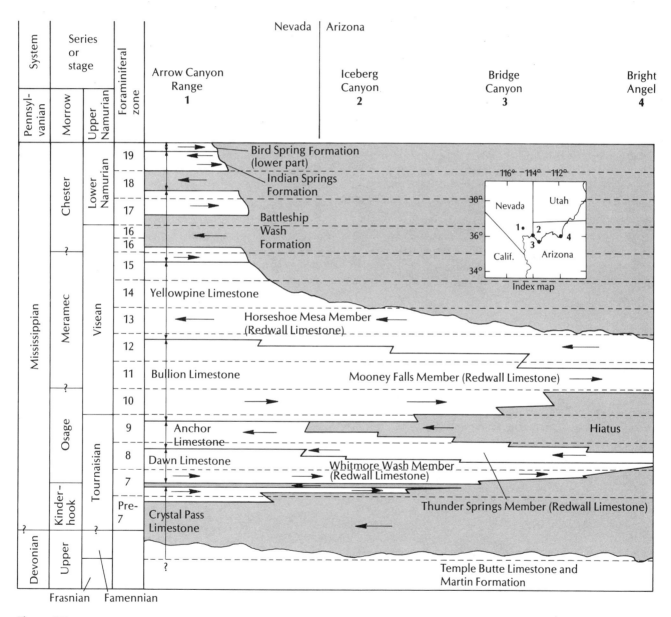

Figure 8.5

The Mississippian Redwall Limestone of the Grand Canyon seems to be uniformly thick and continuous on casual inspection. Detailed biostratigraphic work shows that it is actually composed of a series of transgressive and regressive deposits (arrows) that span only part of the Mississippian. The bases of individual members are actually large unconformities that typically transgress two foraminiferal zones (about 2 million years each) over only a few hundred kilometers. From Skipp (1979).

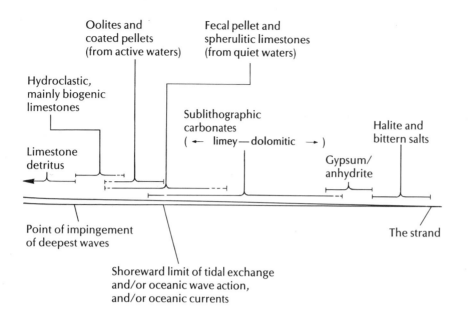

Figure 8.6

Shaw's model of the standard facies belts in a limey epeiric sea, with vertical scale greatly exaggerated. Because each of the lithologic types is contemporaneous and laterally adjacent, each could be expected to form time-transgressive rock units as the seas transgress and regress. From Shaw (1964).

These epeiric seas must have been very broad and very shallow, extending for hundreds of kilometers with slopes of only centimeters per kilometer. Restricted water circulation due to limited wave, current, and tide action would have produced a regular series of facies belts (Fig. 8.6). During transgression or regression, these facies belts would shift back and forth relative to the shoreline, each belt producing a characteristic rock unit that is time-transgressive throughout. From this, Shaw (1964, p. 71) concluded that "all laterally traceable non-volcanic epeiric marine sedimentary rock units must be presumed to be diachronous."

Ager (1981, p. 62) presents a modern analogue of the formation of diachronous deposits. On the coast of the Persian Gulf, the facies belts consist of subtidal lagoonal muds, intertidal algal mats, and supratidal sands and evaporites of the sabkha, or supratidal salt flats (Fig. 8.7). A pit dug through the sabkha about 6.5 kilometers

from the shore revealed an algal mat similar to the one forming today at the tide line. Radiocarbon dates on this mat give ages of 4000 years, or approximately 460 years of diachronism per kilometer. Ager emphasizes that the layer-cake metaphor of sedimentation as a "continuous rain from heaven" must be replaced with the concept of *lateral sedimentation*, or "the moving finger writes." In Ager's (1981, p. 61) words, "if it looks the same it must be diachronous."

Another study of lateral migration of coastal environments (sand spits, bay bottoms, and tidal marshes) showed different results (LeFournier and Friedman, 1974). Based on the changing positions of dikes in France, rates of lateral migration of 1 kilometer per 100 years were estimated. This is about an order of magnitude faster than the rates calculated by Ager (1981). Rates calculated from the migration of sandy spits in the barrier island sequence on the Atlantic shores of Long

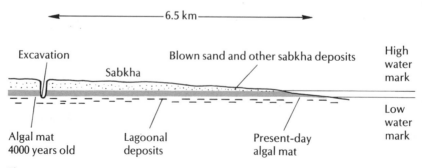

Figure 8.7

Ager used the example of an algal mat in the Persian Gulf to demonstrate diachronism. The mat is forming on the surface today, but its lateral equivalent buried beneath the sabkha can be radiocarbon dated at 4000 years old. The algal mat transgressed only 6.5 kilometers in 4000 years. From Ager (1981).

Island and New Jersey are even faster (Kumar, 1973; Sanders, 1970). The Long Island spit complex migrated at 65 meters per year (see Fig. 4.24), and the point of Sandy Hook in New Jersey has moved about 12 meters per year. The Mississippi delta migrates at rates of 75 meters per year (Gould, 1970). The rates are several orders of magnitude faster than the rates calculated by Ager and 6 to 7 times faster than the migration of French coastal facies.

On the scale of historic time, a lateral migration of millimeters to meters per year seems very slow, and sedimentary units are significantly diachronous. But on the scale of geologic time, these rates are so fast as to be nearly instantaneous. As we saw in the previous chapter, events separated by a few thousand to hundreds of thousands of years can seldom be resolved in the geological record. The geologist must keep two nearly contradictory concepts in mind when thinking about the time significance of a lithologic unit. On the scale of a local depositional environment and a short time span, rock units are markedly time-transgressive. Point bar sequences, delta fronts, prograding barrier islands, shifting tidal channels, and spits are but a few of the many examples of depositional systems that migrate laterally and produce bodies of uniform character that transgress time. However, on a scale of millions of years, this amount of change cannot be resolved. With such poor time-resolution, most rock units seem roughly isochronous.

Most of the ancient examples of diachronism that have been studied, however, show that over long distances rock units transgress large amounts of time, usually millions of years. This apparent discrepancy with modern rates of migration can be explained by the zigzag pattern of transgression. Episodes of transgression can indeed be as rapid as modern rates suggest. As suggested by Barrell's diagram (see Fig. 7.10), however, transgressions need not be continuous. Instead, a short episode of transgression can be followed by a regression that reduces the net lateral shift of the shoreline. As this zigzag "dance" of transgression and regression continues to move landward, it produces the typical intertonguing found at the base of many rock units (e.g., Fig. 8.5). The basal deposits of a single transgressive episode over a very short distance (less than a few kilometers) might indeed by synchronous to the extent that the problem can be resolved. Over any significant distance, however, the base of the unit is significantly younger and therefore time-transgressive.

When approaching a contact between formations in an outcrop for the first time, a geologist cannot be sure whether the contact is roughly synchronous or whether it significantly trangresses time. Ideally, the geologist should use whatever means are available to obtain an estimate of local rates of lateral migration for use on the problem at hand. The beds may be significantly time-transgressive, or the scale of resolution may be so

coarse and the distances so short that their ages are indeed "close enough that the difference doesn't matter" (Shaw, 1964).

Punctuated Aggradational Cycles and Time Correlation

Recently, the concepts of geologically instantaneous facies boundaries and the asymmetry of transgression and regression discussed in Chapter 7 have come together in the concept of *punctuated aggradational cycles* (*PACs*). As proposed by Goodwin and Anderson (1980, 1985), Anderson et al. (1984), and Goodwin et al. (1986), PACs are repetitive cycles of shallowing-upward beds that are punctuated at their bases by rapid rises in sea level (Fig. 8.8). PACs were originally described in the Devonian Helderberg Group of the Catskill Mountains of New York (see Box 6.1). In this area are repeated sequences of shallowing-upward intertidal limestones, in which each cycle begins abruptly with a discontinuity capped by a subtidal limestone. Goodwin and Anderson infer that these basal discontinuities represent rapid, relative sea-level rises and that the rest of each sequence was formed as the intertidal limestones aggraded and filled the basin, lowering relative sea level until the next rapid sea-level rise. In this sense, the PACs are conceptually very similar to the asymmetrical cycles of rapid transgression and prograding basin fill discussed in Chapter 7 (see Fig. 7.6).

Goodwin and Anderson go beyond showing that PACs represent rapid, episodic sedimentation. They assert that the basal discontinuities, representing rapid relative sea-level rises, can be traced for many kilometers across the entire sequence. If these sea-level rises are as rapid as they suggest, then they can be considered geologically instantaneous, and each PAC boundary represents a time plane (Figs. 8.8B,C). This leads to the conclusion that in some cases, rock units *can* be treated as if they were isochronous, and lithostratigraphic boundaries can be time planes.

The concept of PACs is still very controversial (Kradyna and Mehrtens, 1984; Wilkinson et al., 1984). Critics doubt whether these basal discontinuities can be traced as far as Goodwin and Anderson suggest. Instead of representing a single erosional surface covered by a single rapid transgression, each unconformity may be part of a series of short transgressive-regressive cycles. Although these cycles may be geologically instantaneous over short distances, over long distances they could be diachronous, as we saw in the Redwall Limestone example (Fig. 8.5). Brett and Baird (1986) argued that PACs are not pervasive over an entire basin, but occur only where the sediment accumulation rate is very high on the margins of basins. If so, then their significance for time correlation is purely local.

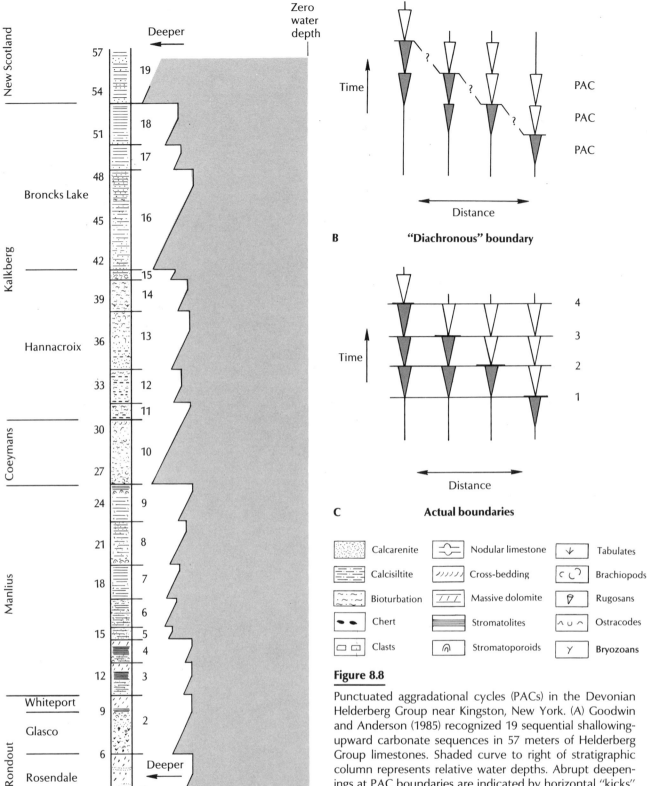

Figure 8.8

Punctuated aggradational cycles (PACs) in the Devonian Helderberg Group near Kingston, New York. (A) Goodwin and Anderson (1985) recognized 19 sequential shallowing-upward carbonate sequences in 57 meters of Helderberg Group limestones. Shaded curve to right of stratigraphic column represents relative water depths. Abrupt deepenings at PAC boundaries are indicated by horizontal "kicks" to the left. Gradual shallowing by aggradation within each PAC follows. (B) Diagrammatic cross section of four hypothetical outcrops with equivalent PACs (indicated by triangles). If each PAC is isochronous across the outcrop area, then their boundaries can be used as synchronous surfaces. The traditional formation boundaries are drawn along time-transgressive facies changes. In (C), however, the more natural boundaries could be drawn connecting each of the isochronous PACs. Under this scheme, the formations would be time equivalent.

THE NATURE OF THE CONTROL

Transgressive-regressive packages bounded by unconformities make up the sedimentary record of most basins on cratons and continental margins. Some of these unconformities are small; others span great distances and long periods of time. Major unconformities divide the stratigraphic record of every continent into discrete packages, which Sloss and others (1949; Sloss, 1963) called *sequences* (Fig. 8.9). These major unconformity-bounded packages of sediment are thought to represent the earth's large-scale tectonic or eustatic events, each

of which persisted for tens of millions of years. The sequences unite natural packages of sedimentary history that may span system boundaries. For example, the Lower Ordovician in most of North America is the culmination of the long Cambrian transgressive episode, which Sloss called the "Sauk Sequence." The rest of the Ordovician record is part of another transgressive episode that continues into the Silurian, called the "Tippecanoe Sequence." The Sauk-Tippecanoe "boundary" represents a major event in North American history, but the Cambrian-Ordovician and Ordovician-Silurian boundaries are not marked by any major tec-

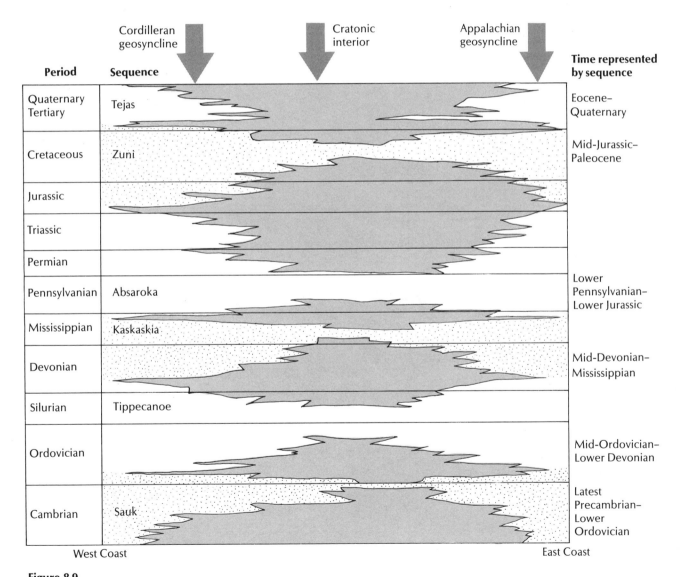

Figure 8.9

Sloss's diagram of the time-stratigraphic relationships of unconformity-bounded sequences in North America. Dark areas represent large gaps in the stratigraphic record, which become smaller toward the continental margins. Light areas represent strata. These "Sloss sequences" have also been recognized in Russia and South America. From Sloss (1963).

tonic or eustatic events in North America. The sequences are therefore more useful for describing the geologic history of the region.

Further analysis of these sequences reveals that they are not restricted to North America. A similar pattern of unconformities is found on the Russian platform (Sloss, 1972, 1978) and in South America (Soares et al., 1978). There is also a spectrum of small-, medium-, and large-scale unconformities preserved in the record of each continent. In some cases (Busch and Rollins, 1984; Ross and Ross, 1985), these can be matched up between continents (Fig. 8.10). According to these authors, the Carboniferous and Permian records of most of the major continents show a similar pattern of cyclic transgression

and regression that can be matched, cycle for cycle, around the globe. These global similarities in transgressive-regressive cycles are unlikely to have been controlled by local tectonics but must also be global in cause. The only reasonable controlling agent for this kind of global change is eustatic sea level.

Similarly, the seismic evidence from the passive margins of the world (discussed further in Chapter 11) shows a similar pattern of small-, medium-, and large-scale sea-level cycles of onlap and offlap. As defined by Peter Vail and others (Vail et al., 1977; Haq et al., 1987, 1988), there are four types (Table 8.1) of cycles: *first-order cycles* (*supercycles* of Fischer, 1981), which span hundreds of millions of years; *second-order cycles* (*synthems* of Chang,

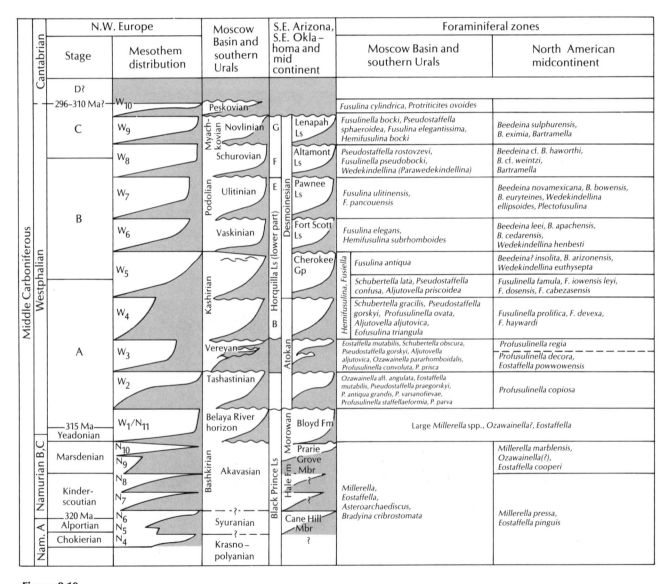

Figure 8.10

Correlation of Middle Carboniferous transgressive-regressive cycles that occurred in Europe, Russia, and North America. Fusulinid foraminiferal biostratigraphy establishes the correlation between the similar patterns of cyclic sedimentation and unconformities. (W, Westphalian; N, Namurian). From Ross and Ross(1985).

Table 8.1

Stratigraphic Cycles and Their Causes

Type (Vail et al., 1977)	Other Terms	Duration (million years)	Probable Cause
First order	Supercycles (Fischer, 1981)	200–400	Major eustatic cycles caused by formation and breakup of supercontinents
Second order	Sequence (Sloss, 1963) Synthem (Ramsbottom, 1979)	10–100	Eustatic cycles induced by volume changes in global mid-oceanic spreading ridge system
Third order	Mesothem (Ramsbottom, 1979)	1–10	Possibly produced by ridge changes and/or continental ice growth and decay
Fourth order	Cyclothem (Wanless and Weller, 1932)	0.2–0.5	Rapid eustatic fluctuations induced by growth and decay of continental ice sheets, growth and abandonment of deltas

1975, and Ramsbottom, 1979), which span tens of millions of years and produce the major unconformities in Fig. 8.10 and Sloss's sequences; *third-order cycles*, which last only a few million years; and *fourth-order cycles*, which have durations of hundreds of thousands of years and correspond to the classic cyclothems of the late Paleozoic. Busch and Rollins (1984) and Busch and West (1987) recognize even smaller-scale patterns. In their scheme, fourth-order cycles—the *mesothems* of Chang (1975) and Ramsbottom (1979)—have durations of 600,000 to 3 million years; *fifth-order cycles* (traditional cyclothems) span 300,000 to 500,000 years; and *sixth-order cycles* (punctuated aggradational cycles) range from 50,000 to 130,000 years in length.

First-Order Cycles

The first-order cycles seem to be related to major plate movements (Worsley et al., 1984). For example, the suturing of all continents into the Pangaea supercontinent in the Permian coincides with a long period of continental emergence, and the rapid breakup and separation of continents in the Cretaceous corresponds to a long period of continental inundation. Fischer (1981, 1982, 1984) suggested that the earth has gone through two complete 300 million-year-long supercycles between what he calls "icehouse" and "greenhouse" states (Fig. 8.11). During icehouse states (late Precambrian, late Paleozoic, and late Cenozoic), there were extensive polar icecaps, lower sea levels, and steep temperature gradients between the poles and the equator. Although there was a thin layer of warm water at the surface, the bulk of the ocean was composed of cold water masses produced at the poles, and the mean temperature of the ocean was about 3°C.

Greenhouse states, on the other hand, had no polar icecaps, and consequently sea level was higher. This implies that the temperature gradient between poles and equator was much less steep and that the ocean was also less stratified, with an average temperature of about 15°C (almost 10°C warmer than today). Fischer suggested that the greenhouse phase is triggered by increased carbon dioxide in the atmosphere (the greenhouse effect), which traps solar radiation and warms the earth (Barron and Washington, 1982, 1985). He pointed out that peaks of vulcanism are correlated with the greenhouse phase of the cycle (Fig. 8.11), which suggests that increased mantle activity during these times releases large amounts of greenhouse gases from the mantle into the atmosphere by means of increased vulcanism. When this mantle "overturn" slows, weathering on the land withdraws carbon dioxide from the atmosphere until its rate of supply by volcanism is balanced by the rate of withdrawal by the lithosphere. This low-level balance corresponds to the icehouse phase of the supercycle. The growth of a glacial armor over the land (preventing further withdrawal of carbon dioxide by lithospheric weathering) and possibly increased mantle convection have apparently prevented the earth from becoming an iceball like some other planets.

Fischer's models have many appealing features, but at the moment they are highly controversial. For one thing, they are based on only the last 700 million years of earth history. With only two complete icehouse-greenhouse cycles, it is difficult to prove that a cyclic causation mechanism really exists. Two greenhouse phases alternating with three periods of glaciation could easily result from coincidental causes that have no underlying cyclic mechanism. Crowell and Frakes (1970) have shown that most of the icehouse phases of earth history coincide with times when the continents were in polar positions. In addition, there are exceptions that

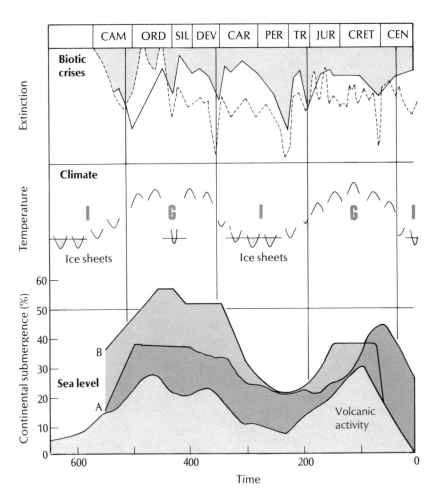

Figure 8.11

According to Fischer (1981, 1982, 1984), the earth goes through 300-million-year supercycles between icehouse (I) and greenhouse (G) states. These first-order cycles are shown by the alternating phases of climate. The greenhouse phases correlate with peaks in vulcanism and eustatic sea-level highs, shown in two versions in curves A and B. Two versions of the curve of extinctions of marine organisms are shown. Many mass extinction events ("biotic crises" along the top) also appear to be associated with the transitions between phases of the supercycles. From Fischer (1982).

are difficult to explain by these cycles, such as the late Ordovician glaciation, which occurred near the peak of a supposed greenhouse phase.

Second-Order Cycles

Many of the second-order cycles are thought to be related to changes in ridge volume and seafloor spreading rates. During periods of rapid sea-floor spreading, the greater volume of recently erupted and intruded oceanic crust that has not yet cooled and contracted increases the ridge volume (Fig. 8.12). During periods of slow sea-floor spreading, the crust has more time to cool and occupies less volume. There seems to be good correspondence between spreading rates and the fre-

quency and magnitude of sea-level changes during the Cretaceous (Hays and Pitman, 1973; Pitman, 1978). However, ridge volume changes are too slow to account for cycles much less than ten million years in length.

Sheridan (1987a, 1987b) suggested that the cycles of fast and slow seafloor spreading may be controlled by cyclic changes of mantle convection, operating on a scale of 30 to 60 million years (faster than the cycles postulated by Fischer). He pointed out that the periods of fast spreading and high sea level are correlated with periods of high magnetic-field activity, which is controlled by the earth's core and mantle (Fig. 8.13A). Most of the periods of high sea level since the Cambrian are correlated with periods of stable magnetic-field polarity (Fig. 8.13B), with a phase lag of about 10 million years. Vogt (1975) and Sheridan (1983, 1987a, b) proposed a

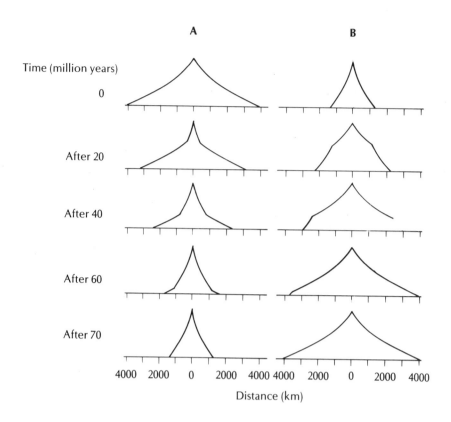

A   B

Time (million years)

0

After 20

After 40

After 60

After 70

4000 2000 0 2000 4000 4000 2000 0 2000 4000

Distance (km)

Figure 8.12

Profiles of fast- and slow-spreading mid-ocean ridges through 70 million years of spreading. (A) Ridge that had been spreading at 6 centimeters per year slows to 2 centimeters per year. After 70 million years, it has one-third its original volume, and epeiric seas return to the ocean basins. (B) Ridge that had been spreading at 2 centimeters per year changes to 6 centimeters per year, greatly increasing its volume and displacing water, causing sea-level rise. After Pitman (1978).

model (Fig. 8.13C) in which the earth alternates between two phases. When its core is relatively cool and quiet, there is a weak magnetic field and few reversals; to compensate for this decreased core energy and to maintain the energy balance of the earth, the mantle undergoes rapid, smooth convection in a series of well-developed convection cells and mantle plumes (Fig. 8.13C, time = t_1). As the core changes to a hotter, more turbulent state, it produces many magnetic-field reversals and a strong magnetic field; the mantle, on the other hand, becomes relatively cooler, and the convection and plate motion slow down, with few discrete convection cells (Fig. 8.13C, time = t_3).

Recently, some geologists (Bally, 1980; Watts, 1982; Thorne and Watts, 1984; Watts and Thorne; Parkinson and Summerhayes, 1985; Summerhayes, 1986; Miall, 1986; Carter, 1988; Hubbard, 1988) have challenged the idea of eustatic control of second-order cycles. They point out that these cycles have been measured exclusively from passive continental margins. Hubbard (1988) studied three of the major basins that were responsible for much of the sea-level curve proposed by Vail et al. (1977), and found that most of the sequence boundaries were not synchronous, but related to local tectonic activity. Only in basins of the same age with identical subsidence and sediment input rates did the second-order cycles appear to be synchronous. Hubbard (1988) suggested that the concept of global second-order cycles may be an illusion created by the fact that the Vail curve was constructed from basins which were

very similar in age and tectonic history. If this is so, then the cycles do not have global chronostratigraphic significance, and should not be used in correlation.

The latest tests of the Vail curve were published in the volume edited by Wilgus et al. (1988). A number of contributors identified many of the key unconformities predicted by the Vail model in sections exposed on land. Poag and Schlee (1984) did the same with well data from the North American Atlantic margin. While these papers clearly established that many of these unconformities and seismic sequences were real, they still do not establish that all the second-order cycles are global and synchronous. Only a few of the studies were able to identify the same unconformities in more than one continent, and only for small parts of the Mesozoic or Cenozoic. Hallam (1988) found that some of the Jurassic unconformities matched the Vail curve, but others did not, and these were clearly influenced by local tectonism. Likewise, Williams (1988) found that the oxygen isotope record of the oceans agreed with the Vail curve for some events, but not with others. In short, the Vail second-order cycles probably have a significant eustatic sea level component, but there is also much local tectonic "noise" that has not yet been filtered out of the eustatic "signal." The Vail curves will probably undergo further refinement, and they should not be used uncritically. Although Fig. 11.17 and Appendix B show the Haq et al. (1987, 1988) "sea level" curves, not all the fluctuations on these curves have been clearly documented as global, eustatic events.

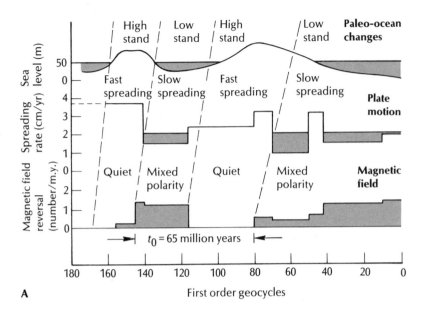

A   First order geocycles

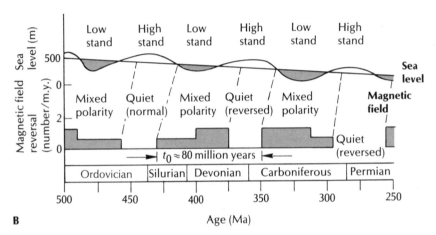

B   Age (Ma)

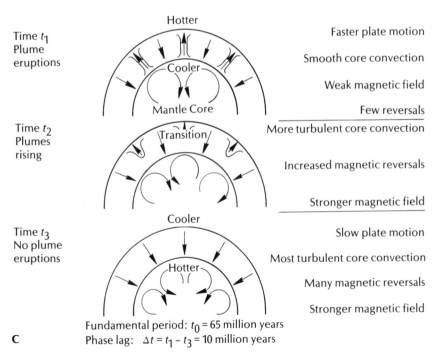

Fundamental period: $t_0 = 65$ million years
Phase lag: $\Delta t = t_1 - t_3 = 10$ million years

C

Figure 8.13

Sheridan (1987a, 1987b) has shown a correlation between the activity in the earth's core (as inferred from magnetic field behavior) and mantle, and the changes in spreading rate and sea level. (A) The correlations among sea level, spreading rate, and magnetic field behavior over the last 180 million years. (B) Although there is no way of calculating spreading rate prior to the Jurassic, the correlation between field behavior and sea level can be traced back to the Ordovician. In both figures, the cycles take about 65 to 80 million years and have a phase lag of about 10 million years between the magnetic field behavior change and the corresponding sea-level change. (C) Sheridan's model for these correlations. At time t_1, the core is cooler, with a weak magnetic field and few reversals. To compensate, the mantle becomes hotter and develops strong convection and faster plate motion. By time t_3, the core heats up and develops strong convection, resulting in many field reversals and a stronger magnetic field. The mantle, by contrast, convects less and plate motion is reduced.

Third-Order and Fourth-Order Cycles

Eustatic sea-level changes on the scale of a few hundred thousand to a few million years (third- and fourth-order cycles) are usually caused by changes in global ice volume. During the maximum Pleistocene ice advance, so much sea water was trapped in the ice caps that sea level was 100 meters lower than it is now. If the present ice caps melted, sea level would rise by 40 to 50 meters (Donovan and Jones, 1979). That adds up to about 150 meters of net sea-level change at rates as rapid as a few centimeters per year. It seems clear that the fourth-order and perhaps some of the third-order cycles are glacially controlled. Indeed, the evidence for glacial control of late Paleozoic eustatic sea level is strong (Crowell, 1978), and the glacial control of Cenozoic sea-level changes is equally well documented (Kennett, 1982). Nevertheless, this does not explain the well-known sea-level cycles of the Jurassic and Cretaceous, when there were no ice sheets so far as we know. Several authors (Barron et al., 1985; Fischer et al., 1985) have proposed models that explain these cycles in terms of alternating wet (possibly cooler) and dry (possibly hotter) climates.

But what controls ice volume and glacial-interglacial cycles? Since 1976 we have known that most of the Pleistocene temperature variations were controlled by changes in sunlight received by the earth, caused by variations in the earth's orbital geometry that affect its distance from and orientation to the sun (Hays et al., 1976; Imbrie and Imbrie, 1979; Denton and Hughes, 1983; Fischer, 1986). The effects of these *Milankovitch cycles* were once known exclusively from the Pleistocene, but recent analysis of the classic Carboniferous coal-bearing cyclothems has shown a periodicity that may be explainable by the Milankovitch cycles (Heckel, 1986). Milankovitch cycles have also been invoked to explain the periodicities found in Triassic lake sediments (Olsen, 1984) and in Cretaceous marine sediments (Fischer et al., 1985; Fischer, 1986). Goodwin and Anderson (1985) attributed the periodicity of punctuated aggradational cycles in the Devonian of New York to the Milankovitch cycles.

Event Stratigraphy

Whether or not Milankovitch mechanisms work for most other short-term cycles, one thing is becoming clear: *cycles that have global eustatic control have time significance.* If transgressions and regressions are induced by simultaneous worldwide changes in sea level, then the peaks (maxima) of transgressions and regressions are more or less synchronous worldwide. Israelsky (1949) first introduced the idea that correlation of lithologies by their position in a cycle can have time significance, and Ager (1981, p. 70) called such correlation *event stratigraphy.* During a typical transgressive-regressive cycle (Fig. 8.14), lithostratigraphic units are clearly diachronous. However, different lithologies that are formed at the same point in a cycle (such as the deposits during peak transgression) are approximately isochronous. Correlation of *different* lithologies that represent the *same* point in the cycle (the event) has a greater chance of indicating time significance than has any other technique. (The fossils in this instance are either scarce or highly facies-controlled). Event stratigraphy offers a

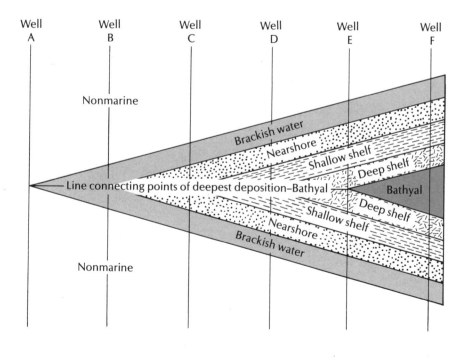

Figure 8.14

Israelsky's (1949) diagram of transgressive-regressive facies cycles, showing that the point of maximum transgression in each facies is essentially time-equivalent. Each lithostratigraphic unit, however, is markedly diachronous. At peak transgression, bathyal deposits on the right are synchronous with nonmarine deposits on the left. Correlation by events such as transgressive or regressive maxima and minima has come to be known as *event stratigraphy* (Ager, 1973, 1981).

method of worldwide correlation of eustatic cycles, with some assurance that they are isochronous. However, correlations of beds other than those at the peaks of cycles is much less certain.

If the correlation of the peak of a single transgression or regression with other such peaks is uncertain, the larger-scale pattern may suggest a correlation. Busch and Rollins (1984) and Busch and West (1987) argue that the pattern of cycles nested within (or superimposed upon) larger-scale cycles can be correlated over large areas. For example, the Carboniferous of the North American midcontinent (particularly Kansas) and that of the northern Appalachians (mostly Pennsylvania, Ohio, and West Virginia) show a similar pattern of fifth-order transgressive-regressive cyclothems clustered into larger fourth-order cycles (Fig. 8.15). The peaks of unusually large transgressive-regressive cycles (such as the Pine Creek–Brush Creek of the Appalachians, the Hertha-Swope-Dennis of the midcontinent, or the Ames and the Stanton) serve as the major tie points for lining up

the pattern of cycles. These have been checked by bio-stratigraphic markers (see Fig. 8.15) and shown to be isochronous.

In summary, we have seen how early stratigraphers tended to think of rock units as having time significance, an inheritance of the old layer-cake ideas that dated back to flood geology. In the 1930s and 1940s, the pendulum began to swing the other way, and the importance of lateral facies changes began to be emphasized. Shaw (1964) and Ager (1973, 1981) represented some of the more extreme proponents of lateral sedimentation: If it looks the same, it must be different in age. Because most sedimentation is controlled by local facies patterns, these ideas about lateral sedimentation are still valid. We cannot consider the lithostratigraphic boundaries between rock units to have time significance, especially over long distances. However, with the recent emphasis on global, eustatically controlled sea-level cycles and their effects on sedimentation, one can find many places in the world where lithostratigraphic units do appear to have time

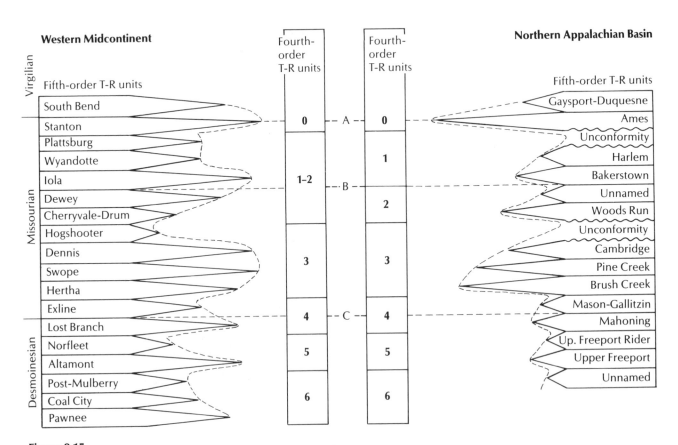

Figure 8.15

Event correlation by hierarchical patterns of cycles within cycles. Both the western midcontinent and the northern Appalachians show a series of fifth-order transgressive-regressive cycles (T-R in the figure) clustered into larger numbered, fourth-order cycles. These can be checked by biostratigraphic markers (levels A, B, and C), which show that the cycles are isochronous over 1200 kilometers. From Busch and West (1987).

significance. It must be emphasized, however, that these correlations are rare exceptions. The assumption of synchrony can be justified only by independent evidence that these cycles are indeed global.

GEOLOGICALLY INSTANTANEOUS EVENTS

How do we establish time-equivalence in the stratigraphic record? In most instances, radiometric dates are rare or unavailable and have estimated measurement errors that are too large for this purpose. The ideal time marker should be *widespread, distinctive* and *geologically instantaneous*. In most cases, events that were separated by years to tens of years (or less) cannot be resolved in the stratigraphic record and are therefore considered instantaneous with respect to geological time scales.

The primary tool for inferring time relations in the geologic record is biostratigraphy (discussed in detail in Chapter 10). However, without fossil evidence, a number of rock types might be considered to be instantaneously formed (in the geological sense). The most common of these are volcanic deposits of a single eruption, such as ash layers and their diagenetically altered equivalents, *bentonites* (Fig. 8.16). Ash deposits fall or flow in a matter of hours or days and cover wide areas during that time. If ash deposits can be precisely correlated, and possibly even dated, they provide unique time planes. The use of ash layers to mark geologic time is called *tephrostratigraphy* or *tephrochronology*.

In areas of frequent volcanic activity, tephrostratigraphy can be a powerful tool. Detailed petrographic

analysis of the ash and its geochemical "fingerprinting" by comparing the amounts of stable, heavy trace elements that are present are needed to determine which ash beds match up. Other methods, such as biostratigraphy or magnetic stratigraphy, can be used to check the correlations. Once the ash layers themselves have been correlated, they offer advantages that few other methods of correlating sections can match. First of all, they are geologically instantaneous and so provide a network of "time planes" among the sections. Second, many ash layers can be radiometrically dated, so it is possible to get a numerical age in a sedimentary sequence.

Third, many explosive volcanic events spread huge clouds of ash over enormous areas, making it possible to correlate sequences across great distances and various sedimentary environments. For example, the Long Valley Caldera near Mammoth Mountain in California erupted about 740,000 years ago (Gilbert, 1938; Dalrymple et al., 1965; Izett et al., 1970; Izett and Naeser, 1976; Izett, 1981) and spread an ash cloud all the way to eastern Nebraska, forming the Bishop Tuff (Fig. 8.17). The eruption was so massive that 150 cubic kilometers of ash were displaced up to 2000 kilometers with pyroclastic deposits up to 125 meters thick near the source and several centimeters thick 1500 kilometers away. This ash layer, along with many others, has been used to correlate Pleistocene sequences all over the western United States (Izett, 1981). By combining the data from this ash with other ashes that happened to be blown west (Sarna-Wojcicki et al., 1987), one can construct a network of ash datum levels that correlate terrestrial and marine sections from the North Pacific to Nebraska. Terrestrial and marine sections have also been correlated in this way on the Pacific Coast of Central America (Bowles et al., 1973) and in the Mediterranean (Keller et al., 1978). Such precise correlations between the terrestrial and marine environments are very seldom achieved because the lithofacies and the fossils of the two environments are completely different. Magnetic stratigraphy is the only other method to achieve this kind of correlation across facies.

Other rock types that are less common and widespread than volcanic ash but are rapidly formed are also useful as time markers. Any deposit formed by a single catastrophic slide or flow, such as a turbidite, olistostrome, or debris flow, can be considered geologically instantaneous. Some lake deposits form by a uniform rain of silt and clay from suspension and therefore are formed along time planes. This is particularly true of climatically controlled laminations and *varve* deposits because each layer reflects a climatic change and is therefore isochronous (see Fig. 2.2). However, the varve deposits of lakes that show migrating facies belts cannot be considered good time markers (Buchheim and Biaggi, 1988).

Certain features within sedimentary beds can also

Figure 8.16

Bentonites (two white layers in lower center-left) are altered volcanic ashes that serve as excellent time-significant stratigraphic markers. These examples are from Cretaceous marine shales. Photo courtesy D. L. Eicher.

A

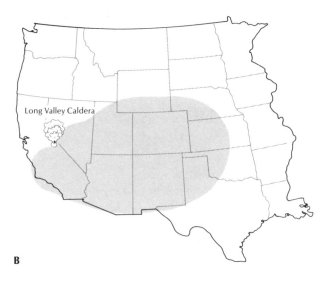

B

Figure 8.17

(A) Outcrop of the Bishop Tuff, an immense ash flow sheet that erupted from Long Valley Caldera, California, about 740,000 years ago. The finely laminated basal pumice layer is shown here; note that the size of pumice fragments increases up the section. Photo by P. C. Bateman, courtesy U.S. Geological Survey. (B) Extent of the Bishop ash cloud, which traveled over 1000 kilometers to eastern Nebraska. From Izett and Naeser (1976).

used as time markers. In Chapter 11, we discuss correlation by magnetic reversals and isotope shifts, which appear to be synchronous. Other geochemical signatures of unique events, such as the iridium anomaly that is believed to represent the impact of an extraterrestrial body (Alvarez et al., 1980), could also be used as time markers. *Tektites* (Fig. 8.18) are glassy spherules that are thought to have been scattered around the earth by the impacts of meteorites. If a series of tektite layers can be

clearly identified as derived from the same event, the layers are useful time markers (Fig. 8.19). These phenomena must be used with caution, however, because there are many pitfalls in interpretation.

Clearly, the geologist must use any and all available tools of correlation and know the strengths of each. When possible, one method of correlation should be checked against another. In Chapter 13, we look at several examples of cross checks. The geologist should keep several caveats in mind. The most of important of these requires abolishing the easy mental habit of layer-cake thinking. Rock units are probably diachronous unless proven otherwise.

TIME, TIME-ROCK, AND ROCK UNITS

Throughout the history described in Chapter 1, there was continual controversy about the relationship between rock and time units and about the time significance of rock bodies. Arduino, Werner, and many other early geologists thought that every rock unit was deposited during a discrete period of time, such as before, during, or after the noachian flood. Werner felt that the formation of a specific rock type, such as schist or limestone, was unique to one of his four great divisions of earth history. In his scheme, one could recognize a particular geologic time by a specific lithology. As the principle of faunal succession became widely used, geologists realized that rocks of similar lithology could differ in age. The classic stratigraphic column of England is full of shales, limestones, and sandstones of various

Figure 8.18

A selection of tektites, aerodynamically shaped because they cooled while flying through the atmosphere. Photo courtesy B. Mason.

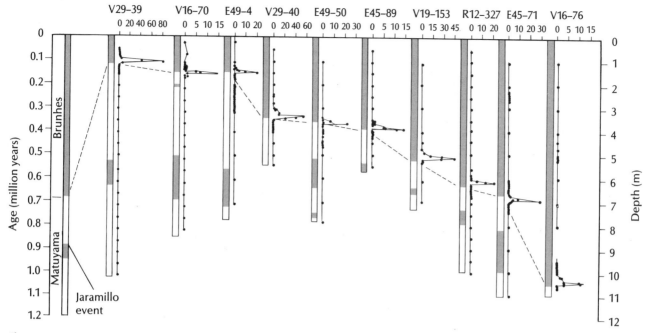

Figure 8.19

Abundance of microtektites (per 8 cubic centimeters of sample) from the Australasian strewn field in deep-sea cores from the South Pacific. Magnetic stratigraphy (solid and open bars) of each core is also shown. Note the peak in microtektite abundance is always within 30 centimeters of the Brunhes/Matuyama boundary, so it appears to represent a single synchronous microtektite shower. From Glass et al. (1979).

ages. These rocks can be distinguished only by their fossil content because their lithologies are so similar. In addition, it was finally recognized that rock units could be of similar age and yet have very different lithologies, as Sedgwick and Murchison showed when they named the Devonian. The Old Red Sandstone is very different in lithology from the graywackes of Devonshire, yet the fossils showed that both units belong between the Silurian and Carboniferous in the rock column.

The battle between Sedgwick and Murchison over the Cambrian and Silurian is a classic demonstration of this conceptual conflict. Sedgwick proceeded in the old-fashioned way, describing local lithologies and attaching names to them. Murchison used faunal succession to define the Silurian so that it could be recognized internationally. Murchison's fossils enabled him to trace his "Silurian" into rocks that Sedgwick had defined as Cambrian on the basis of lithology.

Faunal succession was developed by men who thought that the earth was very young. They thought that fossils had been deposited in successive flood deposits spanning very little time. The publication of Darwin's *On the Origin of Species* was still fifty years in the future, so faunal succession had no evolutionary connotation then. Yet the sequence of faunas and their utility in distinguishing strata of similar lithology suggested that fossils were an independent, more reliable marker

of time than rocks. Geologic time and observable rock sequences had become separate but related concepts. Lyell's use of molluscan percentages as a "clock" for the Tertiary takes this divorce between rocks and time to its extreme.

This uneasy tension between time and rock units still confuses people today. Because observable rock units are not unique to a time, and can often transgress time, formally defined rock units (*lithostratigraphic units*) have no explicit time connotation. Like time on the human scale, geologic time is an abstract concept that can be marked by many different methods: changes in fossil faunas, radioactive decay, changes in oxygen isotopes, reversals of the earth's magnetic field, or whatever is convenient. Therefore, geologic time units (*geochronologic units*) are also purely abstract concepts. Time and rock units are not completely separate, however. Many geologic events have left evidence in the rock record, and the rock record is our only concrete representation of geologic time. For this reason we have developed a hybrid unit, the *time-rock* (or time-stratigraphic, or *chronostratigraphic*) unit. A time-rock unit is the sum total of the rocks formed worldwide during a specified increment of geologic time (as recognized by fossils, radiometric dates, or whatever). A time-rock unit has physical reality though it cannot be seen in its entirety in any one section, and its boundaries are based on the abstract concept of time. Those boundaries are established by biostratig-raphy (see Chapter 9).

Schenck and Muller (1941) formalized the subdivi-sions of the three categories of time, time-rock, and rock units and recognized the hierarchy of units that had developed over the years. Their scheme is shown below:

Time	Time-rock	Rock
Eon...........Eonothem		Supergroup
Era............Erathem		Group
Period.........System		Formation
Epoch.........Series		Member
Age...........Stage		Bed
Zone		

Schenck and Muller emphasized the fact that rock units have no explicit time connotation by showing them at right angles to the time and time-rock units. The equiv-alence of time units and their corresponding time-rock units is shown by the dotted lines. Thus, the Jurassic System consists of all the rocks deposited during the Jurassic Period.

The important distinction between time and rock units is often confused, even by geologists who should know better. It is easy to fall into the layer-cake thinking of our catastrophist forbears and refer to "Tapeats Sandstone time" or "the thickness of the Late Jurassic" or events that occurred "during the Upper Cretaceous." As Owen (1987) points out, geologists who never refer to "upper Tuesday" or "late peninsula of Michigan" will easily confuse time and place words in geology. Such confusion occurs on an even more subtle level. For ex-ample, the term "pre-Dakota unconformity" implies that the Dakota Sandstone is a geochronologic unit; it should be "sub-Dakota unconformity." This may seem like nit-picking, but as Owen (1987, p. 367) puts it, "authors who are careless in usage of time and place words run the risk of implying that their carelessness may extend to data collection, analysis, and conclusions as well."

Each unit in the series is composed of one or more units of the level beneath it in the hierarchy. For exam-ple, the Phanerozoic Eon contains three eras (Paleozoic, Mesozoic, and Cenozoic). The Cenozoic Era contains two periods (Tertiary and Quaternary). The Tertiary Period contains five epochs (Paleocene, Eocene, Oligo-cene, Miocene and Pliocene). The Oligocene Epoch con-tains two internationally recognized ages (Rupelian and Chattian), and the corresponding stages are based on biostratigraphic zones. For example, in the deep sea, the Chattian Stage is presently based on three successive planktonic foraminiferal biostratigraphic zones, the *Glo-bigerina ampliapertura* Zone, the *Globorotalia opima opima* Zone, and the *Globigerina ciperoensis* Zone. Other zona-

tions can be used, but the foraminiferal zones have achieved international consensus in this instance.

Rock units have a similar hierarchy. For example, the Oligocene White River Group is composed of two for-mations in South Dakota: the Chadron Formation and Brule Formation (see p. 229). The Brule Formation, in turn, is made up of the Scenic Member and Poleslide Member. Within the Scenic Member are several local units, such as the lower *Oreodon* beds and the *Meta-mynodon* sandstones.

Instead of building up from the bottom (as in the case of zones), the fundamental unit in lithostratigraphy is the *formation*. A formation is characterized by two im-portant properties: It must have *identifiable and distinctive lithic characteristics*, and it must be *mappable on the earth's surface or traceable in the subsurface*.

Formations can be lumped into groups and super-groups or subdivided into members and beds. However, this is not obligatory. Many formations have no mem-bers and are not part of any larger group. Whether a rock unit merits the rank of member, formation, or group is up to the discretion of the geologist who describes it. For example, in the Paleozoic sequence in the Grand Canyon (Fig. 8.20), the top four units are formations (Kaibab Limestone, Toroweap Limestone, Coconino Sandstone, and Hermit Shale). Below it are four forma-tions that are hard to distinguish and so are usually united into the Supai Group (McKee, 1982). Beneath the Supai is the Redwall Limestone (McKee and Gut-schick, 1969), a formation with four *members* (from bot-tom to top: Whitmore Wash, Thunder Springs, Mooney Falls, and Horseshoe Mesa), which are comparable in thickness to the *formations* above them. The Redwall Limestone (a *formation*) is approximately equal in thick-ness to the Supai *Group*. The decision on whether units deserve member, formation, or group rank is not based on thickness but on other criteria such as their distinc-tiveness, lateral continuity, and ease of recognition.

The criterion of mappability or traceability is one of convenience. In some areas that have thousands of feet of section, formations can be hundreds of feet thick and recognizable on large-scale maps, aerial photographs, and even satellite photos. In other areas, formations are very thin and can only be mapped on very small scales. Clearly, if a formation is mappable, there must be some lithic criterion by which it can be distinguished and mapped. Lithic distinctiveness does not necessarily im-ply lithic uniformity. Formations can include many different rock types and can change lithology over dis-tance. All that is required is some sort of distinctiveness on which the boundaries, or *contacts*, can be based and mapped. Because formations are units of convenience, one stratigrapher's formation might only be a member to a stratigrapher working in an area with different geology.

THE STRATIGRAPHIC CODE

As we saw in Chapter 1, the inconsistencies and conflicts of terminology and concepts were major sources of confusion to early geologists. Eventually, it became necessary to standardize usage and erect a set of guidelines for stratigraphers to follow. The first American Commission on Stratigraphic Nomenclature met in the early 1930s (Ashley et al., 1933). The American code has been revised many times since then, and its most recent revision (1983) appears here as Appendix A. International usage is not as unanimous as North American nomenclature, but the essential features have been codified by the International Subcommission on Stratigraphic Classifica-

tion in the International Stratigraphic Guide (Hedberg, 1976).

The Code states the criteria that are to be used to distinguish a formal rock unit from an informal one. The name of a *formal* geologic unit is capitalized (e.g., Morrison Formation) and must be created under strict rules (Articles 3 through 21). The creation of a formal unit must be published in a recognized scientific medium, and the original description must list a number of important defining criteria. For example, a formation must be named after a local geographic feature, not a person or some combination of Greek and Latin roots. There are rules about priority of names, and compendia to determine if a name has already been used. In the United

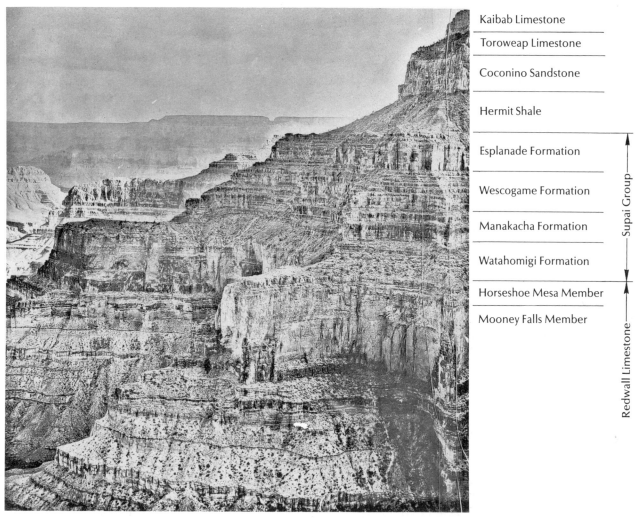

Figure 8.20

Stratigraphic terminology in the upper part of the Grand Canyon. At the base, the Redwall Limestone (a formation) is divided into members (shown here are the Mooney Falls Member and Horseshoe Mesa Member). The next four formations are combined into the Supai Group. The top four formations (Hermit Shale, Coconino Sandstone, and Toroweap and Kaibab Limestones) are not combined into groups or subdivided into members here. Photo by N. W. Carkhuff, courtesy U.S. Geological Survey.

States, this duty is carried out by the frequent publications of the Geologic Names Committee of the U.S. Geological Survey. The boundaries, dimensions, and age of the formation must be clearly given in the original description. Most importantly, a type section, or *stratotype*, must be designated so that later geologists can determine what the name originally referred to and what the author's conception of the name was. The type section should be the best reference section available. Ideally, it should be well exposed, with both the top and bottom contacts of the formation visible. It should be as complete and continuous as possible, without faulting or long covered intervals. Additional *reference sections* are also helpful in establishing the variability of the formation or giving a backup section in case the type is covered, built over, bulldozed, or otherwise lost (or poorly chosen in the first place). In many cases, the formation can be studied in the subsurface as well. It is best to designate the type section on a surface outcrop and to designate subsurface sections as reference sections because the surface data are accessible to more geologists. Some units, however, are known only from boreholes.

When a new or revised description of a unit is published, the information should include the following:

1. Name and rank of unit
2. Locations of type and reference sections, including a map or air photo
3. Detailed description of the unit in the stratotype, including nature and height of the section or well depth of contacts
4. Comments on local or regional extent of unit and its variability
5. Graphic log of unit (including geophysical logs of subsurface units)
6. Location of curated reference material
7. Discussion of relationship to other contemporaneous stratigraphic units in adjacent areas

Informal units are also common, and the Code recognizes their role in "works in progress" and for innovative approaches to stratigraphy. For example, Sloss's "sequences" (discussed above) are informal units. Many of the working units in economic geology, such as coal beds, oil sands, aquifers, and quarry beds, are informal. Informal units serve a valuable role in enabling geologists to articulate new concepts outside the restrictions of the Code. As informal units become more widely accepted, they may eventually become formal. Many of the new units and categories in the Code were once informal units. However, informal units do not have the status of the formal units of the Code and are not capitalized. As more detailed studies are done, many more units are recognized than require formal definition. Most of these units will remain informal because establishing all of them would cause an overwhelming proliferation of names. As the authors of the Code put it, "no geologic unit should be established and defined, whether formally or informally, unless its recognition serves a clear purpose."

One of the commonest errors made by geologists is confusing informal with formal units in print, or in slides and illustrations. As mentioned above, all names in a formally defined unit are capitalized, but only the geographic term is capitalized in informal names. For example, the Tapeats Sandstone of the Grand Canyon is a formally defined unit, but the *Metamynodon* sandstones of the Big Badlands of South Dakota are informal. This distinction becomes tricky when referring to time and time-rock terms. For example, the subdivisions of the Paleozoic and Mesozoic Periods/Systems have been formally defined, but those for the Cenozoic have not. Thus, it is correct to refer to the Late Devonian Epoch or the Lower Jurassic Series, but the upper Eocene must always remain uncapitalized. There are more subtle pitfalls, as well. For example, the Cretaceous has only two formal subdivisions, so it is correct to refer to Early/Lower and Late/Upper Cretaceous, but not Middle Cretaceous (except as an informal term). Owen (1987) clarifies many of these confusing points, and the geologic timescale (Appendix B) is a useful guide to determining which subdivisions have attained formal recognition.

Many striking features of the new North American Stratigraphic Code do not appear in any of its predecessors. The widening usage of the term stratigraphy is reflected in classifications for rock bodies that are not layered. The new units are divided into the following two classes and are defined in Appendix A (new categories are in italics):

Material categories based on content or physical limits: Lithostratigraphic, *lithodemic, magnetopolarity,* biostratigraphic, pedostratigraphic, *allostratigraphic.*

Categories expressing or related to geologic age: Chronostratigraphic, *polarity-chronostratigraphic,* geochronologic, *polarity-chronologic, diachronic,* geochronometric.

The most remarkable of the new categories are the *lithodemic* units, which deal with unstratified igneous and metamorphic rocks. With formal recognition of this category, the rules of stratigraphy now apply to all rocks on the earth's surface. Because lithodemic units are unstratified, their boundaries are based on the positions of lithologic changes. Most such boundaries are intrusive contacts, but there can be gradational contacts. Metamorphic rocks frequently show gradational changes in which the degree of deformation, or the presence of indicator minerals changes gradually away from the source of temperature or pressure. The basic unit, a *litho-*

deme, is comparable in concept to a sedimentary formation. Ideally, a lithodeme has some sort of lithologic consistency and is mappable. The rock name in a lithodeme is capitalized, as in San Marcos Gabbro or Pelona ·Schist. A group of lithodemic units can be united into a formal unit called a *suite*, as long as all the included lithodemes are either igneous or metamorphic. If two or more classes of rock (i.e., igneous, metamorphic, or sedimentary) are lumped together into a formal unit, it is called a *complex*. The Franciscan Complex, for example, is composed mostly of metamorphic rocks but also contains undeformed sediments and igneous intrusions.

Units that are defined and identified on the basis of the discontinuities that bound them are called *allostratigraphic* units. The lithologies of allostratigraphic units are often very similar, but the thin discontinuities between them prevent them from being considered a single formation. The Code cites terrace gravel deposits on opposite sides of valley walls as an example of allostratigraphic units. Such deposits may be either physically overlapping or separated, but they are usually very different in age and are bounded by a distinct unconformity. In the case of terrace deposits, each may have been deposited by a different glacial episode. The unconformity-bounded seismic sequences of Vail et al. (1977) and Sloss's (1963) sequence are also examples of allostratigraphic units, although they are not yet formally defined. Allostratigraphic units have the same criteria for recognition and formal definition as other formal units.

Previously, only time (geochronologic) and time-rock (chronostratigraphic) units existed as units that depend on geologic age. The new Code recognizes *diachronic* units, which are the span of time represented by rock bodies that are known to be diachronous. This is a remarkable reversal from the old concept that rock bodies were usually time equivalent. Now there is a formal method by which geologists can refer to the time represented by rock bodies that are explicitly *not* time equivalent.

The new Code also distinguishes between geochronologic units, which are ultimately based on chronostratigraphic units, and *geochronometric* units, which are expressed in years and have no material referent. This unit has become necessary, especially in Precambrian geology, in which events are referred to by their numerical, not relative, age.

The Code is clearly written, so it is unnecessary to paraphrase it further here. Most geologists have agreed to follow the Code, and most geologic publications insist on adherence to it. If one wants to be understood, and to understand what other people think and write, knowing the Code is a necessity.

FOR FURTHER READING

Owen, Donald E., 1987. Usage of stratigraphic terminology in papers, illustrations, and talks. *Jour. Sed. Petrol.* 52(2): 363–372. A perceptive article that points out many of the common errors made by geologists confusing time, time-rock, and rock units.

Schoch, R. M., 1989. *Stratigraphy, Principles and Methods.* Van Nostrand Reinhold, New York. Advanced-level treatment of the conceptual framework of stratigraphy, emphasizing the theoretical ideas behind the stratigraphic codes.

Shaw, Alan B., 1964. *Time in Stratigraphy*, McGraw-Hill, New York. A stimulating discussion of the nature of epeiric sedimentation and its implications for geologic time.

Wilgus, C. K., et al. (eds.), 1989. *Sea Level Changes—An Integrated Approach.* Soc. Econ. Paleo Min. Spec. Publication 42. The most up-to-date review of the Vail curves and their implications for global correlation. Many of these papers will revolutionize stratigraphy.

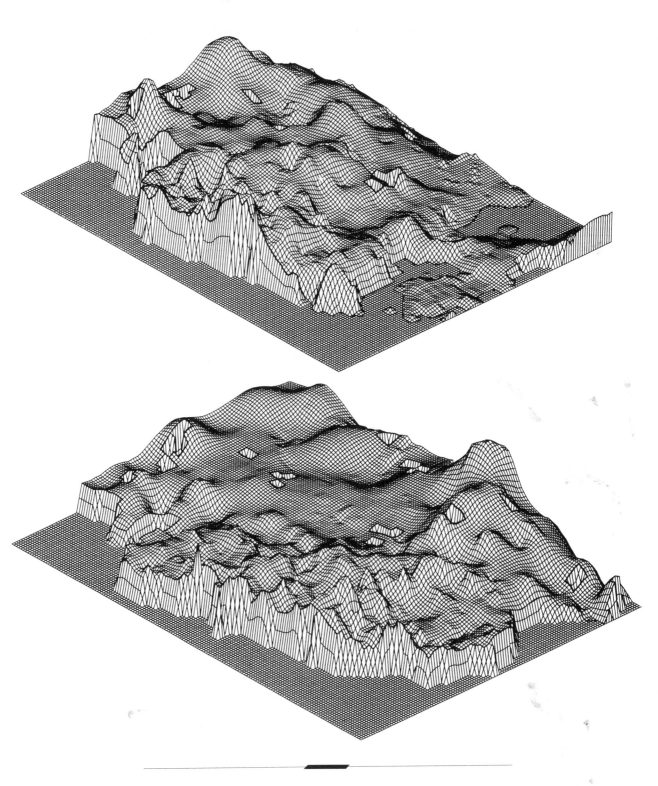

Computer-drawn isometric projection of thicknesses of Cretaceous strata, central Alberta. From Hughes (1984). Courtesy Geological Society of Canada.

9

Stratigraphic Methods

MEASURING AND DESCRIBING STRATIGRAPHIC SECTIONS

The fundamental data for nearly all stratigraphic studies are measured stratigraphic sections. Accurate measurement, precise description, and careful collection of rock samples and fossils are essential before any correlation or interpretation can be made. A stratigraphic column provides a one-dimensional sequence of units; two columns, placed side-by-side and correlated, produce a two-dimensional cross section; three or more columns, placed on a map and correlated in three dimensions, result in a spatial representation of all of the rock units.

Every section presents its own peculiarities. In reconnaissance mapping, it is often necessary to measure a number of sections in minimal detail. For a careful bed-by bed analysis, it may take days to measure one section. The section(s) to be measured must be chosen carefully. In some regions, only partial exposures are available, and every decent outcrop must be measured. Arid regions with high relief, on the other hand, usually have excellent outcrops everywhere. In this case, geologists prefer sections that are the thickest, structurally the most

uncomplicated, somewhat distant from other sections, and the most accessible and climbable. Most learn by experience to recognize outcrops that will enable them to concentrate on measuring and describing rather than on keeping their footing.

Section measurement begins with recording all the pertinent information, such as date and precise topographic location, in the field notebook (Fig. 9.1). Photographs of the outcrop can supplement field sketches. The geologist must also determine if the beds have any dip, and which measurement technique (discussed later) is appropriate to the thickness, exposure, and attitude of the outcrop. Sections are always measured from the bottom of the exposure because, by superposition, the lowest beds accumulated first and are the oldest. The first step is to describe and subdivide the strata in terms of homogeneous, informal units that are natural for the exposures in the outcrop. These may or may not correspond to formally named formations or members. In many cases, the geologist does not know in advance where the formal boundaries lie but must determine this later. Each informal unit is numbered, and a complete sedimentary rock description is recorded in the field notebook. The dominant lithology, texture, color (on

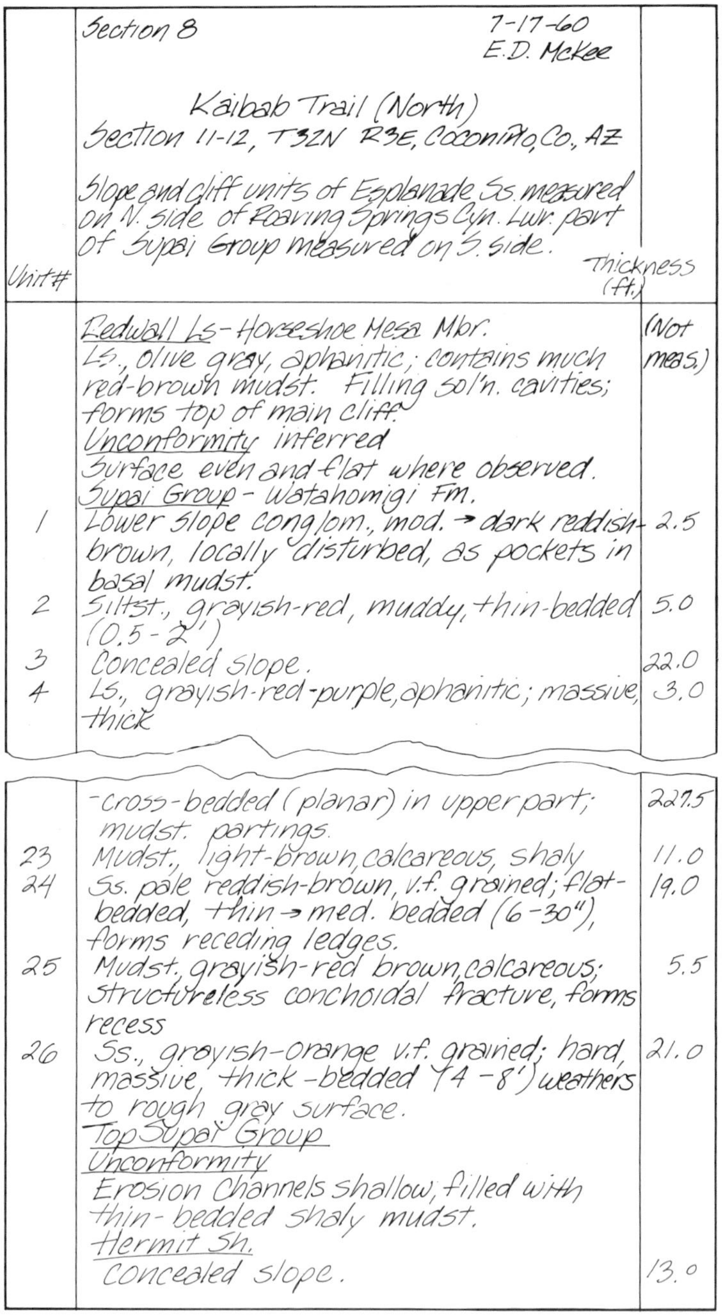

Section 8 7-17-60
E.D. McKee

Kaibab Trail (North)
Section 11-12, T32N R3E, Coconino Co., AZ

Slope and cliff units of Esplanade Ss. measured on N. side of Roaring Springs Cyn. Lwr. part of Supai Group measured on S. side.

Unit #		Thickness (ft.)
	Redwall Ls - Horseshoe Mesa Mbr. Ls., olive gray, aphanitic; contains much red-brown mudst. Filling sol'n. cavities; forms top of main cliff.	(Not meas.)
	Unconformity inferred. Surface even and flat where observed.	
	Supai Group - Watahomigi Fm.	
1	Lower slope conglom., mod. → dark reddish-brown, locally disturbed, as pockets in basal mudst.	2.5
2	Siltst., grayish-red, muddy, thin-bedded (0.5 - 2')	5.0
3	Concealed slope.	22.0
4	Ls., grayish-red-purple, aphanitic; massive, thick	3.0
	-cross-bedded (planar) in upper part; mudst. partings.	227.5
23	Mudst., light-brown, calcareous, shaly	11.0
24	Ss. pale reddish-brown, v.f. grained; flat-bedded, thin → med. bedded (6-30"), forms receding ledges.	19.0
25	Mudst., grayish-red brown, calcareous; structureless conchoidal fracture, forms recess	5.5
26	Ss., grayish-orange v.f. grained; hard, massive, thick-bedded (4 - 8') weathers to rough gray surface.	21.0
	Top Supai Group	
	Unconformity. Erosion Channels shallow, filled with thin-bedded shaly mudst.	
	Hermit Sh. Concealed slope.	13.0

Figure 9.1

Example of field notes taken in measuring section. Data from McKee (1982).

both weathered and fresh surfaces), macroscopic mineralogy, induration, and outcrop exposure and resistance to weathering are the most important properties to be noted in the field. The contact relations (sharp or gradational) between units are also noted, as are any visible structural features (folding, brecciation, or fracturing). Sedimentary structures are informative, and if they give any indication of the locations of tops of deposition beds or of current directions, these should be measured and recorded.

All of these features can be summarized by symbols to produce what is known as a *graphic log* (Fig. 9.2). An example was seen in Fig. 4.27. The graphic log has the great advantage of reducing long verbal descriptions to a visual plot that becomes easy to read with practice. Grain sizes can be shown by the relative width of the stratigraphic profile. Lithology is indicated by a series of standardized patterns. Sedimentary structures and fossils can be shown symbolically either directly on the column or in a series of parallel columns. Paleocurrent directions

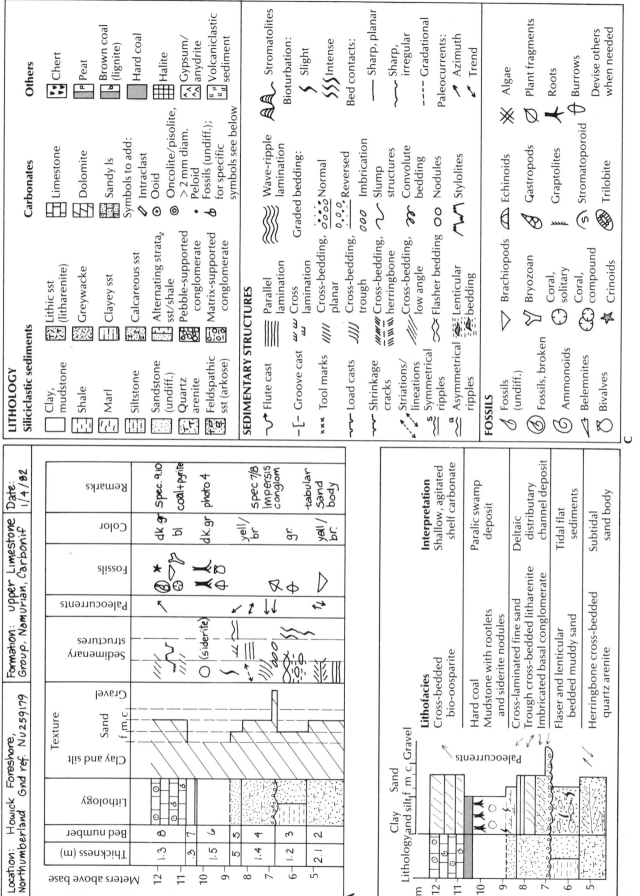

Figure 9.2 (A) Example of field notes taken while measuring a section and constructing a graphic log. (B) Summary draft of graphic log, based on data in A. (C) Standard symbols for lithology, sedimentary structures, and fossils used in measuring sections and constructing graphic logs. From Tucker (1982).

can be plotted and placed in the column where they were measured. Some features still require verbal description, so a separate column for notes is appropriate.

Any fossils that occur should be noted and identified as well as field conditions permit. If the specimens are important or should be studied further for some reason, they should be carefully collected and their *exact* position on the column recorded. Fossils without precise stratigraphic data are almost always useless. Only amateurs and rockhounds pick up fossils in the field without carefully recording where they are found. Trained geologists have no excuse! In most studies, samples of the rock units are needed for laboratory analysis or detailed study. These samples need not be large (a few rocks can make a backpack heavy), but it's better to be safe than sorry. Some problem with the description or correlation may arise later in the laboratory, and it might be impossible to return to the outcrop.

The method of measuring stratigraphic thickness varies with the attitude, exposure, and total thickness of the beds to be measured. The simplest measurements are of horizontal and well-exposed beds, typical of many sequences in the U.S. midcontinent. In such a case, the fastest and easiest method is to use a sighting hand level and measure in increments of one's own eye height (Fig. 9.3A). Starting from the base of the outcrop, the geologist sights to a point that is level with eye height and mentally notes its location. Then the geologist walks to that point and repeats the process, describing the beds between sightings along the way. If the level has a clinometer (such as is found in a Brunton transit or an Abney hand level), it is possible to measure dipping beds using the *Hewett method*. The clinometer is set at the angle of dip, and the geologist sights for the next point to stand on, moving perpendicular to strike (Fig. 9.3B). In this case, the eye height must be multiplied by

the cosine of the dip angle to get the true stratigraphic thickness. If the dip is consistent, this correction can be made once at the beginning, and then each increment on the section in the field notebook is the corrected eye height. Most standard field notebooks contain trigonometric tables for making such corrections.

The best method of measuring dipping beds, especially when the exposures are steep or discontinuous, is to use a simple device called a *Jacob's staff* (Fig. 9.4A). A Jacob's staff is a lightweight pole of some convenient standard length (such as 1.5 meters or 5 feet) with smaller increments marked on the side. A horizontal sighting bar is usually attached to the top, and a good Jacob's staff also has a clinometer. Expensive Jacob's staffs hold a Brunton in a clamp on the top and even telescope shut when not in use. A simple Jacob's staff can be made from a pole with a rotating plate on the top (attached by a wing nut) on which is glued a protractor and an inexpensive line level. It is not necessary to spend a lot of money to get an accurately measured section.

The Jacob's staff can be used to measure horizontal beds in the same manner, except that the increment of measurement is the length of the Jacob's staff (a convenient standard unit) and not eye height. Few geologists have an eye height that is as convenient; thus, measuring a section with the Jacob staff is sometimes faster than hand leveling. The Jacob's staff is by far the easiest and most accurate method for measuring inclined bedding. The staff is held perpendicular to the bedding surface, and the geologist sights along the top, perpendicular to strike, to find the next point of measurement (Fig. 9.4A). The clinometer at the top is set at the dip angle to insure that the staff is exactly perpendicular to bedding. Because good outcrops are seldom exposed in a single straight transect, the geologist must often move

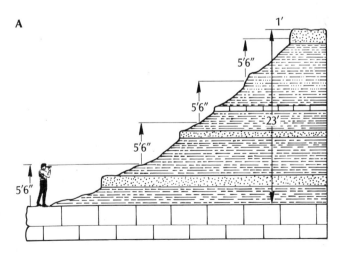

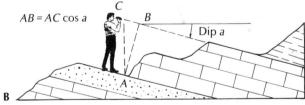

Figure 9.3

(A) Method of measuring horizontal strata with a sighting hand level, using eye height as standard dimension. From Krumbein and Sloss (1963). (B) Hewett method of measuring thickness of dipping strata, using a clinometer set at the angle of dip. In this method, stratigraphic thickness is equal to eye height times the cosine of the dip angle. After Kottlowski (1965).

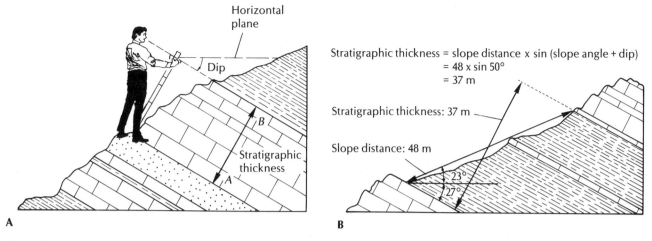

Figure 9.4

(A) Measuring dipping strata with a Jacob's staff. After Kottlowski (1965). (B) Measuring dipping strata using a measuring tape across a sloping surface. In this method, both the dip and slope angle must be measured, in addition to determining the slope distance by measuring tape. From Krumbein and Sloss (1963).

laterally along a bedding plane and continue measuring the section where the exposures are better. Wherever the transect begins, it must be kept perpendicular to the strike of the beds, or else a trigonometric correction is necessary. For this reason, the geologist should continually keep track of the strike and dip of the beds, especially if the attitude is very steep or changes rapidly. A single geologist can measure a section quite rapidly with a Jacob's staff and then go back and fill in the details and make collections. Two geologists, however, can do the job in less than half the time because one can measure while the other writes descriptions in the field notebook and collects samples.

Two geologists are necessary for methods that require measuring tape or a similar device. If the total section is not very thick, or if the individual beds are very thin and each requires detailed study, then a measuring tape is preferable. Using simple trigonometry, one can use a measuring tape to measure inclined beds along a slope (Fig. 9.4B). Many other methods of measuring dipping beds in unusual ways, or with unusual tools, are possible. The excellent little book by Kottlowski (1965) is highly recommended for the details of section-measuring techniques.

STRATIGRAPHIC CROSS SECTIONS AND FENCE DIAGRAMS

Once a series of stratigraphic sections have been measured in an area, they can be drafted side-by-side and correlated by the procedures described in Chapter

8. The result is a two-dimensional representation of the earth known as a *stratigraphic cross section* (See Fig 8.1, for example). It differs from a normal geological cross section in that no topography is shown, and structural deformation is either corrected for or shown diagrammatically. Stratigraphic cross sections are usually aligned along some sort of shared stratigraphic datum or by the topographic elevation of their beds if they are horizontal. After lines of correlation are drawn, patterns of pinchout and other differences in thickness, intertonguing, facies changes, unconformities, and biostratigraphy can be shown, and their relationships can be delineated. A clear stratigraphic cross section often carries more information about the geologic history of a region than any other type of diagram.

However, a stratigraphic cross section is only two-dimensional. One can obtain an even greater understanding of the geometry of lithologic units by drawing a series of cross-sections on an isometric map base to form a type of three-dimensional cross section known as a *fence diagram* (Fig. 9.5). Fence diagrams can make some patterns and relationships very clear, but they must be drawn carefully or they become confusing. Reducing cross sections to fences of the appropriate scale for the isometric diagram also loses much of the detail. If the geology is too complicated, it is impossible to represent it on a few fences; but the more fences that are drawn, the more difficult the diagram is to read. Computer programs are now available that eliminate some of the mechanical problems of drawing cross sections and fence diagrams (particularly in isometric drawing). However, correlation is not an automatic procedure, so geologists must use their own judgment in deciding what is geologically reasonable.

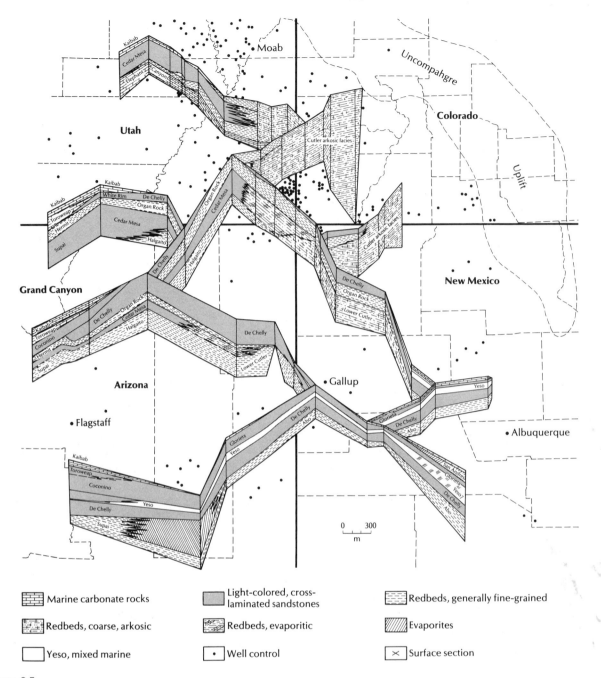

Marine carbonate rocks

Redbeds, coarse, arkosic

Yeso, mixed marine

Light-colored, cross-laminated sandstones

Redbeds, evaporitic

• Well control

Redbeds, generally fine-grained

Evaporites

× Surface section

Figure 9.5

A fence diagram from the Permian rocks of the Four Corners region. By means of this three-dimensional projection of a series of cross sections, complex facies relationships can be clearly seen. The arkoses from the Uncompahgre Uplift in Colorado (Cutler Formation) interfinger with nonmarine shales (Organ Rock, Halgaito) and eolian sandstones (Cedar Mesa, De Chelly, Coconino). In the Grand Canyon region, these interfinger with marine limestones (Kaibab, Toroweap). In southern New Mexico, there are evaporites. From Baars (1962).

STRATIGRAPHIC MAPS
Types of Maps

A map is a two-dimensional representation of points in a three-dimensional space. The purpose of many maps, such as roadmaps, is to show horizontal distances and directional relationships, but maps can also reduce three-dimensional objects to two dimensions by shadowing and shading (which requires considerable artistic talent) or by contouring (which is more mechanical). Topographic maps, the most familiar example of contouring, reduce three-dimensional surfaces to two dimensions. Contouring can also be used to represent three-dimensional surfaces of an underground rock bed (*structure contour maps*) or three-dimensional thicknesses of rock (*isopach maps*). A *stratigraphic map* shows the areal distribution, configuration, or aspect of a stratigraphic unit or surface. The simplest stratigraphic maps, such as isopach and structure contour maps, show the external geometry of a rock body. More complex maps also depict variations in the composition of a rock body, its vertical variability, and its internal geometry. Such maps are usually classed as *facies maps*. Finally, the whole geologic picture can be integrated into interpretive maps, such as paleotectonic maps, and paleogeographic maps.

Structure Contour Maps

The structure contour map is similar to a topographic contour map except that the surface contoured is not topography but is the surface of an underground lithologic unit (Fig. 9.6). Like other contour maps, it is made by placing as many data points on the map base as possible and contouring between those that are different. This requires an extensive array of subsurface points, as well as all the outcrop elevations available. If few data points are available, the contour lines are best drawn by hand using a bit of interpretation to generate as reason-

able a picture as possible. When the density of data points is great enough, then the contours can be drawn mechanically. Computer programs are now available that can accomplish this at minimal cost. However, every geologist should have some practice with manual contouring. There is no better way to understand contour lines and to learn their limitations.

Structure contour maps are not difficult to interpret. Structural highs such as domes and anticlines show up clearly, as do low spots such as basins and synclines. Structure contours that are tightly bunched along a straight line usually indicate faulting, and the contours themselves can show the offset. The structure contour map is the primary tool for finding structural traps for hydrocarbons, such as anticlines and faults.

Isopach Maps

Isopachs are points of equal thickness of a rock unit, and an isopach map contours these points. This requires data points along the top and the bottom of the bed. If there is any dip, the apparent thickness of a bed may be greater than its true thickness. The *apparent* thickness of a unit is shown by an *isochore map*. Many so-called isopach maps are actually isochore maps because they are not corrected for structure. Isopach data usually come from wells scattered around the area of a map base, but they can also be drawn from two structure-contour maps by subtracting the elevations of the lower surface from those of the upper surface. The data base is contoured by the usual methods. The result, however, is not a surface but must be visualized as an irregular lens, or pillow, or bowl, or shoestring, or whatever three-dimensional shape seems appropriate. The contours are not on the top surface of the structure (which is considered horizontal by convention), but along the bottom surface. They indicate where the body bulges up or down, pinches out, or is truncated (Fig. 9.7).

Isopach and isochore maps help locate features that are found in no other type of map. A concentric bull's-eye pattern with increasing thicknesses toward the center clearly shows a basin, with the deepest part having the thickest isopachs. Abrupt changes in isopach spacing can indicate the boundary between thin shelf sequences and thick basin deposits. The boundary between the shelf and the basin is commonly called the *tectonic hinge*. Unusual flexures or breaks in the isopach pattern usually indicate buried folded or faulted uplifts. Where isopachs thin to zero, stratigraphic pinchouts or abrupt fault boundaries are indicated by tight contour spacing. However, zero isopachs may indicate erosion rather than depositional thinning. In such a case the unit has been carved out from above, and the zero isopachs represents "holes" in its upper surface. Further evidence is then needed to determine the true significance of the peculiar patterns on the isopach map.

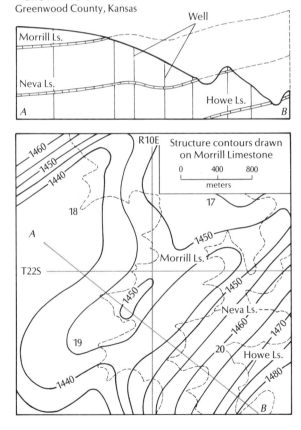

Figure 9.6

Below: Structure contour map, showing the relation of structure contours to outcrop patterns. *Above:* The projection of the contours on the Morrill Limestone beyond its outcrop edge through the use of lower marker beds and known intervals between them. From Krumbein and Sloss (1963).

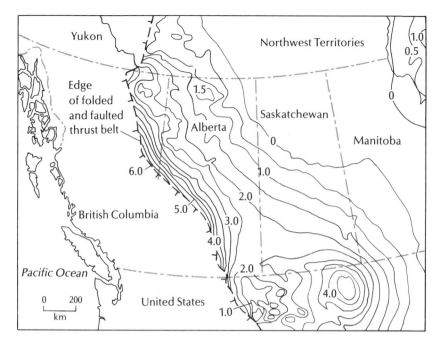

Figure 9.7

Isopach contours of sediments in the Rocky Mountain foreland basin of Alberta, Saskatchewan and Montana. Thickness increases southwestward, with a maximum thickness of 6 kilometers right against the fold and thrust belt, which produced the downwarping and supplied the sedimentary fill. The bull's-eye pattern in the lower right indicates a bowl-shaped closed basin. The irregular contours with the 1 kilometer thickness just to the left of the basin indicate thin spots across basement uplifts. From Porter et al. (1982).

A series of isopach maps indicates the time when basins were subsiding and accumulating sediment and when they were shallow or being eroded. A trend toward zero isopach, which represents a pinchout, implies an ancient shoreline. The thickness and dimensions of a basin are often critical to determining whether it has hydrocarbon reservoir potential or where stratigraphic and structural traps lie. For these reasons, isopach maps have long been used in petroleum geology.

Facies Maps

A facies map shows the areal variation of some aspect of a stratigraphic unit. Variations in lithologic aspects and attributes are shown by *lithofacies maps;* variations in faunal aspects are shown by *biofacies maps.* Facies maps may show the thicknesses of a single component of a unit, such as the sandstone thickness in a formation. An isopach map of a single rock component in a unit is called an *isolith map* (Fig. 9.8B). The actual thickness of a component can be converted to its percentage of the thickness of the whole stratigraphic unit, producing a *percentage map.* More than one component can be shown in a lithofacies map; for example, the ratios of two components (such as sand/shale) can be calculated and contoured. Such a *ratio map* (Fig. 9.8A) gives a good

idea of where a unit is changing in grain size or where various environments are found. If three components are represented, then some sort of triangular diagram for the ratios of the three components must be used as a key (Fig. 9.9). The resulting *triangle facies map* is a bit more difficult to read, but it often gives a good sense of the distributions of sedimentary environments and facies changes. Further, the degree of mixing of the three components can be calculated mathematically, and these values can be contoured to form an *entropy map.* This kind of map is a good indicator of how mixed or homogeneous a rock unit is. Many other types of lithofacies maps have been created for specific purposes, but they are less commonly used.

Most facies maps are based on two-dimensional representations of three-dimensional features. However, computer technology has made isometric projections of isopach, isolith, and other facies data much easier to plot. Figure 9.10 shows a perspective diagram of the isopachs and the sandstone, siltstone, and shale isoliths for late Cretaceous formations in Alberta. The computer perspective diagram is much easier to visualize than a two-dimensional map and conveys information very clearly. However, a high density of data points is necessary for the computer to plot this kind of diagram accurately.

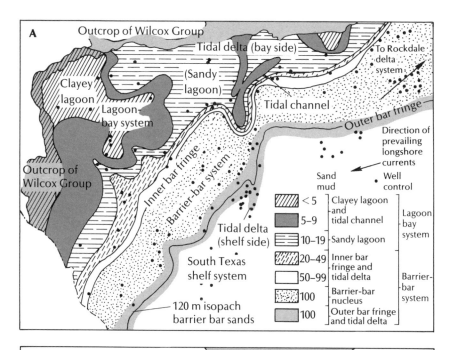

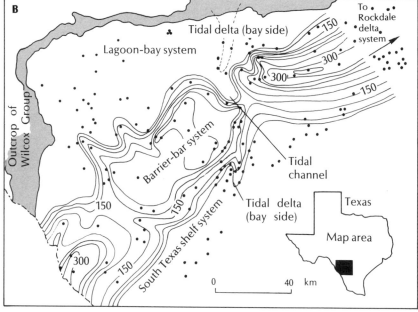

Figure 9.8

Two examples of facies maps of barrier-bar and bay-lagoon systems in the Wilcox Group (Eocene) of the Texas Gulf Coastal Plain. (A) Sandstone/shale ratio map, with the ratios indicated by various patterns of shading. (B) Sand isolith map (isolith interval = 30 meters) showing thicknesses of sands through the unit. From Lofton and Adams (1971).

Paleogeologic Maps

An ordinary geologic map shows the distribution of rock outcrops on the earth's present topographic surface. One can use subsurface information to strip away younger units and to reconstruct the pattern of outcrops before they were buried. This kind of map is called a *paleogeologic map*, or a *subcrop* (i.e., *sub*surface *out*crop) *map*. If one could take a time machine and visit the world as it was before the younger units were deposited, the paleogeologic map could be used as an ordinary geologic map. Conversely, if one mapped the subcrop

pattern of the *base* of a sequence that covers an ancient surface one would have a geologic map of units as seen from below, a so-called *worm's eye map*, or a *supercrop map*.

Paleogeologic maps are particularly useful in areas such as the craton, where broad regional unconformities occur and younger transgressive sequences lap over older beds that have been eroded to expose many other units. The paleogeologic map may show structural features that are not visible because they are covered by younger features that were deposited after structural deformation ceased. A series of paleogeologic maps is often one of the best tools for determining the timing

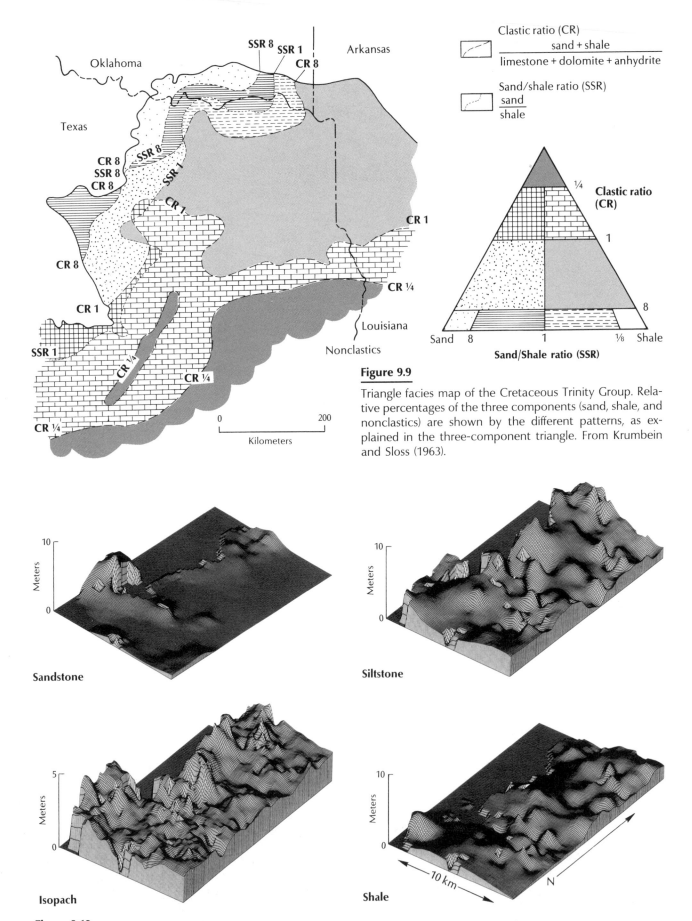

Figure 9.9

Triangle facies map of the Cretaceous Trinity Group. Relative percentages of the three components (sand, shale, and nonclastics) are shown by the different patterns, as explained in the three-component triangle. From Krumbein and Sloss (1963).

Figure 9.10

Computer-drawn isopach and isolith diagrams, showing three-dimensional thickness of the total unit and of its sandstone, siltstone, and shale components. These thicknesses lie between two marker coals in the Cretaceous of central Alberta. From Hughes (1984).

and extent of episodes of structural deformation. Paleo-geologic maps also give a good sense of the relative magnitudes of unconformities, transgressions, and regressions. They can also have economic applications. If, for example, a region is covered by an impermeable cap rock, the paleogeologic map will show where the permeable reservoir rocks occur unconformably beneath it, a classic stratigraphic trap for hydrocarbons.

Paleogeographic Maps

The stratigraphic maps described so far are based on *objective* attributes of lithologic units, such as elevation, thickness, geologic age, percentages of components, and homogeneity. Sometimes it is useful to display subjective *interpretations* of these lithologies on a geologic map as well. From inferences based on lithofacies distribution and paleoenvironmental evidence, the geologist can construct a *paleogeographic map*, which shows the distribution of ancient environments at a given time in geologic history (e.g., see Fig. 3.26 or 4.10). Paleogeographic maps show the shorelines, the areas of marine deposition, the eroding uplands, and sometimes the climatic and vegetational belts at some instant in geologic history. Historical reconstructions of this kind are often the ultimate goal of historical geology. Naturally, such interpretation can be highly speculative, especially if there are missing areas due to unconformities. However, such information can be economically useful in predicting the distribution of shorelines or deltas or other stratigraphic traps, or in getting a good perspective of the total paleoenvironmental picture.

BASIN ANALYSIS

The construction of stratigraphic cross sections, fence diagrams, and stratigraphic maps enables the geologist to analyze the three-dimensional geometry and the depositional history of an ancient sedimentary basin. Many types of evidence can be developed that indicate the source of sediments and their direction of transport, the geometries and thicknesses of rock units and their relative ages, and ultimately, the tectonic factors that controlled the geologic history of the basin. This kind of research is called *basin analysis*. Most of the world's nonrenewable fuel resources are found in sedimentary basins, and predicting the occurrence, abundance, and cost of recovery of oil, gas, and coal requires the most complete understanding possible of these basins.

Most of the fundamental methods of basin analysis are discussed throughout this book. However, particularly common in basin analysis is the determination of the source and direction of transport of the basinal sedi-

ments. One finds this information by examining the basin sediments to determine their *provenance* (source area). In field studies, the geologist must be alert for any rock type or clast that is megascopically distinctive and should also study the area surrounding the basin for possible sources for the sediment.

More often than not, provenance studies require detailed petrographic study. The most useful diagnostic grains are often visible only in thin section. If sediment has few lithic fragments or other polycrystalline grains that are diagnostic by themselves, then the percentage of certain key mineral grains may be important. Most diagnostic are the heavy minerals (with densities greater than the ubiquitous quartz and feldspars), which can be separated from others by settling crushed samples in a dense liquid such as bromoform. Assemblages of heavy minerals can often be traced to distinctive source terranes. For example, some minerals, such as garnet, epidote, staurolite, and kyanite/sillimanite/andalusite, are characteristic of metamorphic terranes. Others may typify the presence of certain igneous rocks. For example, tourmaline, beryl, topaz, and monazite are commonly derived from pegmatites. The most stable heavy minerals, such as zircon, tourmaline, and rutile, can maintain their integrity through many sedimentary cycles. Thus, their abundance increases as sediments mature. The relative abundance of these three minerals is often called the *ZTR index* and is used as an indicator of sediment maturity.

One also can map textural trends of the particles in a basin. For example, the roundness of many sedimentary grains usually increases with distance from the source, so a map of grain roundness (Fig. 9.11A) can indicate source and direction of transport. Similarly, grain size usually decreases with distance from the source, so a map of maximum clast diameter (Fig. 9.11B) can reveal much information about the source area.

The most direct method of determining direction of transport, however, comes from taking measurements of sedimentary structures that show clear directional trends. Cross-beds, ripple marks, and the various directional bottom marks discussed in Chapter 2 all give clues to the flow direction of a depositional event. These structures can be measured in the field and used to plot *paleocurrents*.

Paleocurrent analysis can be a powerful tool, but it has a pitfall: A reliable study may require hundreds of measurements. Many of the sedimentary environments discussed in Chapters 3–6 have highly variable current directions. For example the currents in a meandering river can point in all directions of the compass, depending on which part of the meander loop they come from. Measurements from a restricted outcrop might be from a part of the meander loop that happens to be flowing opposite to the overall flow direction of the river. This could result in a very erroneous paleogeo-

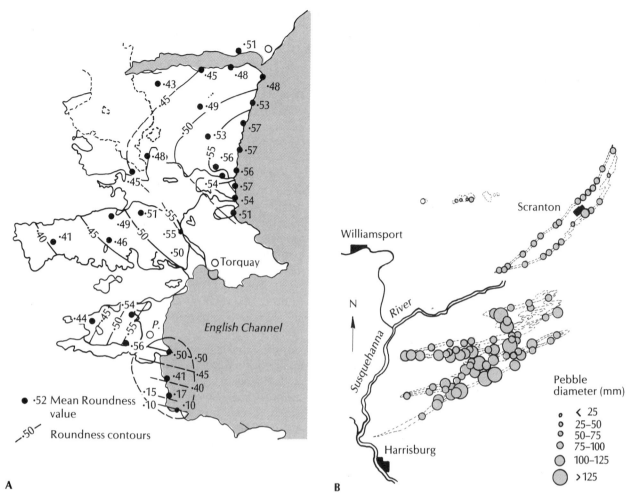

A   **B**

Figure 9.11

(A) Map of variation in roundness of pebbles of the Triassic Lower New Red Sandstone of England. The highest roundness values are found north of Torquay, indicating a souce that is further inland. After Laming (1966). (B) Average maximum clast diameter in the Pottsville Formation of Pennsylvania. After Meckel (1967). Note that the coarsest material is in the southeast, indicating a source area in this direction.

graphic reconstruction of the river system. In beach environments, there can be multiple paleocurrent directions. The foreshore strata could be seaward-dipping, the backshore strata could be landward-dipping, and there could also be shore-parallel structures from longshore currents. All of these could be preserved in the same limited outcrop. A field geologist who lacked understanding of the paleoenvironment would be totally baffled.

In any paleocurrent study, the geologist must take as many measurements as is practical and make field observations about which part of the sedimentary environment they might come from. In addition, paleocurrent measurements from dipping beds must be corrected to the horizontal. This is easy to do on a stereonet though

many calculators and small computers can run programs to do this. To display a large sample of directional data and get a sense of the overall trend, the geologist can plot paleocurrent data on a *rose diagram*. An example of this type of diagram can be found in Box 9.1, which discusses an example of basin analysis. A rose diagram can show a single, unimodal current direction or two current directions (bimodal), which can be perpendicular, parallel, or oriented at any number of angles. The preceding beach example would probably display a polymodal distribution, which might be very difficult to pick out from a noisy random distribution of paleocurrents. The analysis of paleocurrents is a complex procedure, and the reader is referred to an excellent book by Potter and Pettijohn (1977) for further details.

Box 9.1
Basin Analysis of the Ridge Basin, California

A particularly well-exposed and thoroughly studied basin is the Ridge Basin in the Transverse Ranges north of Los Angeles, California (summarized in Crowell and Link, 1982). Small in area, it is now uplifted and deeply dissected, so that most of the features can be studied in outcrop (Fig. 9.12). The Ridge Basin was formed by rifting between two major fault systems, the San Gabriel and San Andreas faults (Fig. 9.13A), which produced a very narrow (10 to 15 kilometers wide) but extremely deep rift that acccumulated 13,500 meters (44,000 ft) or over 8 miles of sediments during about 8 million years of the late Miocene and Pliocene. Because the basin is so narrow and deep, facies changes are very rapid across it (Fig. 9.13B).

The fault-bounded basin margins are characterized by coarse fanglomerates and breccias, which pass laterally into the sandstones and shales of the basin center. The Ridge Basin began filling up in the late Miocene when a marine incursion deposited the Castaic Formation, up to 2800 meters (9200 ft)

of marine mudstones (Figs. 9.13B, 9.14). Although the Ridge Basin itself is thousands of meters deep, the water depth in the Castaic Formation was only in the range of 45 to 90 meters (Stanton, 1960, 1966). Most of the thickness of the stratigraphic units in the Ridge Basin results from continual subsidence along the faults and not from catastrophic deepening. After the Castaic marine incursion, the Ridge Basin continued to fill with lacustrine shales of the Peace Valley Formation. These interfinger with fluvial and deltaic sands and conglomerates of the Ridge Route Formation on the northeastern margin and the coarse Violin Breccia on the southwestern margin (Figs. 9.13B, 9.14). The Peace Valley-Ridge Route sequence is capped by the late Miocene-early Pliocene Hungry Valley Formation, 1200 meters (4000 ft) of fluvial and alluvial sandstones, conglomerates, and mudstones.

Basin analysis of the Ridge Basin included detailed sedimentological studies of each of the stratigraphic units in Fig. 9.13B, and the study of their

Figure 9.12 Aerial photograph of the northern part of the Ridge Basin, showing the thick sequence of lacustrine shales and deltaic and fluvial sandstones dipping away from the viewer. Frenchman Flat and Highway 99 at lower left, Hungry Valley and Frazier Mountain in the background. Photo courtesy J. C. Crowell.

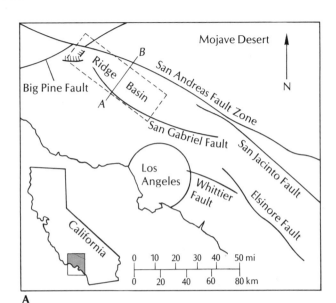

A

Figure 9.13 (A) Location of the Ridge Basin, southern California. Note location of cross section A-B. (B) Diagrammatic cross section of the Ridge Basin, showing the major stratigraphic and structural relationships. Section located along line A-B. From Crowell and Link (1982).

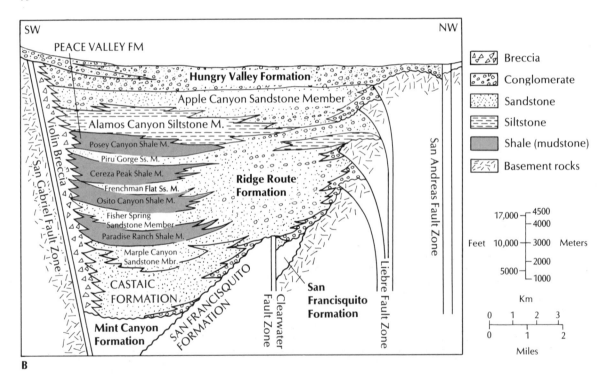

B

depositional environments. In addition, fossils (trace, plant, invertebrate, and mammal) were used to determine the age and depositional environment of each unit. Parts of the sequence have also been dated with magnetostratigraphy (Ensley and Verosub, 1982). Some of the units have also been analyzed geochemically.

Because the basin is so narrow and well exposed, with such obvious source areas, it is ideal for studies of source and direction of transport. The large number of paleocurrent measurements is summarized in Fig. 9.15. Most of the bidirectional paleocurrent marks (such as ripple marks, cross-beds, groove casts) show a strong east-west orientation, roughly perpendicular to the axis of the basin. These marks correspond to currents that transported sediments from the source areas in the east and west. Some of the paleocurrents, particularly those in the center

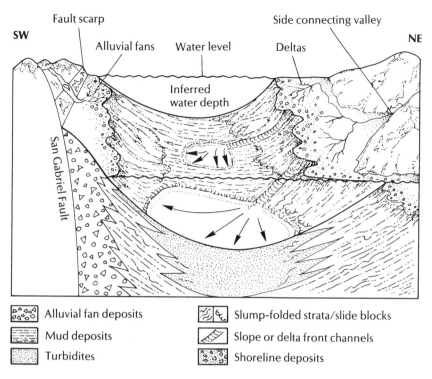

Figure 9.14 Depositional interpretation of Ridge Basin, showing the breccias and alluvial fans produced along the fault scarps. These grade rapidly into fluvial-deltaic deposits, then drop off sharply into deep-water shales and turbidites. From Crowell and Link (1982).

of the basin, show a northwest-southeast orientation, indicating that the transport in the center of the basin was southeasterly. Apparently the basin drained to the southeast, so the paleoslope for the deeper basin sediments was in that direction. The unidirectional indicators in the Violin Breccia show this southeasterly paleoslope direction particularly well.

Provenance studies of the Ridge Basin are complicated by the variety of source areas available on each side of the basin. Pebble counts in the Hungry Valley Formation (Fig. 9.16) showed that the most common rock types are granitics and gneisses which are exposed widely to the west and north. Volcanic rocks are locally abundant near the Big Pine Fault in the northwest and in the southeast margin of the basin. Certain distinct rock types, such as pink granites, augen gneisses, and fault-zone mylonite are derived only from the Alamo Mountain-Frazier Mountain area to the west and do not occur to the north or east. Marbles, on the other hand, are exposed only north of the San Andreas Fault and appear to be from a northeasterly source area. These

studies show that the bulk of the Ridge Basin sedimentary fill came from the north and east (Fig. 9.17), even though the basement west of the San Gabriel fault scarp was a closer source for sediments on the western side of the basin.

Provenance analysis of the cobbles in the Violin Breccia reveal an unusual mode of deposition. The Violin Breccia lithologies are restricted to gneisses, diorites, and granitic rocks, which have a very limited source area on the west side of the fault. No sedimentary clasts are found from prospective source areas just a few kilometers to the west, or from more distant areas. All of these lines of evidence suggest that the Violin Breccia source area was not very extensive and was drained by small streams with limited drainage basins. Yet the Violin Breccia has a total stratigraphic thickness of 13,400 meters (44,000 ft). How could such an enormous thickness of conglomerate have formed from such a small, poorly drained source area for over 7 million years?

Crowell (1982) pointed out that the characteristics of a basin trapped between two strike-slip faults

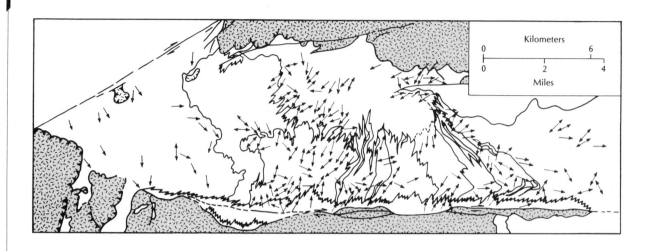

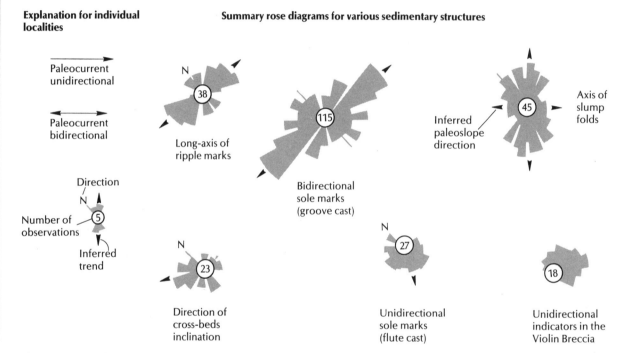

Figure 9.15 Paleocurrent rose diagrams for a variety of directional sedimentary current indicators in the Ridge Basin (circled numbers are number of observations). Notice that the bulk of the unidirectional indicators are out of the north-northeast, even on the southwestern edge of the basin where the souces from the southwest would have been much closer. The basin ultimately drained toward the southeast, as indicated by the paleocurrents. From Crowell and Link (1982).

provide a mechanism for sediment accumulation. The western source area moved northward relative to the basin as the San Gabriel fault continued to move in the late Miocene. In effect, the alluvial fans accumulating in the Ridge Basin were transported southward, away from their source like cars on a southbound coal train (Fig. 9.18A). As each alluvial fan deposit moved south from the source area, a new "coal car" (alluvial fan deposit) was formed behind it to the north. Consequently, the thick Violin Breccia is actually made of a sequence of shingled fan deposits, with the oldest lying farthest south (Fig. 9.18B) and each progressively younger fan overlapping the one before on its northern end.

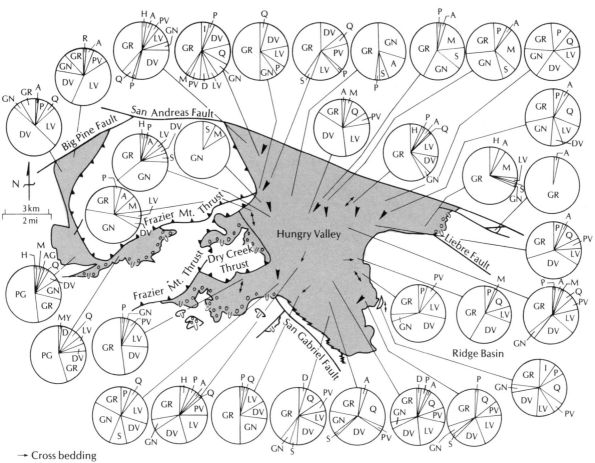

- → Cross bedding
- ↔ Channel orientation
- ► Imbrication

Figure 9.16 Provenance of pebble types from the Hungry Valley Formation, as counted by R. G. Bohannon. Relative percentages of types are indicated by pie diagrams. Abbreviations: A, amphibolite; AG, augen gneiss; D, diorite; DV, dark volcanics; GR, granitics; GN, gneiss; H, hornfels; I, intraclasts, such as rip-up clasts of underlying strata of shale and sandstone; LV, light-colored volcanics; M, marble; MY, mylonite; P, pegmatite; PG, pink granite; PV, purple volcanics; Q, quartz; S, schist. From Crowell and Link (1982).

Thus, the vertical stratigraphic thickness in any one place is less than a third of the total stratigraphic thickness, and the formation is time transgressive over about 7 million years.

The peculiarities of the Ridge Basin are due to its unusual geological setting. Unlike most basins in the craton or mountain belts (discussed in Chapter 14), the Ridge Basin was formed as a chasm or gap opened between two curved segments of a strike-slip fault (Fig. 9.17). This accounts for the extremely narrow yet deep basin, with an enormous thickness of sediment accumulated over 8 million years. These "pull-apart" basins are not common in the geologic record, but they produce a very distinctive result: an extraordinarily thick but narrow sedimentary package with coarse conglomerates on the margins, and very rapid facies changes to lacustrine and marine shales and sandstones in the middle. As the Ridge Basin example becomes better known, more such examples are being found in ancient sediments.

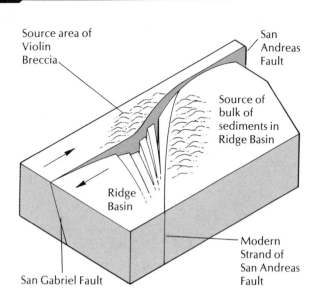

Figure 9.17 Block diagram showing the origin of the Ridge Basin as a sigmoidal bend in the San Andreas Fault. This explains the extremely steep yet narrow geometry of the basin. From Crowell and Link (1982).

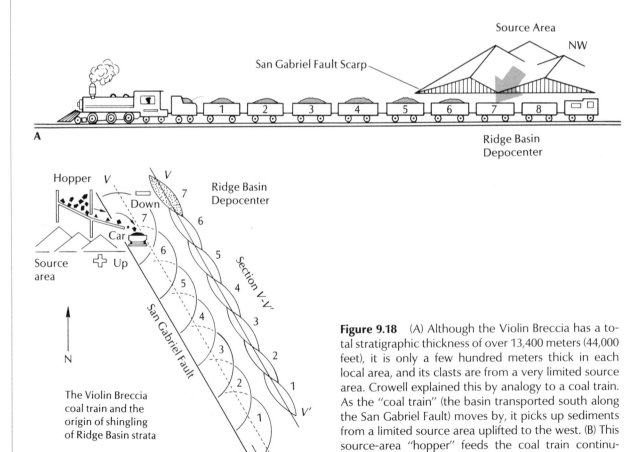

Figure 9.18 (A) Although the Violin Breccia has a total stratigraphic thickness of over 13,400 meters (44,000 feet), it is only a few hundred meters thick in each local area, and its clasts are from a very limited source area. Crowell explained this by analogy to a coal train. As the "coal train" (the basin transported south along the San Gabriel Fault) moves by, it picks up sediments from a limited source area uplifted to the west. (B) This source-area "hopper" feeds the coal train continuously, producing a "shingled" deposit that gets progressively younger (1–7) to the north. From Crowell and Link (1982).

FOR FURTHER READING

Brenner, R. L., and McHargue, J. A., 1988. *Integrative Stratigraphy*. Prentice-Hall, Englewood Cliffs, N. J. Recent textbook which is particularly strong in discussing the practical dimensions of stratigraphic methods, especially as they are used in the oil industry.

Conybeare, C. E. B., 1979. *Lithostratigraphic Analysis of Sedimentary Basins*. Academic Press, New York. An advanced discussion of the many methods and goals of analyzing sedimentary basins.

Crowell, J. C., and Link, M. H. 1982. *Geologic History of the Ridge Basin, Southern California*. Soc. Econ. Paleo. Min., Pacific Section, Field Trip Guidebook. The standard reference on the remarkable Ridge Basin, containing papers on nearly every subject, as well as a field trip guide.

Kottlowski, F. E., 1965. *Measuring Stratigraphic Sections*. Holt, Rinehart, and Winston, New York. The best little paperback guide to the different methods of measuring and describing stratigraphic sections. Unfortunately, it is currently out of print, but it is worth acquiring at any price.

Krumbein, W. C., and Sloss, L. L., 1963. *Stratigraphy and Sedimentation*. W. H. Freeman, San Francisco. A classic text on stratigraphy, containing some of the best chapters on stratigraphic maps available in print.

Langstaff, C. S., and Morrill, D., 1981. *Geologic Cross Sections*. Int. Human Res. Dev. Corp., Boston. A paperback manual explaining the practical side of drawing cross sections and fence diagrams, with many examples.

LeRoy, L. W., and Low, J. W., 1954. *Graphic Problems in Petroleum Geology*. Harper and Brothers, New York. A classic laboratory manual with excellent problems demonstrating stratigraphic columns, cross sections, fence diagrams, structure contours, isopachs, lithofacies maps, and many more. Also out of print but worth obtaining for its wealth of examples alone.

Levorsen, A. I., 1967, *Geology of Petroleum*. W. H. Freeman, San Francisco. The classic text on petroleum geology, with extensive discussion of stratigraphic maps and their interpretation.

Miall, A. D., 1984. *Principles of Sedimentary Basin Analysis*. Springer-Verlag, New York. The most sophisticated and up-to-date book on the study of sedimentary basins.

North, F. K., 1985. *Petroleum Geology*. Allen & Unwin, Boston. Comprehensive modern review of the methods of the oil industry, including stratigraphic maps and basin analysis.

Potter, P. E., and Pettijohn, F. J., 1977. *Paleocurrents and Basin Analysis*, 2d. ed. Springer-Verlag, New York. Outstanding book, with many clear examples, on paleocurrent analysis and its application for the understanding of sedimentary basins.

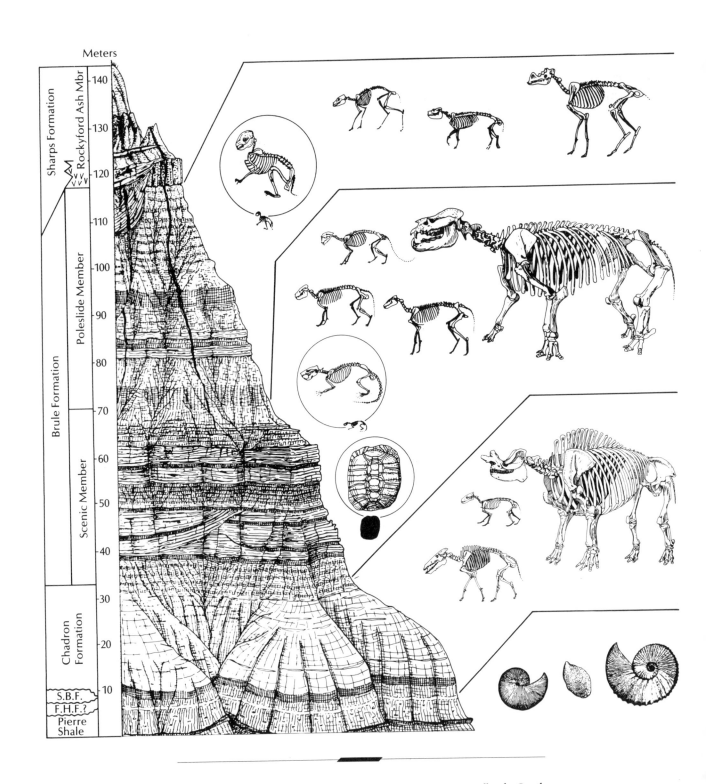

Meters

Sharps Formation — Rockyford Ash Mbr

Brule Formation — Poleslide Member — Scenic Member

Chadron Formation

S.B.F.
F.H.F.?
Pierre Shale

Biostratigraphic succession of fossil mammals from the Big Badlands, South Dakota. Courtesy G. J. Retallack.

10

Biostratigraphy

DEVELOPMENT OF BIOSTRATIGRAPHIC CONCEPTS

After the discovery of the principle of faunal succession, geologists slowly came to understand the importance of fossils in correlating and dating rocks. Realizing this importance, Murchison based his Silurian System largely on fossils and could recognize the Silurian outside the type area in Wales by the fossils in the rock. Sedgwick, who did not use fossils in his definition of the Cambrian System, found his definition nearly abolished until paleontologists studied his area and described a characteristic Cambrian fauna. Lyell tried to use the "modernization" of molluscan faunas as a geologic clock, independent of the rock record. Fossils have indeed proven to be the only widely applicable tool of time correlation in sedimentary rocks although they must be used with caution. Their use for stratigraphic correlation is called *biostratigraphy*.

Once the principle of faunal succession was widely accepted, further refinements of its use were inevitable. Smith's ideas led only to vaguely defined time units, and Lyell's work applied only to the Cenozoic, but the French geologist Alcide d'Orbigny, in his comprehensive study of the Jurassic (*Terrains Jurassiques*, 1842), showed that fossil *assemblages* were the keys to correlation. No matter what the lithology of a rock unit, no matter where it occurred, it could be recognized by its characteristic assemblage of fossils. D'Orbigny called all of the strata defined by one fossil assemblage a *stage*, a term that now has been included in the formal time-stratigraphic hierarchy. Of course, d'Orbigny's stages were catastrophist in meaning, that is, every stage represented a separate creation and flood. As d'Orbigny's 10 stages were studied and subdivided by later stratigraphers, the number of stages increased far beyond what any catastrophist geologist could accept. For other reasons, this multiplicity of stages was also unacceptable to the growing number of uniformitarian geologists. "The naming of new stages became more a sport than a scientific endeavor" (Berry, 1987, p. 125).

In Germany, Friedrich Quenstedt found that d'Orbigny's stages did not work well outside France. Quenstedt had studied the Jurassic succession in minute detail, measuring sections and recording the position of each fossil. He found that d'Orbigny's stages were too broadly and vaguely defined for his use and began to divide his sections according to the stratigraphic ranges of individual fossil taxa. His methods were amplified and popularized by one of his students, Albert Oppel (1831–1865). Oppel traveled widely, applying Quenstedt's range methods to the Jurassic rocks of France, Switzerland, and England. After plotting hundreds of vertical ranges, he realized that the patterns were repetitive on a fine scale all over Europe. These patterns could be broken into distinct aggregates, bounded at the bottom by the appearance of certain taxa and at the top by the appearance of other taxa. Between 1856 and 1858, Oppel described the diagnostic aggregates, or "congregations," of fossils in his scheme of *overlapping range zones*. His zonation scheme was the first that allowed the clear-cut, nonarbitrary recognition and comparison of time units. Oppel also grouped his zones into "stages," but these differed from d'Orbigny's stages by being based on smaller units with distinct boundaries. Eventually, most of d'Orbigny's stages were redefined in this manner.

CONTROLLING FACTORS: EVOLUTION AND PALEOECOLOGY

Oppel's zonation scheme, based on Jurassic ammonites, did not turn out to be applicable worldwide, partly because of changes in taxonomy. Modern ammonite workers tend to combine (or "lump") species that Oppel considered distinct. The major limitation on his scheme, though, was that the ammonites that Oppel studied were not worldwide in distribution. In Europe, for example, ammonites occur in two biogeographic provinces, one in northwest Europe (Oppel's area) and one near the Mediterranean. Efforts to use Oppel's zonation in the Mediterranean province failed because few of the ammonites he recognized were found there. It soon became clear that zonation schemes must first be worked out locally and then compared between regions to see how they correlate.

The failure of Oppel's scheme shows how evolution and paleoecology have controlled biostratigraphic distributions. The *evolution* of organisms is the enabling factor, providing the progressive changes in species through time that makes biostratigraphy possible. Unlike any other means of correlation, biostratigraphy is based on the unique, sequential, nonrepeating appearance of fossils through time. The presence of a single fossil can often be used to determine the age of a rock very accurately. This is not true of the lithology of the rock, its magnetic polarity, its seismic velocity, or its isotopic composition; none of these are unique and cannot be used alone. If the rock can be radiometrically dated, then another criterion of age can be used; except

for volcanic ashes and lava flows, few stratified rocks can be radiometrically dated.

Paleoecology is the limiting factor on biostratigraphic distributions of organisms because no organism has ever inhabited every environment on the face of the earth. The distribution of organisms is also restricted on a local scale. *Facies-controlled* organisms are restricted to particular sedimentary environments (Fig. 10.1). Such organisms manage to migrate as depositional environments change and thus to find their old environment in a new place; they have evolved very slowly, if at all. For example, there is a distinct ecological community dominated by *Lingula* (a genus of inarticulate brachiopod) that has persisted unchanged in Europe since the Cambrian, while other invertebrate communities changed rapidly. This *Lingula* community appears to track its accustomed environment, avoiding evolutionary change, and thus has no time significance for the geologist (Fig. 10.2). An unwary biostratigrapher might use the first appearance of a slowly changing, facies-controlled fossil as a time indicator, when it is really an indicator of time-transgressive facies change.

In the biostratigraphy of the Gulf Coast are assemblages of robust, heavy-walled agglutinated foraminifers that do not occur in the Gulf of Mexico today but are known elsewhere at bathyal and abyssal water depths (Poag, 1977). These foraminifers were used widely in zonation of the Oligocene and Miocene sediments of the Gulf and Caribbean. When offshore drilling began to

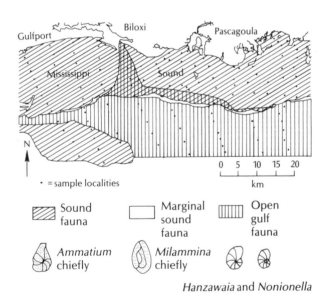

Figure 10.1

Present distribution of benthic foraminifers in the northern Gulf of Mexico, showing their strong ecological separation. This type of facies-controlled distribution is likely to create problems for the biostratigrapher. From Eicher (1976).

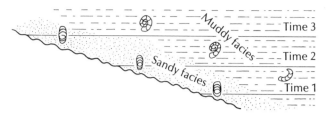

Figure 10.2

Significance of contrasting rates of evolution and shifting environments. The brachiopod *Lingula* lives in the sandy nearshore facies and evolves extremely slowly. It is a poor biostratigraphic indicator because it tracks its environment and thus changes little. Ammonoids that lived offshore, however, are excellent for biostratigraphy. They evolved rapidly and were not tied to any particular facies because they are free swimming. From Dott and Batten (1988).

recover younger deposits, however, these assemblages were found to extend to the top of the Pleistocene. Clearly, this agglutinated foraminiferal assemblage is more useful for paleoecology than for biostratigraphy. The only way to determine whether there are facies-control problems is to compare the biostratigraphy of the suspect group with the biostratigraphy of other groups in the area or to check it with physical stratigraphic methods. Although fossils are better time indicators than any other stratigraphic tool, they too must be used with caution.

Another example of the preceding problem was worked out by McDougall (1980). Benthic foraminifers were used, as in the previous example, to construct a biostratigraphic zonation of the late Eocene marine sequence along the North American Pacific Coast. The zonation in the state of Washington (Rau, 1958, 1966) was worked out independently of the zonation in California (Scheck and Kleinpell, 1936; Mallory, 1959; Donnelly, 1976). When McDougall examined the foraminifers that were found in both environments and analyzed their ecological associations, she found that foraminifers that marked one "time" in Washington occurred at a completely different "time" in California. For example, the *Sigmomorphina schencki* zone, which was used to mark the early Refugian in Washington (Fig. 10.3A) was based on deep-water forms that occurred throughout the Narizian and Refugian in California (Fig. 10.3B). Similarly, the *Cibicides haydoni* subzone, used to mark the early Refugian in California, was based on shallow-water forms that spanned the late Narizian and entire Refugian in Washington. As a consequence, the Narizian, Refugian, and Zemorrian Pacific Coast benthic foraminiferal stages were not sequential chronostratigraphic units, as had been thought. Instead, they overlapped or duplicated each other in various ways, depending on whether one defined them on the foraminifers used in California or in Washington. Problems such as these are enough to give a biostratigrapher gray hairs.

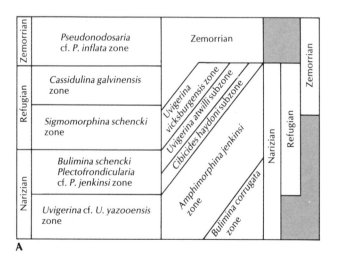

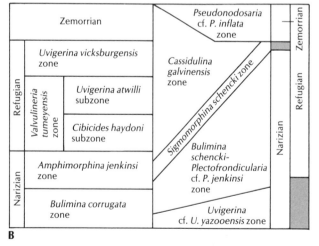

Figure 10.3

Correlation of the late Eocene benthic foraminiferal zonations between California and Washington, showing the problems caused by the strong facies control on these organisms. (A) Correlation scheme assuming the Washington zonation is valid. Under this scheme, the California zonation is time-transgressive, and the stages that are based on California forms overlap. (B) Correlation assuming the California zonation is valid. In this case, the Washington zonation is time-transgressive, and the stages overlap. This is because most of these taxa were strongly facies controlled; they disappeared from one area simply because they had migrated north or south with the facies changes. For example, the *Sigmomorphina schencki* zone appears in the late Narizian of California but migrates north to Washington by the early Refugian. From McDougall (1980).

The distribution of fossils in the rock record is therefore controlled by two primary factors, evolution and paleoecology. The first is the key to telling time, and the second is often useful in determining sedimentary environments. Untangling the two signals is not always simple, but we are fortunate to have both sources of data.

BIOSTRATIGRAPHIC ZONATION

As Quenstedt and Oppel showed, the key to using fossils for telling time is careful plotting of the vertical stratigraphic distribution, or *range*, of every fossil in a local section. Fossils whose original stratigraphic positions were not recorded at the time of collection (usually collected by amateurs and geologists who are not stratigraphers) are useless in this respect. The first step, then, is obtaining a reliable data base: a large suite of various fossils from one or more sections of the local area, with accurate stratigraphic data that were gathered at the time of collection of each specimen.

After all of the fossils have been cleaned up, they must be identified. This is not as trivial a task as it sounds. It is often very difficult to identify a fossil. Sometimes distinctions between one fossil species and another are so subtle that only a specialist can tell them apart. In other cases, a group may not have been studied in years, so there is no clear understanding of which species are valid or when they lived. In many cases, identification is often hampered by poor preservation. An incorrect identification can lead to serious problems if the stratigrapher follows it to incorrect correlations.

The next step is to compile the ranges into a *zonation*. Shaw (1964, p. 102) pointed out that every extinct organism divides geologic time into three parts: the time before it appeared, the time during which it existed, and the time since its disappearance. All of the strata that actually contain a given fossil species are said to be in its *range zone*. In a local section, the observed range of a fossil is its partial range zone, or *teilzone*. Teilzones are the empirical data base from which all biostratigraphy is derived, but they are local and only partial. No teilzone represents the total range in time and space of a species. The various classes of zones are called *biozones*.

The absence of a fossil from an area can be due to a number of factors (Fig. 10.4). A species first appears on earth when it evolves from an ancestral form (*evolutionary first occurrence*), but its first appearance in regions away from its origin is due to *immigration*. Various factors may have limited its immigration so that in some areas a taxon appears in the record very much later than when it first evolved in its home range. Likewise, its last appearance on earth is its final *extinction event*, but it may have been exterminated in many local regions before its final extinction. It may also have *emigrated* out of a region long before it became extinct worldwide. The first appearance of a fossil, whether by evolution or immigration, is called a *first appearance datum (FAD)*, or *first occurrence datum (FOD)*. Similarly, the last appearance, whether by extinction or emigration, is called *last appearance datum (LAD)*, or *last occurrence datum (LOD)*.

If the fossil record were everywhere perfect, biostratigraphic ranges would be simple functions of these four types of events: evolution, extinction, immigration, and

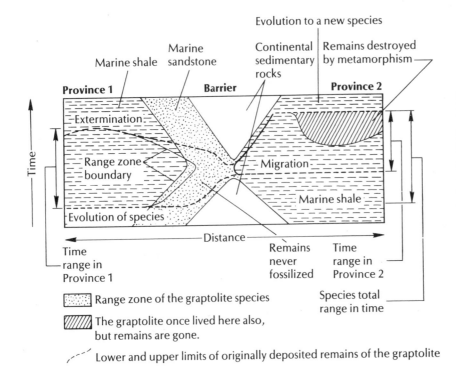

Figure 10.4

Time-space relationships of the range zone of a fossil. It evolves first in Province 1, but the barrier to Province 2 is present, so its migrational first appearance in Province 2 is later than its true first appearance. Likewise, it dies off first in Province 1 but lasts longer in Province 2, where it eventually disappears by evolving into another species. In addition to these factors, which prevent synchronous first and last appearances in different basins, the range zone can be shortened by nonpreservation or by erosion or metamorphism. The total species range in time is shorter than its local range in any one region. From Eicher (1976).

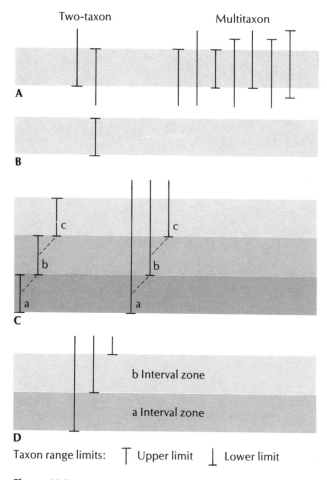

Two-taxon   Multitaxon

A

B

C

c b a   c b a

D

b Interval zone

a Interval zone

Taxon range limits: ⊤ Upper limit ⊥ Lower limit

Figure 10.5

Classes of interval zones. (A) Concurrent range zones, defined by the first and last appearance of two or more taxa with overlapping ranges. (B) Taxon range zone, defined by the first and last appearance of a single taxon. (C) Lineage zones, or phylozones, defined by the evolutionary first appearance of successive taxa in a lineage. (D) Interval zones, defined by two successive first or last occurrences of partially overlapping ranges. From Hedberg (1976).

emigration. However, other factors are involved. Fossils of a species will seldom be found in rocks deposited in environments that were unsuitable for that species (Fig. 10.4); and in many environments where the species did occur, the rocks were not suitable for preserving fossils or were eroded away along with their fossils. In many cases, the rocks may have been heated and deformed by a metamorphic event, so that few or no recognizable fossils remain. The rocks can also be inaccessible for sampling, or a particular fossil might accidentally be missed in the sampling. The final, observed distribution of a fossil group is thus affected by many factors that have nothing to do with its biology. Much of the original "signal" has been destroyed by the "noise" of sampling, nonpreservation, and destruction. Biostratigraphic techniques are designed to reconstruct as much of the signal as possible.

The nomenclature of zonation has been confused since its beginning, but the most recent stratigraphic codes (Hedberg, 1976; Appendix A) have attempted to standardize it. The first class of zones is the *interval zone*, which is bounded by two specific first or last occurrences of taxa. The most important type of interval zone is the *concurrent range zone* (Fig. 10.5A), defined by the overlap of two or more taxa. A *taxon range zone* (Fig. 10.5B) is based on the first and last occurrence of a single taxon. A *lineage zone* (Fig. 10.5C) is based on successive evolutionary first or last occurrences within a single lineage. An *interval zone* (Fig. 10.5D), (not to be confused with interval zones as a class of zones), is defined by two successive first or last occurrences of unrelated taxa.

The second class of zones is the *assemblage zone*, which is based on the association of three or more taxa. Normal assemblage zones are characterized by numerous taxa (Fig. 10.6A); thus, their boundaries are slightly vague and are only recognizable when a sufficient number of their *characterizing taxa* are present. However, this can be an advantage. Zones based strictly on one or two

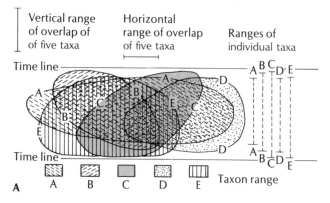

Vertical range of overlap of five taxa   Horizontal range of overlap of five taxa   Ranges of individual taxa

Time line

A B C D E

Time line

A B C D E   Taxon range

A

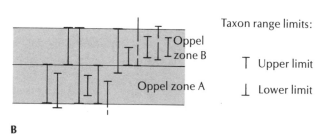

Oppel zone B

Oppel zone A

Taxon range limits:

⊤ Upper limit

⊥ Lower limit

B

Figure 10.6

Classes of assemblage zones. (A) Typical assemblage zones are defined by a suite of taxa, so the upper and lower boundaries are vague, depending on how many taxa are used to define the zone. The actual zone of overlap of all five taxa in this example is considerably less than the total assemblage zone. (B) Oppel zones, defined by the overlap of several taxa. From Hedberg (1976).

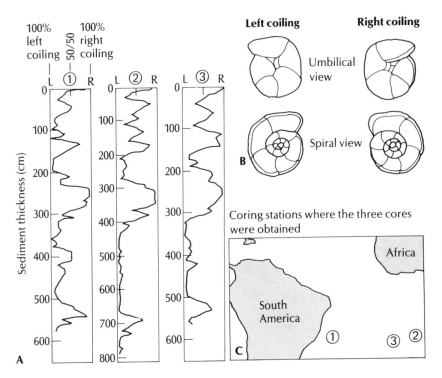

Figure 10.7

Correlation based on climatically controlled coiling changes in the foraminifer *Globotruncana truncatulinoides* in the South Atlantic. This animal is predominantly right coiling during interglacials but switches to left coiling during glacials. From Eicher (1976).

taxa are unrecognizable when those *defining taxa* are absent, whereas if *any* of the characterizing taxa of an assemblage zone are present, the zone can be recognized. An assemblage zone can be given more definite boundaries by using two or more first or last occurrences to define it (it can still be recognized by its characterizing taxa, even if the defining taxa are absent). The North American Stratigraphic Code and the International Stratigraphic Guide call this type of zone an *Oppel zone* (Fig. 10.6B) because it seems to be most similar to Oppel's original usage (though he actually used several different types of zones; see Hancock, 1977; Berry, 1987).

A third class of zone is the *abundance zone*, also known as a "peak" or "acme" zone. Abundance zones are characterized by the relative abundance of certain taxa. Although it seems tempting to correlate peaks in abundance of certain taxa, because they can swamp a fauna or flora at certain times, this kind of abundance change is usually due to local ecological factors and probably has little time significance. When peaks in abundance are due to time-significant changes, such as worldwide climatic change, abundance zones can serve as time markers. For example, in the deep sea an opportunistic coccolith called *Braarudosphaera* bloomed in tremendous numbers in times of climatic stress and formed distinctive *Braarudosphaera* chalks. Because these chalks were climatically controlled, they have been used successfully for correlating rock units.

Other changes in fossil lineages can have biostratigraphic utility. For example, the coiling direction in the planktonic foraminifer *Globorotalia truncatulinoides* shifts from right-coiling in warm waters to left-coiling in cool waters (Fig. 10.7). The rapid temperature shifts in the world ocean during the Pleistocene show up in the changing coiling ratios of this microfossil. Because the temperature shifts are controlled by global climatic change, the coiling ratios are good time markers.

Beyond the theoretical aspects of biostratigraphy there are many practical problems. All formal biostratigraphic zonations should conform to the stratigraphic procedures outlined by the Code. A zone must be formally named, usually by the species names of the two taxa that define its upper and lower limits or the name of one particularly abundant species that first appears in it. There must also be a description of the taxonomic content of the zone (including both the defining and the characterizing taxa). The describer must designate a type section and give the known geographic extent of the zonation.

As stratigraphers have become more careful in documenting ranges and describing type sections, inevitable disputes over boundaries have arisen. The type sections or areas of two successive biostratigraphic units are usually in two separated areas, and there may be no overlap. Often, neither section preserves the boundary between the two units, so biostratigraphers must search for a third area where the transition is recorded. Ideally, this section should be as continuous and fossiliferous as possible, with several taxonomic groups to compare.

After the detailed local biostratigraphy of the available groups has been worked out, the stratigraphers must decide which biostratigraphic event(s) should serve as boundaries between units. At this point, philosophical disputes become acute. Some workers argue that boundaries should be drawn at mass extinctions and other faunal breaks because the high turnover of taxa would make the boundary easy to recognize. Others argue that boundaries should never be based on faunal breaks because the abrupt truncation of many ranges of taxa along a narrow zone probably indicates a major hiatus or unconformity.

Drawing the boundary between two time-stratigraphic units along a major gap is indeed risky. If that gap should be filled later by some other section located somewhere else, the boundary may become blurred. Instead, some stratigraphers advocate drawing the boundary at the position of the evolutionary first occurrence of a single, distinctive taxon. If the evolutionary sequence of that taxon is relatively complete, it is less likely that a section is missing. Once such a boundary is agreed upon, Ager (1964, 1973, 1981) suggests that some physical marker (metaphorically called "the Golden Spike") should be driven into the outcrop at the precise level of the boundary. Although this procedure is more arbitrary than natural breaks, there should be no more argument once it is done. To date, three major boundaries have been internationally defined in this manner (Bassett, 1985): the Silurian/ Devonian boundary (see Box 10.1), the base of the Cambrian (base of the *Paragloborilus-Siphonogonuchites* small, shelly fossil zone in Kunyang Quarry, Yunnan Province, China), and the Ordovician/Silurian boundary (base of the *Parakidograptus acuminatus* graptolite zone at Dob's Linn, Scotland). Many other biostratigraphic boundaries are currently under discussion (reviewed in *Episodes*, vol. 8, number 2, June 1985).

The choice of which boundary types to recognize can also vary with the method chosen. In outcrops, many

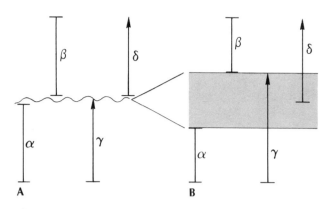

Figure 10.8

(A) Zones defined by both bottoms and tops of ranges (α and β) and "topless" zones (γ and δ). (B) Restoration of the section removed by the unconformity in (A) results in a gap between α and β, but no gap results from topless zones.

stratigraphers mark zones from the bottom (first occurrence) upward because they measure a section from the base. These "topless" zones have the advantage that if a boundary in the stratotype happens to be drawn on a local hiatus, and geologists find a new section that spans this hiatus elsewhere, the new section can automatically be assigned to the zone below (Fig. 10.8). Geologists working with subsurface data tend to use tops of range zones. Not only are they accustomed to drilling downward, but the drilling mud may continue to bring up fossils from the borehole wall long after the base of a particular biostratigraphic zone has been passed. Other geologists have tried to use one of these methods in combination with abundance data, which is particularly striking in some sections (see Fig. 10.9).

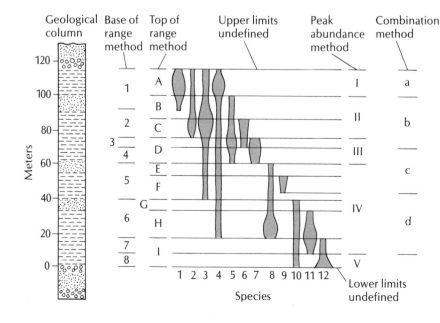

Figure 10.9

Four methods of recognizing concurrent range zones in a single section. Widths of lines indicate relative abundance of specimens. Note that each method gives a slightly different zonation. From Eicher (1976).

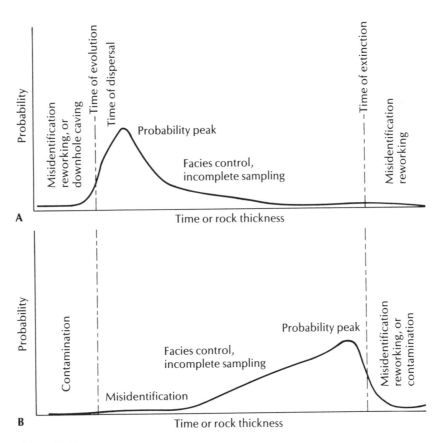

Figure 10.10

A representation of the probable position of local first or last occurrences relative to the true first or last occurrence. (A) The local first occurrence is probably slightly younger than the true time of evolution due to dispersal delay. However, it could be slightly older due to misidentification, reworking, or downhole caving; it could also be much younger due to facies control or incomplete sampling. (B) The local last occurrence is probably slightly younger than the true time of extinction, because extinction in each area has a strong local control. However, it could be much younger due to facies control or incomplete sampling, or even slightly older due to misidentification, reworking, or contamination. From Edwards (1982a).

The decision on whether to use tops or bottoms of ranges hinges on a number of factors (Fig. 10.10). For example, estimating the probability of finding the true first occurrence of a given fossil in a section or core (Fig. 10.10A) suggests that its first occurrence probably does not match its time of evolution. The actual first occurrence is likely to be somewhat later due to the time necessary for the organism to disperse from its site of evolution to the sampling site. However, its first occurrence may be lower in the section than it really should be, due to misidentification, downhole caving, or reworking. Its first occurrence may be much later than its actual evolution because of incomplete sampling or because the local environment was at first unsuitable for preservation. Similar arguments apply to the probability of identifying a fossil's last occurrence (Fig. 10.10B). Its last occurrence might appear to be later than it really is due to misidentification, reworking, or uphole contamination. It may appear to be earlier than it actually is due to incomplete sampling, or local extinction due to facies changes. These theoretical probability curves are likely to be modified by local circumstances. For example, the characteristics of an organism (its ability to disperse or its facies sensitivity) vary the shape of the curve, as does the nature of the sample (core, outcrops, or cuttings). The practicing biostratigrapher soon gains a sense of which problems are most serious in a given situation as he or she learns its idiosyncrasies.

Box 10.1

The "Golden Spike" at the Silurian/Devonian Boundary

The establishment of the Silurian/Devonian boundary is a good example of how such an international effort takes place. The boundary has been defined as the first occurrence of the graptolite *Monograptus uniformis uniformis* in bed 20 (Fig. 10.11) of the section at Klonk, Czechoslovakia (Chlupác et al., 1972; McLaren, 1973, 1977). The name *Klonk* is a fortunate coincidence, because the sound of it suggests the driving of the "golden spike." But the area was not chosen because of its name. A number of areas around the world were considered by a group of

stratigraphers who were familiar with faunas that spanned the Silurian/Devonian boundary. As reviewed by McLaren (1973, 1977), the Silurian/Devonian Boundary Committee of the Commission of Stratigraphy, International Union of Geological Sciences (IUGS) was organized in Copenhagen in 1960 and gave its final report and recommendations in 1972.

Their first problem was that the type areas of both systems were inadequate. As we saw in Chapter 1, Murchison based the Silurian on marine rocks in

Figure 10.11 (A) Type section of the Silurian-Devonian boundary, near Klonk, Czechoslovakia. The boundary is defined as the first appearance of the graptolite *Monograptus uniformis*, which occurs in Bed 20 in this section. From Chlupác et al. (1972). (B) Photograph of the Klonk section. Bed 20 is visible just above the point of the hammer. Photo courtesy I. Chlupác.

the Welsh Borderland found under the nonmarine Devonian Old Red Sandstone. When Murchison and Sedgwick found marine Devonian rocks in southwestern England, they eventually recognized that this sequence was equivalent to the Old Red Sandstone and was younger than the Silurian. Unfortunately, the base of the marine type Devonian is not exposed in Britain, and the top of the marine type Silurian is capped by nonmarine rocks. Because neither Wales nor Devon nor any other area in Britain contained a good sequence spanning the Silurian/Devonian boundary, the old type area had to be set aside and other areas considered.

Another important consideration came from studies of the Bohemian (now Czechoslovakian) sequences first described by the pioneering geologist Joachim Barrande in 1846. These sequences had been important in Murchison and Sedgwick's disputes over the Silurian because they were much thicker, more complete, and more fossiliferous than the type areas in Britain (Secord, 1986). By 1960, it was apparent to the Committee that the "Barrandian" sections included a latest Silurian stage, the Lochkovian, that was missing from the British sequence. The Barrandian was thus one of the first prime candidates for a type section.

Nevertheless, the Committee looked into several other candidate areas and considered many different fossil assemblages that could be used to define the boundary. After discussing graptolites, conodonts, corals, brachiopods, trilobites, ostracods, crinoids, vertebrates, and plants, they settled on graptolites.

Because graptolites were pelagic, their fossils occur worldwide and make the best zonal indicators of all the candidates. By 1968, the Committee had chosen the base of the *Monograptus uniformis* zone as the base of the Devonian.

Now the task was to determine the place where this zone is best exposed. They sought a well-studied rock section with good exposures. The fossils of this section should be abundant at every level, thoroughly studied, and well documented. Ideally, the best section would be the one that is most complete, with no evidence of missing or truncated zones. If it should contain a gap, and a section were found later that spanned the gap, then the sharp boundary could become blurred. Finally, the type section should be easy to reach and should remain accessible to geologists of any nationality who want to study and collect, now and in the future.

In addition to the Barrandian in Czechoslovakia, sections were studied in three places in the Soviet Union, the Carnic Alps of Austria and Italy, the Holy Cross Mountains of Poland, Aragon in Spain, the Roberts Mountains in Nevada, the Gaspé in Quebec, and additional areas in Algiers, Morocco, Australia, Thailand, the Yukon, and the Canadian Arctic. By 1971, the choice was narrowed to four candidates: Podolia in the Soviet Union, Morocco, Nevada, and the Barrandian.

In August 1972, the Committee met during the International Geological Congress in Montreal. They weighed the pros and cons of each section and finally chose the Barrandian by a vote of 31

THE TIME SIGNIFICANCE OF BIOSTRATIGRAPHIC EVENTS

What types of biostratigraphic events make the best boundaries? Many authors (e.g., Murphy, 1977) consider the evolutionary first occurrence (lineage zone) the best because the evolution of a taxon is an unrepeatable event that happens at a unique time and place. However, there are drawbacks to this method. Evolutionary first occurrences are rare, and they may not occur in the sections of interest. If Eldredge and Gould (1972) are right in their theory of punctuated equilibrium, there are few examples of gradualistic change in the fossil record. Gould and Eldredge (1977) claimed that most "gradualistic" evolutionary changes in taxa do not hold up under close scrutiny. Many cases of apparent gradualism may be due to repeated immigration events rather than

to the evolutionary change of a lineage *in situ*. If gradual change is real, and is used for determining boundaries, then problems of definition arise. Does one define the evolutionary first occurrence of a species at the first appearance of the character that defines the species, or when 50 percent of the population has developed this character? Here, there is a conflict between species defined for biostratigraphic purposes and the biological meaning of species. The definition of species by the 50 percent criterion, for example, would be considered unnatural by a biologist.

Some geologists (e.g., Repenning, 1967) have argued that immigrational or emigrational events should be preferred for boundaries. Immigrational first occurences, in particular, can be abrupt and unambiguous. The dispersal of most marine organisms, particularly those with planktonic larval stages, is very rapid by geologic standards (Scheltema, 1977). This is also true of land verte-

to 1. The Barrandian was chosen because it was thoroughly described and had a rich and uninterrupted succession of well-studied graptolites, conodonts, and other fossils. There were enough facies changes within the Barrandian sections to make positive correlations with other benthic fossil assemblages. Two Barrandian sections were voted on, and the one at Klonk was chosen. Another vote was taken on whether to draw the boundary at Bed 20, the base of the *M. uniformis* zone (base of the Devonian) or at Bed 13, the top of the *M. transgrediens* zone (top of the Silurian). There was a 170-centimeter gap between these beds, with no fossils. After discussion, the base was drawn at the bottom of Bed 20 (see Fig. 10.11).

McLaren (1973, 1977) reviewed some of the lessons learned from this exercise. Although the procedure just reviewed sounds very legalistic, it was necessary if stratigraphers are going to communicate effectively. As McLaren (1973, p. 23) put it, "definition is not science, but a necessary prerequisite to scientific discussion." If scientists cannot agree on what a name means or how to recognize it, then they end up talking about different things. This can hinder their ability to describe what they are thinking about and to communicate ideas.

The Committee's work also encouraged international cooperation and discussion and cleared up many misconceptions and miscorrelations that would have taken years to discover. One of the early misconceptions was the idea that rock units could be used as time markers. Some geologists

wanted the boundary to be drawn at the Ludlow Bone Bed in the Welsh Borderland, the traditional top of the type Silurian. As we have seen in Chapter 8, rock units rarely are worldwide and synchronous enough to make good chronostratigraphic boundaries.

McLaren (1973) commented on the remarkable collegiality of the delegates to the Committee, who did not form into warring "schools of thought" representing particular schools or nations (as they easily could have). The democratic procedure used to debate and decide the matter made the results accessible and acceptable to the entire geological community.

Finally, we have seen how the priority of the name or type area, though important, may not always be the deciding factor. The type areas of the Silurian and Devonian in England were both inadequate for the purpose of defining their mutual boundary. Similarly, the type area for the base of the Cambrian is now in China, not in Sedgwick's sections in Wales. The type section for the base of the Ordovician is not in Murchison's sections in Wales but is in Scotland because Lapworth (1879) had found better fossils there and had defined the Ordovician in 1879 with primary reference to Scotland.

brates (Lazarus and Prothero, 1984; Flynn et al., 1984). The main problem with this marker is that organisms immigrate to different places at different times. The notorious "*Hipparion* datum," based on the first appearance of the three-toed horse *Hipparion* in the Old World, was once used as the base of the Pliocene (Colbert, 1935). Recent work (Woodburne, 1989) has shown that the *Hipparion* immigration "event" was at least two different immigrations of different hipparionine horses at different times during the Miocene.

Extinctions are probably less reliable boundaries because it is well known that taxa can linger in a refuge long after they have disappeared elsewhere. Yet extinctions can be attractive boundaries because they often cluster at horizons that represent mass extinctions, which are major episodes in the history of life.

Stratigraphers have long assumed that biostratigraphic events are reasonably synchronous within

the resolving power of the stratigraphic record. This assumption was criticized by Thomas Henry Huxley (1862), who pointed out that biostratigraphers have demonstrated only similiarity in order of occurrence (*homotaxis*), not synchrony. In recent years, other independent techniques have been employed to test the time significance of biostratigraphic events. Hays and Shackleton (1976) used oxygen-isotope stratigraphy to show that the extinction of the radiolarian *Stylatractus universus* was globally synchronous. Haq and others (1980) conducted a similar test using carbon isotopes. Berggren and Van Couvering (1978) state that asynchrony, or *heterochrony*, of biostratigraphic datum planes is rare in the marine realm. Yet Srinivasan and Kennett (1981) showed that the last appearances of some marine microfossils can be heterochronous. Some marine microfossils, such as the benthic foraminifers, are notorious for being facies controlled and heterochronic, as we saw in the example of the Pacific Coast Eocene benthic foraminifers (McDougall, 1980).

In land sections, Prothero (1982) and Flynn and others (1984) tested mammalian biostratigraphy against magnetostratigraphy and found that mammal *assemblages* do not show significant heterochrony as a rule, though the stratigraphic occurrence of *individual taxa* can be significantly time transgressive. These studies demonstrated that the basic assumptions of biostratigraphy must be constantly reexamined when new tools of correlation are developed. Most of the methods and assumptions of biostratigraphy have proven to be quite robust. This is hardly surprising because biostratigraphy has been used successfully since the days of William Smith. For these reasons, biostratigraphy will probably remain the primary tool for correlation and determining relative age.

INDEX FOSSILS

Some fossils are so abundant and characteristic of key formations that they are known as *index fossils*, and many formations in North America can be recognized by index fossils. This information was catalogued in Shimer and Shrock's (1944) *Index Fossils of North America*. Most fossil groups (including index fossils) that make good biostratigraphic indicators are distinctive, widespread, abundant, independent of facies, rapidly changing, and short ranging. From these properties, one can see that the best biostratigraphic indicators are pelagic organisms that evolve rapidly. Pelagic organisms live in the surface waters of the oceans, so they can quickly disperse around the world. They are unaffected by the various bottom facies that control benthic organisms. Practical experience has shown that ammonoids, graptolites, conodonts, foraminifers, and other planktonic microfossils give the best results, though some benthic groups are useful.

There are also some problems with the index fossil approach to correlation. Too often the index fossil is equated with the formation, and no attempt is made to document the *actual range* of the fossil *within* the formation. This results in a loss of resolution of the data. The stratigraphic range of the fossil is often reported to be the same as the total thickness of the formation, which may artificially extend the range. In addition, the use of index fossils can cause confusion of rock units with time-rock units and imply that rock units are time equivalent. Too much reliance on one index fossil may cripple a biostratigrapher who works in areas where the index fossil does not occur. The absence of a key index fossil does not necessarily mean that the rocks are not of the age of that index fossil. Finally, Shaw (1964, p. 99) pointed out that reliance on index fossils can lead to other abuses. For example, rote memorization of index fossils and their formations ("*Spirifer grimesi* is the index

fossil of the Burlington Limestone") can lead a geologist to recognize formations by their fossils rather than by their lithologies. Memorization of lists of index fossils also tends to drive many good, young minds away from paleontology.

NORTH AMERICAN LAND MAMMAL "AGES" AND BIOCHRONOLOGY

Land mammal fossils have long been used as stratigraphic tools in Cenozoic terrestrial deposits. However, most fossil mammals occur in isolated quarries and pockets and are seldom distributed evenly through thick stratigraphic sections, as are marine invertebrates. As a result, mammalian biostratigraphers have not always followed classical Oppelian biostratigraphic procedures. A few thick sections that show superposition of faunas are known, but as a whole, the sequence of mammalian faunas must be worked out indirectly. The result has been called *biochronology*. Williams (1901) defined a *biochron* as a unit of time during which an association of taxa is interpreted to have lived (Woodburne, 1977). A biochron is equivalent to the biozone of an assemblage of taxa, except that it is not directly tied to any actual stratigraphic sections.

Tedford (1970) showed that classical biostratigraphic procedures were used by early North American mammalian paleontologists (e.g., the "life zones" of Osborn and Matthew, 1909). These procedures fell into disuse, and a biochronological sequence was built up by piecing together a sequence of faunas based on their stages of evolution. In 1941, the sequence of mammalian faunas was formally codified by a committee chaired by Horace E. Wood, II (Wood et al., 1941). The Wood Committee set up a series of provincial land mammal "ages" that were based on the classic sequence of mammal faunas in North America. These came to be known as North American Land Mammal "Ages." However, they are not true "ages" in the formal time-stratigraphic sense because they are not built from biostratigraphic zones based on actual rock sections.

The Wood Committee attempted to erect unambiguous definitions for each unit by listing index fossils, first and last occurrences, characteristic fossils, and typical and correlative areas. These multiple criteria have since led to conflicts that would not have occurred if classical biostratigraphic methods had been followed originally. For example, the late Eocene Chadronian Land Mammal "Age" is defined by two criteria: (a) the co-occurrence of the horse *Mesohippus* and titanotheres and (b) the limits of the Chadron Formation. At the time of the Wood Committee's work, the last occurrence of titanotheres was thought to coincide exactly with the top of the Chadron Formation, so there was no conflict.

Since then, titanotheres have been found in rocks above the Chadron Formation (Prothero, 1982). Now we must choose between conflicting criteria. Is the end of the Chadronian marked by the last appearance of titanotheres or by the top of the Chadron Formation?

In recent years, mammalian paleontolgists have made efforts to return to classical biostratigraphic principles (reviewed by Woodburne, 1977, 1987). Much of the mammalian time scale is being restructured in terms of biostratigraphic range zones tied to local sections. There are parts of the Tertiary, however, that may never be zoned, so biochronology is still widely used. Its robustness was demonstrated when radiometric dates first became available for the land mammal record and showed that the biochronological sequence was in the correct order (Evernden et al., 1964). European paleomammalogists, who have fewer good stratigraphic sections to work from, use an explicitly biochronological approach. Every fauna is placed in order according to its stage of evolution. *Niveaux répères*, or "reference levels," of classical faunas are used in place of type sections (Jaeger and Hartenberger, 1975; Thaler, 1972).

QUANTITATIVE BIOSTRATIGRAPHY
Shaw's Graphic Correlation Method

A biostratigraphic data base can get so large that it becomes impossible for a biostratigrapher to find the pattern among so many stratigraphic sections with so many taxonomic ranges. In recent years, techniques have been introduced to quantify and handle large data bases, mostly by computer.

The most popular method, *graphic correlation*, or "Shaw plots," was introduced by Alan Shaw in his book *Time in Stratigraphy* (Shaw, 1964). Much of the work can be done on graph paper with the aid of a pocket calculator, though the entire procedure is usually done on a computer. In essence, it uses the statistician's meaning of the word *correlation*. When two variables are correlated, they form a cluster of points on a bivariate data plot. The statistician then attempts to fit a line of correlation to the cluster. First and last occurrences of fossil taxa become data points along an axis that represents the stratigraphic section. If two stratigraphic sections are placed along the two bivariate axes (abscissa and ordinate), the data points in both sections can be plotted in bivariate space (Figs. 10.12A,B). The line connecting these points is the line of correlation.

The line of correlation is the most powerful aspect of Shaw's method. If all the points fit exactly on the line, there is perfect correlation; outlying points immediately become apparent and can be examined to see if they represent artificial or real range extensions or truncations. Real data sets seldom show perfect correlation, and the scatter of points near the line is a measure of the scatter of the data.

If the two sections are plotted to the same scale, the slope of the line is particularly informative. A line with a perfect 45° slope (Fig. 10.12B) indicates that the two sections have identical rates of rock accumulation

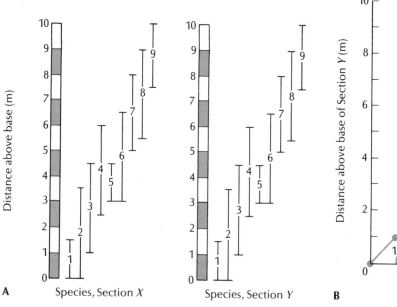
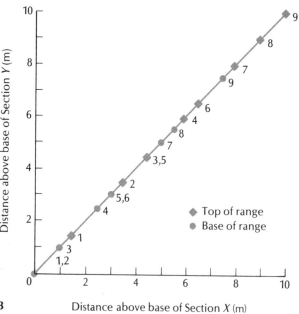

Figure 10.12

Shaw's graphic correlation method. (A) Two sections with the stratigraphic distribution of nine taxa. (B) Correlation between sections *X* and *Y*. Since the ranges in both sections are identical, the line of correlation between the two sections is straight and has a 45° slope. From Shaw (1964).

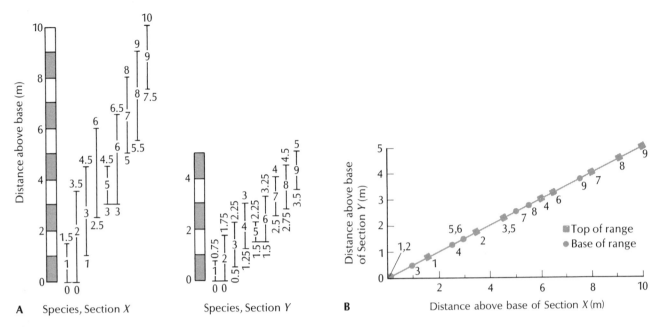

Figure 10.13

(A) Two sections that have the same relative spacing of ranges but different rates of rock accumulation. (B) Graphic correlation of these two sections. The points form a straight line, but the slope is deflected toward the axis of the section with the higher accumulation rate. From Shaw (1964).

and identical distributions of fossils. (Shaw speaks of "rock accumulation" because post-depositional compaction can shorten sections and distort true rates of sedimentation). Commonly, the rates of accumulation in two sections are not identical, in which case the ranges in one section will be proportionately shorter (Fig. 10.13A). The plot of these two sections will not result in a 45° slope but will be a line inclined toward the axis that represents the section with the higher rate of accumulation (Fig. 10.13B). If the rate of accumulation changes during deposition in one section relative to the other, the slope will change, giving a "dogleg" pattern (Fig. 10.14). If there is a hiatus in one of the sections, the ranges will tend to truncate at that level (Fig. 10.15A); the resulting plot will have an obvious "terrace" representing the missing section (Fig. 10.15B). It should be apparent that the graphic correlation method is a powerful tool for spotting bad range data, unconformities, and missing sections, and for interpreting rates of rock acccumulation.

Shaw's method has applications beyond the correlation of two sections. It can be used as the basis for a large-scale correlation scheme among multiple sections. Shaw recommends that the stratigrapher begin by correlating the two best sections in the study area. Any outliers on the line that represent true range extensions are then added to the best reference section, producing a *composite standard*. The composite standard can then

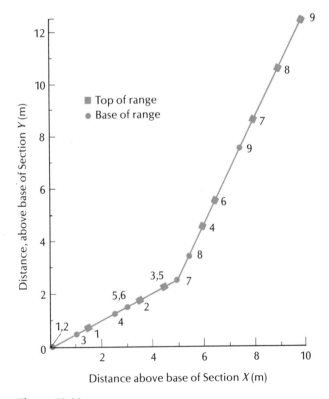

Figure 10.14

Graphic correlation of two sections that show a change in rate of rock accumulation. This forms the characteristic "dogleg" kink in the slope of the line. From Shaw (1964).

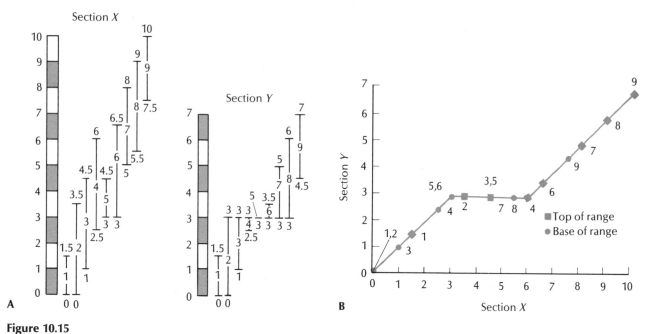

Figure 10.15

Two sections with the same taxa, but one section has an unconformity that truncates the ranges of some taxa. (B) Graphic correlation plot of these two sections. The unconformity causes a "step" or "plateau." The section with the unconformity (Y) shows no change whereas the complete section (X) continues to spread out the points. From Shaw (1964).

be correlated with each additional section, one at a time, and any further range extensions can be added to it. The result is a composite standard that is not a real section but a synthesis of all the information from a group of sections. The *composite standard ranges* are not teilzones but are the maximum ranges of each taxon in the area. If this were done for every section in the world, the total range zone would be approximated. The subdivisions of the composite standard are no longer measured in thicknesses of section but are abstract units that approximate geologic time planes. These can be used to correlate and align sections, producing results that show patterns more clearly than do correlations based on lithologic boundaries or on individual biostratigraphic planes.

Because Shaw's methods are easily learned and are applicable to a variety of biostratigraphic problems, they have had widespread use in the oil industry (Miller, 1977). Simple graphic correlation plots can be routinely used when plotting together any two sets of stratigraphic data, including magnetostratigraphic data and isotopic data.

Probabilistic Stratigraphy

Another method of quantifying biostratigraphic data was invented by William W. Hay (1972; Hay and Southam, 1978). Hay's method does not require the measured thicknesses of section that are essential to

graphic correlation. Instead, the data base consists of the order of appearance of biostratigraphic events in multiple parallel sections; these sequences differ slightly because of locally shortened range data (Fig. 10.16A). The biostratigrapher is concerned only with determining which sequence is *most probable*, hence the name *probabilistic stratigraphy*. This type of probability calculation is a simple statistical problem, easily done by computer. The computer generates a series of matrices (Figs. 10.16B,C) that calculate how many times each possible sequence arranges events in the proper pairwise order. Finally, a matrix emerges that has the least number of events out of order and is therefore most probable (Fig. 10.16C). The probabilistic method is designed to give only the most likely order. Unlike graphic correlation, it does not address problems of correlation between sections or varying rates of accumulation.

Multivariate Methods

Computers make it possible to handle many more variables than ever before. A great variety of techniques have been designed to reduce the complexity of multiple variables to a few simple variables that explain the underlying structure of a large data set. Multivariate methods have been used in many fields and have recently been applied to stratigraphy.

Biostratigraphers have concentrated on methods that detect similarities between data sets. Cluster analysis,

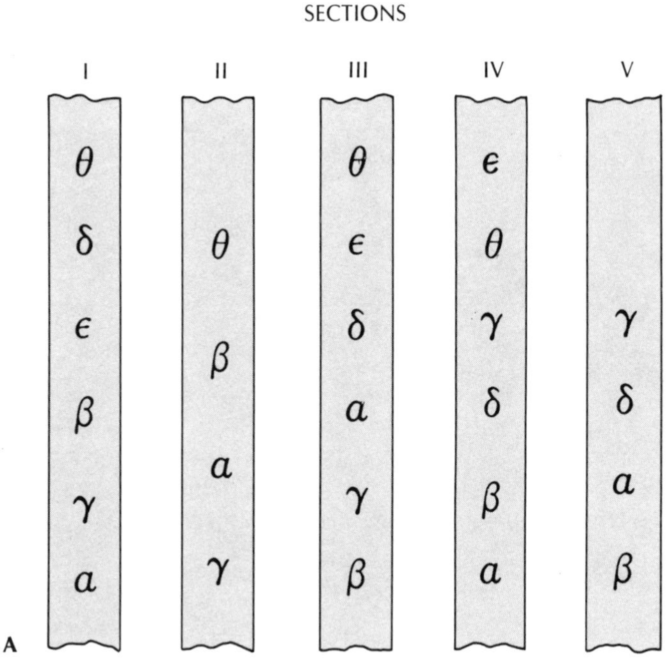

SECTIONS

Lower occurrence (B)

	α	γ	β	ε	θ	δ
δ	4/4	2/4	4/4	1/3	0/3	
θ	4/4	4/4	4/4	2/3		
ε	3/3	3/3	3/3			
β	3/5	2/5				
γ	3/5					
α						

Higher occurrence

Lower occurrence (C)

	α	β	γ	δ	ε	θ
θ	4/4	4/4	4/4	3/4	2/4	
ε	3/3	3/3	3/3	2/3		
δ	4/4	4/4	3/4			
γ	3/5	3/5				
β	3/5					
α						

Higher occurrence

Figure 10.16

Probabilistic stratigraphy. (A) Five hypothetical stratigraphic sections (I–V) with the sequence of biostratigraphic events found in each. (B) Matrix of the first attempt at determining the most-probable sequence. The numerator of the fraction in each box is the number of times that the event in the vertical column was found above the event in the horizontal row. The denominator indicates the total number of sections in which both events occur. This sequence does not have the maximum probability because δ occurs above γ only two out of four times, δ above ε only one out of three times, and δ above θ in none of the three sections; β is above γ only twice in five sections. (C) Highest-probability arrangement of this example. Note that all the fractions are 50 percent or better. Adapted from Hay (1972).

principal components analysis, and principal coordinates analysis are all multivariate techniques that can be used for assessment of similarity. Each of these techniques begins by sample-by-sample comparison of matrices that contain numerical expressions of similarities or differences between all pairs of samples. Multivariate analysis is then used to find groups of similar samples and to generate some sort of grouping by similarity. In *cluster analysis* (Fig. 10.17), the similarity between samples is expressed as a branching *dendrogram* (Fig. 10.17B). Sample points from one biostratigraphic zone cluster together, and dissimilar samples appear on different branches. The zonal boundaries (Fig. 10.17D) can then be drawn between the samples that show the least similarity.

Principal component analysis and *principal coordinate analysis* both generate axes that explain the variability of the data set. For example, the first principal component is the axis that explains the most variability within the multivariate data set. The second principal component explains the second-largest component of variability, and so on. If the samples are analyzed by these methods and plotted with respect to any two of the principal components (Fig. 10.17E), they will again cluster by similarity. Zonal boundaries can be drawn between dissimilar clusters. When used in biostratigraphy, all of these multivariate techniques make the *range-through* assumption, in which all samples between the first and last occurrence of a taxon are assumed to contain that taxon, whether or not it actually occurs in each sample. Multivariate techniques are well suited to analyzing large masses of data, especially for locally short sections and single samples that might not work well in graphic correlation methods. The use of the range-through method implies that the actual presence of a

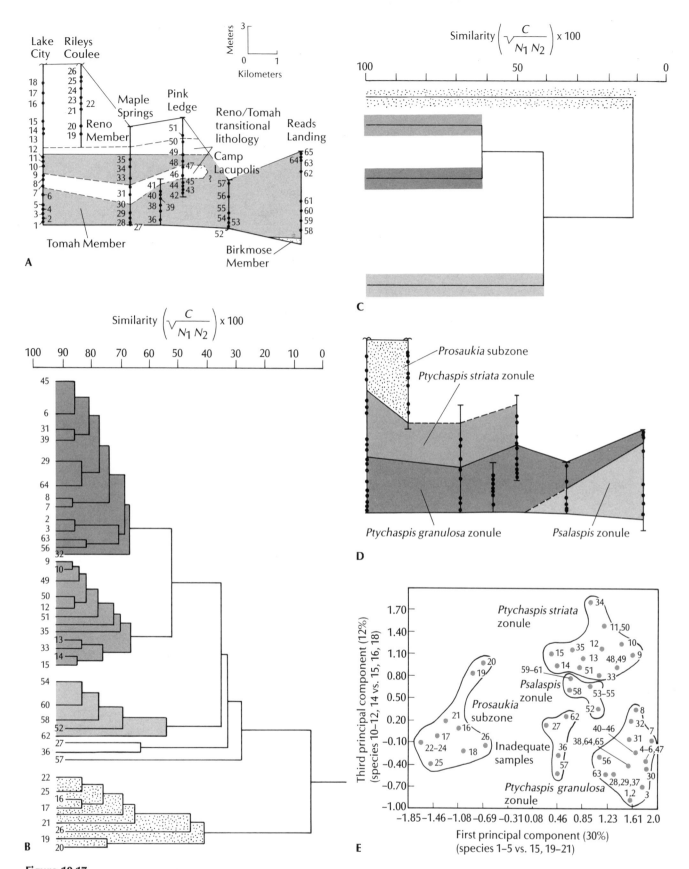

Figure 10.17

Multivariate cluster analysis method of correlation. (A) Lithostratigraphy and position of the trilobite samples from the upper Franconia Sandstone (upper Cambrian) of southeastern Minnesota. (B) Q-mode dendrogram resulting from comparison of 65 samples containing 24 trilobite species. (C) Summary of the larger dendrogram, showing four major clusters. (D) Biostratigraphic interpretation of results. (E) Principal components analysis of the trilobite data, projected on the first and third eigenvectors. From Hazel (1977).

taxon is no longer important in the analysis. Samples containing first and last occurrences are weighted particularly heavily. On the other hand, the method avoids investigator bias toward particular taxa because all taxa can be weighted equally.

Relational Strategies

Another approach to quantification of biostratigraphy is to analyze the relationships between the ranges of all of the taxa two at a time (Guex, 1977; Davaud and Guex, 1978). Three possible relationships exist: the range of species A is totally below the range of species B, totally above it, or overlaps it. By extending this *relational* procedure pairwise through the entire data set, the computer can generate an order of occurrences of all the taxa in the data set. This method is particularly robust in taking an unknown sample and placing it in the ordinal sequence.

Strengths and Weaknesses of Each Method

Edwards (1982a, 1982b) analyzed the four kinds of methods previously described and showed that each is appropriate for a different type of data set. Table 10.1 shows her comparison of the major strengths and weaknesses of each method. Before choosing a technique the biostratigrapher should ask several questions about the type of data to be analyzed. First, if not all the data are from stratigraphic sections, then only multivariate or relational strategies can be used. If all of the data are from sections, but not all of them are *measured* sections, then the graphic correlation method and Hohn's (1978, 1982) principal components analysis are eliminated. The number of sections or samples is also critical. If there are too many samples, the data set will be too large for multivariate methods. Probabilistic methods, relational methods, and graphic correlation work better with more

Table 10.1

Comparison of Quantitative Biostratigraphic Techniques

Feature	Technique			
	Probabilistic	Multivariate	Graphic	Relational
Degree of automation	Can be fully automated	Can be fully automated to produce groups; requires user input to interpret groups	Requires user input for each section and each iteration	Can be fully automated to produce range charts; requires user input for correlation
Number of stratigraphic sections	Works best with many (more than 10) sections; cannot handle isolated samples	Can exceed computer storage with many samples; can handle isolated samples	Any number of sections; cannot handle isolated samples	Any number of sections, but works best with many; can handle a few isolated samples
Range hypothesis produced	Most probable ranges (shortened relative to total ranges)	Can be extended relative to total ranges	Total ranges	Can be extended relative to total ranges
Major assumptions (all require superposition, consistent identification, unique evolution, and extinction)	Spacial homogeneity and independence between all pairs of events	Representation of coefficients in reduced-dimensional space	Best interpretation of the data determined by user judgment and causes the least disruption of the best known ranges	Chronostratigraphic subdivision based on co-occurrences and not on absences of taxa
Comments	May require additional assumptions (e.g., normal or logistic distributions); independence violated if first and last occurrences are treated jointly	Is completely distinct and has additional assumptions and restrictions and produces shortened ranges[a]	Requires measured sections and assumption of linear relative rates of rock accumulation[b]	Will leave chronostratigraphic gaps at boundaries between "associations unitaires"

[a] Hohn, 1978.
[b] Shaw, 1964.
From Edwards (1982b)

sections and require less computer space. If the data set consists of both tops and bottoms of ranges, then graphic and relational strategies will emphasize the differences between them.

Probabilistic and multivariate methods ignore the inherent differences between bases and tops. In many cases, it might be important to distinguish between the tops and bottoms because the probability of error is not equal between them. If, for example, the biostratigrapher is working with well data, the probability that the ranges will be artificially extended downward by downhole caving makes bottoms of ranges more prone to error than tops of ranges. Finally, the biostratigrapher must decide whether the goal is to determine maximum ranges or average ranges. If upward or downward errors are commonly introduced by caving, contamination, or reworking, then maximum ranges are probably larger than true ranges, and an average-range method is better. If there is less chance of artificial range extension, then maximum ranges are closer to the idealized true range. Probabilistic techniques tend to produce average ranges. Graphic correlation, as we have seen, generates maximum ranges. Relational strategies produce ranges that are between these two extremes.

Because many of these methods are still quite new, there have not been many tests of all four strategies on the same data set. Graphic methods are still popular because they are easy to understand and simple to plot and manipulate. They also generate additional information (e.g., relative rates of rock accumulation and the presence of unconformities) that the other methods do not address. However, the graphic method requires knowledgeable input on the part of the biostratigrapher, and this may introduce bias. Some workers argue that a little human intervention is necessary to avoid results

that are geologically absurd. Other methods, particularly multivariate methods, prevent human bias but must be interpreted with caution. Graphic methods also fail when the data base is not from measured stratigraphic sections. In short, there is no single best method. The method chosen should suit the type and quality of the data to be analyzed and the ultimate goal of the biostratigrapher.

FOR FURTHER READING

Cubitt, J. M., and R. A. Reyment (eds.), 1982. *Quantitative Stratigraphic Correlation.* John Wiley & Sons, New York. Excellent collection of papers on quantitative methods; some of the papers were sources for this chapter.

Eicher, Don L., 1976. *Geologic Time*, 2d ed. Prentice-Hall, Englewood Cliffs, New Jersey. An excellent short introduction to the concepts we have examined in the last four chapters. Eicher's chapter 5 on biostratigraphy is particularly good.

Gradstein, F. M., J. P. Agterberg, J. C. Brower, and W. J. Schwarzacher, (eds.), 1985. *Quantitative Stratigraphy.* D. Reidel Publ. Co., Dordrecht, Netherlands. The most up-to-date collection of papers on quantitative biostratigraphic methods.

Kauffman, Erle G., and Joseph E. Hazel, 1977. *Concepts and Methods of Biostratigraphy.* Dowden, Hutchinson, and Ross, Stroudsburg, Pennsylvania. An excellent compendium by many distinguished authors of the frontiers of biostratigraphy.

Shaw, Alan B., 1964. *Time in Stratigraphy.* McGraw-Hill, New York. The classic work which started a revolution in stratigraphy. Clear, very readable, and provocative, but unfortunately out of print.

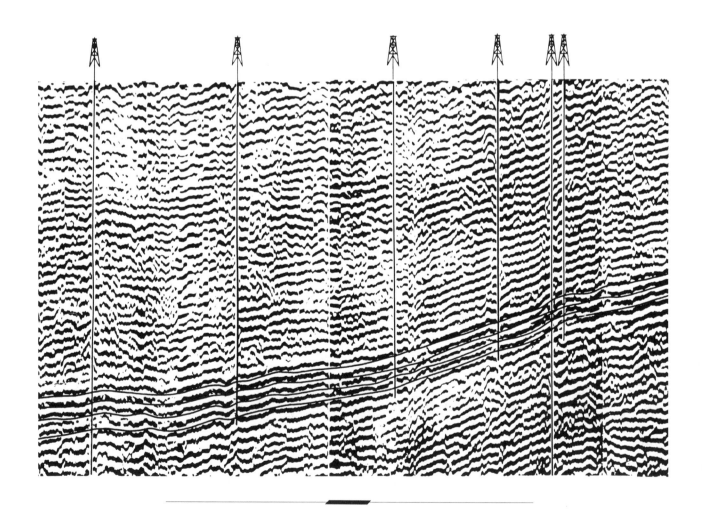

Seismic section from a Tertiary basin in South America, showing well control. From Vail, Todd, and Sangree (1977). Courtesy American Association of Petroleum Geologists.

11

Geophysical and Geochemical Correlation

WELL LOGGING

Method

Previous chapters discussed the correlation of sections based on the lithologic features of the rocks or on their fossil content. In this chapter, methods of correlation that depend on the geophysical or geochemical properties of rocks are discussed. Many of these techniques have been developed in the last decade and are just proving their stratigraphic potential. Each method has certain strengths and limitations. One of the oldest and most widely used methods that depends on the geophysical properties of rocks is subsurface *well logging*.

Many geologic problems cannot be solved on the basis of surface outcrops. This is particularly true in the areas where outcrops are limited, such as the midcontinent of North America. In some places, meters or more of glacial debris or soil horizons cover nearly all outcrops. Seismic reflection can be used to determine the subsurface structure of these covered areas, but a direct sample of the formation is needed to be sure of the lithology from which the seismic reflections are coming. The most practical way to obtain this subsurface infor-

mation is to drill a well and record all the useful information possible in the form of a core of the drilled sequence.

Continuous core recovery from most wells is far too expensive (thousands to hundreds of thousands of dollars) for more than occasional use, except when great detail is wanted. In modern drilling a lubricating mud is pumped down through the drill string (Fig. 11.1) to cool and lubricate the drilling area. The drilling mud (usually a mixture of bentonitic clay and oil or water, plus barite to regulate the density) is then forced up the well to the surface so that some mud constantly flows toward the mud tanks, carrying the cuttings, or "chips," away from the drilled formation to prevent the well from clogging. These chips are a primary source of information about the subsurface unit. The site geologist usually keeps a continuous log and sample of the chips as they come up and are screened out of the drilling mud. The chips are the only direct record of the lithology of the sequence as it is encountered by the drill bit. However, if the density of one lithology greatly differs from the next, the chips may rise to the surface at different rates and give an erroneous impression of the sequence. The driller also keeps a log of the drilling rate, which indicates the degree of induration of the formation.

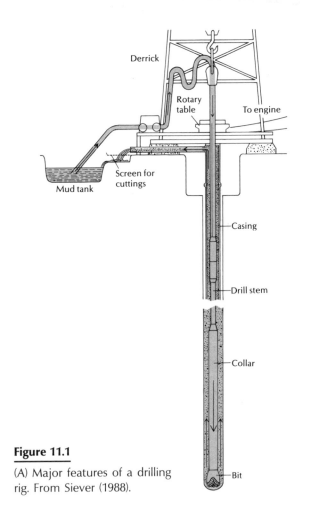

Figure 11.1

(A) Major features of a drilling rig. From Siever (1988).

The second and more important source of stratigraphic information comes after the well has been drilled and the drill string removed. A *caliper,* lowered down the hole, measures hole diameter as it is pulled upward. These measurements not only indicate the presence of shales, which are prone to caving, but also are important for calculating the response of the logging tools. An exploring device, called a *sonde,* is then lowered down the hole, and as it is raised, various electrical and other properties of the hole are measured (Fig. 11.2). A whole suite of measurements can be logged, depending on what information the driller is seeking. Several measurements are usually made simultaneously, and the results are used in combination for a better understanding of the nature of the logged lithostratigraphic sequence. Grouped by the nature of the properties that they measure best, the various types of well logs include those that measure the following:

Bulk properties of rock	Dipmeter and sonic logs
Compositional aspects	Gamma ray log
Fluid properties of pores	Spontaneous potential (SP), resistivity (R), and compensated neutron log (CNL)
Gases in the well	Gas detector and gas chromatograph

Figure 11.2

The electric logging sonde and the configuration of electrodes for recording short normal, long normal and laterolog resistivity curves. The short normal curve measures the resistivity difference between the current electrode A and point M_1, about 40 centimeters away. The long normal curve measures the resistivity between A and M_2, about 160 centimeters apart. The laterolog measures the resistivity difference between A (at the surface) and O. From Krumbein and Sloss (1963).

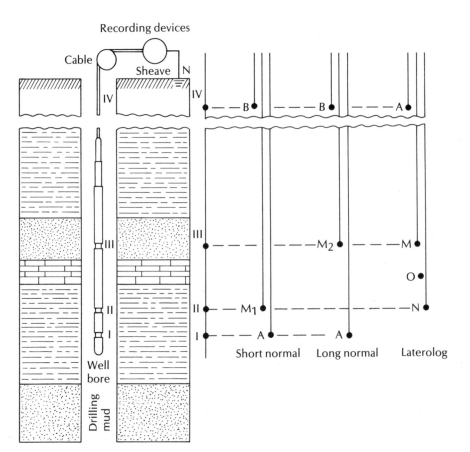

Spontaneous Potential and Resistivity Logs

Of the methods just listed the two most commonly used are the spontaneous potential and resistivity logs. These properties are usually measured together, with spontaneous potential conventionally plotted on the left and resistivity on the right (Fig. 11.3).

The *spontaneous potential* (*SP*) log measures the difference in electrical potential between two widely spaced electrodes, a grounded electrode at the top of the well and an electrode on the sonde. Drilling mud invades the pore spaces of the rock adjacent to the well. When the drilling mud and the natural pore fluid come into contact, they set up an electrical potential. Movement of ions from the drilled formation to the borehole accounts for 85 percent of the measured voltage difference and invasion of drilling mud from the borehole into the **formation accounts for 15 percent. For this reason, the** SP log is a measure of permeability. Shales are impermeable, so their reading is near zero (the "shale line") along the right edge of the log (see Fig. 11.3). Limestones are low in permeability unless they are porous or fractured; but sandstones usually show a large deflection toward the negative pole because of their high permeability. If the sonde encounters a fluid that is a better conductor than the drilling mud (such as salt water), the curve will deflect to the left; if the fluid is a poor conductor (such as fresh water or oil), it will deflect to the right.

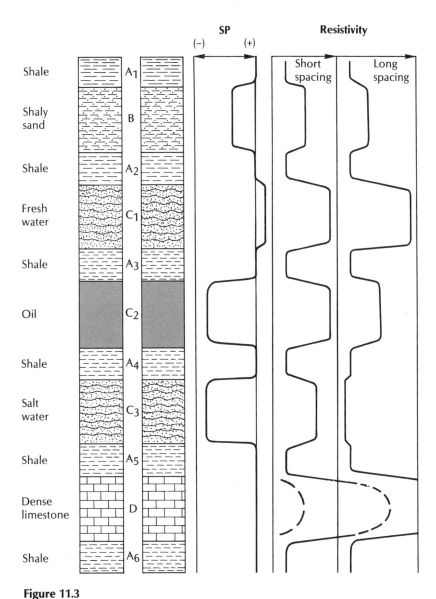

Figure 11.3

Idealized spontaneous potential and resistivity curves for various combinations of rock types and contained fluids. From Krumbein and Sloss (1963).

The *resistivity* (R) log measures the resistivity of the fluids contained in the surrounding rock to an applied electrical current. This indicates the amount of fluid in the rock and therefore the pore space. Because the drilling mud invades the porous rock from the borehole, this measurement is actually of the difference in resistivity between the invaded zone (mud has a high resistivity) and the uninvaded zone (natural pore fluids have a lower resistivity). Resistivity increases with decreasing pore space, so 10 percent porosity is about 10 times more resistant than 30 percent porosity.

Dense rocks with no pore space (such as limestones and quartzites) have very high resistivities, so they usually deflect the record to the right, even off scale. Off-scale peaks are typically truncated, and their tops begin again at the zero line. In addition, the R profile for limestone is typically jagged because of solution cavities or fractures. Sandstones filled with a nonconducting fluid (such as oil or fresh water) also deflect to the right. Shales, on the other hand, have low resistivities and deflect to the left.

The spacing of the electrodes is critical. For example, because drilling muds are usually less saline and more resistive than pore waters, a saltwater-bearing unit infiltrated by drilling mud shows a high resistivity and can be mistaken for an oil sand. To get the true resistivity beyond the invaded zone, the spacing between electrodes must be increased. A typical R log has two curves, one for *short normal* spacing (about 40 centimeters of separation) and one for *long normal* spacing (about 160 to 180 centimeters of separation). Even longer spacing can be obtained by placing a third electrode (O in Fig. 11.2) between the measuring electrodes and recording the resistivity between O and the surface (A in Fig. 11.2) This extra-long spacing (called *laterolog*) measures the resistivity at a long distance from the bore and is least sensitive to mud invasion. The difference between the short and long spacing of R logs is shown in Fig. 11.3. Fresh water gives a large deflection to the right with long spacing, but oil shows almost no change. A permeable saltwater-bearing unit, on the other hand, deflects to the left on the long spacing and to the right on the short spacing.

Other Logs

Other types of logs measure different properties of the borehole. The *dipmeter* has four electrodes spaced 90° apart that measure the resistivity simultaneously in four directions. A rock bed has the same resistivity on all sides of the borehole. Therefore, if the bed is horizontal, there will be no difference in the resistivity in the four directions (Fig. 11.4A). However, if the bed dips, then one or more of the electrodes may encounter a different bed with some other resistivity value. The computer

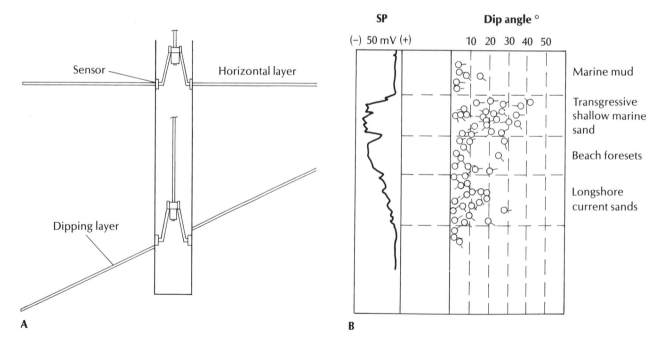

Figure 11.4

(A) A dipmeter, used to determine the dip of rock layers in a well. If the bed is horizontal, all the electrodes detect it simultaneously; if the bed dips, then one electrode detects it before the others. From Hyne (1984). (B) Environmental interpretation using a high-resolution dipmeter. The consistency or variability of the dips and the steepness of their dip angle can be characteristic of certain facies. After Gilreath et al. (1969).

Table 11.1

Wire-Line Well Logs

Name	Measures	Primary Uses	Comments
Electrical log			Older type of log
Spontaneous potential (self-potential), SP	Permeable vs. shale beds	Correlation, location of reservoir rocks	Reservoir rocks kick to left
Resistivity, R	Electrical resistivity		
Short normal	Adjacent to well bore	Tops and bottoms	
Long normal	Away from well bore	Identification of reservoir fluids	Oil and gas gives kick to far right
			Higher oil saturations kick further to right
Induction log, dual induction log	Measures SP, R, and conductivity	Same as electrical log; shallow induction same as short normal and deep induction same as long normal	Used in well filled with any type of drilling mud or air
			No SP measurement in air or oil drilling mud
			Modern type of log
Gamma-ray log	Natural radioactivity of rocks	Correlation, tops and bottoms, location of reservoir rocks, shale content	Shale kicks to right; used in cased and uncased wells
Neutron porosity log, neutron log	Hydrogen atom density	Porosity	Reads low on gas reservoirs; used in cased and uncased wells
Formation density log, density log	Density of rock	Porosity	Must know lithology (matrix)
Acoustic velocity log, sonic log, velocity log	Sound velocity through rock layer; measures interval transit time Δt	Porosity and correlation	Must know lithology (matrix)
Caliper log	Size of well bore	Engineering calculations Calibration of other logs	Thick filter cake (small hole) indicates permeable zone
Dipmeter, dip log	Orientation of subsurface rocks	Interpretation of structure and depositional environment	Uncased well

selects similar resistivities on the four different curves to calculate a dip. The continuous log of changes in dips is used not only to determine regional structure and unconformities but also to detect changes in bedding attitudes of sedimentary structures. Certain assemblages of dips indicate a diagnostic suite of sedimentary structures that aid in paleoenvironmental interpretation (Fig. 11.4B).

The *gamma ray* log measures the natural radioactivity of the strata. Unlike other methods, it can be run even after the borehole has been cased in cement to prevent it from collapsing. Most of the gamma radiation comes from the decay of ^{40}K in the minerals of the surrounding rock, so this method is sensitive to rocks that are high in potassium. Potassium feldspars, micas, clays, and organic material have the highest potassium content whereas quartz and calcite have none. Thus, shales and arkosic, lithic, or muddy sandstones will give high gamma-ray values, and limestones and quartz-rich sandstones will give very low ones.

Sonic, or acoustic, velocity logs use the methods of seismology on a very small scale. By measuring the velocities of sound waves traveling through rock, geologists can estimate the density and porosity of the rock. This logging tool has a sound transmitter at one end and two receivers spaced along it. Both receivers pick up pulses sent out by the transmitter, with the more distant receiver picking them up slightly later. From the difference in receiving times and the distance between the receivers, one can calculate the sonic velocity through the rock. Shale has the lowest sonic velocity, followed by sandstone, limestone, and dolomite, which has the highest. As porosity increases, sonic velocity decreases because the gas or liquid in the pore spaces has a much lower sonic velocity than rock. If the lithology is known, the porosity can be calculated.

There are a number of other well-logging methods of lesser importance, and new methods and refinements on old methods are developed continuously. Table 11.1 summarizes some of the more important methods and

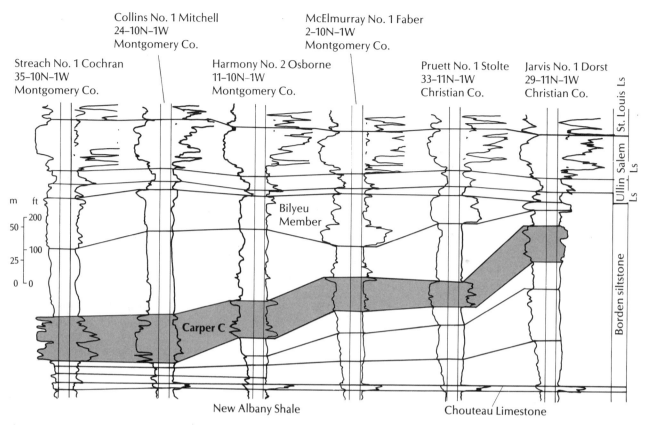

Figure 11.5

Electric log correlation between shelf limestones, deltaic siltstones and shales, and delta-front turbidite sands (Carper C) from the Mississippian of central Illinois. From Lineback (1968).

their relative strengths and weaknesses. After a geologist uses these methods, they become almost second nature. For example, the lithologic interpretation shown in Fig. 11.5 shows how limestone, shale, and sandstone curves can be matched for a considerable distance. In many cases, the shapes of the curves indicate not only the lithology and pore water but also the geometry of the unit. Many of the classic stratigraphic sequences that are characteristic of particular depositional environments have characteristic shapes on the SP log (Fig. 11.6). However, well-log interpretation is not always "cookbook" or straightfoward. Many subtle variations are possible in rocks and in the mechanical behavior of the logging method that can mislead. It always pays to have a good understanding of *why* a particular method works in case the rules of thumb break down.

SEISMIC STRATIGRAPHY
Waves in the Earth

One of the most powerful and rapidly growing fields of stratigraphy and geophysics is *seismic stratigraphy*. Long used for determining the deep structure of the earth and

shallow surface structure for oil exploration, seismic stratigraphy has reached new levels of sophistication and information content during the last decade. Now it is possible not only to recognize stratigraphic horizons but also to recognize the shape of stratigraphic sequences and interpret their depositional history, to recognize unconformities and reconstruct the transgressional-regressional history of an area, and even to detect the fluid content of rocks and identify hydrocarbon accumulations. Seismic stratigraphy allows two- and three-dimensional analysis of subsurface geology with resolutions of tens to hundreds of meters. The technique has also become less expensive as it has become more refined with the result that running a seismic line is much cheaper than drilling a dry hole.

In simple terms, reflection seismology is the process of making a loud bang and then listening for echoes. It is similar to a ship's sonar, which listens for the returning echoes of its sonar "pings" to estimate the depth of the seafloor. On land, the loud bang is usually produced by an explosion of buried dynamite or by some sort of mechanical pounder or thumper. The most widely used thumper is a method called Vibroseis® (developed by Conoco), which uses an array of large trucks to produce vibrations in the ground (Fig. 11.7). At sea, a device

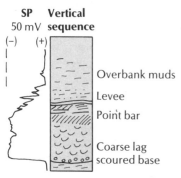

SP   Vertical
50 mV sequence
(−)   (+)

Overbank muds

Levee

Point bar

Coarse lag
scoured base

Common thickness: 8–35 ft (2–10 m)
Geometry: multistory sheet or shoestring
Width and length: extremely variable
Trend: roughly perpendicular to
    depositional strike

A

SP   Vertical
    sequence

Variable succession
Erosional surface
Low angle and planar
cross-stratification
Trough cross-
stratification

Ripple laminated sands
Pro-delta mud

Common thickness: 10–40 ft (3–13 m)
Geometry: sheet or shoestring
Length: up to tens of kilometers
Width: variable
Trend: parallel to depositional strike

B

Figure 11.6

Idealized examples of stratigraphic motifs and self-potential
(SP) logs for (A) a meander fluvial,
(B) a coastal barrier, (C) a barrier
island, and (D) a distributary
channel mouth. Note that the log
shape is an approximation of the
grain size and sand content. For
example, the fining-upward fluvial sequence also tapers upward
on the SP curve; the coarsening-
upward barrier-island sequence
also widens upward. After Galloway (1978).

SP   Vertical
    sequence

Lagoonal muds
Dunes (rarely preserved)
Beach, low-angle cross-laminated

Upper shoreface
trough, cross-stratification

Lower shoreface
bioturbated, rippled

Shelf muds

Common thickness: 20–60 ft (7–20 m),
    but may exceed 80 ft (25m)
Geometry: elongate
Length: up to 60 km
Width: 1–20 km (depending on preservation)
Trend: parallel to depositional strike

C

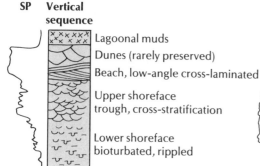

SP   Vertical
    sequence

Delta plain, mud peat
Distributary levee,
rippled and laminated sand
Distributary channel mouth
laminations, rippled,
trough cross-bedding
Laminated and rippled
sandstone
Laminated prodelta mud

Common thickness: 10–40 ft (3–13 m)
Geometry: podiform to elongate
Length and width: variable
Trend: associated with delta lobes

D

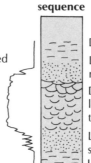

Shaly marine
succession
Erosion
Distributary
channel

Prodelta

Real example of preserved
distributary channel mouth:
Wilcox Group, Gulf Coast

Figure 11.7

COCORP seismic reflection
trucks in Wyoming. Each truck
has a large plate in back that is
lowered and pressed to the
ground until it supports the truck.
Hydraulic motors vibrate the
truck on the plate, creating a
rhythmic pulse that is ideal for
deep seismic reflection. Photo
courtesy S. Kaufman.

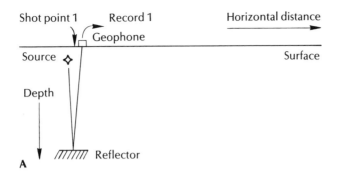

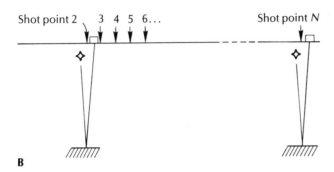

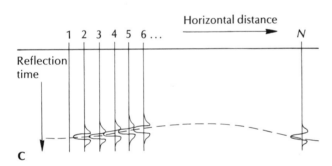

Figure 11.8

The basic principle of seismic reflection. (A) When a seismic wave is generated, the geophone picks up the wave's two-way travel time down to a reflecting layer and back. (B) Moving the shot point and geophone generates a series of reflections of the layer. (C) These reflections show up as the wiggly trace on the seismic record, which can be correlated across the profile. From Anstey (1982).

called an air gun explodes a bubble of gas underwater to create a noise.

The sound waves travel in all directions, but only those that travel almost directly downward can be reflected off structures underneath the explosion (Fig. 11.8). On land, the returning sound wave is detected by a listening device called a *geophone* (or "jug"; the geophone crew members are "jug hustlers"); at sea, the reflected sound is recorded by a hydrophone. A geo-

phone is a remarkably simple device. It consists of a casing that encloses a spring-mounted magnetic coil wired to the sound truck. When the ground vibrates, the casing vibrates with it, but the internal magnet, hanging from its spring, remains motionless due to inertia. The relative motion between the magnet and the geophone casing is recorded as electrical current fluctuations, which become the signals transmitted to the receiving equipment.

A single shot recorded by a single geophone is recorded on a strip chart as a long line, with a pulse representing the time when the reflected sound wave returned. If more than one reflecting surface is encountered, there will be more than one pulse on the trace. The vertical scale on most seismic lines is therefore two-way travel time in seconds. (Knowledge of the average seismic velocity of various rock types enables geologists to calculate the actual depth of the reflecting layer in meters rather than in seconds, but this is needed only when they are trying to match seismic records with well logs or surface outcrops.) A single strip chart is as one-dimensional as a borehole record. To get two-dimensional layers and other structures, an *array* of recording points is needed (Fig. 11.8). Every point records the shot separately and produces a strip chart. When all the charts are lined up, the pulses, or reflections from the same layer also line up from one chart to the next, and the first approximate measurement of the layer is made. This is called "picking" the marker reflections.

The geophones in an array are laid out in a traverse at regular intervals away from each recording truck, which contains the electronics to record the signals and the computer to process and interpret them (Fig. 11.9). (At sea, a ship tows a string of regularly spaced hydrophones underwater behind it to achieve the same effect.) The sound is then generated. Every geophone in the linear array picks up the direct reflection at a time that depends on its distance from the acoustic source. To improve the signal-to-noise ratio, after a shot is made, the vibrator truck and its array are moved a distance downline, so that the reflections from the same layer will be picked up by the geophones from slightly different positions (Fig. 11.9). This is called the *common-depth-point* method. By repeating the shots and measuring from slightly different positions along a linear transect, the signals are recorded many times and can be collected and stacked. This amplifies the signals and screens out the noise because the noise has no regular pattern along the line. Finally, all the signals from the single traverse are collected, screened by the computer, and printed out as a *seismic profile*. The dense clusters of vertical traces from the many subsurface reflector horizons form an almost continuous pattern of lines and pulses. If all the complicated signal processing is done correctly, the reflections from distinctive horizons align and look like real cross sections of units.

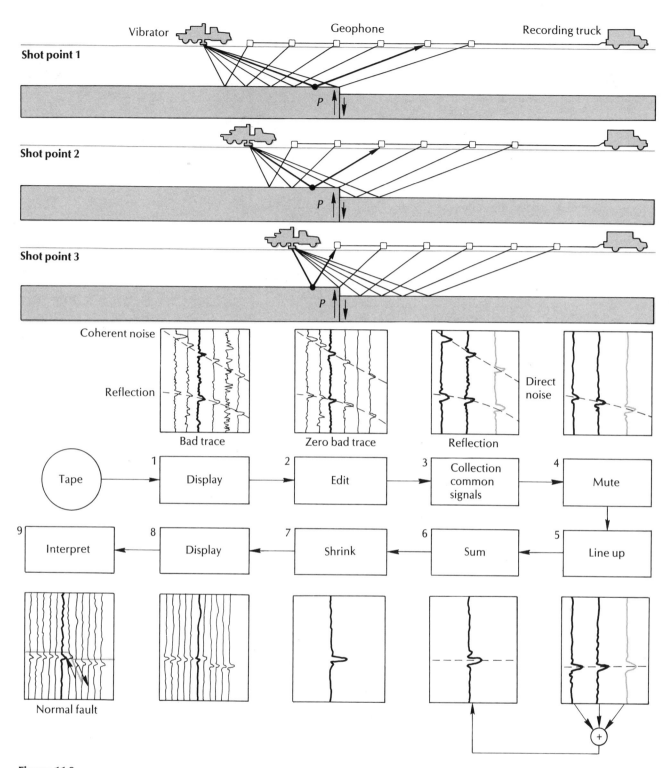

Figure 11.9

Generation and processing of seismic data require many steps, as shown in this schematic diagram. The vibrator trucks (see Fig. 11.7) generate a low-frequency signal that is reflected off an underground layer. A series of geophones picks up the arrivals of waves reflected at different angles from the source. The trucks and geophones are then moved, enabling every geophone to pick up reflections from various angles (note angles of reflection from point P). These reflections are recorded on tape and the recordings (1) displayed. Computer editing (2) removes any noisy traces, and the reflection signals from a common depth point are collected (3). High-amplitude noise is muted (4), and common signals are lined up (5) to compensate for the various angles of the geophones. The pulse is then summed (6), shrunk (7), and displayed (8) along with records from other positions on the seismic line. These can then be interpreted (9) as a geological feature (in this case, a normal fault). From Cook et al. (1980).

Seismic Profiles

Seismic profiles are not stratigraphic cross sections in the strict sense. The vertical scale is two-way travel time, not actual thickness. If the seismic velocity is the same in all the underlying rocks, then the travel time and thickness are linearly related, but variations in lithology and seismic velocity almost always destroy this relationship. Moreover, the reflector horizons are not necessarily lithologic boundaries. A seismic reflection is produced by any abrupt change in seismic velocity, usually denoting a sharp contrast in density. Most bedding horizons produce reflections, but if the boundary is subtle and lies between two lithologies of the same density, the signal may be very weak. On the other hand, lithologic features that have little stratigraphic significance can produce strong reflections. Dense chert horizons, for example, produce reflections as strong or stronger than major unconformities in the continental margins. For a long time, these chert horizons, which do not correspond to any formation boundary, confused marine geologists. Many other lithologic phenomena can also produce spurious reflections, and these are often difficult to screen out. Great care must be used in interpreting seismic profiles because they look deceptively easy to read.

Nevertheless, the structures of many subsurface layers and their approximate depths can be revealed by seismic profiling. If outcrop or well data are available, then the prominent reflectors can be matched up with known lithologic horizons. In doing this, however, the relative scales must be kept in mind. A single seismic deflection usually represents tens to hundreds of meters of section, which is orders of magnitude coarser than the resolution of a well log (Fig. 11.10). Only the most prominent

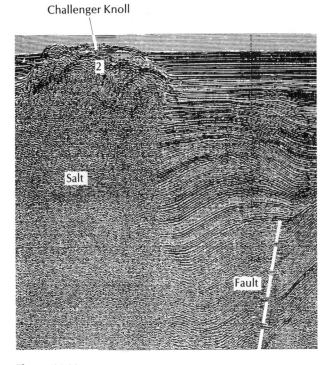

Figure 11.11

The reflection-free interior of a salt dome. From Anstey (1982), courtesy U.S. Geological Survey.

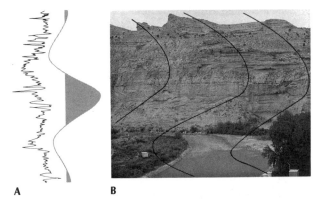

A   **B**

Figure 11.10

(A) The relative dimensions of typical stratification (as seen on an electric log) and the size of a seismic pulse. From Anstey (1982). (B) The scale of the seismic pulse compared with an outcrop about 150 meters thick. Obviously, seismic stratigraphy is suitable only for coarse-scale resolution of stratigraphic features. Based on an idea from Miall (1984).

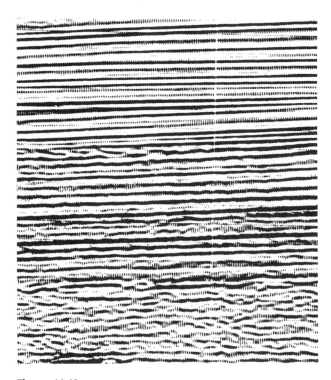

Figure 11.12

The contrast between calm marine deposition (upper third) and nonmarine deposition (lower third). The middle third shows some open marine and some shallow-water deposition. From Anstey (1982), courtesy Merlin Geophysical Company.

lithologic breaks appear on the seismic record, and the actual deflection may represent a zone that is meters or more in thickness (typically, 10 meters or more per impedance "wiggle") where the density contrast is greatest. Layered stratigraphic sequences are the most obvious and easy to interpret, and structural features such as dipping beds, folds, and faults are usually apparent. Unlayered structures do not show any coherent pattern of reflections, but instead are characterized by an almost random scatter of reflections. Massive features such as salt domes, reefs, and hardrock basement usually show this pattern (Fig. 11.11). Even subtle differences in the texture of the reflections can be meaningful. For example, nonmarine beds typically have jagged reflections, with many discontinuous horizons, whereas quiet marine deposition produces reflections that are smooth, continuous, and homogeneous (Fig. 11.12). The presence of gas (which has a very low seismic velocity) inside a reservoir rock changes the seismic reflectance considerably, producing "bright spots". This is one of several instances in which hydrocarbon accumulations produce direct seismic traces. In most instances, however, economic hydrocarbon deposits must be found by analyzing the total stratigraphic pattern.

COCORP

In recent years, seismic stratigraphy has made discoveries that have implications far beyond local geology and economic concerns. One of the most important development has been the Consortium on Continental Reflection Profiling (COCORP) project, developed at Cornell University with the aid of a consortium of oil companies. The COCORP project has concentrated on producing detailed, high-resolution profiles with considerable depth penetration (up to 50 kilometers); their ultimate aim is to produce complete, deep-continental profiles across most of North America. They use a large array of Vibroseis®trucks and stack a much larger sequence of seismic arrays than was possible before the advent of high-powered computers. The COCORP project has made several key traverses through mountain ranges and basins, with some startling results. For example, their profile of the Wind River thrust of Wyoming showed it to be much more deeply seated than previously thought. Even more surprising are the profiles across the Appalachians, which have revealed that they are underlain by deep, horizontal thrust faults (Fig. 11.13) that have hundreds of kilometers of offset. Con-

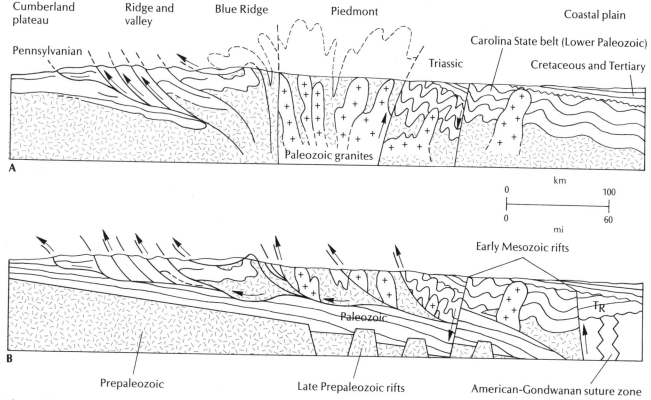

Figure 11.13

(A) The traditional interpretation of the basement structure of the southern Appalachians. The Piedmont and Blue Ridge were thought to have deep crustal roots, with intrusions coming from deep in the crust. (B) The COCORP seismic reflections showed that the southern Appalachians are actually made of thin-skinned thrusts; thus the "crystalline basement" is part of a thin thrust slice and has no roots. Instead, it is thrust over Paleozoic sediments. From Dott and Batten (1988).

trary to the old notion of the shallow, steep thrusts that die out at depth, it appears that the Appalachians are produced by "thin-skinned" thrusting that has transported Precambrian basement hundreds of kilometers over Paleozoic shelf sediments. This radically changes the proposed tectonic models for this area. It is also possible that hydrocarbons are trapped below these thrusts in regions that have never before produced oil.

The Vail Curves

Probably the most widely discussed result of seismic stratigraphy has been the interpretation of unconformities. These show up on seismic records as places where beds abruptly truncate or gently pinch out, and frequently they produce strong reflections. Each unconformity-

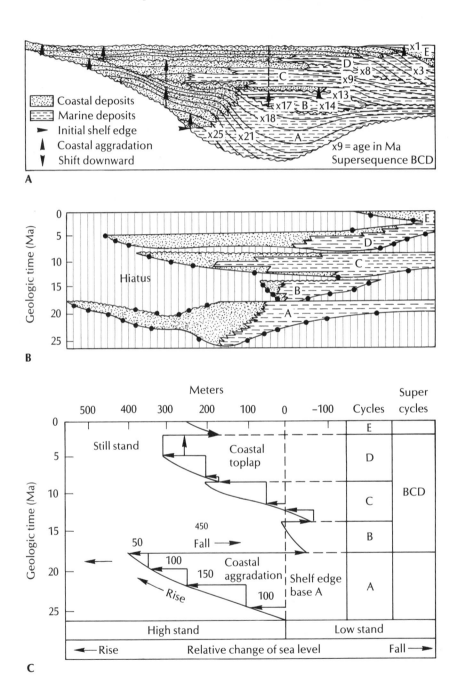

Figure 11.14

(A) Seismic sequences are analyzed by marking the unconformities in a seismic section and determining their degrees of onlap or offlap and their ages. (B) These can then be plotted as a series of unconformity-bounded sequences with a time axis. The lateral extent of the units is shown, and the vertically striped regions are absent units. (C) From this diagram, a curve of relative onlap and offlap can be constructed for this particular section. From Vail et al. (1977).

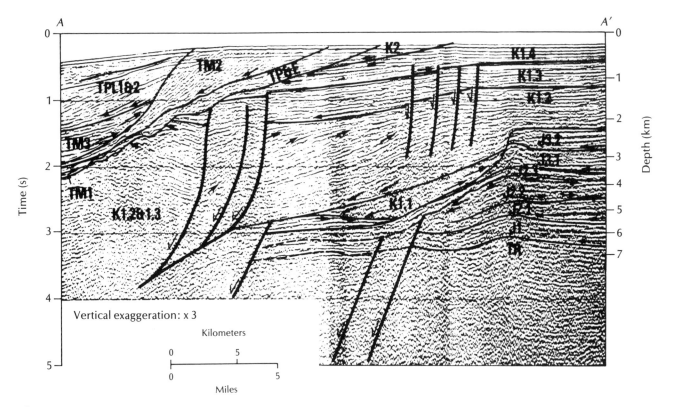

Figure 11.15

Seismic profile across the passive margin of northwestern Africa, showing sequences as defined by seismic reflections. In this view, the fault-bounded Triassic basement (TII) can be clearly seen and is overlain by several small Jurassic (J1–J3.2) packages. The bulk of the section is composed of thick Cretaceous shelf sequences (K1.1–K2), which are cut by normal faults. There are thin Paleocene and Eocene deposits (TP&E) and Miocene deposits (TM1–TM3), but the Oligocene is completely absent due to an enormous offlap event (see next two figures). Major unconformities also separate the Jurassic from the Cretaceous, the K1.1 from the K1.2 sequences, and the Cretaceous from the Paleocene. Based on this and similar sequences from other passive margins, Vail et al. (1977) constructed their famous "sea level" curves. From Mitchum et al. (1977).

bounded package on a seismic profile is called a *seismic sequence* (Fig. 11.14A). The sequences are bounded by curved lines on the seismic profile, but with well control, it is possible to determine the ages of the beds. Assuming that these minor reflectors within sequences are isochronous, one can estimate the ages and magnitudes of the unconformities that bound them. The reflectors then can be placed in a chronologic framework (Fig. 11.14B). From the ages and dimensions of these plots, the ages and magnitudes of changes in coastal onlap and offlap that formed these seismic packages can be interpreted (Fig. 11.14C).

The seismic profiles of a number of passive continental margins have been studied in this fashion, and each was analyzed for the ages and magnitudes of its major unconformities. In the northwest African margin, for example, there are major unconformities between the Jurassic and Cretaceous (J3.2 and K1.1 in Fig. 11.15), between the Cretaceous and Tertiary (K2 and TP&E), between the Eocene and Miocene (TP&E and TM1, elim-

inating the entire Oligocene), and between the middle and upper Miocene (TM2 and TM3). There are minor unconformities between units within the Jurassic and within the Cretaceous. A significant number of normal faults are also apparent in the figure.

After a number of these curves were interpreted, it became clear that most passive margins have major unconformities of approximately the same magnitude and the same age. From this, Vail, Mitchum, and Thompson (1977) suggested that sea level had fluctuated on a worldwide basis, producing unconformities of approximately the same magnitude on all continental margins at the same time. The product of their work was at first called the *Vail sea level curve* (Fig. 11.16), which purports to show the changes in relative eustatic sea level through the last 200 million years.

Since the initial publication of the Vail curve, however, there has been considerable controversy over what controls the shape and apparent magnitude of a cycle in the curve. As we saw in Chapter 8, the large-scale

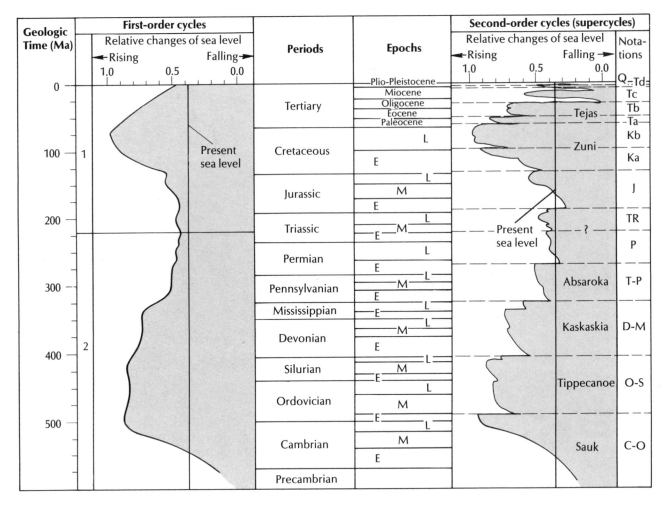

Figure 11.16

First- and second-order changes in global onlap and offlap (called "sea level changes" in this diagram) through the Phanerozoic. Note that there have been two first-order peaks of major transgression (during the early Paleozoic and the Cretaceous) and peaks of regression during the late Precambrian, Permo-Triassic, and late Tertiary. These are punctuated by second-order cycles that span tens of millions of years. The Paleozoic second-order cycles correspond to Sloss's (1963) sequences (see Fig. 8.9). From Vail et al. (1977).

first-order cycles (Fig. 11.17) appear to be real and to have been caused by major continental movements. But the jagged, sawtooth shape of the second- and third-order cycles, with their curved bottoms and flat, abrupt tops, has led to controversy. No explanation for this asymmetry involving the known behavior of sea level made sense, and now it appears that the sea-level fall was not that rapid. When sea level falls, it is normal for the shallow marine coastal sequence to prograde out across the shelf, building up a sedimentary package even though sea level is dropping (Fig. 11.18). This "fills in" the peak during the regressive phase and gives the appearance that sea level falls more abruptly than it rises. Because sea level can be falling while the prograding wedge is still building up, it is necessary to distinguish between the descriptive terms *onlap* and *offlap* (which describe the degree to which the package laps onto the continent) and true rise and fall of sea level. Recently,

Vail and his colleagues (1984; Haq et al., 1987, 1988) have differentiated between their sawtoothed curve of coastal onlap and offlap and a true sea level curve, which has a smoother, more rounded shape (see Fig. 11.17 and Appendix B).

The traditional terminology of disconformity and angular unconformity is insufficient to describe the complex relations seen in seismic sequences. For this reason, Vail and his colleagues have adopted a slightly modified terminology for subsurface unconformities (Fig. 11.19). The upper boundary (Fig. 11.19B) can be erosionally truncated, concordant with overlying sequence, or initially inclined and covered by the overlapping sequence (*toplap*). The lower boundary can also be concordant, or it can *onlap* against an inclined depositional surface or be deposited at an angle (as in delta foresets) against the depositional surface (*downlap*). *Baselap* can refer to either onlap or downlap.

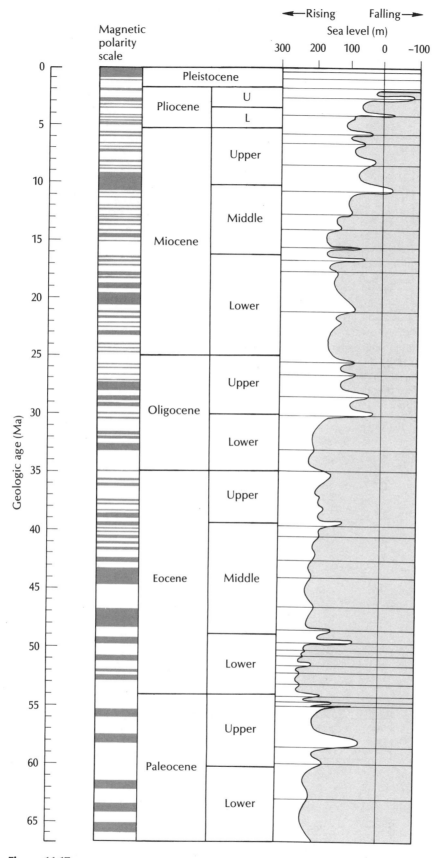

Figure 11.17

Updated version of the Cenozoic portion of the Vail curve generated by Haq, Hardenbol, and Vail (1987, 1988). This curve differs from the 1977 Vail "sea-level" curve in showing a smoother true eustatic curve rather than the sawtoothed onlap-offlap curve (see Fig. 11.16 and Appendix B).

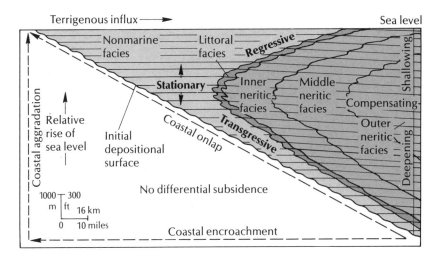

Figure 11.18

The sawtoothed onlap-offlap curve, with its slow transgressions and apparently abrupt regressions, can be explained by the fact that nonmarine facies continue to prograde out and build up the seismic sequence long after eustatic regression has taken place. The true beginning of eustatic sea-level fall is much earlier than the abrupt truncations shown on the original Vail curve. From Vail et al. (1977).

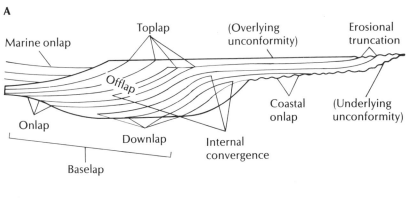

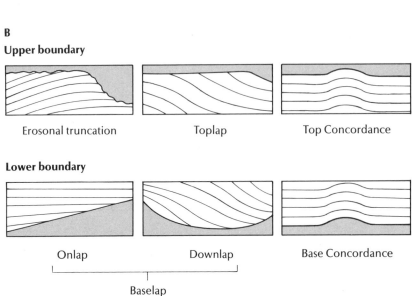

Figure 11.19

The complicated internal relationships of seismic sequences and their unconformities require a special terminology beyond the usual descriptions of unconformities, transgressions, and regressions. (A) Terminology of the typical seismic sequence. (B) Defining examples of each term. *Erosional truncation*: Strata terminate against the upper boundary, mainly due to erosion. *Toplap*: Initially inclined strata terminate against an upper boundary created mainly as a result of nondeposition (e.g., foreset strata terminating against an overlying horizontal surface where no erosion or deposition took place). *Top concordance*: Strata at the top of the sequence do not terminate against the boundary. *Onlap*: Strata terminate updip against an inclined surface. *Downlap*: Initially inclined strata terminate downdip progressively against initially horizontal or inclined surface. *Base concordance*: Strata at base of sequence do not terminate against lower boundary. From Vail et al. (1977).

Some geologists (reviewed in Chapter 8) doubt whether the second-order cycles of the Vail curve are truly eustatic in nature. Watts and Thorne (1984) produced a model that simulates the interaction of thermal contraction of the lithosphere, continental margin subsidence, and the slow first-order sea-level changes. These interact in such a way that they produce a jagged, asymmetrical curve very similar to the original Vail curves. Hubbard (1988) and Hallam (1988) have shown that many of the supposedly "eustatic" events correspond much better to local tectonic events. As we discovered in Chapter 8, the Vail curve is probably a composite of eustatic sea level and local tectonism, and should not be used naively. To their credit, Posamentier, Vail, and their colleagues (Posamentier et al., 1988; Posamentier and Vail, 1988) have refined their models to incorporate the effects of local tectonism. These new models separate the parts of the Vail curve that are eustatically controlled from those which are due to local tectonism or progradation. Further research on the sequences themselves should begin to filter out local tectonics, and produce a true eustatic sea level curve.

MAGNETOSTRATIGRAPHY

Rock Magnetism

Another geophysical correlation technique that has achieved great importance in the last 20 years is magnetic polarity stratigraphy, or *magnetostratigraphy*. Unlike the methods described previously, magnetostratigraphy has primarily been used to correlate surface outcrops, though it also has been successfully used to correlate subsurface cores. Magnetostratigraphy correlates rocks by means of the similarities of their magnetic patterns. It has some unique advantages, including the possibility of recognizing time planes that are worldwide and independent of facies. However, its main limitation is that it seldom works independently but must be used with some other geochronologic method.

Most rocks contain minerals that are naturally magnetic, such as the iron oxides magnetite (Fe_3O_4) and hematite (Fe_2O_3). In cooling igneous rock, these minerals do not lock into a permanent magnetic *remanence* until they cool to a temperature called the *Curie point*. For hematite, the Curie point is about 650°C, and for pure magnetite (with no titanium) it is 578°C. There are three types of remanent magnetization:

Thermal remanent magnetization Once the mineral cools below the Curie point, its magnetization is locked into the crystal, aligned in the direction of the earth's magnetic field at the time. Magnetization formed by cooling below the Curie point is called *thermal remanent*

magnetization (TRM). Some igneous rocks, especially basaltic lava flows, contain large amounts of magnetic minerals and are strongly magnetized.

Detrital remanent magnetization When an igneous rock is weathered and eroded, its magnetic minerals, along with the rest of the rock, become sedimentary particles. The small magnetic grains (only a few micrometers in diameter) behave as tiny bar magnets, aligning themselves with the magnetic field of the earth when the grains are deposited as sediments. The magnetization of these sediments when they have lithified is called depositional, or *detrital remanent magnetization (DRM)*. DRM is usually two or three orders of magnitude weaker than TRM, depending on how much magnetic mineral the sedimentary rock contains.

Chemical remanent magnetization After weathering, iron is dissolved out of rock and moves through the groundwater. Eventually, it can precipitate in another place, usually as hematite. As chemically precipitated iron minerals nucleate and grow, they acquire a remanence that is parallel to the magnetic field direction of the earth at the time they form. This is called *chemical remanent magnetization (CRM)*. Because the hematite weathering product could have been acquired at any time, it often has little to do with the magnetic field prevailing when its rock was formed. In this case, it is irrelevant to the stratigraphic age of the unit and may only hinder magnetostratigraphic investigations. However, there are examples of sedimentary rocks whose chemical remanence can be shown to have formed at the same time as the rock unit itself.

Sampling, Measurement, and Analysis

One or more of the preceding types of remanence may be present in a rock as it occurs in the field. However, to measure the magnetization, samples must be taken back to the laboratory. In well-indurated rocks, such as lava flows, samples must be collected with a portable coring drill (Fig. 11.20). The drill produces a small core, which must be oriented with respect to true north before it is taken from the outcrop. Samples of softer sedimentary rocks can be collected with simple hand tools, but with careful orientation, and even soft muds can be sampled if they are enclosed in some sort of oriented container to prevent them from falling apart. Three or more samples are usually collected at each stratigraphic level, or site, so that they can be averaged. A larger number of samples also allows the calculation of simple statistics. The most important statistical calculation is the variation around a mean direction. This is a measure of how well the samples cluster and can be used to determine whether the magnetic vectors differ significantly from a random scatter.

A *magnetometer* is a device that measures both the intensity and the direction of the magnetic vector of the

A

B

Figure 11.20

(A) Paleomagnetic sampling of lava flows with a coring drill. Photo courtesy S. Bogue. (B) Orientation device is placed over the drilled core to transfer the compass azimuth and dip to the core before it is broken free and removed. Photo by the author.

sample. Samples that are strongly magnetized can be measured with a spinner magnetometer (Fig. 11.21A). This device rotates the sample rapidly within a coil around an axis and measures the electrical current created in the coil by the spinning sample. By spinning the sample in various orientations, the direction and intensity of the magnetic vector can be determined. Most samples are too weakly magnetized or too poorly cemented together to be measured by the spinner magnetometer. In the mid-70s, the cryogenic magnetometer (Fig. 11.21B) was developed to measure such specimens. The cryogenic magnetometer uses liquid helium at 4° Kelvin (K) (four degrees above absolute zero) to create a superconducting region around the sensors. Electrical currents move with almost no resistance at such low temperatures. When a magnetized specimen is brought into the superconducting sensor area, the specimen's magnetic field sets up a current in the superconducting coil, which then can be measured. The cryogenic magnetometer has many advantages over older methods. It is almost three to four orders of magnitude more sensitive and thus is capable of measuring even the most weakly magnetized lithologies. It responds almost instantly, as opposed to the many minutes required to measure each sample with a spinner magnetometer. In addition, the sample does not have to be spun, so even liquid suspensions and live animals can be measured. Currently, many laboratories have cryogenic magnetometers that allow the sample to be rotated while it is in the sensing area. All measurements are taken, averaged, and corrected by a microcomputer that interfaces with the magnetometer electronics.

When a sample is first measured in the laboratory, all the magnetization it has acquired since it was first formed is still present. This initial magnetization is called the *natural remanent magnetization* (NRM). Frequently, the primary (or original) magnetization of the sample is overprinted by a younger magnetization direction, picked up in the earth's current magnetic field. To get rid of younger overprinting, each sample must be partially demagnetized, or "cleaned," until only the primary component remains. The younger overprinting is usually the easiest to erase because it was acquired more recently and usually is not strong enough to realign the primary magnetic fields of the mineral grains.

Demagnetization can be done in two ways, and sometimes both are used on the same sample:

Alternating field (AF) demagnetization If a sample is placed in a strong alternating field, the weaker components of magnetization are eliminated as they oscillate back and forth in response to the rapid change of field direction. At stronger alternating fields, more and more of the magnetization is destroyed until a cleaning field is reached in which only the primary component remains.

Thermal demagnetization The sample can also be heated to higher and higher temperatures, allowed to cool in an area shielded from the earth's magnetic field, and then measured. The weaker, overprinted components of magnetization should disappear at lower temperatures as a sample is heated, but the primary component should be eliminated only when the sample reaches the Curie point of the magnetic mineral.

Demagnetization can be a complicated, tedious procedure, and sometimes good results are never obtained. Problems with demagnetization can lead to erroneous results. For example, most early paleomagnetic studies were done with AF demagnetization alone because it was the easiest to use. It was later discovered that many

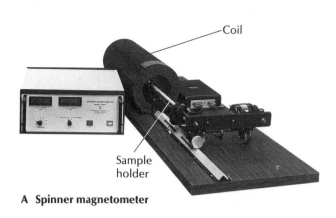

A Spinner magnetometer

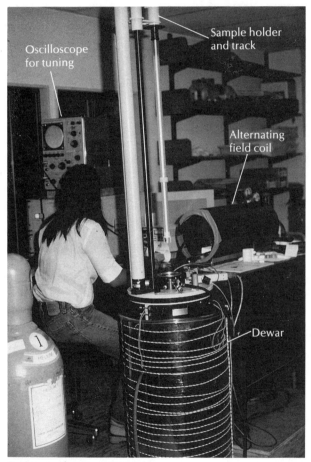

B Cryogenic magnetometer

Figure 11.21

(A) Spinner magnetometer. The long white rod spins the sample within the sensing coil (black cylinder). Photo courtesy Schonstedt Instrument Company. (B) Cryogenic magnetometer. The sample is placed at the end of the long white plastic tube and lowered on the track to the left into the dewar (black cylinder wrapped with wires), where it is measured in the superconducting sensing area in the bottom. The black cylinder to the right is the coil of the alternating field demagnetizer. Photo by the author.

terrestrial mudstones have an overprinting due to chemical remanent magnetization from goethite and other iron hydroxides, which do not respond to AF treatment. Thermal treatment at 200°C, which dehydrates the goethite and eliminates its overprinted component, radically changed the results of several geologists, who had to retract their earlier interpretations.

Field Reversals and the Polarity Time Scale

The result of all this sampling, cleaning, and measurement is a series of rock samples that show the direction and approximate intensity of the earth's magnetic field at the time they were lithified. As early as 1906, the French physicist Bernard Brunhes noticed that some volcanic rocks were not magnetized in the direction of the earth's present magnetic field, but 180° in the opposite direction, or *reversed* from the present *normal* direction. Our compass needles would have pointed south 750,000 years ago, when the earth's field was last reversed. Brunhes' and other observations of reversed rock magnetization were not followed up until the early 1960s, when a group of scientists led by Allan Cox, Richard Doell, and G. Brent Dalrymple at the U.S. Geologic Survey (U.S.G.S.) began to systematically study the magnetization of ancient rocks. Ironically, one of the earliest samples studied, the Haruna dacite of Japan, has the virtually unique property of being self-reversing. As a result, many geologists doubted that the new evidence of reversely magnetized rocks meant that the entire magnetic field of the earth had reversed.

Between 1963 and 1969, Cox, Doell, and Dalrymple continued to provide new data, in friendly competition with the McDougall and Chamalaun group at the Australian National University. To convince skeptical geologists, these paleomagnetists needed to show not only that the magnetization of their samples was reversed, but also that all rocks of the same age were of the same polarity worldwide. Only worldwide data would not be explained by local self-reversal. They sampled lava flows of various ages from all over the world, not only because such rocks are strongly magnetized but also because they could be dated directly by the newly refined method of potassium/argon dating. As the samples accumulated, the pattern became clear. Regardless of where they came from, rocks of the same age had the same magnetic polarity (Fig. 11.22). Not only had the earth's magnetic field reversed in the geologic past, but it had done so repeatedly at irregular intervals. The data produced a distinctive pattern of long and short reversed and normal episodes that could be correlated worldwide. We now call this the *magnetic polarity time scale*. The method by which the magnetic polarity time scale was extended back to the Jurassic is described in detail in Chapter 13.

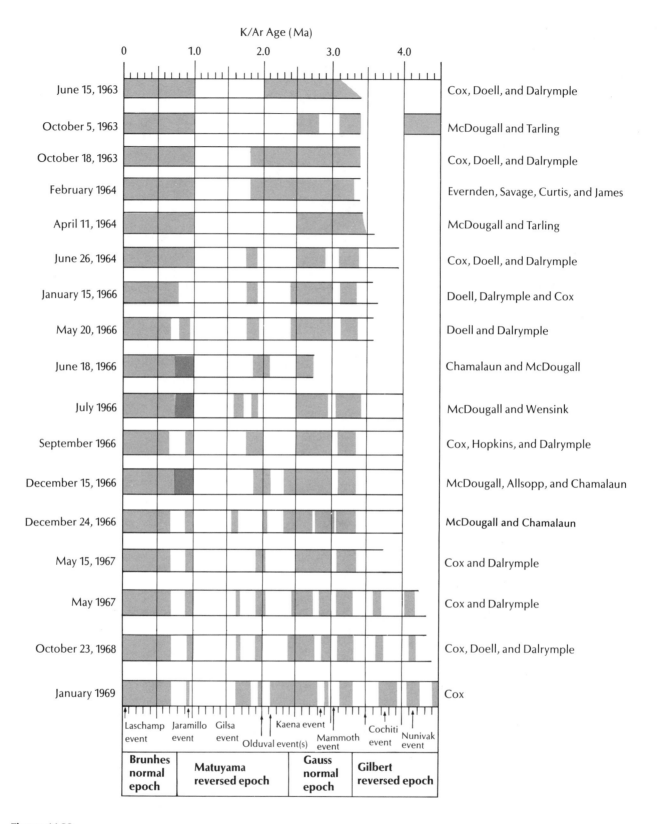

Figure 11.22

Successive versions of the geomagnetic reversal time scale as determined from the potassium/argon dating of volcanic rocks from various continents. Shaded areas represent periods of normal polarity, and unshaded areas are of reversed polarity. The original 1963 time scale of Cox, Doell, and Dalrymple had only three polarity episodes in the last 3 million years, but the competition between them and McDougall and Chamalaun in Australia produced so much data that the 1969 version of the time scale contained more than 20 events in the last 5 million years. From Dalrymple (1972).

Magnetostratigraphic Correlation

The polarity history of the earth presented stratigraphers with a powerful tool for interregional correlation. The pattern of magnetic polarity reversals is irregular and nonperiodic, so distinctive long episodes can be recognized. Once a local polarity record has been matched to the global polarity time scale, it is possible to make correlations of unmatched precision. First, polarity reversals happen *worldwide* and are *independent of facies or lithology*. No other method of correlation can make this claim. The magnetic record of a rock, on the other hand, can be measured whether it is a basalt, a siltstone, a limestone, or whatever. This allows correlation between the deep sea and the terrestrial record where no such correlation was possible before. Second, polarity reversals are *geologically instantaneous*. Studies of sequences that were deposited at high sedimentation rates have shown that a typical polarity reversal takes 4000–5000 years to occur. For all intents and purposes, polarity zone boundaries are isochronous.

Magnetostratigraphy has its own peculiar limitations as well. Polarity events are not unique. Rocks of normal and reversed polarity occur throughout the geological record, so a single sample cannot be dated by its polarity. Only a long sequence of rocks with polarity zones of distinctive lengths can be correlated. Even then, if there are significant hiatuses or rapid changes in sedimentation rates, the interpretation can be erroneous. Even if the polarity reversal pattern appears distinctive, it cannot be correlated by itself unless the top of the section is recent and the first zone down marks the present episode of normal polarity. In all other sections, some other form of independent time control (usually biostratigraphy or radiometric dating) is necessary for placing the section in its approximate geochronologic position. After this is done, then the pattern can be matched to the polarity time scale at that age to see what interpretation gives the best fit. In some cases, there may be several reasonable interpretations, and so the correlation is ambiguous. Thus, magnetostratigraphy, despite its strengths, has important limitations. It is not independent but requires other dating techniques. It requires a long sections of favorable lithology that produce a correlative pattern. If there are unconformities or fluctuations in sedimentation rate, the pattern may be distorted.

As the polarity time scale continues to develop, general patterns are emerging (Fig. 11.23). Most of the Cenozoic and late Cretaceous was a period of rapidly changing, or mixed, polarity that makes it suitable for magnetostratigraphy. The same is true of much of the Triassic and Jurassic, and the early Carboniferous. However, some parts of geologic history were characterized

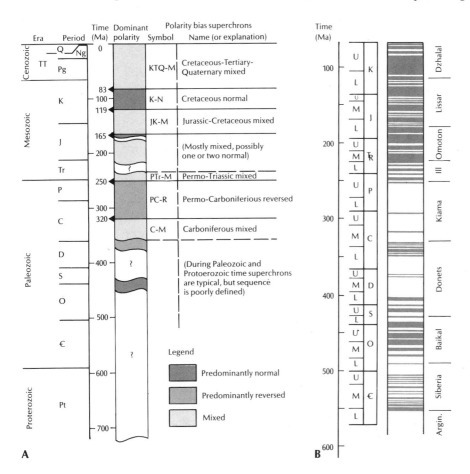

Figure 11.23

The large-scale pattern of the earth's polarity history. (A) Time scale according to Cox (1982). Mixed polarity predominates except during the long Cretaceous normal and the Permo-Carboniferous reversed episodes. The pre-Carboniferous polarity history was poorly known at the time. (B) Time scale according to Molostovsky et al. (1976), based on data from the Siberian Platform. This time scale goes back to the Cambrian, though the relative lengths of zones and the dating are not as well constrained as for the Mesozoic-Cenozoic time scale.

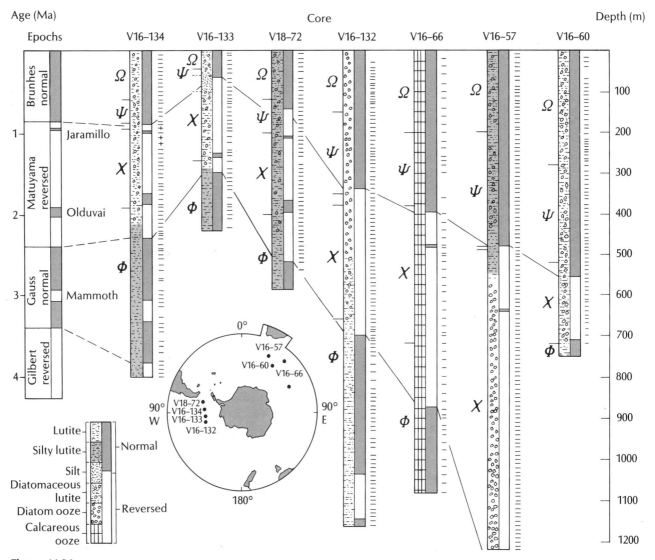

Figure 11.24

One of the earliest examples of magnetostratigraphic correlation of sediments (Opdyke et al., 1966). The polarity zones found in Antarctic deep-sea cores clearly agree with the biostratigraphy of the radiolarians (Greek letters denote biostratigraphic zones).

by long periods of stability. During the late Cretaceous, the earth's field was continuously normal for 36 million years. During the late Carboniferous and Permian, the earth's field was reversed for 70 million years. Obviously, during long periods of field stability, there is little potential for magnetostratigraphy. Pre-Carboniferous magnetic stratigraphy is still too poorly known to generalize about the dominant patterns. Work on the Siberian Platform (Molostovsky et al., 1976) has produced a tentative Paleozoic magnetic polarity time scale, but it is not as easy to calibrate as the Mesozoic-Cenozoic seafloor spreading record (Fig. 11.23B).

Despite these limitations, many of the classic studies have revealed the power of the method. For example, the relative thicknesses of polarity zones may fluctuate somewhat, but in many cases they are remarkably consistent. For example, the polarity zones in Antarctic deep sea cores (Fig. 11.24) are all different in total thickness, but each core has zones of the same relative length. Even in the more episodic sedimentary conditions of terrestrial fluvial sequences in Pakistan, there have been good results. For example, the sections in the Potwar Plateau region of Pakistan (Fig. 11.25A) produce polarity zones of consistent relative thicknesses. The mammalian biostratigraphy and dated volcanic tuffs place the sequence in the magnetic polarity time scale. A Shaw graphic correlation plot of one of the sections against the polarity timescale (Fig. 11.25B) shows that the slope is relatively straight, with linear changes. If any of the polarity zones had irregular thicknesses, they would stand out as outlying points far from the line.

The new North American Stratigraphic Code (Appendix A) has standardized the terminology of magnetostratigraphy. In the early days of magnetic stratigraphy

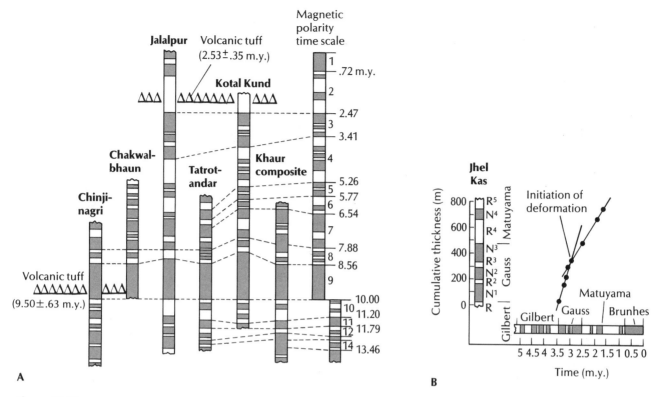

Figure 11.25

(A) Correlation of terrestrial sections from the Miocene Siwalik Group of Pakistan with the magnetic polarity time scale (shown at right). The long normal interval with the tuff dated at 9.50 ± 0.63 Ma ties the section to Epoch 9 (old terminology; now called Chron C5). The date of 2.55 ± 0.35 Ma at the base of a reversed interval ties the upper part of the section to the Matuyama reversed chron (Epoch 2 in the old terminology). From these correlations, the rest of the pattern in the area can be internally correlated and each interval matched to the polarity time scale. (B) The relative spacing of events in individual sections (ordinate) matches the polarity times (abscissa) quite well, as shown by the graphic correlation plot. From Johnson et al. (1982).

(see Fig. 11.22), polarity events were called *epochs* and were named after great scientists associated with magnetism (Brunhes, Matuyama, Gauss, and Gilbert). Shorter changes in polarity within the epochs were called *events* and were named after the place they were found (e.g., Olduvai Gorge in Tanzania, Mammoth Mountain in the Sierra Nevadas, and Jaramillo Creek in New Mexico). Paleomagnetists soon ran out of names of great magnetists, so they began numbering instead. Using the first polarity interval prior to the Gilbert as Epoch 5, they numbered both normal and reversed intervals back to Epoch 23 in the early Miocene (Hays and Opdyke, 1967; Opdyke et al., 1974). Stratigraphers soon pointed out that these intervals were not the same as epochs of the geochronologic hierarchy (see Chapter 8). In addition, the numbering scheme was somewhat arbitrary, because not all normal and reversed boundaries were "epoch" boundaries.

Eventually, magnetostratigraphers began to work with polarity events before the Miocene and needed more labels. They adopted the numbers of the positive magnetic anomalies assigned by Heirtzler et al. (1968) to the seafloor spreading profiles. Each episode of normal polarity was numbered the same as its positive anomaly on the seafloor. Events were called magnetic polarity *chrons* to prevent the confusion with the epochs of geochronology. According to LaBrecque et al. (1983), a chron begins at the top of the normal interval that corresponds to the numbered positive magnetic anomaly and ends just above the beginning of the next oldest normal interval below it. To prevent confusion with the older numbering scheme, the chrons of LaBrecque et el. (1983) are labeled with the prefix *C*. The suffixes *N* and *R* have also been used to subdivide the normal and reversed part of the same chron, and some chrons have been further subdivided with additional numbers and letters (Cox, 1982). Part of the Old Magnetic Epoch 7 became Chron C4, Epoch 13 became Chron C5A, and so on (Tauxe et al., 1987). The current terminology is shown in Appendix B.

In the new Code, the chrons can be subdivided into subchrons or lumped into superchrons. The traditional terminology for the last 5 million years of polarity history is still retained, but now they are called the Matuyama Chron, the Olduvai Subchron, and so on. The Code distinguishes between magnetostratigraphic units (bodies of rock distinguished by any difference in magnetic properties) and magnetopolarity units (bodies of rock distinguished by difference in magnetic polarity). By this definition, magnetostratigraphic units include not only magnetopolarity units but also units distinguished by slight changes in field direction, as happens during reversal transitions or in secular variation studies. Like other formal stratigraphic units, the magnetic polarity units require a type section and adherence to all the other recommended procedures of formal designation and publication.

STABLE ISOTOPE STRATIGRAPHY

Although isotopes formed by radioactive decay have become important in dating (see Chapter 12), there are many isotopes that are stable and do not decay over time. These isotopes exist in well-defined ratios in the ocean and atmosphere. When oceanographic and climatic changes occur, some isotopes can become more or less abundant with respect to others, due to *fractionation*, or separation of isotopes by their differences in atomic weight. These fluctuations of certain isotope ratios have become a powerful tool for stratigraphy in recent years.

Oxygen Isotopes

Most of the oxygen in the earth (99.756 percent) is in its normal form, ^{16}O, which has eight protons and eight neutrons; a slightly heavier isotope, ^{18}O, has two more neutrons and is relatively rare (0.205 percent). Usually, these two isotopes are present in the ocean in this ratio. In 1947, Harold Urey and Cesare Emiliani found that these two isotopes fractionated with changes in temperature. By examining the oxygen in the calcite of the shells of foraminifers, Emiliani found that the oxygen isotope ratio seemed to fluctuate in response to the changes of temperature caused by the ice ages.

Later work showed that the picture was more complicated. In addition to the temperature effect, there was also an effect due to ice volume. Water that is richer in ^{16}O, which makes it lighter, evaporates more readily than water with more ^{18}O. In the unglaciated world (Fig. 11.26), this water rains into the ocean or onto the land and travels back to the ocean via the rivers, so

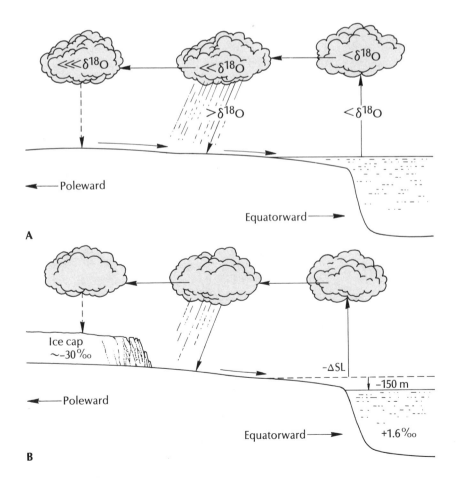

Figure 11.26

Oxygen isotope fractionation during glacial-interglacial cycles. (A) Water carrying the lighter isotope ^{16}O is preferentially evaporated to form clouds. As the clouds move landward and rain out, they become even further depleted in ^{18}O. During interglacials, however, this ^{18}O-poor water returns to the sea, and there is no net change. (B) During glacials, the ^{18}O-depleted water is trapped in the ice caps, which have $^{18}O/^{16}O$ ratios of -30 parts per thousand (‰). The ocean, as a consequence, is relatively enriched in ^{18}O $(+1.6‰)$. From Matthews (1984).

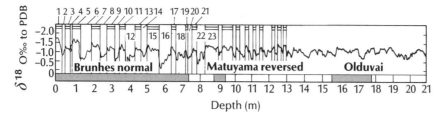

Figure 11.27

Oxygen isotope fluctuations in the carbonate of planktonic foraminifers for the last 2 million years. The more negative values (upward fluctuations) are interglacials; the more positive values (downward fluctuations) are glacials. The major glacial-interglacial isotopic stages were given numbers by Emiliani (1955, 1966) and Shackleton and Opdyke (1973), and their correspondence to the magnetic polarity time scale is shown. From Shackleton and Opdyke (1976).

there is no change in the ratio of the oxygen isotopes. During glaciation, the clouds precipitate their ^{16}O-rich water on the ice caps as snow, where it remains locked up for a long time. As a result, the oceans become relatively depleted in ^{16}O and enriched in ^{18}O during periods of glaciation. It is now apparent that the change in oxygen isotope ratio during the ice ages is primarily on ice volume signal, with a minor effect due to temperature.

· Oxygen isotopes are measured with respect to an arbitrary laboratory standard called *PDB*, after the Pee Dee belemnite. Calcite from this abundant cephalopod in the Cretaceous Pee Dee Formation of South Carolina is used to calibrate the mass spectrometer. The ratio is calculated by the following equation.

$$\delta^{18}O$$

$$= \frac{[(^{18}O/^{16}O)\text{sample} - (^{18}O/^{16}O)\text{standard}] \times 1000}{(^{18}O/^{16}O)\text{standard}}$$

A shell that has a $\delta^{18}O$ value of 3 parts per thousand (parts per mil, or 0/00) to PDB means that the CO_2 derived from that shell is three parts per mil richer in that isotope than PDB. Positive $\delta^{18}O$ values are enriched in ^{18}O, indicating increased ice volume and cooling; negative values are enriched in ^{16}O, indicating decreased ice volume and warming.

The result of these changes in ice volume and temperature is a distinct pattern of fluctuating oxygen isotopes in the Pliocene and Pleistocene (Fig. 11.27). This oxygen isotope record has been divided into distinct "stages," which are numbered back from the present. Because oxygen isotopes in the ocean respond to climatic changes, these changes are worldwide. Ice caps advance and retreat so rapidly that oxygen isotope changes can be considered geologically instantaneous. Like paleomagnetic samples, the ratio of a single sample is not sufficient to determine its age because each value of $\delta^{18}O$ has occurred many times in geological history.

A series of samples is needed to match a pattern to the global oxygen isotope record (Fig. 11.28). Biostratigraphic control is necessary to determine which part of the total pattern is being matched. Like magnetic stratigraphy isotope stratigraphy is not independent but depends on another source of time control. Unlike magnetic stratigraphy, however, oxygen isotope stratigraphy is dependent on lithology. Oxygen isotopes are measured in the shells of marine organisms, because the isotope record occurs primarily in oceanic water. Not all marine organisms give good results, either. Some

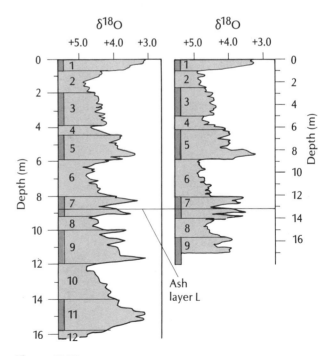

Figure 11.28

Correlation of oxygen isotopic oscillations between two deep-sea cores from the eastern equatorial Pacific. The two cores are scaled by the correlation along ash layer L. Isotopic stages 1 through 12 are indicated. From Ninkovitch and Shackleton (1975).

organisms fractionate oxygen isotopes in their own individual ways by physiological means and thus are useless for this kind of analysis. A large number of organisms must be sampled to screen out those that might give spurious results.

Carbon Isotopes

Like oxygen, carbon has more than one stable isotope in the earth's oceans and atmosphere. 98.89 percent is in the form of ^{12}C, which has six protons and six neutrons. However, 1.11 percent is the heavier isotope ^{13}C, which has an extra neutron. These two isotopes circulate through the ocean and are incorporated into the calcite of organisms in much the same way as are oxygen isotopes. The two are typically measured together during an isotopic analysis, and the formula for calculating $\delta^{13}C$ is the same as the formula for $\delta^{18}O$ with the appropriate substitutions.

The carbon system is not controlled by ice volume and temperature but by oceanic circulation. Organic materials tend to be low in ^{13}C, so when they decay, they release not only excess ^{12}C but less ^{13}C, so the amount of $\delta^{13}C$ decreases in the water. Deep ocean waters, in particular, are traps for organic nutrients and CO_2, which are relatively depleted in surface waters due to photosynthesis, sinking of organic debris, and respiration by bottom-dwelling organisms. This means that deep-ocean waters trap carbon that is depleted in ^{13}C and rich in ^{12}C. When major changes occur in oceanic circulation, these bottom waters exchange with the rest of the surface ocean and release their ^{12}C, which drives the $\delta^{13}C$ more negative.

In general, the ratios of the carbon isotopes are primarily a reflection of oceanographic and climatic changes, which happen on a global scale in a geologic instant. Thus, carbon isotopes, like oxygen isotopes, can be good time markers. Distinctive $\delta^{13}C$ "spikes" (Fig. 11.29) have been used in correlation, though there is no global $\delta^{13}C$ curve like that for oxygen isotopes. Like

oxygen isotopes, carbon isotopes require a continuous record with a distinct pattern and an independent form of time control (usually biostratigraphy). Carbon isotopes are also dependent on the nature of the organisms that trap them and can be used only in marine sections. Until recently, neither oxygen nor carbon isotopes could be used in fresh water or on land because the systems depend on the organism being in physiological contact with normal ocean water. However, recent studies have shown that terrestrial oxygen isotopes can also be used.

Strontium Isotopes

Recently, refinements in the measurement of strontium isotopes have made a new form of isotope stratigraphy possible. Strontium, along with calcium, is an alkali earth element on the periodic table, so it readily forms divalent cations. Its ionic radius is only slightly larger than that of calcium, so it frequently fills the calcium sites in trace amounts in many minerals. The common isotope of strontium in the oceans is ^{86}Sr. ^{87}Sr occurs in lesser abundance, the normal $^{87}Sr/^{86}Sr$ ratio in modern oceanic waters is around 0.7090. As ^{87}Sr is produced by the decay of ^{87}Rb, this ratio fluctuates through geologic time, with a steady, almost linear increase since the Jurassic. Recent work (DePaolo and Ingram, 1985; Elderfield et al., 1982; Palmer and Elderfield, 1985; Elderfield, 1986; Hess et al., 1986) has shown that the rate of change is continuous and linear since the late Eocene (Fig. 11.30). If the $^{87}Sr/^{86}Sr$ ratio of calcite in a marine organism can be measured precisely enough, it can be placed on the curve and dated.

This method is capable of dating single samples, unlike other stable isotope methods. Like the other methods, however, it is presently limited to normal marine organisms. In addition, it does not appear that it will work for samples older than Eocene, so some sort of time control is necessary to get a preliminary estimate of age. However, it is a technique with great potential, particularly if other dating techniques are not applicable.

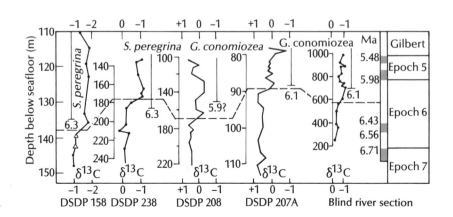

Figure 11.29

Correlation by carbon isotopes. Each section shows a strong negative (rightward) shift of $\delta^{13}C$ in the late Miocene (about 6.1 Ma), which is used as a marker for this time interval. From Kennett (1982).

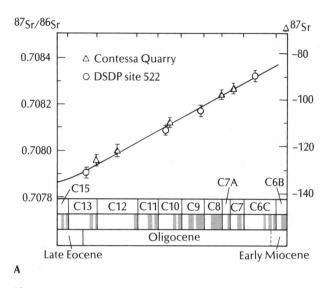

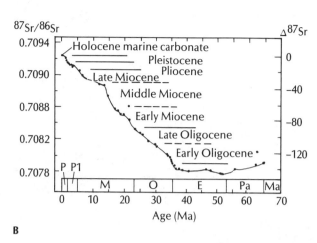

Figure 11.30

(A) Measured $^{87}Sr/^{86}Sr$ values of late Eocene and Oligocene chalks calibrated against the magnetostratigraphy. The vertical error bars are remarkably small, allowing for very tight correlation. (B) The change in strontium isotopic ratios has been almost linear since the early Tertiary, allowing a single unique determination of strontium isotopes to date a sample with fairly high precision. From DePaolo and Ingram (1985).

FOR FURTHER READING

Anstey, N. A., 1982. *Simple Seismics.* Int. Hum. Res. Dev. Corp., Boston. Excellent, short paperback on seismic stratigraphy written "for the petroleum geologist, the reservoir engineer, the well-log analyst, the processing technician, and the man in the field."

Asquith, G. B., 1982. *Basic Well Log Analysis for Geologists.* Amer. Assoc. Petrol. Geol. Methods in Exploration Series, Tulsa, Oklahoma. A detailed text describing and illustrating most of the modern methods of well logging.

Cox, A. V. (ed.), 1973. *Plate Tectonics and Geomagnetic Reversals.* W. H. Freeman, San Francisco. Contains many of the classic papers of the plate tectonic revolution, including most of the important papers on the early development of magnetostratigraphy.

Dobrin, M. B., 1976. *Introduction to Geophysical Prospecting,* 3d ed. McGraw-Hill, New York. Classic text on geophysical methods, emphasizing the equipment and interpretation rather than the theory.

Hyne, N. J., 1984. *Geology for Petroleum Exploration, Drilling and Production.* McGraw-Hill, New York. Excellent introductory text on petroleum geology, containing simple discussions of well logging and seismic stratigraphy.

Kennett, J. P., 1980. *Magnetic Stratigraphy of Sediments.* Dowden Hutchinson, & Ross, Stroudsburg, Penn. A collection of the classic early papers of magnetostratigraphy.

McElhinny, M. W., 1973. *Palaeomagnetism and Plate Tectonics.* Cambridge Univ. Press, Cambridge, England. A classic, although slightly dated, text on paleomagnetism, with good discussions of techniques and of magnetostratigraphy.

Payton, C. E. (ed.), 1977. *Seismic Stratigraphy—applications to Hydrocarbon Exploration.* Amer. Assoc. Petrol. Geol. Memoir 26. Contains the controversial Vail sea-level curves and many other useful papers.

Sheriff, R. E., 1978. *A First Course in Geophysical Exploration and Interpretation.* Int. Human Res. Dev. Corp., Boston. Simple introduction to seismic stratigraphy and other geophysical methods.

Sheriff, R. E., 1980. *Seismic Stratigraphy.* Int. Human Res. Dev. Corp., Boston. A well-written introductory discussion of the applications of seismic stratigraphy to correlation, facies analysis, and hydrocarbon detection.

Tarling, D. H., 1983. *Palaeomagnetism.* Chapman and Hall, London. The most up-to-date text on the subject.

Williams, D. F., I. Lerche, and W. E. Full, 1988. *Isotope Chronostratigraphy, Theory and Methods.* Academic Press, San Diego, Calif. The first book-length treatment of the use of stable isotopes (particularly carbon and oxygen) for stratigraphic correlation. It is valuable for the wealth of technical details that were once only available to specialists familiar with the primary literature.

Several generations of cross-cutting igneous dikes in the Sierra Nevada, California, from Bateman (1965). Photo courtesy U.S. Geological Survey.

12

Geochronology

RADIOACTIVE DECAY

All of the dating methods discussed so far are capable of producing *relative* ages of rocks, fossils, and geologic events. There was no way of knowing the *numerical* ages of events until the discovery of *radioactive decay* in the early twentieth century. Radioactive decay is one of the only predictable, clocklike process in nature, and understanding its mechanism is essential to obtaining and interpreting numerical ages.

During the process of radioactive decay, unstable parent atoms change into more stable, daughter atoms by emitting subatomic particles and energy. If a neutron is lost, then the *atomic weight* (given as a superscript before the symbol of the element, e.g., ^{238}U) changes, producing a different isotope of the same element (e.g., ^{235}U). If an atom loses a proton, then the *atomic number* (usually given as a subscript before the symbol of the element, e.g., $_7N$) and the atomic weight change and a different element results (e.g., $_6C$). Sometimes both protons and neutrons are given off. For example, in *alpha decay* two protons and two neutrons are given off as an alpha particle, thereby changing the atomic number

by two and the atomic weight by four. An example might be:

$$^{238}_{92}U \longrightarrow \, ^{234}_{90}Th + {}^{4}_{2}He \quad \text{(an alpha particle, } \alpha\text{)}$$

In *beta decay*, the parent nucleus emits an electron, and an uncharged neutron in the nucleus changes into a positively charged proton. The latter event increases the atomic number by one, changing the element but not the weight. For example:

$$^{40}_{19}K \longrightarrow \, ^{40}_{20}Ca + e^- \quad \text{(a beta particle, } \beta\text{)}$$

In electron capture, a proton in the nucleus picks up an orbital electron and becomes a neutron, giving the atom a lower atomic number, but the mass again remains unchanged. For example:

$$^{40}_{19}K + \text{orbital electron} \longrightarrow \, ^{40}_{18}Ar + \text{gamma particle } (\gamma)$$

A typical radioactive decay series, showing changing atomic numbers and masses and alpha and beta decay products, is shown in Fig. 12.1. Because radioactive decay

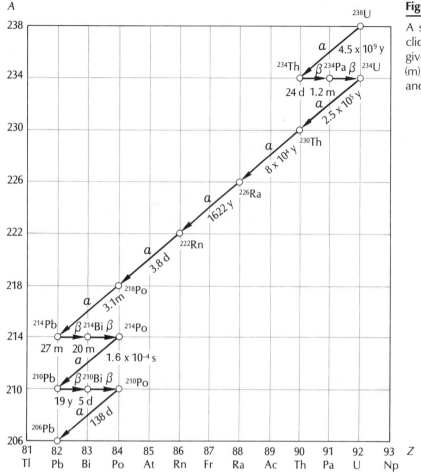

Figure 12.1

A series of naturally occurring unstable nuclides having ^{238}U as parent. Half-lives are given in years (y), days (d), hours (h), minutes (m), or seconds (s); α denotes alpha emission and β denotes beta emission.

takes place entirely within the nucleus of an atom, it is unaffected by chemical changes (such as oxidation and reduction), which affect only the orbital electrons. The ratio of parent to daughter atoms of a radioactive element present in a crystal is determined only by the elapsed time since the radioactive element and its decay products were locked into the crystal (assuming neither have escaped). If we take N to represent the number of radioactive nuclei present in a sample, then

$$\frac{-\delta N}{\delta t} \propto N$$

Rearranging gives $-\delta N/N = \lambda \, \delta t$, where λ is a previously determined *decay constant*. By integrating both sides, one obtains

$$-\int \frac{\delta N}{N} = \lambda \int \delta t$$

$$-\ln N = \lambda t + C$$

where C is the constant of integration and can be evaluated at $t = 0$. Therefore, if $N = N_0$, then $C = -\ln N_0$ (since $t = 0$), and

$$-\ln N = \lambda t - \ln N_0$$

$$\ln N - \ln N_0 = -\lambda t$$

$$\ln \frac{N}{N_0} = -\lambda t$$

$$\frac{N}{N_0} = e^{-\lambda t}$$

$$N = N_0 e^{-\lambda t}$$

The preceding equation, $N = N_0 e^{-\lambda t}$, gives the radioactive decay curve shown in Fig. 12.2. The decay process is exponential, so the amount of parent material decreases exponentially, and the amount of daughter product increases exponentially. Another way of looking at this process is shown in Fig. 12.3. At each equal increment of time (called a *half-life*), half of the parent material decays to daughter atoms. If there are eight parent atoms initially, there are four parents and four daughters after the first half-life, two parents and six daughters after two half-lives, and one parent and seven daughters after three half-lives. The half-life is thus defined as *the increment of time needed for half of the parent atoms to decay to daughter products*. The half-life has a constant value for each different set of radiogenic parents and daughters. It can be calculated from the preceding formulas. If we represent half-life by $t_{1/2}$, when $t = t_{1/2}$, $N = \frac{1}{2}N_0$, thus

$$\frac{1}{2}N_0 = N_0 e^{-\lambda t_{1/2}}$$

$$-\ln\left[\frac{1}{2} = e^{-\lambda t_{1/2}}\right]$$

$$\ln 2 = \lambda t_{1/2}$$

that is, half-life $= t_{1/2} = \ln 2/\lambda = 0.693/\lambda$.

In other cases, we may want to arrange the equation so that we can solve for t, the age, in terms of the known half-life, decay constant, and amounts of parent and daughter atoms. Since

$$N_0 = p + d = pe^{\lambda t}$$

and since $N/N_0 = e^{-\lambda t}$, then $Ne^{\lambda t} = N_0$ and $p = N$. In this case,

$$\frac{p + d}{p} = e^{\lambda t}$$

$$t = \frac{1}{\lambda}\ln\left(\frac{p + d}{p}\right) = \frac{1}{\lambda}\ln\left(\frac{d}{p} + 1\right)$$

The latter is the simplest equation for calculating the age of most radiogenic systems.

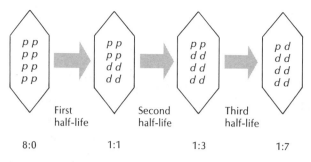

Figure 12.3

Schematic diagram of radioactive decay of a hypothetical crystal with eight atoms of an unstable parent isotope p. After one half-life, half of the parent material has decayed to the daughter product d, resulting in four parent and four daughter atoms. After the second half-life, half of the remaining parent material again decays leaving two parents and six daughter atoms. After the third half-life, half of the two remaining parent atoms decays, leaving only one parent and seven daughters. The change in the p:d ratio from 8:0 to 1:1 to 1:3 to 1:7 is a measure of the age of the crystal in terms of half-life.

MEASUREMENT

The standard tool of the geochronologist and isotope geochemist is the *mass spectrometer* (Fig. 12.4), a device that separates and counts atoms of different masses or charges in a sample substance. The mass spectrometer,

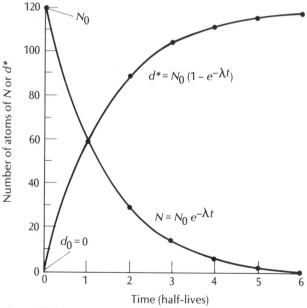

Figure 12.2

Decay curves of a hypothetical radionuclide N to a stable radiogenic daughter d^* as a function of time measured in half-lives. Note that $N \to 0$ as $t \to \infty$ whereas $d^* \to N_0$ as $t \to \infty$. From Faure (1986).

Figure 12.4

The modern Nier-type mass spectrometer, standard tool for nearly all studies of isotope geochemistry. The ion source (cylindrical device on the top left of the cabinet) bombards the sample with electrons to ionize it. The ions travel the curved path through the magnetic field (center), where they are separated according to their mass. The ion collector at the right side of the curved path detects the ions of isotopes of various masses. The results are then analyzed by computer to determine the amounts of each isotope. Photo courtesy VG Isotopes Ltd.

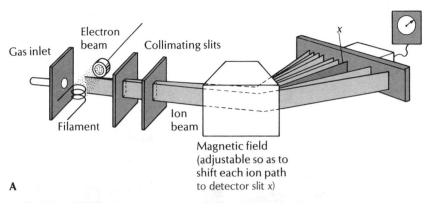

A

B

Figure 12.5

(A) The working parts of the mass spectrometer. (B) Close-up of the curved ion path between the poles of the magnet. Photo courtesy VG Isotopes, Ltd.

first developed by Nier in 1940, consists of three essential parts: (1) a source of a positively charged, monoenergetic beam of ions; (2) a magnetic analyzer; and (3) an ion collector. The entire system is evacuated to pressures of 10^{-6} to 10^{-9} millimeters of mercury. The solid sample is heated in the ion source until it emits ions, which then are accelerated and collimated into a narrow beam. This ion beam passes through the magnetic analyzer (Fig. 12.5), a curved "track" surrounded by a magnet. The magnetic field forces the ions to move in curved paths through the analyzer. The heavier the mass of an ion, the less it responds to the magnetic field and the less curved its path. By the time the ions have gone through the analyzer, they have been sorted by their masses into separate beams. The lightest ions show the most magnetic deflection, and the heaviest ions show the least. The collector picks up these separated ion beams and records the relative strengths of their peaks (Fig. 12.6), which measure the relative abundance of each isotope in the sample.

Even the most sensitive modern mass spectrometers have limitations, the most important of which is *measuremental error*. This is usually about ± 0.2 to 2.0 percent, which means ± 2 million years for an age of 100 million years or ± 20 million years for an age of 1 billion years.

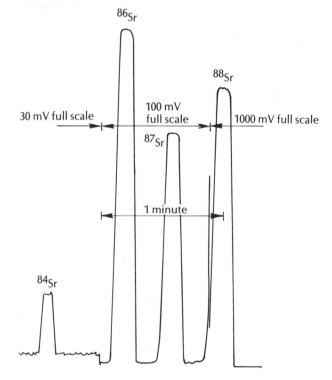

Figure 12.6

Display of a mass spectrometer. Each isotope (of strontium, in this case) produces a peak, and their amounts are gauged by the height of the peak. After Gast (1962).

Table 12.1

Decay Schemes

Parent Isotope	Daughter Isotope	Parent's Natural Abundance (%)	Half-life (10^9 years)	Parent's Decay Constant		Practical Dating Range
^{40}K	^{40}Ar	0.01167	1.250	$\lambda_\beta = 4.692$	$\times 10^{-10}$/yr	1–>4500 Ma
	^{40}Ca			$\lambda_\varepsilon = 0.581$	$\times 10^{-10}$/yr	>1500 Ma
^{87}Rb	^{87}Sr	27.835	48.8	1.42	$\times 10^{-11}$/yr	10–>4500 Ma
^{147}Sm	^{143}Nd	14.97	1.060	6.54	$\times 10^{-12}$/yr	>200 Ma
^{176}Lu	^{176}Hf	2.59	3.50	1.94	$\times 10^{-11}$/yr	>200 Ma
^{232}Th	^{208}Pb	~100	14.010	4.9475	$\times 10^{-11}$/yr	10–>4500 Ma
^{235}U	^{207}Pb	0.72	0.7038	9.8485	$\times 10^{-10}$/yr	10–>4500 Ma
^{238}U	^{206}Pb	99.28	4.468	1.55125	$\times 10^{-10}$/yr	10–>4500 Ma
^{14}C	^{14}N	—	5370 yr	1.29	$\times 10^{-4}$/yr	<80,000 yr

From Parrish and Roddick (1985).

Thus, under the best conditions (fresh specimens, great analytical care, and material that is old enough to contain a measurable amount of daughter product) a sample said to be 100 million years old is really 98 to 102 million years old. The age cannot be resolved better than that, due to the limits of the instrument. The magnitude of this error limits spectrometry to estimating the age of a sample; the relative size of the error tells how reliable the estimation is. Because of measurement error and bad samples, we are hesitant to refer to a date as "accurate" or "close to the truth." We *can* make some statement about its *precision*, or reproducibility, which is indicated by the size of the error and by subsequent analyses.

Criteria for Usefulness

The short-lived, unrenewable radiogenic isotopes that formed at the beginning of the earth's history 4.5 billion years ago have all decayed. Only isotopes with long half-lives (Table 12.1) are still present on earth to serve as geological clocks. Many of these have limited geological application and are not discussed here. A handful of isotopes occur in great enough abundance in the right types of rocks to be geologically useful (Table 12.2).

Besides an appropriate half-life, there are many other criteria for whether a system is geologically useful. The measurement of a sample gives the time of its cooling, when solidification locked the parent and daughter atoms into the mineral. Whether this "date" is also a meaningful *age* must be determined from geological relationships. For example, a detrital particle records the date of its cooling in its igneous parent but gives no indication of the date at which it became a sedimentary particle or the date when it was incorporated into sedimentary rock. This can lead to serious problems. A volcanic ash flow may pick up a small amount of detrital feldspars or biotite that is much older than the ash flow.

If such particles are inadvertently included in the analysis, an erroneous age estimate can result. Except for volcaniclastic rocks, *very few sedimentary rocks can be directly dated*. They can be dated only by indirect methods, such as cross-cutting dikes or by their relationships to volcanic flows. Finding enough places where these relationships are met has taken decades, but each success has refined the numerical dating of the geological time scale (see Box 12.1).

The loss of parent or daughter atoms due to heat or weathering creates other problems. For example, during metamorphism, the original mineral can be recrystallized, and some of the parent or daughter atoms may escape,

Table 12.2

Materials used for Geochronology

Dating Method	Materials That Can be Dated
K/Ar	Hornblende, muscovite, biotite-phlogopite, feldspars, glauconite, whole rock volcanics, some glasses
Rb/Sr	Micas, K-feldspar, cogenetic whole rocks that have a dispersion of Rb/Sr ratios; apatite, sphene for initial Sr^{87}/Sr^{86}
Sm/Nd	Pyroxene, plagioclase, garnet, apatite, sphene, other phases; whole rocks with a dispersion of Sm/Nd ratios
Lu/Hf	Much the same as Sm/Nd, zircon for initial Hf isotopic composition
U/Th/Pb	Zircon, monazite, xenotime, baddeleyite (ZrO_2), sphene, apatite, allanite, pyrochlore, U or Th minerals
Pb/Pb	Galena or other Pb minerals, K-feldspar, tellurides, carbonates in carbonatites.
^{238}U fission track	Zircon, apatite, sphene, garnet, epidote, volcanic glass

From Parrish and Roddick (1985).

Box 12.1
Bracketing the Age of the Silurian/Devonian Boundary

In Chapter 10 (Box 10.1), we saw how the Silurian/Devonian boundary was established biostratigraphically. Although this gives us a standard for its recognition and relative age, it says nothing about its numerical age. The type section in Czechoslovakia has no igneous intrusives or extrusives, so the boundary cannot be directly dated there. Geologists have had to search all over the world for places where rocks of late Silurian or early Devonian age (as established by their fossils) occur in relation to datable igneous rocks. Very few of the rocks found in such relations have preserved the actual Silurian/Devonian boundary, and the dates of the igneous rocks that are in relation to boundary rocks may not be close to the boundary. The entire process requires considerable interpolation between dated horizons and those that have the diagnostic fossils.

We are dealing with ages in the order of 400 million years, so the typical error for these samples is in the order of ± 5 to 15 million years. Even if we had a lava flow perfectly interbedded between the Silurian and Devonian, its error would allow us to give the date only to the nearest 5 million years or so. Experience has also shown that there may be several conflicting dates, even when the sampling and measurements are done as carefully as possible. Consequently, geochronologists (Odin, 1982; McKerrow et al., 1985; Gale, 1985) have used a suite of dates from different areas (Fig. 12.7). Instead of treating each date as a single point, they represented each date by a rectangle. The horizontal dimension of a rectangle represents the measuremental error ("plus or minus") in millions of years. The vertical dimension represents the stratigraphic error, or the degree to which the date can be placed precisely in the fossil sequence. Some dates have huge stratigraphic errors because they occur well above or below the sequence of interest and thus provide only a maximum or minimum age. Dates with no upper or lower stratigraphic limit may be indicated by open rectangles.

Only a few of the dates in Fig. 12.7 are relevant to the age of the Silurian/Devonian boundary. Date 5 (the Shap Granite in England) intrudes upper Silurian rocks but may be as young as late Devonian. Date 6 (the Skiddaw Granite, also in England) is post-late Silurian, but there is no younger limit on its stratigraphic age. Both dates only crudely fix the curve, and their relation to the line of correlation through them suggests that they are early or middle Devonian. Date 8 (the Lorne Lavas of England, interbedded with lowest Devonian rocks) gives a Rb/Sr age of 399 $\pm$ 5 Ma for the earliest Devonian. Date 9 comes from the Hedgehog Volcanics of Maine, which are interbedded with lower Devonian rocks and produce Rb/Sr and whole-rock ages of 400 $\pm$ 10 Ma for the earliest Devonian. Date 10 is from the Gocup Granite of Australia, which could range in age from middle Silurian to early Devonian. It gives a date of 409 $\pm$ 5 by K/Ar on muscovite.

Dates 11 and 12 occur in the late but not latest Silurian (Ludlovian), so they help fix the line but are not close to the Silurian/Devonian boundary. Date 11 is from the Laidlaw Volcanics of Australia, which are interbedded with deposits containing early Ludlovian fossils. However, the spread on the dates from this area is discouraging: 409 $\pm$ 5 Ma by K/Ar on sanidine; 420 $\pm$ 5 by K/Ar on biotite; 425 $\pm$ 17 by Rb/Sr on whole rock; and 421 $\pm$ 5 by Rb/Sr on biotite. Date 12 is a little more useful because it comes from the type Ludlovian in the Welsh Borderland. It gives a date of 419 $\pm$ 10 by K/Ar on biotite. Not shown in Fig. 12.7 are new dates from the Wormit Bay Lavas of Scotland, which are interbedded with rocks that may be Ludlovian or Pridolian in age. These give a date of 408 $\pm$ 5 Ma by Rb/Sr on biotite; this is the first date listed that is truly latest Silurian.

From the preceding numbers, it is apparent how difficult it is to get a decent date with good stratigraphic constraints anywhere near the desired target of the Silurian/Devonian boundary. The only rea-

thus completely eliminating any possibility that the original parent rock can be dated. However, recrystallization resets the clock so that the decay process begins all over again. The dates then will reflect the time of metamorphism. This process is called *metamorphic resetting*. Weathering can cause a similar problem by allowing some of the parent or daughter atoms to escape,

and once again any date derived from the remaining atoms is meaningless. These kinds of problems are not easy to overcome, so they are best avoided. Great care must be taken to get the freshest samples possible and to check every date several times.

Another major limitation of radiometric methods is that there must have been enough of a radiogenic

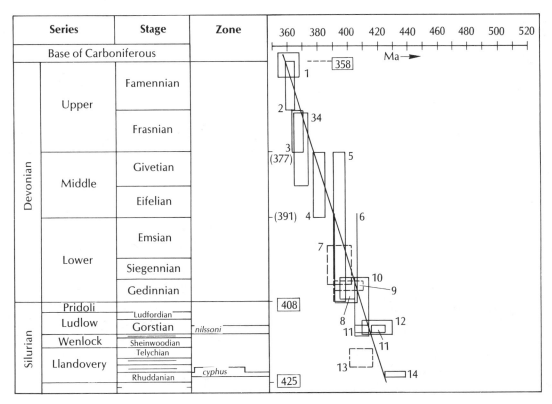

Figure 12.7 Radiometric constraints on rocks of late Silurian and Devonian age. Each date has some error due to the imprecision of radiometrics (shown by the horizontal widths of the boxes, typically ±4 to 5 million years) and some error due to difficulties in locating the date in a stratigraphic horizon (vertical height of the box). Some of the numbered boxes are discussed in the text. Because there is a large margin for error in every date, McKerrow et al. (1985) chose to fit a line through the centers of most of the boxes to interpolate the ages of the boundaries.

sonable procedure is to fit a line through the spread of the rectangles, as shown in Fig. 12.7. *The age of the boundary is not directly dated anywhere, but it is estimated by interpolation.* Naturally, the method of interpolation can vary from author to author, and each estimate may be modified slightly when new dates are available. Consequently, different authors give various estimates for the desired age. A typical range of estimates for the age of the Silurian/ Devonian boundary from the last decade include:

411 Ma   McKerrow et al., 1980

401 Ma   Jones et al., 1981
400 Ma   Odin, 1982, 1985
408 Ma   Harland et al., 1982
412 Ma   McKerrow et al., 1985
408 Ma   Gale, 1985

This illustrates the uncertainty inherent in the whole problem. Whenever you see a numerical time scale or run across a geologist confidently giving a numerical age, it pays to keep the real limitations of those "solid" numbers in mind.

element in the rock for there to be measurable amounts of parent and daughter atoms. Most radiogenic elements are too rare in terrestrial rocks to be useful, with the exception of potassium, rubidium, and uranium. Also, the half-life of the parent element must be short enough for there to be a measurable amount of daughter product; this places a lower limit of about 100 million years on

the usefulness of the Rb/Sr, U/Pb, and Th/Pb systems (see Table 12.1). On the other hand, systems with half-lives that are too short are radioactively "dead". Carbon 14, for example, has such a short half-life that the duration of its usefulness is about 60,000 to 80,000 years. In specimens older than this, there is no parent material left to measure.

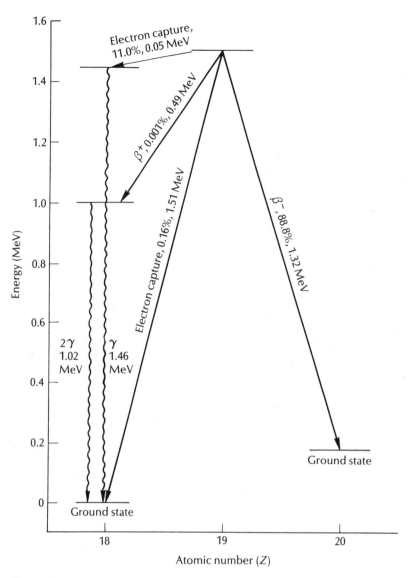

Figure 12.8

Decay scheme diagram for the branched decay of ^{40}K to ^{40}Ar by electron capture and by positron (β^+) emission, and to ^{40}Ca by emission of negative beta particles (β^-). The percentage of the parent material in each decay path and the energy changes are also shown. After Dalrymple and Lanphere (1969).

Potassium/argon Dating

The decay series most commonly used for dating is potassium/argon. Potassium, the seventh most abundant element in the earth's crust, is especially common in potassium feldspars, micas, hornblende, and clays, which are found in many geological situations. The most common isotope of potassium is ^{39}K (93% of total potassium), and the next most common is ^{41}K (7.9%). The rarest isotope is the radioactive form ^{40}K (0.01167%); because it is an unstable atom, most of the earth's original ^{40}K has already decayed, enhancing its rarity.

Potassium-40 has a peculiar branching decay path (Fig. 12.8). Eleven percent of any given amount of ^{40}K decays to ^{40}Ar by electron capture and gamma decay. The rest decays to ^{40}Ca by beta decay. Large amounts of non-radiogenic ^{40}Ca also occur in crustal rocks, so this decay path is not often used for dating. On the other hand, ^{40}Ar is formed only by radioactive decay, and so this is the system that is used. Because there are two decay paths, there are two decay constants, λ_ε and λ_β, for the K/Ar and K/Ca decay paths, respectively.

The main problem with the potassium/argon system is that the daughter product, argon, is an inert gas that can leak out of a crystal. If this has happened in a sample, one will find too little daughter product, and the date obtained will be too young. Argon-40, which also occurs in the atmosphere, can adhere to the sample and contaminate it, making the date too old. To check for this, it is necessary to analyze also for ^{36}Ar, which has a known ratio to ^{40}Ar in the atmosphere ($^{40}Ar/^{36}Ar = 296$). The formula for correcting for atmospheric argon is:

$$^{40}Ar \text{ (measured)} - 296 \, (^{36}Ar) = \,^{40}Ar \text{ (radiogenic)}$$

To obtain useful dates, the crystal lattice of a mineral must be tight enough to retain the argon gas. At depths of 5 kilometers or more or at temperatures above 200°C, most minerals leak argon, so a date from a rock that has been at these pressures or temperatures, or greater, has probably been metamorphically reset. If the mineral has been altered or weathered, it almost certainly has leaked some argon. As the sample is broken down and prepared in the laboratory, it must also resist dissolution, or more argon will leak out.

However, the potassium/argon system is very useful for many geological problems because potassium is abundant in many rocks. There is no upper age limit on this system because its half-life of 1.3 billion years is long enough for use on Precambrian rocks and nearly everything younger. The lower limit depends on atmospheric argon contamination and the sensitivity of the mass spectrometer used. Biotite that is younger than a million years cannot usually be dated, but sanidines as young as 10,000 years have been successfully dated.

The age equation for potassium/argon dating is more complex than the standard formula because there are two decay constants:

$$t = \frac{1}{\lambda_\varepsilon + \lambda_\beta} \ln \left[\frac{(^{40}Ar_{rad})}{(^{40}K)} \left(\frac{\lambda_\varepsilon + \lambda_\beta}{\lambda_\varepsilon} \right) + 1 \right]$$

The decay constants λ_ε and λ_β have been corrected since 1976 (Steiger and Jäger, 1977), so any date printed before then must be corrected. Table 12.3 provides an easy method to correct any older date for the new constants. For a potassium/argon date to be reliable, a number of criteria must be met:

1. No argon should have escaped or leaked.
2. The crystal must have cooled rapidly so that argon is trapped.
3. No extra argon should have been added. Correction must be made for ^{40}Ar contamination from the atmosphere.
4. The system must have been closed to potassium gain or loss during its history.
5. There should be no unusual isotopic fractionation of potassium.

The suitability of various minerals for potassium/argon dating is highly variable.

Micas contain 7 to 9 percent potassium and are common in felsic igneous rocks and in metamorphic rocks. Biotite and muscovite are both reasonably reliable but are very susceptible to weathering and reheating. They are common as detrital grains, so one must be wary of detrital contamination.

Hornblende contains 0.1 to 1.5 percent potassium and is very common in felsic igneous rocks and in certain high-grade metamorphic rocks (especially amphibolite grade). Hornblende holds tenaciously to argon, so it is good for dating metamorphic terranes.

Potassium feldspars contain 7 to 12 percent potassium and occur in a wide variety of rocks. Plutonic igneous feldspars show large variations in argon loss and are seldom reliable. Sanidine, the high-temperature potassium feldspar common in volcanic rocks, cools rapidly and has a different structural state from low-temperature potassium feldspars. As a result, it traps argon well and produces dates with small error bars. It is especially good for dating young, fresh volcanics, but it alters easily and so seldom survives in its fresh state from rocks as old as the Mesozoic. Contamination from detrital potassium feldspars can also be a big problem.

Plagioclases contain 0.01 to 2 percent potassium and are abundant in nearly all igneous rocks, but their potassium content is so low that they are rarely usable.

Pyroxenes contain 0 to 1 percent potassium and are common in basic igneous rocks, but their low percentage of potassium limits their use. At depth, there is apparently significant contamination of pyroxenes by ^{40}Ar. However, pyroxenes in shallow eruptive basalts have been successfully used for dating.

When a rock is too fine-grained for a geologist to pick out individual crystals, he can break down the whole rock and then date it (called *whole-rock analysis*). If the rock is fresh and has a tight, nonporous groundmass that traps all the argon, it may be suitable for dating. Whole-rock dating is not suitable for plutonic igneous rocks that have been heated above 200°C and so have lost argon. However, it is very good for fresh basalts, especially because the individual minerals in basalts tend to be extremely fine grained.

Glauconites (called "glauconies" by Odin, 1982) are found in many shallow marine environments, where they grow authigenically at low temperatures in sediments. Ideally, they would seem to be the exception to the rule that sediments cannot be dated directly. However, experience has shown that they are susceptible to both argon loss and potassium gain, making their dates 10 to 20 percent too low (Thompson and Hower, 1973; Berggren et al., 1978; Berggren, Kent, and Flynn, 1985; Obradovich, 1988). Some scientists have chosen to rely heavily on glauconites for dating the geological time scale and have come up with calibrations that differ by several million years from those derived from high-temperature potassium/argon dates. When this conflict arises, the glauconite dates should be viewed with suspicion because they have a poor track record (see Chapter 13).

In general, then, potassium/argon dating is excellent for most geologic problems, but it must be used with common sense. A single age determination is usually insufficient. Samples should be run several times to reduce the error estimates. The date is even more reliable if two minerals from the same rock, or two specimens from the same unit, or two radiogenic minerals are used. Agreement of dates within 5 percent is generally considered concordant. Anomalous ages can occur for a number of reasons. If they are too young, the problem is probably bad samples, reheating, alteration, weather-

ing, or potassium gain. If they are too old, the problem could be atmospheric ^{40}Ar contamination, or inherited ^{40}Ar. Random errors can be produced by analytical problems such as those discussed earlier, as well as faulty sampling, mislabeled samples, or an incorrect preliminary age estimate based on mistaken field relations.

Argon/argon Dating

Another technique that is increasingly being used is $^{40}Ar/^{39}Ar$ dating, in which potassium-rich samples are bombarded with neutrons in a nuclear reactor to convert ^{39}K to ^{39}Ar. Argon-39 is unstable, with a half-life of 269 years, and decays back to potassium-39 by beta decay. After irradiation, the argon gas is extracted from the sample as in conventional K/Ar methods and is analyzed in the mass spectrometer. Using stable argon-

Table 12.3

Critical Table for Conversion of K/Ar Ages from Western Constants to New IUGS Constants

Age	F	Age	F	Age	F	Age	F	Age	F	Age	F	Age	F
0	1.0268	182	32	385	96	606	1.0160	847	24	1111	88	1404	52
5	66	193	1.0230	397	94	619	58	861	22	1127	86	1421	1.0050
15	64	204	28	409	92	632	56	875	1.0120	1142	84	1439	48
25	62	215	26	421	1.0190	645	54	889	1.0118	1158	82	1456	46
35	1.0260	226	24	433	88	658	52	903	16	1174	1.0080	1474	44
45	58	237	22	445	86	671	1.0150	918	14	1190	78	1491	42
56	56	248	1.0220	457	84	684	48	932	12	1206	76	1509	1.0040
66	54	259	1.0218	469	82	697	46	947	1.0110	1222	74	1527	38
76	52	271	16	481	1.0180	710	44	961	08	1238	72	1545	36
87	1.0250	282	14	493	78	724	42	976	06	1254	1.0070	1563	34
97	48	293	12	506	76	737	1.0140	990	04	1270	1.0068	1581	32
108	46	305	1.0210	518	74	750	38	1005	02	1287	66	1599	1.0030
118	44	316	08	530	72	764	36	1020	1.0100	1303	64	1618	28
129	42	327	06	543	1.0170	778	34	1035	1.0098	1320	62	1636	26
139	1.0240	339	04	555	1.0168	791	32	1050	96	1336	1.0060	1655	24
150	38	350	02	568	66	805	1.0130	1065	94	1353	58	1674	22
161	36	362	1.0200	580	64	819	28	1081	92	1370	56	1693	1.0020
172	34	374	1.0198	593	62	833	26	1096	1.0090	1387	54	1712	1.0018

40 to correct for atmospheric and other interferences, one can calculate an age once the conversion rate of potassium-39 to argon-39 is known. This is determined by irradiating a standard of known age along with the sample. This technique has many advantages:

1. It does not require a separate analysis of potassium, so it is unnecessary to split the sample to retrieve solid potassium; only argon gas need be measured.

2. It is so sensitive that very small samples can be used.

3. Because only one element is measured, the uncertainties of sample weighing and concentration measurement are eliminated.

4. With the proper techniques, atmospheric argon contamination can be removed; in fact, the method of measurement called step heating not only eliminates this problem but also allows correction for partial loss of argon from a sample that has been disturbed since cooling.

Figure 12.9 shows how the argon-argon system works. If before decay, the argon-39 was uniformly distributed at potassium sites within the mineral, then it should diffuse out of all parts of the mineral at the same rate. Therefore, measurements of the $^{40}Ar/^{40}K$ ratio taken after step heating from the center to the edge of the sample should produce constant values and a flat curve (Fig. 12.9A). If the sample has been heated during its history, however, some of the argon-40 will have diffused from the edges, but the argon-40 in the center, which was sealed in, will have kept the original $^{40}Ar/^{40}K$ ratio. Thus a $^{40}Ar/^{40}K$ plot of an altered

Age	F	Age	F	Age	F	Age	F	Age	F	Age	F	Age	F
1731		2056		2422		2837		3316		3878		4551	
	16		84		52		0.9920		88		56		24
1750		2078		2446		2865		3348		3916		4597	
	14		82		0.9950		0.9918		86		54		22
1770		2099		2471		2893		3381		3955		4645	
	12		0.9980		48		16		84		52		0.9820
1789		2121		2495		2922		3414		3994		4693	
	1.0010		78		46		14		82		0.9850		0.9818
1809		2143		2520		2950		3448		4034		4741	
	08		76		44		12		0.9880		48		16
1829		2165		2546		2979		3482		4074		4790	
	06		74		42		0.9910		78		46		14
1849		2188		2571		3008		3516		4115		4840	
	04		72		0.9940		08		76		44		12
1869		2210		2597		3038		3550		4156		4891	
	02		0.9970		38		06		74		42		0.9810
1889		2233		2622		3068		3585		4198		4942	
	1.0000		0.9968		36		04		72		0.9840		08
1909		2256		2648		3098		3620		4240		4994	
	0.9998		66		34		02		0.9870		38		06
1930		2279		2675		3128		3656		4283		5049	
	96		64		32		0.9900		0.9868		36		04
1950		2302		2701		3159		3692		4326			
	94		62		0.9930		0.9898		66		34		
1971		2326		2728		3189		3728		4370			
	92		0.9960		28		96		64		32		
1992		2350		2755		3221		3765		4414			
	0.9990		58		26		94		62		0.9830		
2013		2373		2782		3252		3802		4459			
	88		56		24		92		0.9860		28		
2035		2397		2809		3284		3840		4505			
	86		54		22		0.9890		58		26		

Note: To convert an age based on old Western constants to one based on new IUGS constants, multiply by the indicated correction factor F. Ages are in 10^6 yr. Old western constants: $\lambda_\varepsilon + \lambda_\varepsilon^1 = 0.585 \times 10^{-10}$ yr^{-1}, $\lambda_\beta = 4.72 \times 10^{-10}$ yr^{-1}, $^{40}K/K_{total} = 1.19 \times 10^{-4}$ mol/mol; new IUGS constants: $\lambda_\varepsilon + \lambda_\varepsilon^1 = 0.581 \times 10^{-10}$ yr^{-1}, $\lambda_\beta = 4.962 \times 10^{-10}$ yr^{-1}, $^{40}K/K_{total} = 1.167 \times 10^{-4}$ mol/mol. From Dalrymple (1979).

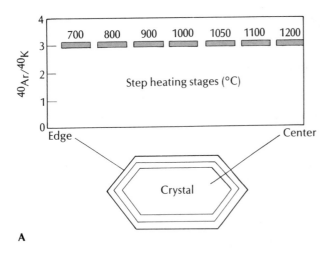

A

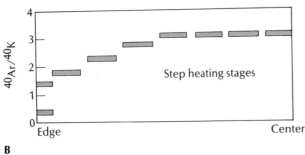

B

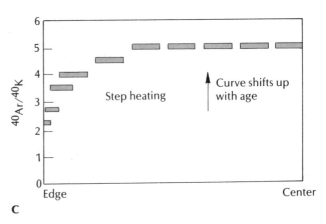

C

Figure 12.9

The argon/argon dating method. (A) Plot of the $^{40}Ar/^{40}K$ ratio with step heating of an unaltered crystal. If there has been no leakage, then the outer rim of the crystal will release the same amount of material at low temperatures as the center releases at high temperatures. (B) An altered crystal, however, will show lower ratios at the lower temperatures due to leakage of argon from the edge. The plateau at the higher temperatures should give the true age of the unaltered center of the crystal. (C) As time passes and the crystal gets older, the entire curve shifts upward, but the plateau remains as long as no further alteration takes place.

sample (Fig. 12.9B) will be low in argon-40 near the margin relative to the center.

As argon continues to decay, it will be trapped throughout the crystal as long as the crystal remains a closed system. This results in an increased $^{40}Ar/^{40}K$ ratio in both the center and the edge of the crystal (Fig. 12.9C). The $^{40}Ar/^{40}K$ ratio in the center of the crystal reflects its original age because the center was relatively immune to alteration. The age estimate from the edge of the crystal gives the time elapsed since partial outgassing due to geologic heating. Thus this method is one of the few that give dates for both the original cooling and later thermal events.

During analysis, this cooling history is reconstructed by stepwise heating of the crystal. At the lowest temperature, the argon from the edge diffuses out first and is measured. At each higher temperature step, argon from deeper into the crystal is released and measured. If the fraction of argon-39 released stays on a plateau through most of the heating steps, then this plateau represents the age of the argon in the unaltered center of the crystal (Fig. 12.10). If the plateau is relatively smooth all the way across (Fig. 12.10A), then an age can be estimated by averaging all the values for the argon-39 on the plateau to produce an *integrated age*. The more irregular the steps in the plateau (Fig. 12.10B), the greater has been the loss of argon and therefore the greater the measurement error.

Another method of obtaining argon/argon dates is

the laser-fusion method. Instead of slowly heating a crystal and measuring its argon to obtain a plateau, the crystal is melted by a laser. As the crystal fuses, it releases its argon, which is measured as described above. The advantage of this method is that the laser can be aimed very precisely, so that the argon can come from a single crystal, or even from part of a single crystal. By doing a large number of carefully chosen crystals, the investigator can determine which crystals appear altered or contaminated and which ones give good agreement on their ages. Thus, both high reproducibility and precision are possible. Unlike the step-heating method, however, one cannot plot a "plateau" that separates the altered edges from the original date obtained from the plateau. Nevertheless, by fusing enough carefully inspected crystals, the investigator can determine if the fresh crystals cluster around a concordant value and assess the reliability of the date in terms of its scatter (the usual "plus or minus" value).

The main limitation of the argon/argon system is that minerals can absorb atmospheric argon. This contamination is difficult to detect except by the irregularities in the plateau of the curve. Best results have been obtained on hornblende and potassium feldspars; biotites tend to produce very irregular age spectra. However, the major advantage of this system over potassium/argon is its potential for detecting and dating thermal disturbance with a single sample. For these reasons, it will probably be used more frequently in the future.

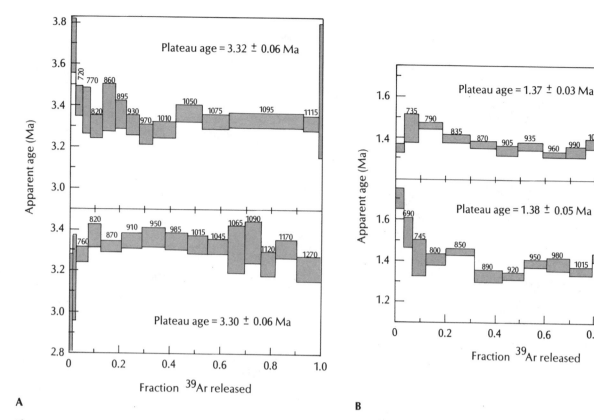

Figure 12.10

Examples of real argon/argon data. (A) Age spectra of feldspars from pumice clasts in the Toroto Tuff, Lake Turkana (see Fig. 13.5). Uncertainty of the age for each heating step at the level of one standard deviation is shown by the thickness of the bar. Temperature at which gas was released is shown by numbers over the bar. Both samples have very well-defined plateaus, with only minor edge alteration at 760° and below. (B) Age spectra from feldspars from the Chari Tuff, Lake Turkana. The plateaus are less well defined than in the previous example. From McDougall (1985).

Rubidium/strontium Dating

Rubidium-87 decays to strontium-87 in a single step by beta decay. Twenty-eight percent of the rubidium in the earth's crust is ^{87}Rb, but rubidium is not a common element in the crust as a whole, occurring in abundances of about 3 to 140 parts per million (ppm). Rubidium is an alkali element that forms a univalent cation with an ionic radius slightly larger than that of potassium, so it substitutes readily for potassium in many minerals. Strontium, however, forms a divalent cation, so it prefers calcium sites. As a result, when ^{87}Rb decays to ^{87}Sr, there is a strong tendency for the daughter strontium to migrate out of the host mineral. If whole rock dating is used, however, the radiogenic strontium can be found in the calcium sites. Rubidium, like potassium, occurs primarily in micas, feldspars, and amphiboles. In most minerals, some inherited ^{87}Sr has been trapped in the crystal. To correct for this ^{86}Sr in rock samples is also measured because its ratio to ^{87}Sr is known for most of earth history. Therefore, the age formula has the correction factor built in:

$$t = \frac{1}{\lambda} \ln \left\{ 1 + \left(\frac{^{86}Sr}{^{87}Rb} \right) \underbrace{\left[\left(\frac{^{87}Sr}{^{86}Rb_{present}} \right) - \left(\frac{^{87}Sr}{^{86}Rb_{initial}} \right) \right]}_{\text{Radiogenic Sr correction}} \right\}$$

Rubidium/strontium dating works best in older samples because rubidium is so rare that a long decay period is necessary to accumulate measurable quantities of strontium. The correction for the $^{87}Sr/^{86}Sr$ ratio can introduce huge errors because ^{86}Sr is so abundant, especially in younger rocks. For example, corrections for $^{87}Sr/^{86}Sr$ are good to one part in 100 million, but in many samples, one part in 100 million is all the ^{87}Sr that three million years of decay of ^{87}Rb will produce. A young sample could have 100 percent error, but Precambrian samples have enough radiogenic strontium that the error is in the order of 0.3 percent.

The most reliable technique of rubidium/strontium dating is called the *isochron* method. A single rock contains several minerals with different initial concentrations of rubidium. Yet at the time of crystallization, the $^{87}Sr/^{86}Sr$ ratio throughout the rock was constant (Fig. 12.11). As the ^{87}Rb in the rock decayed, the $^{87}Sr/^{86}Sr$

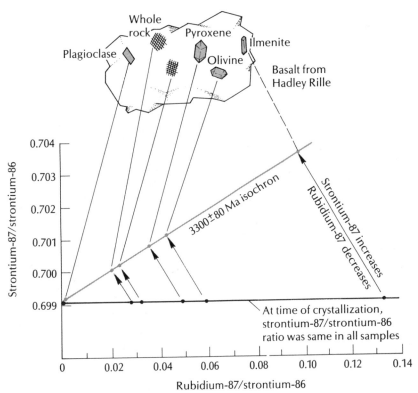

Figure 12.11

Rubidium/strontium isochron dating of a lunar basalt. The isochron method depends on analysis of samples with differing rubidium content. At the time of crystallization, the $^{87}Sr/^{86}Sr$ ratio was the same for all minerals in the sample. In minerals with low rubidium content, such as plagioclase or the whole rock, this ratio changed relatively little; but in minerals with high rubidium content, such as ilmenite, it changed much more over the same amount of time. Slope of the isochron increases with time and provides the date. After Eicher (1976).

ratios in the minerals that contained more of the parent ^{87}Rb changed more than the ratios in minerals with less. Therefore, measurements of the strontium isotopes present in the various minerals of a sample differ and fall along different paths in the plot shown in the figure. At any given time, the isotope ratios of all of the minerals should fall along a common line, called an isochron. The slope of the isochron line is the ratio of radiogenic ^{87}Sr to ^{87}Rb in the sample and thus is a measurement of age of the sample; older samples have a progressively steeper slope.

The rubidium/strontium method is useful in both igneous and metamorphic rocks that are old enough. It has proven especially useful for lunar basalts, meteorites, and metamorphic rocks, providing both the age of original crystallization and the date of metamorphism. Because the migrating ^{87}Sr is trapped in calcium sites, the whole rock age represents the original date of crystallization. Individual minerals will have lost strontium during metamorphism and their rubidium/strontium age is therefore a measure of their metamorphic resetting date. The blocking temperatures also vary in the different minerals, giving a rough index of the maximum temperature reached by the rock. The whole rock has a blocking temperature of 700°C, but the muscovite in it is stable only at 500°C and the biotite at 280° to 320°C. If resetting has occurred in the biotite but not the muscovite, for example, the maximum temperature can be bracketed.

Uranium/lead Methods

Uranium/lead dating is often the most reliable of the techniques, especially for old igneous rocks, because uranium is much more abundant than rubidium. The system is made even more robust by the fact that three uranium/lead isotopic decay series are available for cross-checking. The earth's uranium is 99.27 percent ^{238}U and 0.72 percent ^{235}U. These uranium isotopes decay to different isotopes of lead, ^{206}Pb and ^{207}Pb, respectively. A third system, ^{232}Th, decays to ^{208}Pb, so all three related systems can be used for cross-checking. Each of these parent isotopes decays to lead through a series of alpha and beta decays, producing a number of short-lived intermediate products and helium at each alpha decay step (see Fig. 12.1). The ratio of $^{207}Pb/^{206}Pb$ in a rock also gives an age estimate, which can be used to check for concordance. Inherited lead can throw measurements off, so the nonradiogenic ^{204}Pb must also be measured. The normal ratio of ^{204}Pb to ^{207}Pb and ^{206}Pb is assumed to be equivalent to the primordial ratio measured in meteorites, and this amount can be subtracted to correct for inherited lead.

Minerals that are rich in uranium are rare, but many minerals contain traces of it. Less common minerals, such as zircon, sphene, and monazite are often used because they are impervious to weathering or other alteration and are extremely stable. They are also very

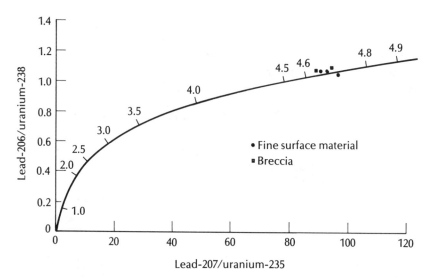

Figure 12.12

Uranium/lead concordia plot, showing essentially concordant ages of lunar fine surface material and breccia collected by Apollo 11. Dates between 4.6 and 4.7 billion years agree with the data for the age of the solar system. After Eicher (1976).

easy to isolate by standard separation techniques, followed by immersion in a bath of hydrofluoric acid until the entire rock except the zircons has dissolved. However, these minerals are also very stable detrital particles, so care must be taken to avoid detrital grains in a sample. Some zircons are known to have been picked up by magmas and crystallized into much younger rock, but these can be recognized by their reworked texture. A reworked zircon seldom shows a euhedral habit.

Because three systems are in operation, the best way to determine the age is called a *concordia plot* (Fig. 12.12). On a plot with the $^{207}Pb/^{235}U$ ratio as abscissa and the $^{206}Pb/^{238}U$ ratio as ordinate, all the sample data points should fall along a smooth curve called the *concordia curve*. This is the theoretical curve that data from the sample should continue to follow as the uranium decays. When the material of the sample cooled and crystallized, its lead–uranium ratios would have been at the origin of the plot, and as the sample aged, the ratios would have moved up the curve. The age of the sample can be read directly from the position of the sample on the curve. If the ratios of isotopes in the sample fall off the curve, they are *discordant* (Fig. 12.13). This may be due to metamorphism, variable lead loss, or continuous

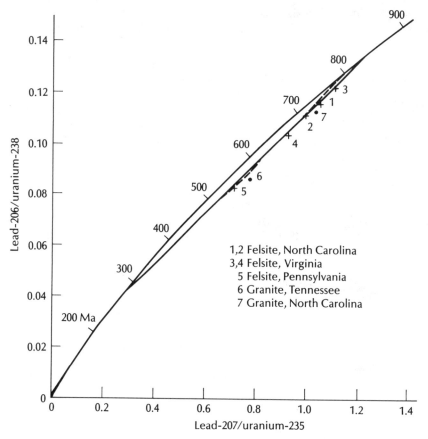

1,2 Felsite, North Carolina
3,4 Felsite, Virginia
5 Felsite, Pennsylvania
6 Granite, Tennessee
7 Granite, North Carolina

Figure 12.13

A concordia plot of a series of discordant uranium/lead ages from the Appalachians indicates a major igneous episode at 820 Ma (upper intersection of straight line with concordia curve) and metamorphic reheating during the Appalachian Orogeny at 240 Ma (lower intersection of straight line with concordia curve). After Eicher (1976).

lead loss by diffusion. Discordant results plot below the concordia curve, usually in straight lines. A regression line fit to these discordant points intercepts the concordia curve in two places. The upper intercept should give the original age of cooling, and the lower intercept (chronologically younger) gives the time of metamorphism or lead loss. Even discordant data yield valuable information.

Fission-track Dating

Spontaneous radioactive fission of ^{238}U produces not only daughter isotopes but also two high-energy charged nuclei that shoot out in opposite directions from the decaying nucleus. As they do so, they may disrupt the crystal lattice of their host mineral and leave "tracks" that can be seen under a microscope (Fig. 12.14). These *fission tracks*, about 10 to 20 micrometers long, can be seen in many crystals. Because the number of tracks is a function of the original amount of radioactive parent in the crystal, counting the tracks should give the age of the material. The older a crystal (given equal amounts of parent material), the more fission tracks will be formed. The method works best on substances with hard, resistant crystal lattices that are translucent and can be studied easily. Fresh volcanic glass, zircon, sphene, apatite, muscovite, biotite, and tektites have been successfully dated in this way. The sample is cut to expose a cross section, etched with hydrofluoric acid or other appropriate etchant to make the tracks visible, and then

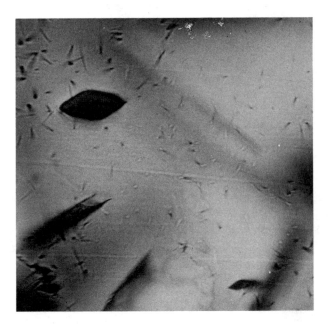

Figure 12.14

Fission tracks in a zircon crystal. Photo courtesy C. W. Naeser.

polished. The tracks are visible under a petrographic microscope at 800 to 1800 power and are counted using an eyepiece graticule to get a *spontaneous track density*.

The tracks are products of the fission of ^{238}U. To find out how much parent ^{238}U is present in the crystal, the sample is then placed in a reactor and bombarded with neutrons. This generates new particle tracks from thermal-neutron-induced decay of ^{238}U. The more original ^{238}U there is, the more tracks will be produced. These are counted in the same way as the spontaneous tracks, giving an *induced track density*. The age can then be calculated as a ratio of the spontaneous-to-induced track density as follows:

$$A = \ln\left[1 + \left(\frac{\rho_s}{\rho_i} \times \frac{\lambda_D \phi \sigma I}{\lambda_F}\right)\right]\frac{1}{\lambda_D}$$

where ρ_S is the spontaneous track density, ρ_i the induced track density, ϕ the neutron dose in neutrons/cm^2, and the constants currently in use (Naeser, 1979) include $I = 7.252 \times 10^{-3}$, $\lambda_D = 1.551 \times 10^{-10}$/year, $\lambda_F = 7.03 \times 10^{-17}$/year, and $\sigma = 580 \times 10^{-24}$ square centimeters.

Fission-track dating has its stringent requirements. The tracks must be stable and etchable. The uranium must be uniformly distributed through the crystal and must be concentrated enough to give a minimum of 10 tracks per square centimeter. The crystal must also be free of lattice defects, distortions, and other flaws. The main problem with the procedure is that heating can anneal fission tracks, eliminating some of them and giving ages that are too young. However, various minerals have different temperatures at which their tracks anneal. In a whole rock, the discordant fission track ages can be used to estimate the ages of both the primary cooling and any later metamorphic events. For example, apatite has a blocking temperature of only 74° to 120°C, but sphene anneals at 197° to 350°C, and zircon at 150° to 210°C. Thus the apatite might give the age of latest cooling, and the sphene is more likely to retain the original age.

As long as the mineral to be dated can be recovered, weathering is not a problem with fission tracks. In general, fission-track ages yield error estimates that are larger than those produced by potassium/argon dating. These larger error estimates are due to problems with counting the large number of tracks necessary to achieve errors of 1 to 2 percent. To achieve errors of less than 2 percent, more than 10,000 tracks would have to be counted each for the fossil, the induced, and the neutron dose determination. Less than 2000 tracks are usually counted for practical reasons. Fission track dating is excellent for young Tertiary and Quaternary volcanic rocks or for determining the provenance of detrital zircons. It is also less expensive than most other dating techniques.

Carbon-14 Dating

Carbon-14 dating was first developed in the 1940s for use in archeology. It is the only isotopic decay system that occurs naturally in sedimentary rocks and fossils, so it is another exception to the rule that sedimentary rocks cannot be directly dated. The half-life of the system is so short (5730 years) that all the primordial carbon-14 decayed long ago. However, ^{14}C is continually produced in the atmosphere by bombardment of ^{14}N by cosmogenic neutrons. The ^{14}N loses a proton, thereby becoming ^{14}C. This atmospheric radioactive carbon is assimilated into the carbon cycles of plants and animals while they live. The decay clock is started at the moment the tissue is no longer living and exchanging carbon with the environment.

The nuclear reaction is

$$neutron + {}^{14}_{7}N \longrightarrow {}^{14}_{6}C + {}^{1}_{1}H$$

and the corresponding decay reaction is

$${}^{14}_{6}C \longrightarrow {}^{14}_{7}N + \beta + energy$$

The age equation is

$$t = 19.035 \times 10^3 \log\left(\frac{A}{A_0}\right) \text{ in years}$$

where A is the measured activity of the sample in disintegrations per minute per gram of carbon (dpm/g), and A_0 is the initial activity. The currently accepted value for A_0 is 13.56 ± 0.07 dpm/g.

Carbon-14 dating is based on two critical assumptions: first, that the rate of ^{14}C production in the upper atmosphere is constant; second that the rate of assimilation of ^{14}C by organisms is rapid relative to its rate of decay. Both of these assumptions appear to be valid.

The procedure for measuring carbon-14 in a sample is quite different from that of any other isotopic system. First, the sample is burned in a vacuum chamber to produce CO_2, part of which is composed of ^{14}C. The carbon dioxide gas is run through a chamber from which most of the other atmospheric gases have been removed. It is then concentrated in a copper counting tube, where a Geiger counter measures the decay activity. During measurement, it must be shielded from atmospheric gamma rays and neutron bombardment. In many cases, only a few grams of material are needed, but in others, several kilograms are required (Table 12.4). Even with sufficient material, it takes 12 to 24 hours of counting per sample to get a measurement. For precision, multiple samples are counted and recounted to screen out background noise and to get consistent measurements with small error estimates.

The main limitation of carbon-14 dating for geology is its short half-life. The practical upper age limit is about 40,000 years, though with extraordinary care and extremely sensitive equipment, marginal dates of 60,000 to 80,000 years have been produced. Another problem has been uncovered by analyzing the carbon in dated tree rings. These studies show that the relative amount of atmospheric ^{14}C has fluctuated tremendously in historic times, decreasing significantly in the last 100 years due to the burning of trees and fossil fuels (Fig. 12.15). Nuclear explosions have also added slight amounts of atmospheric ^{14}C. A correction must be added to the date to compensate for these effects.

There are other pitfalls as well. Marine shells are notorious for picking up carbon that has been dissolved in seawater from ancient carbonate rocks. This ancient carbon is radioactively "dead" and depresses the ^{14}C count, giving the impression that the shell is much older than it really is. This is why an occasional living shell gives a very ancient date. In general, however, carbon-14 is the primary dating tool for Quaternary geology and archeology. It is possible to directly date any organic material that was once living, such as shells, wood, peat, bones, baskets, cloth, paper, food materials, and even some pottery and iron objects. Carbon-14 dating proved critical in dating the ice ages as well; undoubtedly, new uses will continue to be found for it.

Other Methods

Some relatively minor isotopic decay systems have significant applications. *Thorium-230* has been used to date deep-sea cores that are too old for carbon-14 (40,000 years), too young for the first magnetic reversal boundary (730,000 years), and have no potassium for potassium/argon dating. Thorium-230, part of the ^{238}U decay series (see Fig. 12.1), has a half-life of 75,000 years. When it forms from the decay of ^{238}U, it is fractionated in seawater, with the uranium remaining in solution and the thorium incorporated in the bottom sediments. If the sedimentation rates and the ^{230}Th precipitation rates are approximately constant, the thorium can be measured from the top of a long core downward to obtain the age.

Another system used in similar situations is *thorium-230/protactinium-231*. Produced from the decay of ^{235}U, ^{231}Pa is also found in marine sediments. It has a half-life of only 37,000 years. If this isotope is found in combination with ^{230}Th, the two have a combined half-life of 57,000 years. The ratio of these two isotopes thus changes constantly through time, because they decay at different rates. By combining the two isotopes, one can obtain a date that is independent of sedimentation rates. Therefore this method is more versatile than using ^{230}Th alone. Due to the short half-life, however, its maximum limit is about 150,000 years.

Table 12.4

Material Suitable for Dating by the Carbon-14 Method

Material	Amount Required (g)[a]	Comments
Charcoal and wood	25	Usually reliable, except for finely divided charcoal, which may adsorb humic acids; removable by treatment with NaOH. Subject to "post-sample-growth error," i.e., difference in time between growths of tree and use of the wood by humans.
Grains, seeds, nutshells, grasses, twigs, cloth, paper, hide, burned bones	25	Usually reliable. These materials are "short lived" and have negligible post-sample-growth errors.
Organic material mixed with soil	50–300	Should contain at least 1% organic carbon in the form of visible pieces. Efforts should be made to remove as much soil as possible in the field.
Peat	50–200	Often reliable, but intrusive roots of modern plants must be removed. The coincidence of peat formation with the occupation or archeological sites requires careful consideration.
Ivory	50	Often well preserved and reliable. Interior of tusks is younger than the exterior. Some ivory tools may have been carved from old rather than contemporary material.
Bones (charred)	300	Heavily charred bones are reliable. Lightly charred bones are not because exchange with modern radiocarbon is possible.
Bones (collagen)	1000 or more	Organic carbon in bones, called collagen, is reliable. However, the organic carbon content is low and decreases with age to less than 2%.
Shells (inorganic carbon)	100	The carbon in calcite or aragonite of shells may exchange with radiocarbon in carbonate-bearing groundwater. Shell carbon may be initially enriched in ^{14}C relative to wood due to isotope fractionation. It may also be depleted in ^{14}C due to incorporation of "dead" carbon derived by weathering of old carbonate rocks. The reliability of shell dates is therefore questionable.
Shell (organic carbon)	Several kilograms	Organic carbon is present in the form of conchiolin which makes up 1 to 2 of modern shells. Dates may be subject to systematic errors due to uncertainty of initial ^{14}C activity of this material.
Lake marl and deep-sea or lake sediment	Variable	Such materials are datable on the basis of the radiocarbon content of calcium carbonate. Special care must be taken to evaluate errors due to special local circumstances.
Pottery and iron	2 to 5 kilograms	Pottery sherds and metallic iron may contain radiocarbon that was incorporated at the time of manufacture. Reliable dates of such samples have been reported.

[a] The approximate amounts of material required for dating are based on the assumption that 6 g of carbon should be available, which is sufficient to fill an 8-liter counter with CO_2 at a pressure of 1 atm.
From Ralph (1971).

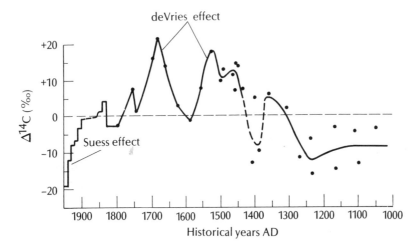

Figure 12.15

Deviation of the initial radiocarbon activity in wood samples of known age. The decline in radiocarbon content starting at about 1900 results from the increase in all CO_2 from burning fossil fuels (the Suess effect). The anomalously high radiocarbon activity around 1500 and 1710 AD is known as the de Vries effect, but its causes are not understood. From Faure (1986).

Another system that has been used much in recent years is *samarium/neodymium* (Sm/Nd). Both are rare earth elements with high atomic numbers on the periodic table. The ^{147}Sm changes to ^{143}Nd by alpha decay with a half-life of 106 billion years. Both elements are found widely in the earth's crust in trace amounts of less than 10 parts per million. Phosphate minerals, such as apatite and monazite, and alkalic igneous rocks tend to have higher concentrations. However, mafic and ultramafic igneous rocks have the highest Sm/Nd ratios and are most easily dated. Whole rock dating, or the isochron method of dating various minerals in the same rock, are most commonly used. Sm/Nd dates are less susceptible to alteration by metamorphism than are Rb/Sr dates. They are especially useful for dating rocks that were initially low in rubidium, such as mafic and ultramafic rocks.

FOR FURTHER READING

Dalrymple, G. B., and M. A. Lanphere, 1969. *Potassium-Argon Dating,* W. H. Freeman, San Francisco. The best single book on this widely used method, covering nearly all the practical and theoretical aspects.

Eicher, D. L., 1976. *Geologic Time,* 2d ed. Prentice-Hall, Englewood Cliffs, N.J. An excellent short paperback on many aspects of stratigraphy and geochronology. Chapter 6 covers numerical dating.

Faure, G., 1986. *Principles of Isotope Geology,* 2d ed. John Wiley & Sons, New York. The classic text on isotope geology, with detailed chapters on all the standard dating techniques; now updated with a new edition.

McDougall, I., and T. M. Harrison, 1988. *Geochronology and Thermochronology by the ^{40}Ar/^{39}Ar Method.* Oxford University Press, New York. Detailed treatment of argon/argon dating, one of the hottest new methods in geochronology.

Parrish, R., and J. C. Roddick, 1985. *Geochronology and Isotope Geology for the Geologist and Explorationist.* Geol. Assoc. Canada, Cordilleran Section, Short Course 4. Excellent short, readable summary of the most common and practical methods, emphasizing the strengths and limitations. Highly recommended.

York, D., and R. M. Farquhar, 1972. *The Earth's Age and Geochronology.* Pergamon Press, Oxford, England. An excellent short paperback covering most aspects of geochronology.

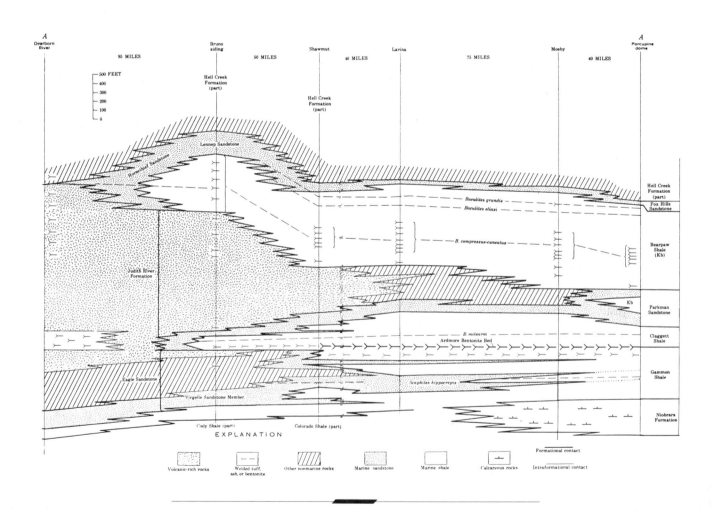

The web of correlation in the late Cretaceous Montana Group, central Montana. Bentonites and ammonite biostratigraphy provide time markers in the rapidly transgressing and regressing marine and nonmarine units. From Gill and Cobban (1973). Courtesy U.S. Geological Survey.

13

Chronostratigraphy

INTEGRATED CORRELATION

Because most real problems present us with a frustrating and incomplete mix of data from various sources and because every method of dating and correlating has its strengths and weaknesses, most stratigraphic problems are tackled using a variety of techniques. Usually, this requires collaboration among specialists in various branches of geology. Once this information is gathered, however, it is possible to integrate the results of the various methods into a combined picture, cross-checking one set of correlations or dates against the others to spot possible problems (Fig. 13.1). The result is a correlation and a chronology of geologic events that are more rigorously tested and have greater precision and resolution than any single method could produce. This "web of correlation" (to paraphrase Berry, 1987) is often complex and difficult to construct, but the value of the system is correspondingly greater. Establishing the time relationships among geologic units by means of integrated methods is called *chronostratigraphy*.

Sometimes, two data sources (e.g., a biostratigraphic identification and a radiometric date) may disagree. The specialists who produced each result must reexamine their data to see if the conflict can be resolved. This

process is healthy because the flaws in the data often become apparent, and the result is greater refinement. Ideally, every episode of conflict and resolution produces a smaller and smaller range of disagreement, and the corresponding chronology of events needs fewer and fewer changes. The resulting chronology has greater *accuracy*; it is closer to that abstract, unattainable ideal we call "truth."

As we check the results of one method against another, we must avoid circular reasoning. Each method must retain its own integrity, or else it tells us only what we already know. A classic example concerns a certain biostratigrapher who implicitly incorporated the stratigraphic ranges of his fossils into his species definitions. In others words, he defined species not only by their anatomical features but also by the formations in which they occurred. Each formation had a different species, even if the fossils had not actually changed in any way. The result was that the formations were correlated by their fossils, and the fossils were recognized by the formation they came from. Another paleontologist called this man's bluff by handing him a loose specimen of one of the fossils. Without knowing the formation it had come from, he could not identify its species.

The methods of dating and correlation that we have

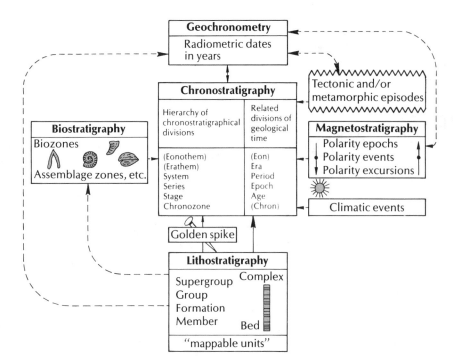

Figure 13.1

Flow chart of the procedures of stratigraphy and their relationships. The chronostratigraphical record is the central repository for the data derived from the procedures and phenomena indicated around them. After Dineley (1984).

examined fall into two categories that correspond broadly to the concepts that underlie Gould's (1987) book, *Time's Arrow, Time's Cycle.* Some methods of dating are based on irreversible processes that operate continuously in one direction ("time's arrow"). These include biostratigraphy, which is based on the unidirectional evolution of organisms through time, and geochronology, which is based on the unidirectional process of radioactive decay. Both of these methods can be used independently of any other method though they seldom provide maximal information alone.

Other methods are based on processes that have several possible states and fluctuate or cycle among them ("time's cycle"). Correlation is based on the overall *pattern* of fluctuations of these states and is dependent on biostratigraphy or geochronology to place this pattern within the matrix of geologic history. Most other correlation methods fall into this second category. For example, most lithologic correlation is based on a pattern of fluctuations among certain rock types: sandstone, limestone, shale, and so on. Recognition of the pattern comes from the *sequence* of rock types and the *relative thickness* of each type. A particular unit (or sample) cannot be dated by itself but depends upon the sequence, and ultimately on biostratigraphy and/or geochronology. By contrast, a single key fossil or a single sample with decaying radioisotopes in the minerals can be dated in isolation, after the necessary background work is done (e.g., the biostratigraphic sequence of an area or the decay constants of an isotopic system).

Other stratigraphic methods also depend on pattern recognition. For example, magnetostratigraphy is based

on recognition of a pattern of fluctuation between two stable states: normal and reversed polarity. A sample can be measured by itself, but it tells us only that the rock was cooled or deposited during a time when the earth's polarity was in one of the two possible states. A reversely magnetized rock could have been formed during the most recent episode of reversed polarity, the Matuyama, which ended about 730,000 years ago, or in any one of the hundreds of reversed episodes that preceded it. A magnetostratigraphic sequence with an identifiable pattern of polarities is needed to tie it to the polarity time scale. Usually, such a sequence also needs some independent means of dating (typically biostratigraphy or geochronology) to give us some idea of how old the pattern is. Then it can be matched with the appropriate part of the magnetic polarity time scale to see if there are characteristic polarity intervals that match in duration and sequence.

A similar approach is needed for most other means of correlation. Seismic correlations are based on changes in density of the material through which the seismic wave has passed and make sense only in the context of the entire pattern. Stable isotope correlations are based on fluctuation of carbon or oxygen isotopes and make sense only if a complete sequence has been run to determine the characteristic peaks and valleys in the pattern. The new method of strontium isotopes, however, is an exception: It appears to be directional rather than cyclic (see Chapter 11). Since the Eocene, the unidirectional change in the $^{87}Sr/^{86}Sr$ ratio from about 0.707 to 0.709 means that any given sample can be dated within the level of precision of measurement.

RESOLUTION AND PRECISION

Other factors that become important when combining stratigraphic methods are their relative *precision* (inherent error of a system, which indicates how repeatable measurements are) and *resolution* (the ability of a system to discriminate between two closely spaced events in geologic time). Every stratigraphic method makes compromises between the two qualities. The precision of radiometric dating is stated in terms of the percent analytical error (given by the "plus or minus" after a date). This precision remains fairly constant in terms of the error percentages, but the resolution decreases with increasing sample age (Fig. 13.2). For example, a date of 10 million years with an analytical error of ±1 percent means that radiometry can resolve events that are separated by about 100,000 years. However, the same 1 percent error on dates of around 100 million years means that the method can only resolve events that are 1 million years or more apart. More typical values of 5 percent error can only resolve events 5 million years apart for the Cretaceous.

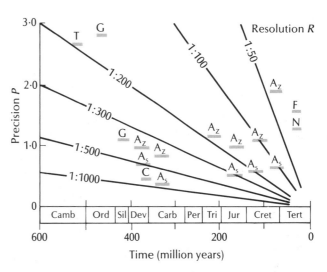

Figure 13.3

Resolution and precision in some fossil groups of biostratigraphic importance. Abbreviations: A, Ammonoidea; C, conodonts; G, graptolites; N, nannoplankton; T, trilobites. Subscripts: z, zones; s, subzones. Precision P is given on the vertical axis and refers to discrimination in years of average zonal or subzonal units. If t is the span of years by numerical dating, and z is the number of zones or subzones over that interval, then $P = t/z$. Resolution R represents the resolving power that a given value of P has for the numerical time span of the zone considered. If T is the mean numerical age for the time span t of z zones, then $R = T/P$. Radiating lines of equal resolution are shown in terms of $1/R$. From House (1985).

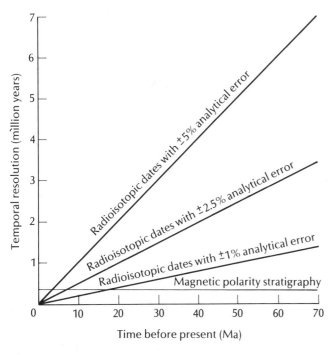

Figure 13.2

Comparison of the temporal resolution for radiometric dating and magnetic polarity chronology for the Cenozoic. Analytical errors of 1, 2.5, and 5 percent are shown for radiometric dating. The maximum temporal resolution is twice the value of the standard error. The average frequency of reversals for the Cenozoic is 3 per million years. Notice that the resolution of magnetic stratigraphy is constant through time whereas the resolution of radiometrics gets poorer with increasing age. From Flynn et al. (1984).

Precision in biostratigraphy is limited by a number of factors (discussed in Chapter 10), such as upward or downward mixing or transport of fossils, improper identification or poor preservation of fossils, and unfossiliferous intervals. In the best cases, however, the precision is extremely good and limited only by the rate of sediment accumulation. For example, some deep-sea cores have nearly continuous records of microfossils, with no large gaps, so the first appearance of a key microfossil can be very precisely located if the preservation and identification of the fossils are good. If the core also happens to represent a time of rapid sedimentation, then a precision of thousands to tens of thousands of years is possible. If there were annual climatic varves or some similar phenomenon, then a precision of hundreds of years to less than a year is possible.

Resolution in biostratigraphy, on the other hand, is limited primarily by the rate of evolution of the fossil assemblages or individual lineages. If they evolve slowly, then every zone is long, and resolution is low. If they evolve rapidly, the zones can be short, and excellent resolution is possible. Resolution varies from group to group among the biostratigraphic indicator fossils (Fig. 13.3). For example, 117 ammonite zones and subzones are currently recognized in the 70-million-year span of the Jurassic (House, 1985), so each zone or subzone averages only 600,000 years in length. Ramsbottom (1979) reported a zonation of the Namurian (Early Carboniferous) down to 25,000-year increments using goniatitic ammonoids. As is apparent from Fig. 13.3, not all fossil groups have resolution this fine, but many have zones that are shorter than 3 million years in average duration. Because radiometric precision decreases with increasing age, the relative resolution of the biostratigraphy compared to its numerical age gives a rough measure of the equivalence of biostratigraphic resolution. In Fig. 13.3, for example, Cambrian trilobites have zonal lengths of about 3 million years, but the radiometric precision is typically ± 5 million to ± 10 million years. Cretaceous zonal lengths are only a million years or less, but dates typically have errors in the ± 2 million to ± 4 million year range. Scaling the zonal lengths to the error in numerical age puts Cretaceous ammonite subzones on approximately the same 1:200 line on which Cambrian trilobites fall. From Fig. 13.3, it appears that the finest resolution available is in the Devonian and Carboniferous, in which the zones are extremely short despite the age of the system. For most of these examples, good biostratigraphy has much better resolution than radiometric dates.

Let us consider one more system. The precision of magnetic stratigraphy is limited by the quality of the magnetic signal, the shortness of the sampling interval, the completeness of the stratigraphic record, and the time it takes to make a transition from one polarity to the other. However, in exceptional cases, a detailed record of the transition interval itself can provide finer precision than the 4000 to 5000 years needed for a polarity transition. The resolution, on the other hand, is limited by the length of the polarity interval. For the Cenozoic, the frequency of polarity changes ranges from 3.84 per million years in the Neogene to 1.85 per million years in the Paleogene (Flynn et al., 1984). The average length of a polarity zone is about 3 million years or less. *This does not change with increasing age* (see Fig. 13.2), so the resolution relative to the numerical age is much better than even the best radiometric dates for anything older than the Quaternary. During the mixed polarity of the Jurassic and early Cretaceous, the paleomagnetic resolution is again in the order of less than a million years whereas the resolution of radiometric ages is typically 3 to 8 million years, or worse.

Magnetostratigraphy, then, offers resolution on the same order of magnitude as biostratigraphy (or better in some cases), but it is applicable to just about any rock type, terrestrial or marine. High-resolution biostratigraphy, on the other hand, is possible only in environments in which rapidly evolving organisms are preserved. In the Jurassic, for example, the marine zonation based on ammonites has very high resolution, better than the paleomagnetic record. There is no comparable terrestrial biostratigraphy, however, so magnetic stratigraphy offers the potential of greater resolution in that environment.

The relationships among precision, accuracy, and resolution are summarized in Fig. 13.4. On the vertical axis, the age of the event increases logarithmically upward. On the horizontal axis, the time separation of events increases logarithmically to the right. Naturally, events of very short durations, from 1 hour (tsunamis and gravity flows) to about 1 year, cannot be resolved except in the most unusual circumstances. They fall in the simultaneous-instantaneous zone, the zone in which events can be considered geologically instantaneous. The range of accuracy of ^{14}C is from tens of years to about 80,000 years, whereas the accuracy of potassium/argon dating begins in the range of 500,000 years and older. However, the zone of resolution for these methods is much more limited. Potentially, ^{14}C can resolve events spaced only years apart, though it is much more practical for events space ten to tens of thousands of years apart. Because the outside limit is less than 100,000 years, however, it cannot be used for events further apart than 10,000 years.

Potassium/argon dating, on the other hand, cannot discriminate events separated by tens of thousands of years except in the Pleistocene. Resolution decreases with increasing age, so for the Tertiary events separated by a million years are at the lower limit of resolution (as we have already seen). Notice that at this point the zone of biochronological discrimination begins to overlap that for potassium/argon (Fig. 13.4). Although the resolution of potassium/argon continues to get worse in older systems (hence the slope), that for biochronological discrimination remains in the order of one million to ten million years (hence the vertical boundary). This reinforces the point that biostratigraphic resolution is dependent on the rate of turnover of the fossil species in question and not their numerical ages.

In this chapter, we shall look at a few examples of integrating stratigraphic methods to produce the most complete geologic history possible. Each example is chosen to demonstrate the given-and-take among several methods and to show the pitfalls of relying too heavily on one method.

ON THE LOCAL SCALE: THE PLIO-PLEISTOCENE OF EAST AFRICA

The continental deposits of the East African rift grabens have become world famous as sites of important discoveries of human fossils. These include Olduvai Gorge in northern Tanzania, the areas around Lake Turkana (formerly Lake Rudolf) in northern Kenya, the Omo

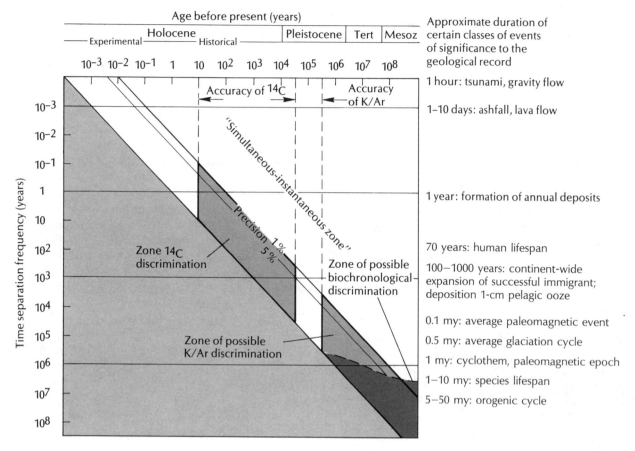

Figure 13.4

Resolving power of geochronologic systems in the Cenozoic. The horizontal axis shows age increasing exponentially to the right, from hours to hundreds of millions of years. The vertical axis shows the time separation of events, from hours to hundreds of millions of years. The range is from events of short duration (tsunamis, turbidity currents, ash falls, lava flows, to events of long duration (species life spans, orogenic cycles). Carbon-14 dating is useful only in the area represented by the parallelogram, ranging from tens of years to less than 100,000 years. In the most recent materials, ^{14}C dating can resolve events that are years to tens of years apart. At its older limit, it can only resolve events that are hundreds of years apart. Events more closely spaced than this cannot be resolved, and fall in the simultaneous-instantaneous zone. Similarly, K/Ar dating can date materials less than a million years in age, but its resolution is at best only tens of thousands of years. The light gray area indicates the zone in which Cenozoic biochronology is effective. Note that although biostratigraphy can seldom resolve events spaced less than a million years apart, this resolution does not decrease with increasing age. From Berggren and Van Couvering (1978).

River drainage in southern Ethiopia, and the Hadar area in the Afar Triangle of northern Ethiopia. This last area is the home of our oldest known hominid relative, *Australopithecus afarensis*. The most famous specimen of this species is a nearly complete skeleton of a female nicknamed "Lucy."

Because of the interest in these hominid fossils and their ages, there have been many efforts to determine the stratigraphy as precisely as possible. These rift graben valleys provided suitable homes for our early ancestors and fossilization sites for their bones; fortunately, they also provided the depositional environment for exceptional stratigraphic detail. Rift valleys tend to subside continuously for long periods of time, producing unusually thick terrestrial sequences. Their frequent fault

ing and tilting means that many of these sequences are well-exposed today, though they require detailed mapping to work out the faulting. The rift valley was and is a site of active vulcanism, so there are also many layers of airfall and ashflow tuffs, suitable for both radiometric dating and for tephrostratigraphic correlation. Where the sequences are thick and fine-grained enough, they are also suitable for magnetic stratigraphy. Finally, these deposits entombed not only the hominids but the entire evolving East African Miocene-to-Recent mammal fauna, some of which are excellent biostratigraphic indicators. All of these techniques have been applied to the East African sequence, and the cross-checking, correction of bad correlations and dates, and ultimate refinement of the system is one of the most thorough in the world.

The discovery of Olduvai Gorge in 1911 began the study of the geology of the region, and the fossil faunas were studied by a number of people for the next 50 years (reviewed by Leakey, 1978). The area received worldwide attention in 1959 with the discovery of the first good hominid specimens and again in 1961 with the publication of a potassium/argon date of 1.8 million years on Olduvai Bed I (Leakey et al., 1961; Evernden and Curtis, 1965). This was one of the first potassium/ argon dates to be run, and it radically altered the prevailing view of the age of human origins by proving them to be much older than had been thought. In subsequent years, the stratigraphy of the Olduvai section was worked out in great detail, using dates from other ashes and flows (Hay, 1975). Work in the Omo Valley produced a longer and more complete sequence with many radiometric dates going back over 3 million years.

Spectacular hominid fossils also came from the Koobi Fora area on the east shore of Lake Turkana. Of these, Richard Leakey's best *Homo habilis* skull, known by its catalogue number, 1470, was the most significant. Not only was it much more advanced than the australopithecines that had been previously found, but it was associated with stone tools. The specimen and the tools were found below the KBS Tuff, which was dated by the ^{40}Ar/^{39}Ar method (Fitch and Miller, 1970) at 2.61 ± 0.26 million years.

As described in Johanson and Edey's (1981) book *Lucy* and in Roger Lewin's (1987) *Bones of Contention*, this date caused severe problems for the paleoanthropologists. The 1470 skull was not only very advanced but almost a million years older than would have been predicted from other areas. The chief evidence that the date was anomalously old came from the mammalian biostratigraphy. As worked out by a number of scientists, the fauna found with 1470 were much more like those known from about 2 million years ago at Omo and Olduvai. Of the mammals, the elephants and especially the pigs (Maglio and Cooke, 1972; Cooke, 1976) proved to be the most useful. Maglio's (1972) "Faunal zone a," which included the pig *Mesochoerus limnetes* and the elephant *Elephas recki*, occurred in association with dates of around 2 million years at Omo but below the KBS Tuff at Koobi Fora. At a conference in London in 1975, the dispute between Leakey and the adherents to the old KBS Tuff date, and those who doubted it, reached a climax. As described by Johanson and Edey (1981, p. 239), one of Leakey's associates arrived wearing a hat he called a "pig-proof helmet" to protect against the "pig men." Basil Cooke, the chief "pig man," wore a tie with "MCP" woven into it; most of the audience assumed that this abbreviation meant "male chauvinist pig." After Cooke made a strong case for the pig biostratigraphy and against the old KBS Tuff date, he said, "You may think you know what MCP stands for, but you don't. It really stands for *Mesochoerus correlates properly*."

The mammalian biostratigraphy was double-checked by Tim White and John Harris (1977), who also came to the conclusion that the date on the KBS Tuff was too old. Their detailed studies of the fauna even showed that there was a significant gap in the Koobi Fora sequence that was not apparent from the lithostratigraphy or troublesome K/Ar dates. Several more attempts to date the KBS Tuff produced new dates on *two different* KBS tuffs of 1.6 and 1.8 million years by K/Ar (Curtis et al., 1975), and then a better date of 1.8 ± 0.1 by K/Ar (Drake et al., 1980). The ^{40}Ar/^{39}Ar dates on the KBS Tuff which first produced the notorious date of 2.61 (Fitch and Miller, 1970), later gave a date of 2.42 million years (Fitch et al., 1976). When rerun by others (McDougall et al., 1980; McDougall, 1985) the KBS Tuff produced a ^{40}Ar/^{39}Ar date of 1.88 ± 0.02 million years. The early fission track dates on the KBS Tuff first gave dates of 2.42 million years (Hurford et al., 1976) but eventually produced ages of 1.87 ± 0.04 million years (Gleadow, 1980). Ironically, because of the initial bad

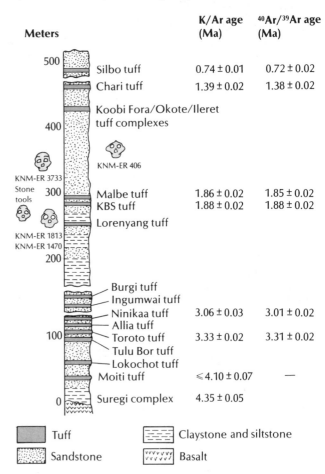

Figure 13.5

Current K/Ar and Ar/Ar dates for the tuffs in the Koobi Fora region, showing the excellent agreement of both methods after much trial and error. Location of some key hominid fossils and artifacts are also shown. From McDougall (1985).

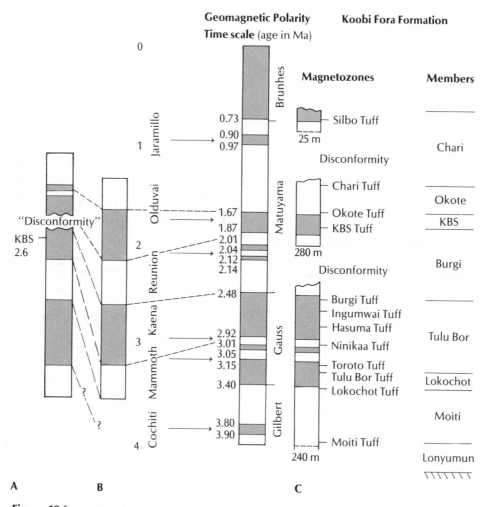

Figure 13.6

Magnetostratigraphy of the Koobi Fora Formation and its correlation with the magnetic polarity time scale, which changed with the changing interpretation of the KBS Tuff date. (A) Interpretation of Brock and Isaac (1974), which placed a disconformity *above* the KBS Tuff, so that the normal polarity in which it lies could be correlated with the upper part of the Gauss (in agreement with the date of 2.6 Ma then prevailing). (B) Interpretation of Hillhouse et al. (1977), which the KBS Tuff date had been revised to 1.6 to 1.8 Ma (C) Current interpretation of Hillhouse et al. (1986), with much more complete sampling than previous studies. This extended sampling spanned the time scale from the Brunhes to the Gilbert Chrons and revealed a disconformity *below* the KBS Tuff.

date on the KBS Tuff and its tremendous scientific importance, the KBS Tuff is now one of the best dated units in the world.

In concert with the dates on this most controversial unit, many of the other tuffs at Omo, Koobi Fora, and other areas produced dates that provided a total framework for the area (Fig. 13.5). Another approach to these volcaniclastic units was tephrostratigraphy by chemical fingerprinting (Cerling and Brown, 1982). Eight of the tuffs at Koobi Fora could be matched with those at Omo based on their trace and minor elements. Chemical fingerprinting showed that the Tulu Bor Tuff in the Koobi Fora Formation had been misidentified. This method made it possible to use the tuffs as synchronous surfaces

to separate the Koobi Fora Formation into a set of sequential, isochronous units (Findlater, 1978) and to match these with the units at Omo.

Magnetostratigraphy of the Koobi Fora Formation was first attempted by Brock and Isaac (1974), but many of the specimens apparently had a normal overprint that was not removed during demagnetization. This, compounded with the fact that Brock and Isaac were trying to calibrate their magnetostratigraphy with the erroneous dates of the KBS Tuff, meant that the magnetostratigraphy had to be reinterpreted. Originally, they correlated the Koobi Fora Formation with the polarity time scale from just above the Olduvai event to the top of the Gilbert reversed interval (Fig. 13.6A). Because of the anomalously old date on the KBS Tuff, they placed a disconformity above it to account for the "missing" lower Matuyama and upper Gauss intervals. As revised by Hillhouse et al. (1977), the sequence spanned from above the Olduvai down to the middle of the Gauss, eliminating the disconformity that was demanded by the bad date on the KBS Tuff (Fig. 13.6B). The most recent interpretation (Fig. 13.6C) added much additional sequence to the section and showed that the sequence spans part of the Matuyama, with a big disconformity *below* the KBS Tuff and above another sequence that runs from the top of the Gauss to the middle of the Gilbert. Clearly, the paleomagnetic interpretation is susceptible to the revisions in radiometric dating and does not reveal the disconformities by itself. On the other hand, the paleomagnetic reversal boundaries provide additional time planes for even better resolution of events than was possible with the ashes or biostratigraphy alone.

The example of the Koobi Fora and related formations of East Africa points out a number of problems typical of this kind of study. The radiometric dates provide the numerical ages, but they are subject to many types of error and so had to be redone many times in several laboratories by three methods over 15 years before all the results were consistent and undisputed. Because the KBS Tuff date supported Leakey's contention that the genus *Homo* was very old, there was much political and sociological maneuvering before the problem was corrected. Leakey and his associates placed too much importance on the 2.6 Ma dates, which caused much bitterness in paleoanthropology for almost a decade (Lewin, 1987).

As discussed by Lewin (1987), the mystery of why the original KBS feldspar crystals gave the anomalously old date has never been solved. Every time they were rerun by Fitch and Miller, their ages came out around 2.4 to 2.6 Ma. Lewin (1987, p. 252) recently found some of the original crystals and had them rerun by the K/Ar method; their dates came out 1.87 ± 0.04 Ma. Even methods that are supposedly objective and absolute, such as radiometric dating, can be biased by the expectations of the experimenter. For example, Lewin (1987, p. 246) quotes Hurford on running the fission-track method, as follows:

> You can bias your results ten percent either way, easily. You go crystal by crystal, and you begin to see where the rolling average is going. If you need the count to be higher with the crystal you're working on, so that it will fit in, you might include something that is a doubtful track. If you want the count to be lower, you don't include it. That was poor practice.

Biostratigraphy proved to be the most reliable method, least subject to errors. It also revealed gaps in the sequence that were not made apparent by other methods. However, it is only locally applicable to that part of East Africa and does not correlate with other continents or the deep sea. Tephrochronology also improved correlation within the Turkana Basin but, like the evidence of the mammals, was not useful outside the areas of the volcanic ashfall.

Finally, magnetostratigraphy provided the finest resolution of time and the best correlation with other parts of the world, but it was subject to problems with the magnetic signal and had to be interpreted in the framework of the available radiometrics and biostratigraphy. As each of these sources of data was reinterpreted, the magnetic story changed too. As of this writing, however, most of the controversies have apparently subsided, and now the hominid-bearing sequences of East Africa are among the best-dated terrestrial sequences in the world.

ON THE CONTINENTAL SCALE: THE WESTERN INTERIOR CRETACEOUS OF NORTH AMERICA

In Chapter 5, we studied the epeiric seaways of the Western Interior Cretaceous of North America as examples of marine deposits. This region also serves as an excellent example of the web of correlation by many stratigraphic techniques in the marine environment. In fact, the integrated stratigraphy of the Western Interior Cretaceous appears to have the highest resolution reported for any part of the geologic record prior to the Pleistocene and spans almost 40 million years.

The Cretaceous seaway was subjected to frequent transgressions and regressions of the epeiric seaway, which formed a cyclic set of deposits from nonmarine to open ocean (Fig. 5.28). From this, it is apparent that a first level "event" correlation can be made on the basis of this cyclic pattern. The peaks of transgression in all areas are probably equivalent, as are the peaks of regression. That is merely the beginning, however. The Cretaceous seaway was rich in marine life, especially ammonites, bivalves, gastropods, and benthic foraminifers. Every fossiliferous level is collected and its fossils identified, starting with local levels. Although there may be locally unfossiliferous levels, when samples from enough sections are collected and combined, their range produce a high-resolution subdivision of the late Cretaceous (Fig. 13.7). The formal zonation is locally named after the oysters (particularly *Inoceramus*), but the ammonites are more important regionally as zonal markers because they are pelagic and therefore less provincial. Cobban and Reeside (1952) originally recognized 32 ammonite zones from the Albian to the Maastrichtian, but later revisions resulted in 72 zones with an average duration of 450,000 years. The local bivalve zonation

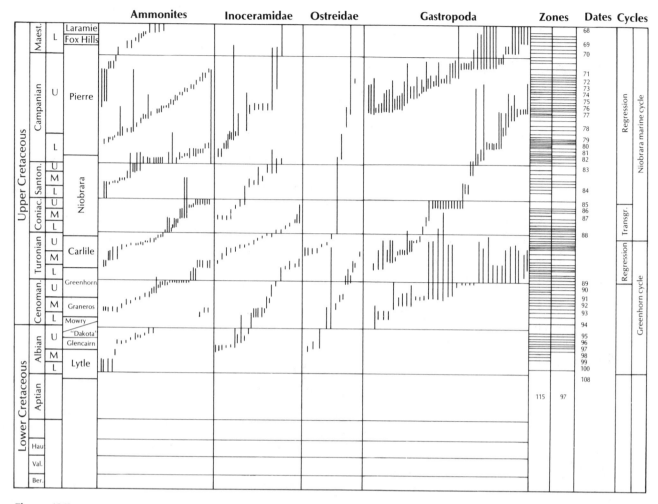

Figure 13.7

Range zone data of some key molluscan taxa (ammonites, inoceramid bivalves, oysters, and gastropods) for the Western Interior Cretaceous. When the overlapping ranges of so many taxa are combined, the composite biostratigraphy (right column) has 119 zones spaced over 40 million years, giving a resolution of 120,000 to 280,000 years per zone. From Kauffman (1970).

provide a resolution of 780,000 years, and combining the two subdivides the zones in such a way that a resolution of 250,000 to 330,000 years is possible (Kauffman, 1970, p. 635). When all the taxa are combined (Fig. 13.8, right column), the overlap of ammonite, bivalve, gastropod, and foram biostratigraphy gives 119 separate sequential levels with an average duration of 120,000 to 280,000 years per zone (Kauffman, 1970, p. 651).

Although this is excellent biostratigraphic control by any standard, there are other sources of data for testing the biostratigraphy and providing additional datum levels. During the late Cretaceous, frequent volcanic eruptions from the uplifting Rocky Mountains to the west spread clouds of ash over the seaway. These ashes sank to the bottom and became widespread bentonite horizons (see Fig. 8.12). Approximately 400 of these bentonites are scattered through the Albian to Maastrichtian section, each of which provides a clear synchronous surface. Some spread over 2000 miles (the

X-bentonite) whereas others were local and cannot be traced through every section (Fig. 13.8). Nearly all the concurrent range zones of the ammonites and bivalves prove to be isochronous when tested against these bentonites. The network of bentonites can be traced over long distances (p. 297), producing an amazing "web of correlation." There are also limestone-calcarenite beds that can be traced over long distances, and these can be used as secondary correlation planes. Some of them show great lateral persistence and consistency of thickness, with the same number of limestone beds between each bentonite (Figs. 13.8, 13.9A). These alternating thin limestone-shale units represent fluctuations of some oceanographic parameter, probably related to sea-level fluctuations. In some areas (Fig. 13.9B), there are distinct bundles of these limestone beds arranged in short cycles that appear to span 20,000 to 25,000 years, longer cycles that span 40,000 to 50,000 years, and larger cycles of 100,000 to 125,000 years (Fischer et al., 1985). These

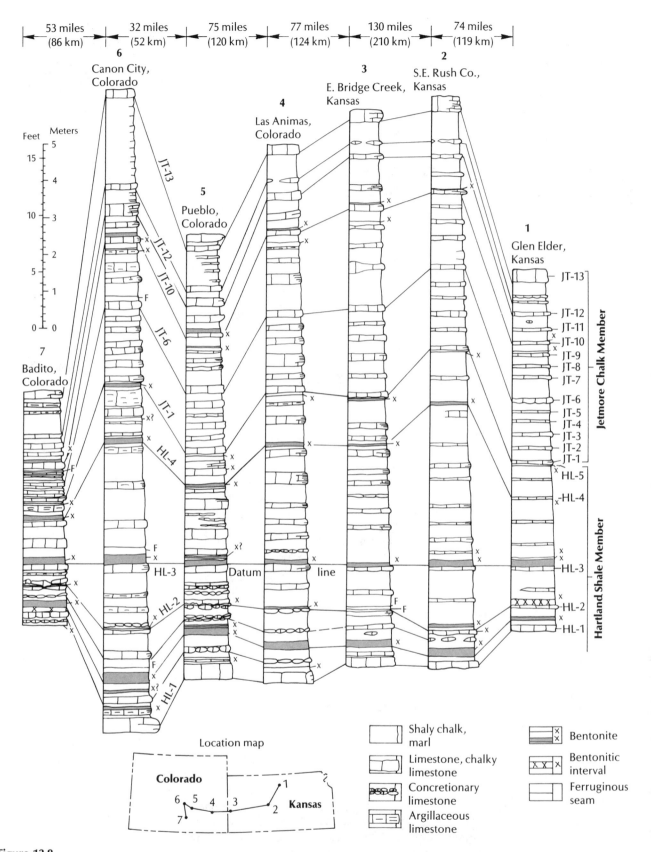

Figure 13.8

Correlation of sections in the Hartland and Jetmore Members of the Greenhorn Limestone from central Colorado to central Kansas. Sections can be correlated by distinctive bentonites (black layers and X along sections), and many of the chalk layers in the shale are so widespread and distinctive that they also have been named and correlated. From Hattin (1985).

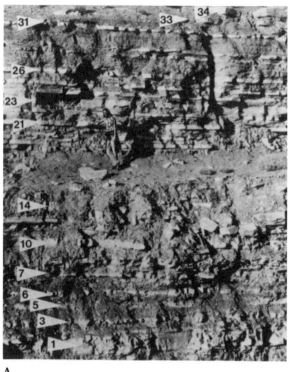

A   B

Figure 13.9

(A) Middle and lower Bridge Creek Limestone Member of the Greenhorn Limestone, showing the regular alternation between resistant limestone beds and less resistant shales. These limestone markers (numbered in sequence) can be widely correlated across the Western Interior, as seen in Fig. 13.8. (B) Finely laminated Fort Hays Limestone Member of the Niobrara Formation, showing thick chalks interrupted by shale layers: The regular pattern of shale-limestone couplets appears to show the superimposed cyclic pattern that may correspond to the Milankovitch variations of the earth's orbit. On the left are Kauffman and Pratt's interpretations of the 20,000 to 25,000 year, 40,000 to 50,000 year, and 100,000 to 125,000 year Milankovitch periodicities as seen in the outcrop. From Kauffman and Pratt (1985).

may correspond to the three Milankovitch cycles of the earth's orbital variation that have already been shown to control the Pleistocene fluctuations in ice volume, temperature, and sea level (Barron et al, 1985). If so, then these climatic cycles could be correlated down to the 20,000-year level, and the duration of events between cycles could be estimated on an even finer level of resolution.

Yet another method is now being tried on these sections. All of the beds are now being examined geochemically (Arthur et al., 1985) to determine the conditions of deposition. Although the primary goal of this work is to determine the paleoceanographic conditions responsible for the geochemical fluctuations, the pattern could also serve as yet another correlation tool. For example, Pratt (1985) found that there is a global geochemical event associated with the mass extinctions at the Cenomanian/Turonian (mid-Cretaceous) boundary. This event shows up in the $\delta^{13}C$, the $\delta^{18}O$, and the organic carbon levels in sections from Montana to Kansas to Arizona (Fig. 13.10). Geochemical events are also reflected by the boundary surfaces between layers representing unusual conditions, such as the reducing

anaerobic conditions that produce organic-rich black shales.

Unfortunately, magnetic stratigraphy is of limited use in the late Cretaceous. The long Cretaceous normal interval spans the entire Aptian through Santonian (about 118 to 83 Ma), so there are no changes in polarity for correlation during 35 million years of the late Cretaceous. Magnetic stratigraphy has proven useful in the early Cretaceous, however (Lowrie et al., 1980).

Clearly, there are many sources of high-resolution datum levels in the Western Interior Cretaceous. What about the numerical ages of these events? Unfortunately, the geochronology leaves much to be desired. Many of the bentonites provide sufficient potassium-rich minerals for K/Ar dating, and there are some glauconites as well. The bentonites are very weathered, however, so the biotites are altered, and the sanidines can be contaminated with much older detrital feldspars. Fortunately, both of these minerals are found in the same bentonite and can be used to check one another. With extreme care in sample preparation, the best radiometric dates for the late Cretaceous usually have analytical errors in the range of ±700,000 to as much as ±2.6 million

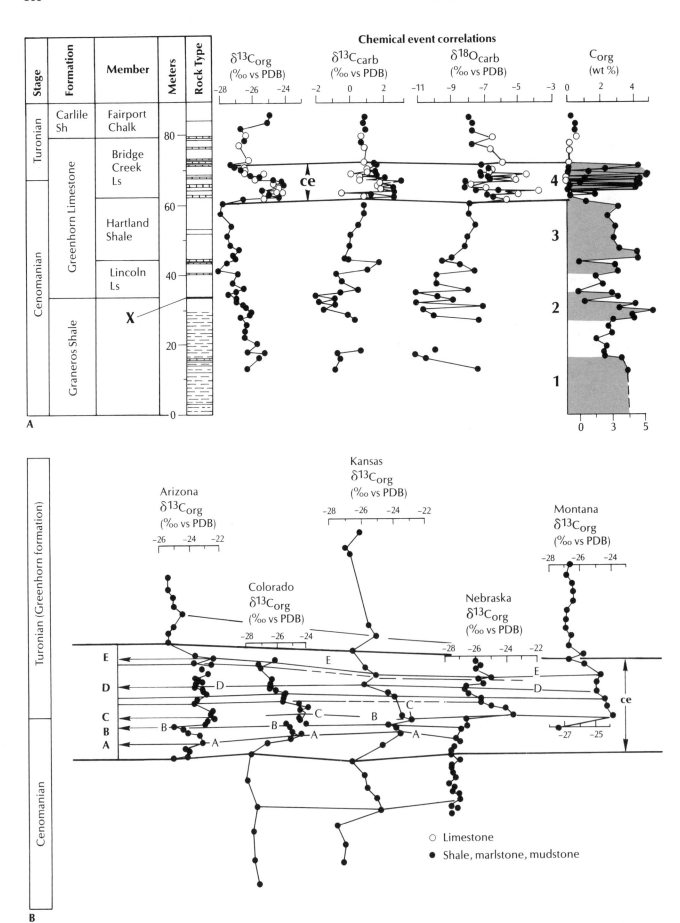

Figure 13.10

(A) High-resolution chemostratigraphy, as shown by the organic carbon, ^{13}C, and ^{18}O analyses of the Cenomanian/ Turonian boundary interval from Arizona to Montana to Nebraska. The chemical event (ce) shows up clearly in all the chemical signals at a fixed distance above the X-bentonite bed (X), a marker bed that is found from Alberta to Texas. (B) Details of the Cenomanian/Turonian chemical event in the ^{13}C signal. Notice that even the small-scale fluctuations can be correlated from state to state (fine lines). From Pratt (1985) and Kauffman (1988).

years (Obradovich and Cobban, 1975). This is much poorer resolution than the other methods, but it is the only way to produce numerical ages. If the bundled cycles shown in Fig. 13.9B are indeed caused by the well-known cycles of the earth's orbital variations, then they could produce numerical estimates of time in the order of 20,000 years or less. This is much better than the ± 700,000 years (or worse) resolution of radiometric dates.

As a consequence, most Cretaceous workers have adopted these numerical ages as rough estimates of endpoints, but they have used the fine-scale stratigraphy to subdivide the interval more finely than a series of dates could. This has been called *high-resolution event stratigraphy* (HIRES) by Kauffman (1988). In this method, all of the tools of correlation (event correlation of ocean-ographic events, bentonites, biostratigraphic events, and geochemical events) can be combined from area to area, using the graphic correlation method of Shaw (1964) discussed in Chapter 10 (Fig. 13.11). With such a multi-tude of synchronous correlation points between sec-

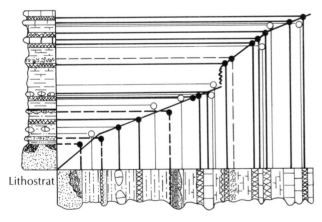

Lithostrat

Figure 13.11

Graphic correlation of stratigraphic events (bentonites, cli-matic cycle beds, biostratigraphic markers) between two Cretaceous sections. Event correlations are shown by heavy lines; biostratigraphic markers are shown by light lines. Notice that the biostratigraphic boundaries do not fit the line of isochron correlation as well as do the event markers. From Kauffman (1988).

tions, it is possible to constrain the line of correlation very precisely. Once numerical age estimates are made for this line, then the numerical age of any point along it can be interpolated with great precision. Kauffman (1988) suggests that typical resolution is on the order of 50,000 years or less for most of the Cretaceous.

In summary, the Western Interior Cretaceous provides an exceptional example of integration of multiple bio-stratigraphic datum levels, bentonites (and their ra-diometric dates), limestone-shale cycles that may be climatically controlled, and possibly even geochemical correlation. Even though this sequence ranges in age from about 68 to 108 Ma, resolution of 50,000 years or less is possible in some instances. Unlike our East African example, these correlations can be traced for thousands of kilometers across the entire central region of North America. Not many places yield such fine-scale correlation over such wide areas.

ON THE GLOBAL SCALE: THE EOCENE/OLIGOCENE BOUNDARY

So far we have looked at local or regional units that have been correlated to a standard geological time scale. But how is that time scale constructed? How can we say, for example, that the Eocene/Oligocene boundary is 36 or 34 or 33 million years old? How do we recog-nize it in our local outcrops?

The global geologic time scale has been constructed by integration of stratigraphies and dates from all over the world. Combining data of unequal quality from so many sources makes worldwide correlation much trickier than local or regional correlation because there are many more sources of possible error. In some cases, the data have yielded a relatively uncontroversial results; in other cases, the data are so poor that there is little confidence in the numbers and little grounds for strong argument. We have seen such a case with the dating of the Silurian/ Devonian boundary (Box 12-1). In a few other cases, however, there are multiple sources of data, but their interpretation is subject to alternative points of view and strong controversy. These last examples are the most revealing for us because they show most clearly the steps by which the rest of the time scale has been con-structed. For this reason, let us look at one part that has been particularly controversial: the Eocene/Oligocene boundary.

As we saw in Chapter 1, the name *Eocene* was coined originally by Lyell (1833) to divide the Tertiary ac-cording to faunal percentages. Unlike most stratigraphic names, it was based on an abstract concept, the "ticking" turnover of the molluscan "clock," rather than on a typical area or type section. Clearly, Lyell conceived of time units as independent of local stratigraphy, but most of Lyell's contemporaries and successors were more interested in designating stages based on local lithostratigraphic sequences, that may or may not represent all of Tertiary time. In spite of his interest in abstract time terms, Lyell designated the Paris and London basins as the typical areas for his Eocene. Lyell (1833, pp. 57–58) explicitly recognized that there might be further subdivisions of his units but did not try to designate them himself.

Naturally, geologists began to find areas with faunas that were intermediate to Lyell's Eocene and Miocene. The upper part of the Eocene is sparsely fossiliferous in the Paris Basin, but in northern Germany and Belgium an extensive marine transgression produced many fossils that were younger than those in the Eocene of the Paris Basin. These northern rocks were called "Lower Miocene" by some and "Upper Eocene" by others, until von Beyrich (1854) coined the term *Oligocene* for them. In 1856, von Beyrich was more precise about the units included in the Oligocene, and its upper and lower boundaries.

Long before this, however, geologists had named parts of their local successions of fossiliferous strata. As reviewed in detail by Berggren (1971), these names and their type areas proliferated and soon became a confusing mess. In many cases, two successive stages were not in the same basin, or otherwise did not lie in superposition one to another, so it was impossible to determine whether they were really successive or whether they overlapped because of lateral facies changes. In other cases, it became clear that a stage, as originally defined, left a gap between it and the next stage; then arguments ensued as to whether to draw the boundary at the base of the upper stage or at the top of the lower stage, or to create another stage in between.

The primary cause of the whole controversy was that these Tertiary stages were created for local sequences within various European basins, which had undergone complicated local histories of transgression and regression throughout the Tertiary (Fig. 13.12). The type area of each stage is thus part of a time-transgressive sedimentary cycle that is separated from adjacent cycles by unconformities above and below it. When a geologist tried to trace any of these stages laterally from its type area, problems were certain to be encountered. These problems were compounded by the fact that most of the faunas were endemic, shallow marine molluscs. These were often difficult to match from one European basin to another, let alone with other parts of the world. The result was almost a century of argument, much of it trivial or fruitless. When someone wanted to call some rocks in North America "Eocene" or "Oligocene," it was almost impossible for anyone else to substantiate or refute the claim. Not only were there few fossils in North America that could be correlated to the type areas in Europe, but there was considerable confusion as to what

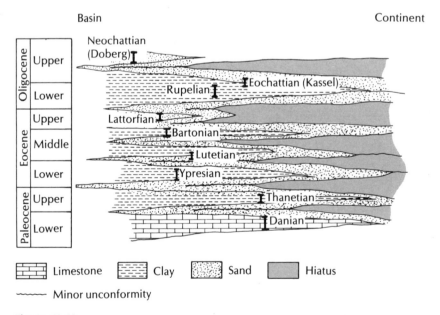

Figure 13.12

Depositional history of the type areas of the Paleogene stages and ages in northwest Europe. Notice that the section is incomplete and composed of irregular transgressive and regressive sequences with large hiatuses. Each of the stratotype sections spans only a small part of the total age or epoch. This incompleteness is responsible for much of the confusion and dispute about stratigraphic terminology in the Paleogene over the last century. From Hardenbol and Berggren (1978).

was Eocene or Oligocene even within the type areas in Europe.

By the 1960s and 1970s, a solution began to appear. What was needed were the most continuous possible sequences with few gaps containing a continuous record of fossils that could be correlated around the world. The shallow marine basins of Europe failed to provide this, but the deep-sea record, with its globally distributed, rapidly evolving planktonic microfossils, did. Deep-sea sequences were studied first from uplifted exposures on land in New Zealand and Italy and then from oceanic cores recovered by the Deep Sea Drilling Project. The changes in planktonic foraminifers in particular provided clear, nearly global datum levels that could be correlated from core to core with great confidence. Because these changes were generally controlled by climatic changes in global water masses, they proved to be synchronous within the resolution of the system in nearly all examples. The planktonic foraminiferal and calcareous nannofossil zones became the worldwide standard of correlation for the Tertiary for lower and middle latitudes. In areas of siliceous sedimentation, diatom and radiolarian zones were used. A few key cores had both calcareous and siliceous sediments and provided the neccessary tie between the two stratigraphies.

The micropaleontologist, therefore, was in the key position to correlate sequences in the Tertiary. Berggren (1971) and many other scientists examined samples from the type areas in Europe to find the microfossils needed to tie the Tertiary epochs and stages there to the deep-sea record. This was not an easy task because the shallow-water European sequences contained few open-ocean planktonic microfossils, and often these were poorly preserved. This research showed that some long-accepted beliefs and standards were wrong and that some entrenched names were useless or overlapped other names. Naturally, this led to more controversy, but most of it has died down since the studies of the early 1970s.

The deep sea, which provided the continuous, fossiliferous record needed to correlate the European type areas on a worldwide basis, could not provide numerical ages for these epochs and stages. Radiometric dates were needed, but unfortunately, they were not ideal, either. Almost no ash falls occurred in the deep marine sequences, and few occurred in the shallow marine sequences. Most dates were obtained from glauconites, which grew diagenetically in the shallow marine environment. As we saw in Chapter 12, dates from glauconites are subject to problems that do not affect dates from high-temperature minerals from volcanic events, Glauconites are notorious for leaking argon, and for this reason, most of the first glauconite dates to be run have been rejected. Other early glauconite dates were rejected because of equipment insensitivity to atmospheric argon. Even the freshest glauconites have problems, however. Because they form diagenetically, they can crystallize long after they are buried and can thus lock in a much younger age. They occur as rounded pellets on the seafloor, so they are very easily transported and reworked. Many glauconite-rich deposits also yield unsuitable ages because of recycled older glauconite. Other problems have been documented as well. In short, many isotope geologists regard glauconite dates with suspicion, especially when they clash with high-temperature dates that are not subject to these problems. Nevertheless, glauconites are abundant and have necessarily been used extensively for dating the Tertiary (Hardenbol and Berggren, 1978, Fig. 6; Odin, 1978, 1982).

Besides the steady rain of microfossils onto the deep-sea floor, there is another steady deep-sea process that records events. This process is seafloor spreading. As the ridges spread apart, they are intruded by magmas from the mantle which cool and record the prevailing direction of the magnetic field. This cooled midocean ridge crust is then split apart by further spreading, and the magnetic record is transported away from the ridge in both directions. As the earth's magnetic field flips back and forth between normal and reversed polarity, the continuous process of spreading and cooling of magma records the polarity history away from the ridge. Thus the spreading seafloor acts like a magnetic tape recorder of the earth's field.

How does one calibrate this "tape recording"? The first test was Leg 3 of the Deep Sea Drilling Project. The researchers drilled holes on both sides of the mid-Atlantic ridge in the South Atlantic and found that the cores got progressively older with distance from the ridge. The basalts of the deep-sea floor were so altered that they could not be radiometrically dated, but the sediments overlying the basalts had microfossils that yielded a stratigraphic age. To get reliable numerical estimates, the youngest part of the paleomagnetic time scale was constructed by dating many fresh lava flows all over the world and then measuring their polarities. As described in Chapter 11, these measurements eventually produced a magnetic polarity time scale for the last 5 million years. Basalts on Iceland can be dated as far back as 13 Ma (Harrison et al., 1979), providing both polarity and direct basalt dates, but in the older dates, the radiometric error has already exceeded the precision achievable by other methods. The time scale earlier than 13 Ma must be constructed by extrapolation of the seafloor spreading record itself.

The first historic attempt at this extrapolation was made by Heirtzler et al. (1968), who compared the spreading profiles from the South Atlantic, South Pacific, North Pacific, and Indian Oceans and found that the South Atlantic had the least variation in spreading rate and that its spreading history extended back to about 80 Ma. Then they did something daring. Using a spreading rate calculated from the dated time scale from 0 to 4 Ma, they extrapolated back to 80 Ma and produced the *entire* time scale. We now know that their extrapolated time scale is within 10 percent of the currently accepted version, a remarkable achievement considering the lack of control points. Their assumption of the constancy of the rate of seafloor spreading must have been good for them to come so close on the first attempt.

The process has been refined many times since the Heirtzler et al. (1968) time scale. Alvarez et al. (1977) analyzed a sequence of Cretaceous pelagic limestones near Gubbio, Italy, that produced a continuous fossil and paleomagnetic record for the late Cretaceous and Paleocene. Using this tie point and a date on the late Cretaceous, LaBrecque et al. (1977) produced a new version of the time scale that was tied at the late Cretaceous and at the Quaternary. All of the time scale between these tie points had to be interpolated. Ness et al. (1980) revised the time scale yet again because new K/Ar decay constants (Steiger and Jäger, 1977) had just been published (see Chapter 12). Unfortunately, Ness et al. (1980) simply revised the ages of the interpolated boundaries rather than going back to the original dates to recalculate them to see how they changed the time scale.

Lowrie and Alvarez (1981) once again modified the time scale according to the Gubbio sequence (Lowrie et al., 1982), which now had almost continuous magnetics and marine microfossils for the late Cretaceous through Miocene. They took the paleomagnetic signature of eight stratigraphic boundaries that were based on microfossil evidence to add more calibration points between the original two points of LaBrecque et al. (1977). Unfortunately, they assumed that the ages of these biostratigraphic boundaries were well constrained, and so they fit the spreading profile between them. This forced fit caused the spreading profile to change by rapid jerks between each pair or data points, which goes against everything that geophysicists know about seafloor spreading. Although the spreading rate varies from ocean to ocean, and does change slowly over time, there is no geophysical mechanism that could change it by rapid spurts and stops every few million years. These rapid fluctuations were an artifact of the uncertainties of the numerical ages of the eight paleontological datum levels, which were less reliably known than the rate of seafloor spreading.

The Gubbio section of Lowrie et al. (1982) and Leg 73 of the Deep Sea Drilling Project (Poore et al, 1982) both provided continuous pelagic records with good microfossils and magnetics. It was clear from these records that the microfossils used to recognize the Eocene/Oligocene boundary occurred between normal Chrons C13 and C15 of the polarity time scale, in the upper third of the reversal between them, known as C13R. At that time, however, neither of these sections could be radiometrically dated. What was needed were sections with good high-temperature dates and magnetics that could be tied directly to the polarity time scale. This was first provided by Prothero et al. (1982, 1983; Prothero, 1985a) in a section at Flagstaff Rim, near Casper, in central Wyoming (Fig. 13.13). This section spanned over 200 m (700 feet) of volcaniclastic sediments which were full of Chadronian (then thought to be early Oligocene) mammals. More importantly, there were already K/Ar dates in four places in the section. An unusually long episode of reversed polarity was found between dates of 32.5 (Ash J) and 36.1 (Ash B). The only reversal on the Oligocene part of the polarity time scale that approaches this almost 3-million-year length was the reversal between Chrons C12 and C13, which is over 2 million years long (Fig. 13.13). Other dates (Ashes F and G) established intermediate points in this correlation. The mammal record showed no evidence of a major unconformity that might reflect missing section and therefore missing polarity intervals. On this basis, direct radiometric calibration points (Prothero et al., 1982, 1983; Prothero, 1985a) were provided between the calibration points from the Neogene and late Cretaceous. Other calibration points (Prothero, 1985a; Prothero and Armentrout, 1985; Flynn, 1986) came from terrestrial and shallow marine sequences in North America with good magnetics and direct high-temperature radiometric dates.

With this new data, another version of the time scale (Berggren, Kent and Flynn, 1985) was established using these direct calibrations of the magnetic polarity record. Avoiding the sudden, unwarranted jerks in spreading rate of Lowrie and Alvarez (1981), Berggren, Kent, and Flynn fit three line segments through the calibration points which were based on high-temperature (not glauconite) dates found in sequences with clearly identifiable polarity records (Fig. 13.14). The two inflection points of these three line segments correspond to periods of well-established changes in spreading rates, based on comparison with other spreading profiles. The Berggren, Kent and Flynn time scale does not assume constant spreading for the whole time interval, but only between times where it has been independently established that there were changes in spreading rates. This is preferable to the Lowrie and Alvarez (1981) approach, which assumes that the biostratigraphic levels were well dated (the biostratigraphers knew they were not) and then produced geophysically impossible fluctuations in spreading rate. Berggren, Kent, and Flynn, on the other hand, assumed that spreading rate is relatively constant (except where demonstrated otherwise) compared to the large known errors of biostratigraphic age assignments and radiometric dates.

Some scientists, however, have constructed their own time scales, which give radically different versions of the Eocene/Oligocene boundary and other important events. The most persistent of these is Odin (1978, 1982), who uses glauconites from the European marine

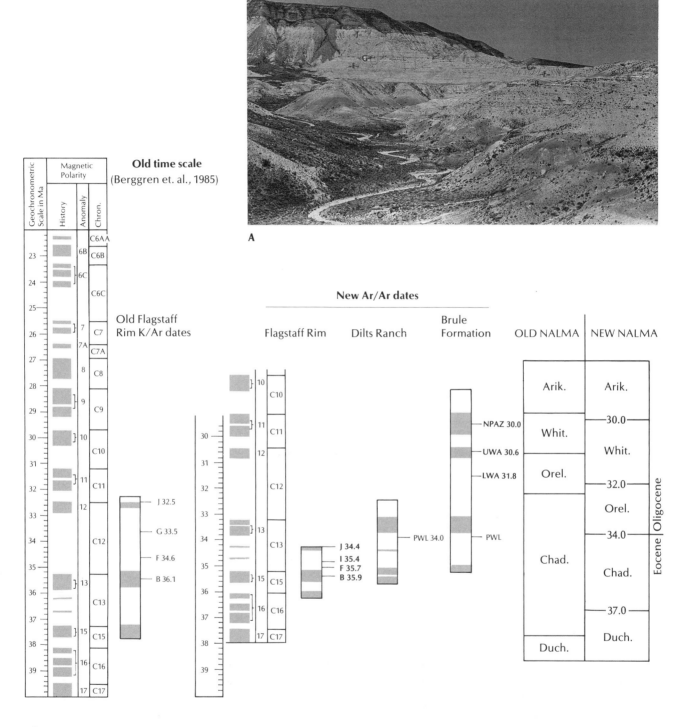

Figure 13.13

(A) The classic section of the White River Formation at Flagstaff Rim, near Alcova, Wyoming. The section contains mammal fossils that span most of the Chadronian land mammal "age" (Emry, 1973). It also contains ash layers (labeled B, F, G, and J) that were K/Ar dated by Evernden et al. (1964), and recently redated by Ar/Ar methods. (B) Old and new interpretations of the time scale, based on the change in dating methods. On the left is the time scale of Berggren, Kent, and Flynn (1985), which was partially based on the old Flagstaff Rim K/Ar dates. In the center is the new adjustment to the time scale suggested by the new Flagstaff Rim Ar/Ar dates, plus Ar/Ar dates from previously undated ashes at Dilts Ranch and elsewhere in the Brule Formation. At the right are the revisions in the North American land mammal "ages" (NALMA) suggested by the new dates. Photo courtesy R. J. Emry.

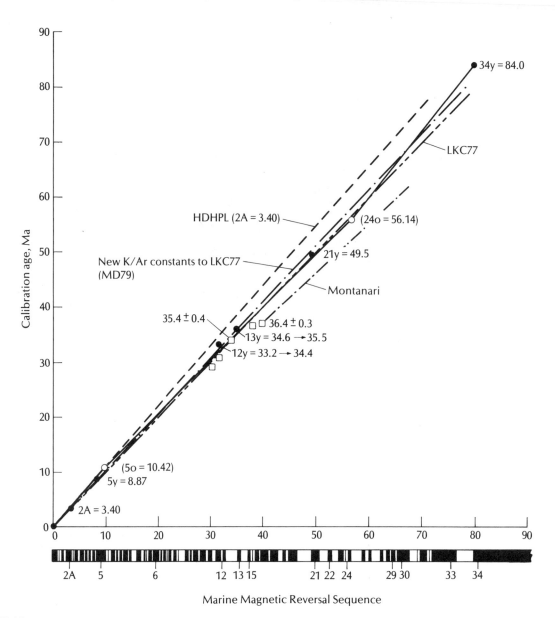

Figure 13.14

Calibration of the magnetic polarity time scale (Berggren, Kent, and Flynn, 1985). The solid line is fit through a set of control points (solid dots) representing high-temperature K/Ar dates and leaving out glauconite dates. On the assumption that seafloor spreading is fairly constant, the slope is straight except in the early Paleogene, for which other evidence suggests a slight change in the seafloor spreading rate. The dotted lines represent earlier versions of the time scale (HDHPL, Heirtzler et al., 1968; LKC, LaBrecque et al., 1977; MD, Mankinen and Dalrymple, 1979), which are quite close to the Berggren, Kent and Flynn version in most areas. □ represents new Ar/Ar dates (see Fig. 13.13), and the new Ar/Ar adjustments to the old 12y and 13y calibration points are also shown. From Berggren et al. (1985).

sequences. Odin has produced a time scale which consistently gives dates much younger that those from high-temperature methods (Fig. 13.15, open circles). Curry and Odin (1982) and Odin and Curry (1985) placed the Eocene/Oligocene boundary at 32–34 Ma or younger, and their Paleocene/Eocene boundary differs from the estimate of Berggren, Kent and Flynn by almost 4 million years! Such a huge difference cannot be dismissed due to the normal error in the radiometric dates, which is typically less than a million years for the early Tertiary.

For anyone trying to use a time scale and make calculations based on these numbers, this discrepancy is so big that it cannot be ignored. Each reader must look over the extensive documentation of each time scale, and judge the assumptions that were made and the quality of data that was used to construct it. Aubry et al. (1988) did so, and found some severe problems with Odin's glauconite dates. Many of Odin and Curry's (1985) dates were mislocated stratigraphically (Aubry et al., 1988, Fig. 4). When they are replotted, they are more consistent. In other cases, there were a wide range of glauconite dates from the same bed, and Odin and Curry (1985) arbitrarily picked certain ones and ignored others that tended to give older ages. In other cases, they selectively ignored dates which did not agree with

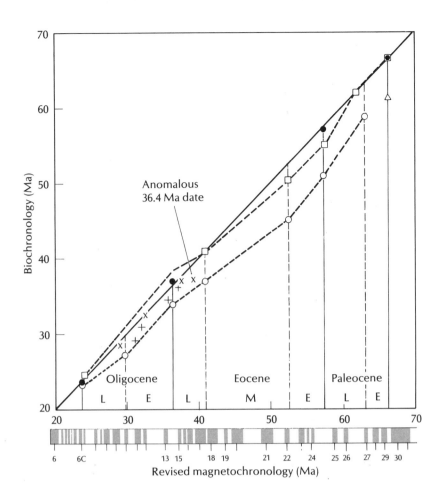

Figure 13.15

Comparison of the Berggren, Kent, and Flynn (1985) version of the time scale (solid lines and circles) with other versions of the time scale. Open squares, Ness et al. (1980); open circles, Odin and Curry (1981), who used glauconite dates almost exclusively; +, new Ar/Ar dates (see Fig. 13.13); ×, new dates from Montanari et al. (1988). Note that the oldest late Eocene date of Montanari et al. (1988) falls furthest off the line suggested by the other dates. From Berggren et al. (1985).

their preconceptions, even though these came from areas with well-established stratigraphy. Obradovich (1988) showed that, although some glauconites gave ages which appear to be reliable, others did not, and there was no way of knowing in advance which were more reliable.

A more fundamental problem is the nature of glauconite dates themselves. As discussed above, there are many reasons why their reliability is generally inferior to that of high-temperature dates. Aubry et al. (1988, Fig. 6) replotted the dates from Odin and Curry (1985), and found that while all the high-temperature dates gave a consistent line of correlation, some of the glauconite dates agreed with it and most were too young. There seems to be systematic bias toward ages from glauconites that are too young, so that when glauconite dates are used in the time scale (see Figure 13.15), the time scale is also too young. As Obradovich (1988) pointed out, what is being measured is actually the date when the glauconite crystals are "closed" and no longer exchange materials with the surrounding sediment. Even Odin (1982, p. 728) admits that this closure may occur millions of years after the glauconite grain was deposited, so that most glauconite ages are minimum estimates. Odin and Curry (1985) dispute the relative constancy of spreading rate where their dates disagree, but given the poor track record of glauconites, and the fact that many were mislocated stratigraphically, or located in shallow marine European sequences of question-

able stratigraphic level, the burden of proof seems to be on them.

Nevertheless, a number of divergent estimates have appeared for the age of the Eocene/Oligocene boundary. Harris and Zullo (1980) and Harris et al. (1984) dated the Eocene/Oligocene boundary at 34 Ma, based on an Rb/Sr date from glauconites in the Castle Hayne Limestone in North Carolina. However, Berggren and Aubry (1983) and Berggren et al. (1985) have shown that this date was misplaced stratigraphically; it was not even in the right microfossil zone, but significantly older. Recently, Harris and Fullagar (1989) have dated a bentonite, rather than unreliable glauconites, and their dates (both K/Ar and Rb/Sr) suggest that the Eocene/Oligocene boundary is younger than 36 Ma.

Glass and Crosbie (1982) estimated the age of the Eocene/Oligocene boundary at 32.3 Ma based on dates on K/Ar and fission-track dates of microtektite layers from the Caribbean. However, Berggren et al. (1985) have shown that there are problems with the biostratigraphic age assignment of this date. More recently, Glass et al. (1986) dated microtektites with laser-fusion argon/argon dating, and got an age for the late Eocene of 35.4 ± 0.6 Ma. By extrapolation, they placed the Eocene/Oligocene boundary at 34.4 ± 0.6 Ma. However, it is not certain how early or late in the Priabonian (late Eocene) their microtektites are located, so this extrapolation does not place very precise limits on the Eocene/Oligocene boundary. It could be only slightly younger, or much younger, than their 35.4 Ma date.

The Italian Apennine sections analyzed by Lowrie et al. (1982) have recently provided high-temperature dates that are closely tied to marine biostratigraphy and magnetostratigraphy. In several places in the classic Gubbio sections, there are ash partings that were dated by both K/Ar and Rb/Sr methods by Montanari et al. (1985). Their first publications gave an estimate of 35.7 ± 0.4 for the Eocene/Oligocene boundary. Although this is a bit younger than the 36.5 Ma estimate of Berggren et al. (1985), most of their dates fit quite well on the Berggren time scale line (see Fig. 13.15). However, additional dates from Gubbio and another section near Massignano have caused Montanari et al. (1988) to reject one contaminated date of 35.4 ± 0.4 for the base of Chron C12R (early Oligocene), and accept dates of 36.0 ± 0.4 and 36.4 ± 0.3 Ma for the late Eocene. By refitting their line to these dates, they estimated the age of the Eocene/Oligocene boundary at 33.7 ± 0.5 Ma.

Although most of these dates appear to be reliable, it is not clear that the oldest date of 36.4 ± 0.3 Ma is really from the early Priabonian (early late Eocene) and magnetic Chron C17N, as Montanari et al. (1988) argued. The magnetic polarity of this part of the section is difficult to interpret, and there is not much biostratigraphic change among the Priabonian microfossils to determine just how early in the Priabonian this date lies. If this one anomalous date is taken at face value, then the entire Priabonian is less than a million years in duration, when most other indicators suggest that it spans about 5–6 million years (Berggren, Kent, and Flynn, 1985). In addition, this date falls far off the line (see Fig. 13.15) established by dates from the Cretaceous through Eocene. If one accepts the slope of the line determined by this anomalous date, then the line will not pass anywhere near such well-established points as the 66 Ma date for the Cretaceous/Tertiary boundary. Thus, the new Montanari et al. (1988) calibration suggests that the Eocene/Oligocene boundary is younger than 36 Ma, but cannot be relied upon for the late Eocene.

A breakthrough in this controversy occurred just as this book was going to press. The volcanic ashes from Flagstaff Rim, Wyoming (see Fig. 13.13), which had been originally dated by K/Ar, have just been redated by laser-fusion Ar/Ar methods. Carl Swisher of Berkeley Geochronology Center found that many of the old K/Ar numbers were unreliable, probably due to contamination of the samples when they were originally run in 1964. By isolated single, unweathered crystals of biotite and sanidine, and running dozens of them with laser-fusion methods, Swisher produced dates which differed radically from the old estimates (see Fig. 13.13). For example, the old Flagstaff Rim dates ranging from 32.5 (Ash J) to 36.1 (Ash B) has been replaced by dates of 34.4 (J) to 35.8 (Ash B). By shortening the span of the Flagstaff Rim section from over 4 million years to less than 2

million years, it is clear that the reversed interval between Ashes F and J cannot be the long Chron C12R reversed interval, as Prothero et al. (1982, 1983; Prothero, 1985a) argued. Instead, it appears that Flagstaff Rim spans Chron C13R, C15N, and possibly down to C16N (Fig. 13.13).

Where did the long C12R reversal go? Fortunately, the new Ar/Ar technique has made it possible to date many other ashes that were previously undatable by K/Ar methods. The Dilts Ranch section, near Douglas, Wyoming (see Fig. 13.13), produced a number of Chadronian dates that agreed with the new Flagstaff Rim dates. A date of 34.0 Ma on the Persistent White Layer (PWL) occurs just below the Chadronian/Orellan boundary. In addition, new dates on the Lower and Upper Whitney Ashes produced the first reliable dates from the Orellan or Whitneyan. From these new dates, it is apparent that Chron C12R must be the long reversal that spans the early Orellan to the middle Whitneyan. It is bracketed by dates of 34.0 (the PWL) and 30.8 Ma (the Upper Whitney Ash), giving a duration of about 2.5–3.0 million years, in good agreement with the known duration of Chron C12R.

The implications of these new dates for the time scale are staggering. The old Berggren, Kent and Flynn (1985) time scale is shown at the left in Fig. 13.13, and the new adjustments are shown in the middle. The new Ar/Ar dates suggest a shift of almost 2 million years for Chrons C12 through C16! More importantly, this places the Eocene/Oligocene boundary (upper third of Chron C13R) around 34.0–34.5 Ma, in good agreement with many of the new dates discussed above. The new dates are also plotted in Fig. 13.15 to show how much younger they are than the Berggren et al. (1985) fit (but still older than the glauconite dates of Odin and Curry). Clearly, the new time scale will have to be fit through these points, as well as those of Montanari et al. (1988) and the others mentioned above, but still fit through known control points in the Miocene, and in the Cretaceous through middle Eocene. Such a revision of the time scale is underway as this book goes to press, but every time scale now available (including the one reprinted in Appendix B) is now out of date.

The dispute over the age of the Eocene/Oligocene boundary does not appear to be over yet, but we are clearly seeing a convergence around dates of 34.0 to 34.5 Ma. Yet this is no mere tempest in a teapot, interesting only to specialists. The Eocene/Oligocene transition was one of the most dramatic climatic and faunal extinction crises in the entire Tertiary (Cavelier et al., 1981; Prothero, 1985b), so dating it is very important. For example, Prothero (1985b) labeled the Chadronian/Orellan extinctions the "mid-Oligocene event," and thought that the Terminal Eocene Event occurred at the beginning of the Chadronian. The new calibrations (see Fig. 13.13) make it clear that the Chadronian/Orellan transition is the Terminal Eocene Event, and the mid-Oligocene event is much later. A time scale is not an easy thing to construct, and there are a whole host of assumptions and decisions made in producing the clean, simple results that most people use. A good geologist or stratigrapher, like an intelligent shopper, must do the homework. The old motto, "Caveat emptor" (Let the buyer beware) applies to all scientific work, including time scales.

FOR FURTHER READING

Berggren, W. A. 1971. "Tertiary Boundaries and Correlations." In B. M. Funnell and W. R. Riedel (eds.), *The Micropaleontology of Oceans.* Cambridge Univ. Press, Cambridge. The classic work that summarized and synthesized over 150 years of stratigraphic dispute in the Tertiary.

Cohee, G. V., M. F. Glaessner, and H. D. Hedberg (eds.), 1978. *Contributions to the Geologic Time Scale.* Amer. Assoc. Petrol. Geol. Studies in Geology, 6. A series of classic papers on calibration of the time scale from the mid-1970s. Some papers have become dated, but most have held up well.

Harland, W. B., A. V. Cox, P. G. Llewellyn, C. A. G. Pickton, A. G. Smith, and R. Walters, 1982. *A Geologic Time Scale.* Cambridge Univ. Press, Cambridge. Another complete time scale, with discussion of much of the theory behind it. Excellent listings of the stratotypes areas and the geochronology behind them.

Johanson, D., and M. Edey, 1981. *Lucy, The Beginnings of Humankind.* Simon and Schuster, New York. The lively best-selling account of the discovery of the oldest hominids, which vividly portrays much of the scientific and sociological background to the disputes about the stratigraphic age of our ancestors.

Lewin, R., 1987. *Bones of Contention: Controversies in the Search for Human Origins.* Simon & Schuster, New York, Lively account of the controversies and prejudices in the study of human evolution, including a blow-by-blow description of the KBS Tuff controversy.

Odin, G. S. (ed.), 1982. *Numerical Dating in Stratigraphy,* 2 vols. John Wiley & Sons, New York. A complete listing of all of the radiometric dates used by Odin and others up to 1982, with some important theoretical papers. Unfortunately, it is marred by excessive reliance on questionable glauconite dates.

Pratt, L. M., E. G. Kauffman, and F. B. Zelt (eds.), 1985. *Fine-grained Deposits and Biofacies of the Cretaceous Western Interior Seaway: Evidence of Cyclic Sedimentary Processes.* Soc. Econ. Paleo. Min. Field Trip Guidebook 4. Excellent collection of papers documenting the incredible stratigraphic resolution possible in the Western Interior Cretaceous.

Snelling, N. J. 1985. *The Chronology of the Geological Record.* Geol. Soc. London, Memoir 10. The latest work on the time scale. Note especially the paper by Berggren, Kent, and Flynn.

Rifting between Africa and the Arabian Peninsula, forming the Red Sea (foreground) and the Sinai Peninsula. From Stanley (1989).

14

Tectonics and Sedimentation

THE "BIG PICTURE"

We have discussed stratigraphy on the scales of texture, fabric, structures, lithologic units, facies and environments of deposition, and basinal sequences. A further hierarchical level that encompasses all the aspects of basin analysis is what might be called the "big picture"— the ultimate causes of accumulations of sedimentary rocks and stratigraphic sequences throughout geologic history. Sedimentary rocks are not randomly scattered over the surface of the earth but are concentrated in deep basins and shallow shelf sequences. For almost two centuries we have known that particular sedimentary rocks are associated with particular types of tectonic situations. But not until the discovery of plate tectonics was there an adequate explanation of *why* these sedimentary rocks and sequences formed during certain times and places. In the 1970s, enormous scientific breakthroughs related sedimentary sequences to specific motions of the earth's crustal plates. Now it is possible to explain sedimentary sequences with plate tectonic models and even to infer the motions of ancient plates from the sedimentary record they leave behind.

CRATONIC SEDIMENTATION

As early as 1859 the pioneering American geologist James Hall recognized that there was a fundamental difference between the sedimentation in the center of a continent and that on its margins. The center of a continent appears to be relatively stable, with thin sedimentary sequences punctuated by unconformities. As one approaches the continental margins, however, one sees an increasing association between deep sedimentary troughs that have accumulated thousands of meters of sediment and the mountain belts that exist on many continental margins. The stable center of the continent was named the *craton* by Stille in 1936. *Kraton* is the Greek word for *shield*, and the cores of the continents are indeed broadly convex upward like the shields of ancient warriors. The marginal troughs that parallel the mountain belts were called *geosynclines* by James Dwight Dana in 1873. Geosynclines are discussed in the next section.

The craton is characterized by thin sequences of sedimentary rocks that unconformably overlie the Precambrian basement. Most continents have a Precambrian

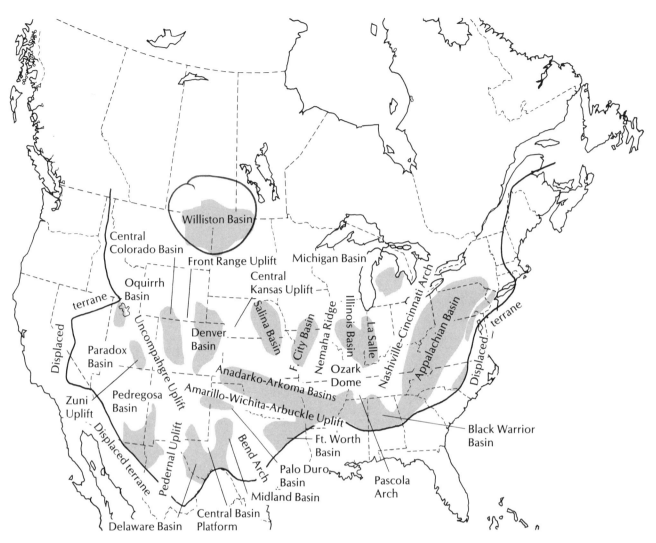

Figure 14.1

Location of major cratonic basins and arches in North America during the late Paleozoic. From Sloss (1982).

core or shield area to which mountain belts have been added by later accretion along the margins. Because cratons are usually positive features (they are frequently above base level), much of their post-Precambrian history is represented by erosion surfaces and unconformities. Limited vertical movement in the craton has formed shallow *basins and arches*, which have a few tens of meters of relief and which have sunk or risen very slowly and steadily over long spans of time (Fig. 14.1). The cratonic sedimentary cover on most continents is mostly Paleozoic and Mesozoic, averages about a kilometer in thickness, and is composed almost entirely of shallow marine sands, limestones, and shales, with occasional fluvial and deltaic sequences. The sandstones are usually clean and very mature, indicating that the grains have gone through multiple cycles of reworking before final deposition.

Cratonic basins are typically shallow and bowl-shaped, with the units thickening gradually toward the center (Fig. 14.2). The isopachs of many basins show a bull's-eye pattern, indicating a very regular thickening toward the center of subsidence. Judging from the sedimentary records of such basins, they remained structural and topographic lows for long periods of time. Yet their fill is not continuous. During some periods these basins undergo gentle downwarping, and during others they are relatively stable. As a consequence, their sedimentary record consists of unconformity-bounded packages that represent intervals of major transgression across the entire craton. These packages were called *sequences* by Sloss, as discussed in Chapter 8. Even during the peak of a transgression, however, the cratonic basin fill is thin and discontinuous compared with the sediment found in the marginal basins.

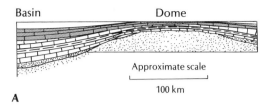

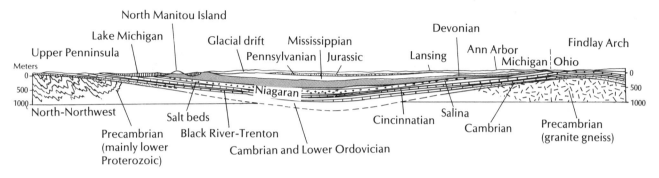

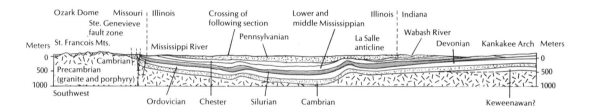

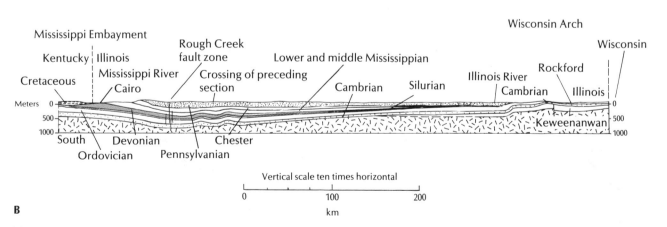

Figure 14.2

(A) Idealized section of a dome and basin in the craton, showing thinning of all units away from the basin and toward the dome. This results partly from deposition of a greater thickness of each unit in the basin and partly from truncation and overlap along unconformities. (B) Cross sections of two representative cratonic basins, showing the characteristic geometry. The top diagram shows the Michigan Basin, and the lower two diagrams show east-west (middle) and north-south (bottom) cross sections of the Illinois Basin. From King (1977).

The cause of this gentle cratonic basinal downwarping has been argued for a long time. Lately, it has become apparent that cratonic basins are not in fact very different in origin from those on margins. Many apparently have resulted from deep intracontinental rifts that were interrupted and became deeply buried. Even though these ancient rift valleys did not split the craton, they remain zones of crustal thinning and weakness, and therefore are subject to further subsidence. An example is the bull's-eye pattern seen in the Michigan Basin, which is centered in lower Michigan. Older maps of this basin show smoothly concentric isopachs, indicating an almost perfect bowl-shaped depression. Recent gravity data, however, show a pronounced linear trend in the depression suggesting a deeply buried northwest-southeast trending fault graben underneath the Paleozoic cover (Byers, 1982). It appears that the Michigan Basin is not a simple bowl but a narrow rift graben that is part of an ancient Proterozoic rift system underlying Lake Superior, Minnesota, and much of the midcontinent. Discoveries such as this promise to radically change many of our traditional notions about cratonic sedimentation and tectonics.

GEOSYNCLINES

In contrast to the gentle downwarping of the craton, the margins of a continent show a pattern of deeply subsiding troughs, which may have accumulated thousands of meters of sediment. These sedimentary piles have been frequently folded, faulted, metamorphosed, and then, at some later time, thrust upward as mountain belts. Hall (1859) first recognized the association between the deep subsidence and the uplift and mountain

building, and thought the subsidence was caused by the weight of the accumulated sediments. Dana (1873), who coined the term *geosyncline*, suggested that the subsidence and subsequent deformation were not due to sediment loading but to lateral compression. The American concept of geosynclines was based on the idea that they formed on the margins of the continental crust and were asymmetrical and wedge-shaped.

By 1900, European geologists had applied Dana's concept of geosyncline to the thick, compressed sedimentary sequences that were uplifted to become the Alps. Alpine geosynclines, however, contain deep marine sediments in the middle and are symmetrical, tapering away from the central trough in both directions. As geosynclinal theory developed, European and American geologists argued about the causes and characteristics of geosynclines. Unfortunately, much of the argument was due to the fact that they were using the same name to talk about two different beasts. In 1936, Stille subdivided the classic geosyncline (Fig. 14.3): (a) The sequence of shallow marine sandstones and limestones that tapered gradually into the craton was named the *miogeosyncline* ("near geosyncline"); (b) away from the center of the craton, the miogeosyncline ended at an abrupt break and dropped off into thick, deep marine shales, sandstones, volcanic rocks, and cherts, which Stille called the *eugeosyncline* ("true geosyncline"). Many eugeosynclines included not only deep marine sediments, but also submarine volcanics and volcaniclastic debris. Another anomalous feature was that much of their sediment had been shed into the basin from some no longer existing uplift further out in the ocean, *away* from the craton, and this he called "tectonic land." The eugeosyncline usually had been subjected to intense compression and tectonic deformation, and in some continental margins it had been thrust over the miogeosyncline.

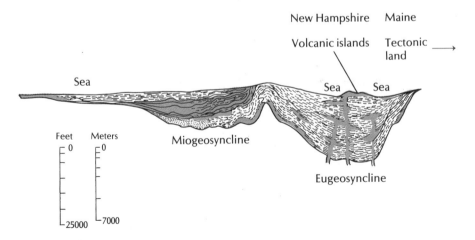

Figure 14.3

Classic profile of a geosyncline, based on the Ordovician of the Appalachians.
Adapted from Kay (1951).

Table 14.1

Kay's Classification of Tectonic Elements

ORTHOGEOSYNCLINES   Linear, deeply subsiding, shallow to deep water; located between cratons
    Eugeosynclines: actively subsiding, with associated volcanics
    Miogeosynclines: less active; no volcanics
PARAGEOSYNCLINES   Commonly ovate, less actively subsiding, shorter-lived than orthogeosynclines; located within craton or adjacent to craton
    Exogeosynclines: tonguelike extensions from orthogeosyncline, detritus mainly from orthogeosyncline
        Example: Catskill-Chemung (Devonian) "delta"
    Autogeosynclines: isolated depositional areas within craton, detritus from distant cratonic sources
        Example: Michigan Basin
    Zeugogeosynclines: subsiding areas adjacent to complementary uplifts in craton, detritus mainly from uplifts
        Example: Denver Basin Area in Pennsylvanian Period
GEOSYNCLINES OF LATER CYCLES   Subsiding areas that developed later in the sites of older geosynclines
CRATON   The consolidated, relatively neutral area that comprises the main part of the continent or of oceanic basins
    Positives: complementary rising areas adjacent to intracratonic geosynclines

From Kay (1951).

Along with the concept of the geosyncline, a related concept of tectonic cycles evolved. The development of the deep eugeosynclinal trough usually preceded an episode of compression and mountain building. The term *flysch* was applied to the thick sequence of deep marine shales, graywackes, turbidites, and cherts that filled the eugeosyncline before and during the episode of mountain building. The term *flysch* is not purely descriptive but carries the genetic implication that the sequence was deposited at the same time as the mountain building (*synorogenic*). After the mountains were deformed and uplifted, they usually shed a sequence of fluvial and lacustrine sandstones, redbeds, and shales off their flanks. This sediment was called *molasse*, and the term almost always implies that it is a *postorogenic* deposit. (These terms have been used in many senses over the years, so their original meanings have been distorted and must be determined from context.)

Geosyncline theory reached its greatest development in 1951 with the publication of Marshall Kay's classic monograph *North American Geosynclines*. Kay systematized the many confusing terms and subdivisions of geosynclines that had been proposed over the years. His classification of geosynclines (Table 14.1) also included descriptions of block-faulted "parageosynclines," which did not form mountain chains. Parageosynclines

included "exogeosynclines" (found on the continental margin but receiving detritus from uplifted areas outside the craton), "autogeosynclines" (isolated cratonic depositional basins such as the Michigan Basin), and "zeugogeosynclines" (fault grabens yoked to adjacent uplifts on the craton). Other terms were also proposed, and many more arguments arose when geologists tried to fit basins other than North American basins into Kay's framework. The chief failure of geosyncline theory was that, though it described *how* basins subsided, it had no explanation for *why* they did so. The naming and describing of geosynclines became sterile exercise because it led to no understanding of fundamental causes.

THE PLATE TECTONIC REVOLUTION
The Basic Model

In the late 1950s and 1960s, geology underwent a scientific revolution as profound as those of Copernican astronomy, Einsteinian physics, or Darwinian biology. The discovery of seafloor spreading showed that the earth's crust is not the static foundation we had assumed it to be. Instead, it is composed of a group of lithospheric plates that move relative to one another over the surface of the globe. Plate tectonic theory developed rapidly after it was widely accepted in the late 1960s and early 1970s, but stratigraphers were relatively slow to realize its implications for the formation of sedimentary basins. A landmark paper by Dewey and Bird in 1970 first related the formation of mountain belts and geosynclinal basins to plate tectonic models, and by the late 1970s most mountain belts and sedimentary basins were being explained by plate tectonics.

Plate tectonic theory contained many revolutionary implications. Geosyncline theory became obsolete, and its terms survive today with only part of their original meanings. Even more startling are the plate tectonic implications for the tempo and scale of geologic events. All of the modern seafloor has been created since the Jurassic, and all older seafloor has been subducted and lost beneath the trenches. Ocean basins up to 500 kilometers wide can form and disappear in 50 to 100 million years. This cycle of opening and closing of an ocean basin is called a *Wilson cycle*, and there have apparently been many Wilson cycles in the history of every continent. This means that no sedimentary basin with a long history of deposition is likely to have remained static or in its original position, except on cratons.

On continental margins that are associated with subduction zones, it appears that much of the geology consists of terranes from elsewhere (*exotic*, or *allochthonous* terranes), which have been accreted by the collisions of plates. Seventy percent of the Western Cordillera of

North America, for example, is estimated to be made of blocks that were not part of North America in the Precambrian, but came from other parts of the Pacific basin during the Phanerozoic. Most of California, Oregon, Washington, and some of Idaho and Nevada are essentially foreign—a fact that has profound implications for geologic studies in this region. Every time you cross a major geologic boundary, such as a fault zone or volcanic arc, you may be stepping onto rocks that have no genetic relationship to the terrane you just left. This sense of lateral discontinuity radically changes the way stratigraphers view these regions (Byers, 1982).

Plate tectonics has left us with a fairly coherent picture of both *how* and *why* the major crustal plates move. Most of the major plates have a core of *continental crust*, which is sialic in composition, much less dense than the mantle, and about 30 to 50 kilometers thick. This fragment of continental crust is surrounded by a thin (5 kilometers) layer of basaltic *oceanic crust*, which is much denser than continental crust. Oceanic crust arises as magma along the midocean ridges as seafloor spreads when continents drift apart and then, because ocean crust is thin and dense, it slides under the edge of the buoyant continental crust and sinks down into the earth's mantle. When continental plates collide, however, neither can sink under the other, so they produce an uplifted mountain belt such as the modern Himalayas.

Three types of margins between lithospheric plates are possible:

1. If two plates are separating, they form a *divergent* (passive) margin. Divergent margins are characterized by extensional features, especially seafloor spreading, extensional grabens, and normal faulting.

2. If two plates are moving toward one another, they form a *convergent* (active) margin. Such margins are characterized by compressional tectonics, especially folding and thrusting. Except when the colliding plates are continental crustal plates, there is subduction of one plate (usually oceanic crust) beneath another. Thus a convergent margin includes a subduction zone and all its associated features.

3. If two plates are sliding past one another, neither separating nor converging, they form a *transform* margin, which is characterized by horizontal shear and strike-slip faulting. Along a transform margin, crustal material is transferred from its origin at the ridge to its destruction in the trench without significant net compression or extension.

Many plate boundaries are combinations of all three of these types of margins or have changed from one type to another during their geologic history, so these categories are necessarily oversimplified. However, they are useful models for understanding the nature of mountain building and sedimentary basin development.

Divergent Margins

The evolution of divergent margins might be subtitled "how to build an ocean." The *first stage* (Fig. 14.4A) occurs when continental crust begins to rift, possibly due to mantle plumes upwelling underneath. Upwarping, caused by hot spot or mantle upwelling, produces domal uplifts that shed coarse, immature alluvial and fluvial deposits onto their flanks. Mantle upwelling can also result in volcanic eruptions which contribute alkalic volcanic ash flows and falls and volcaniclastic sediments to the sedimentary piles on either side of the rift.

In the second stage, the crust breaks due to uplift and extension, and a crustal block drops down to form a fault graben called a *rift valley*. This stage is occurring today in the East African rift valley (Fig. 14.5). Continued spreading results in the development of a progressively wider rift graben, with numerous downdropped normal fault blocks along both flanks of the rift (see Fig. 14.4A). These fault graben valleys form basins for the sediment that erodes from the upthrown areas. Most of this sediment is coarse, immature alluvial debris and lesser fluvial deposits. In the center of the basin, small lakes may form lacustrine shales or limestones, and even evaporites. Frequently, the faults become vents for alkalic volcanics and produce ash falls and ash flows, as well as abundant volcaniclastic sediment. The famous hominid-bearing beds in East Africa (discussed in Chapter 13) are a good example of this stage of passive margin development. A classic ancient example of a rift graben sequence is the Triassic/Jurassic Newark Supergroup of the eastern margin of North America (Fig. 14.6). Fault graben valleys were formed by the Triassic opening of the Atlantic, and these basins were filled by a thick sequence of arkoses, lithic arenites, lacustrine shales, and volcanics such as the Palisades Sill and the Watchung lava flows.

Not all rift valleys become ocean basins. When hot spots begin to force a continent apart, they arise at isolated points and produce a triradiate set of rift valleys, which is one class of *triple junction* (Fig. 14.7A). Two of the three arms of the triple junction link with rifts of nearby triple junctions to form the eventual continental margin. The third arm opens for a while but stops spreading when the other two arms link. This failed third arm of a triple junction, called an *aulacogen*, eventually fills with sediments. Aulacogens form at steep angles to the eventual continental margin and are zones of subsidence. River systems become entrenched in the abandoned trough and eventually fill it with deltaic deposits. Nearly all of the larger rivers that drain into the modern Atlantic, including the Mississippi, the Amazon, the Congo, the Niger, and a number of lesser rivers, have developed in aulacogens formed during Triassic and Jurassic rifting. After the deltaic and coastal plain sequence buries the aulacogen, a deep linear basin remains

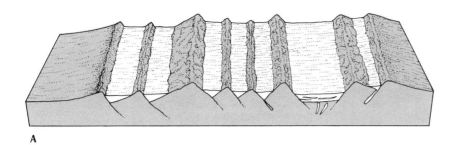

A

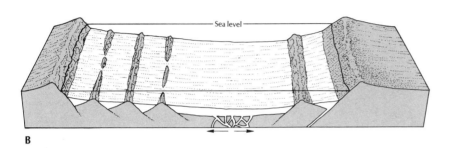

B

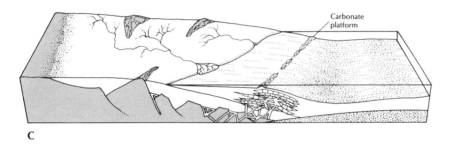

C

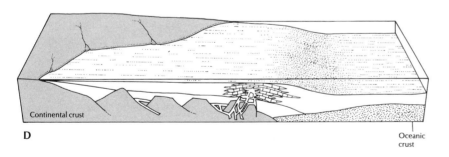

D

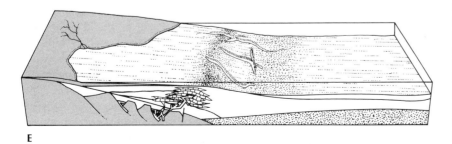

E

Figure 14.4

Schematic cross sections of a passive margin showing four phases of development. (A) Rift valley phase. (B) Proto-oceanic gulf phase. (C) Early post-rift phase. (D) Late post-rift phase, when clastic sediments overwhelm the carbonate platform and build up the shelf-slope-rise complex. (E) Late post-rift phase, when low sea level incises submarine canyons into the shelf edge. Modified from Schlee and Jansa (1981).

Figure 14.5

Aerial photograph of the East African rift valley in the Afar Triangle, Ethiopia. Note the massive, tilted, normal fault blocks. The depressions are coated with white evaporites from dry lakes. From Tazieff (1970).

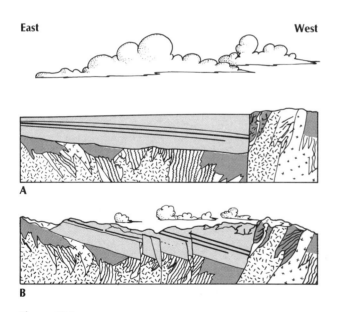

Figure 14.6

Diagrammatic cross sections of the large Triassic fault graben that formed during the early opening of the Atlantic. It is filled with the redbeds and lacustrine shales of the Newark Supergroup. (A) Early stage of filling of the basin. As the basin subsided, lavas welled up to form dikes and sills, and alluvial fans spread from the uplands to the east. (B) Eventual destruction of the basin by further normal faulting. From Stanley (1989).

that can be detected in the subsurface. For example, the Anadarko Basin and the Mississippi Embayment are believed to be ancient aulacogens now buried under the coastal plain cover (Fig. 14.7B). Many other deep, narrow continental rifts, such as the Precambrian Keeweenawan basins beneath Lake Superior and central Michigan, are also believed to be aulacogens.

The *third stage* might be called the *proto-oceanic gulf stage* (see Fig. 14.4B). The two successful arms of the triple junction continue to spread and eventually are invaded by marine waters. However, they have not yet spread far enough for oceanic crust to appear at the center. The fault grabens of the rift valley margin are buried under a sequence of shallow marine shales, limestones, and even pelagic oozes. Many of these basins happen to be located in tropical areas and so produce sabkha dolomites and evaporites. Because the rift is still narrow at this stage, the marine waters are probably restricted in their circulation, which may result in thick sequences of marine evaporites. The best modern analogue for this stage is the Red Sea (Fig. 14.8), which is 2000 kilometers long, 200 to 350 kilometers wide, and only 1 kilometer deep. Here, the faulted rift graben sequences are capped by alkalic volcanics, shallow marine shales and limestones, supratidal dolomites and sabkha deposits, and thick evaporite sequences. In the center of the Red Sea, marine oozes accumulate. The rifting of

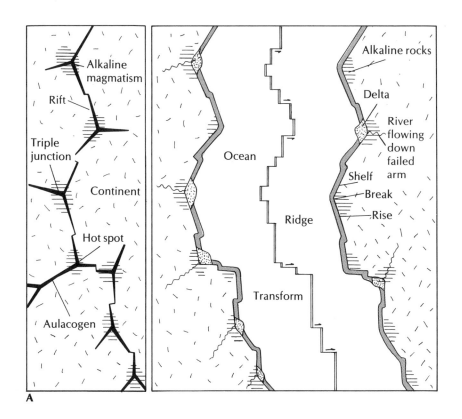

A

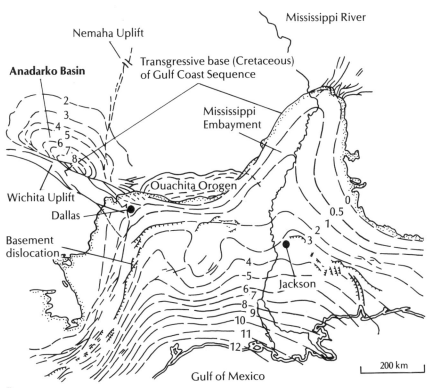

B

Figure 14.7

(A) Two stages in formation of aula-cogens. In the early stages, hot spots under the continental crust cause domes that are intruded by alkalic magmas. The domes eventually fracture into three rift valleys at approximately 120° angles. Eventually, two of the three rifts join up to form a rifting continental margin. The abandoned rift valley becomes the aulacogen. As the sides of the rift pull further apart, the aulacogens become deep linear basins that often dictate the positions of future drainage basins, long after they are filled and buried. From Dewey and Burke (1974). (B) Isopach map of the Gulf Coastal Region showing the southward-thickening wedge of Gulf Coast passive margin sediments with two aulacogens at right angles to the margin. The Anadarko Basin, over 8 kilometers deep, is one of the deepest Paleozoic basins in North America. The Reelfoot Aulacogen underlies the Mississippi Embayment and was responsible for the faults that caused the gigantic New Madrid, Missouri earthquakes of 1811–12. Contours in kilometers on base of Paleozoic sequence. From Burke and Dewey (1973).

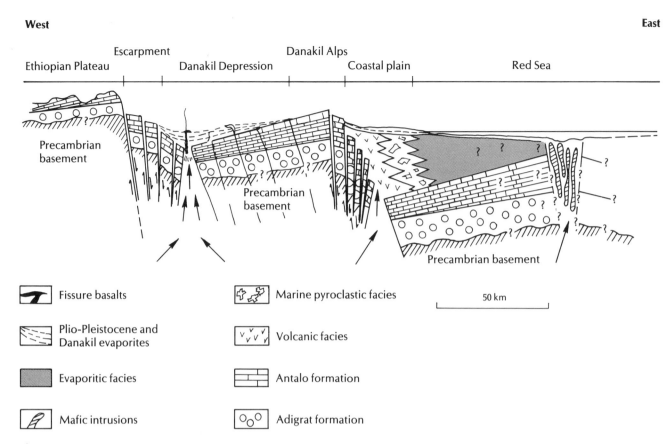

Figure 14.8

Cross section across the Red Sea and Danakil Depression, showing a classic example of a rift valley filling with a proto-oceanic gulf sequence. From Hutchinson and Engels (1970).

the North Atlantic reached this stage in the Jurassic, and the Gulf of Mexico became a restricted evaporitic basin that produced a thick unit known as the Louann Salt, which is responsible for most of the salt domes and much of the oil wealth of the Gulf Coast.

The *fourth stage* of development is *normal oceanic rifting* (see Figs.14.4C–E). Continental separation reaches the point at which a midocean ridge appears and begins producing oceanic crust at the spreading center (see Fig. 14.7A). As the spreading continues, the continental margins separate and sink down the ridge flanks. Pelagic oozes and shales accumulate over most of the seafloor, and the turbidite prism of the continental slope and rise forms near the continental margin. Above the shelf-slope break, the proto-oceanic gulf sediments become the basement for a thick sequence of shelf sediments and deltaic deposits, building up the continental terrace and eventually the continental embankment (see Figs. 14.4C–E). This sequence of clean marine sandstones and limestones progrades outward from the continent and is sometimes bounded by a reef at the shelf-slope break (Fig. 14.9A). The modern passive margin sequence of the North Atlantic shows many variations on this geometry (Figs. 14.9A,B), but the basic pattern is the same.

The total sequence is almost 10 kilometers thick with a basement of Triassic-Jurassic rift grabens, proto-oceanic Jurassic evaporites, and early Cretaceous shallow marine sediments. The rest of the shelf built outward during the Cretaceous and Tertiary, providing the bulk of the sequence that now underlies our present Atlantic continental shelf and the coastal plain. This is the climax of the evolution of passive margins, and this geometry continues to develop unless the plate motions change so that it becomes a convergent margin.

Convergent Margins

When two plates collide, several types of mountain belts can be produced. If both plates are continental crust, neither can be subducted, so they become uplifted and deformed and eventually suture along the line of collision, like the modern Himalayas. The suture belt itself is an area of erosion, but it sheds coarse clastic debris and fluvial deposits off its flanks into adjoining plains. The best known example of this is the molasse sequences of the Siwalik Hills of Pakistan and India, which are a product of Himalayan uplift since the Miocene.

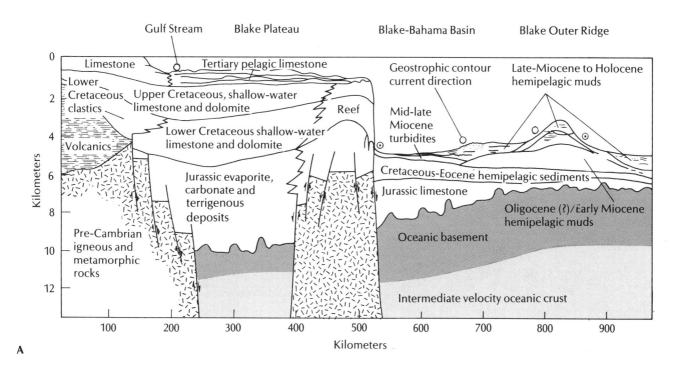

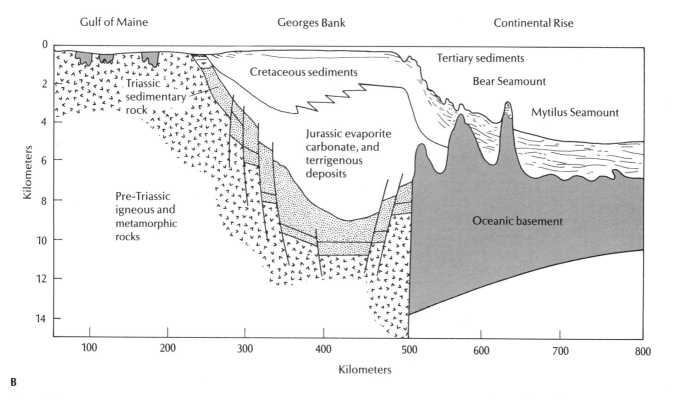

Figure 14.9

(A) Cross section of the Florida platform, Blake Plateau, and Blake-Bahama Basin, showing a classic example a thick passive margin continental shelf-slope-rise sequence. Notice the buried rift grabens, the Jurassic proto-oceanic gulf evaporites, and the Cretaceous shelf-edge reefs and shallow shelf limestones. (B) Cross section of the Atlantic margin through the Gulf of Maine. Similar Triassic grabens, Jurassic evaporites, and Cretaceous shelf-slope-rise sediments can be seen. From Sheridan (1974).

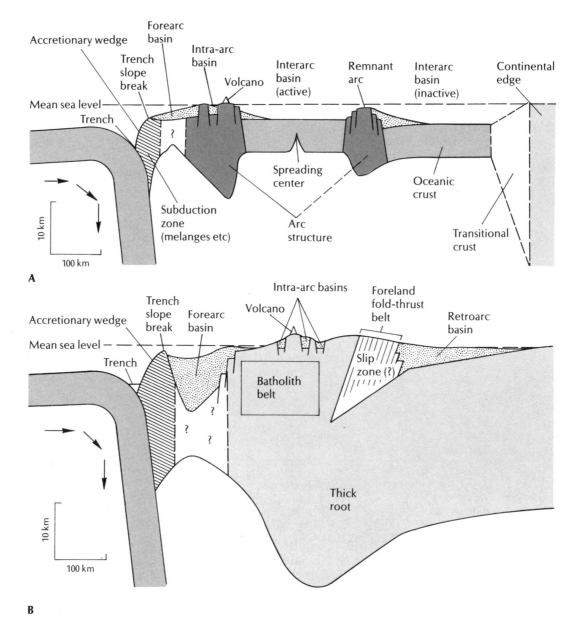

Figure 14.10

Schematic diagrams of sedimentary basins associated with arcs. (A) Intraoceanic (Japan-type) arc with backarc spreading. (B) Marginal (Andean-type) arc. From Dickinson and Yarborough (1976).

If two oceanic plates collide, either can be subducted underneath the other forming the classic *island arc complex* (Fig. 14.10) on the overriding plate. These *intraoceanic arcs* (Fig. 14.10A) are well known along the Pacific Rim today and include many familiar modern analogues, such as Japan, Indonesia, and the Philippines. Because of the predominance of oceanic crust in these islands, most of the volcanics are very mafic, predominantly basalts. When the downgoing slab is melted in the subduction zone, it forms a small island arc by volcanic eruption on the overriding plate. The products of submarine vulcanism (pillow lavas) and submarine sedimentation (turbidites, graywackes, shales, pelagic oozes) are dominant because this arc setting is mostly under water. Some intraoceanic arcs exhibit *backarc spreading*, producing an interarc basin behind the active arc (Fig. 14.10A). In this case, the spreading separates the active arc from the continent, and the spreading center tends to change position. If it changes position, it leaves a *remnant arc* bordered by both active and inactive interarc basins.

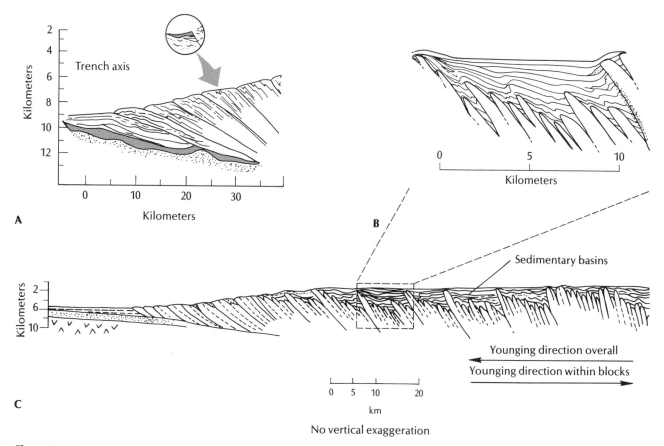

Figure 14.11

(A) Geometry of an accretionary wedge showing the thick slabs of crustal material that are sheared off and stacked up from the bottom. (B) Enlargement of the circled inset in (A), showing the shingling of the accreted slabs, which get younger toward the bottom of the pile. (C) Enlargement of the accretionary wedge basins showing the fill of pelagic sediment when they become tectonically deformed with further accretion. From Karig and Sharman (1975) and Moore and Karig (1976).

When oceanic crust collides with continental crust, the oceanic crust is always subducted beneath the less dense continent (Fig. 14.10B). This forms a type of island arc called a *marginal arc*. The term "island arc" is misleading in this context since there are no islands, but by convention an island arc is produced by the eruption of volcanics from melting in a subduction zone, whether or not a physiographic island results. The best modern example of a marginal arc is the Andes mountains, so the marginal arc complex is sometimes called an Andean-type arc (in contrast to the intraoceanic, or Japan-type, arc).

Because the rising magma in a marginal arc complex must penetrate a thick layer of continental crust, it becomes differentiated in the classic *calc-alkaline trend* of andesite-dacite-rhyolite. The batholith underneath the arc is also felsic and follows the calc-alkaline trend, with quartz monzonites and granites dominant. Most of the volcanics are very silicic, so they tend to erupt explosively and to form ignimbrites and ash flows. Because the region is typically epicontinental and subaerial, the

sediments filling the basins are mostly immature alluvial and fluvial sandstones and shales. The marginal arc has some unique features, such as the foreland fold-thrust belt and the retroarc basin which are discussed later.

The two types of arcs have certain features in common. Where two plates meet, a *trench* is formed. Most of the material lying on the subducted plate goes down the "hole" without leaving a trace. Sometimes, however, pelagic sediments are scraped off the subducted plate onto the overriding plate, and these sediments accumulate along the arc to form an *accretionary wedge* just below the trench slope break (see Fig. 14.10). The accretionary wedge is built up by continual underplating of material scraped off the downgoing slab, so new material is added to the *bottom* of the pile constantly. This process is the exact opposite of Steno's law of superposition (Fig. 14.11). The accreted slices are mostly oceanic sediment and pieces of oceanic crust (pillow lavas, sheeted dikes, and layered gabbros) known as *ophiolites*, which were formed by eruption at the midocean ridge.

A

The most characteristic rock type of the accretionary wedge is *melange* (French, "mixture"), a mass of chaotically mixed, brecciated blocks in a highly sheared matrix (Fig. 14.12A). This deformation and pervasive shearing and brecciation are due to the tremendous compressional and shear forces generated by the downgoing slab (Fig. 14.12B). Melange is so mixed that it shows no stratigraphic continuity or sequence, and blocks and boulders from everywhere are mixed together. Some are exotic blocks from terranes no longer present in the vicinity. Melanges can look something like submarine landslide deposits (or olistostromes, discussed in Chapter 5). Unlike olistostromes, however, melanges are not associated with undeformed deep marine turbidites and shales but with regional deformation. A melange terrane is regional in scale, with various degrees of deformation in various parts of the wedge. Melanges are usually much more sheared and fractured than olistostromes, and the included blocks are themselves usually fractured. In some cases, the sheared matrix of a melange has actually undergone ductile flow. A summary of the major differences between melanges and olistostromes is shown in Table 14.2 The distinction between them is important

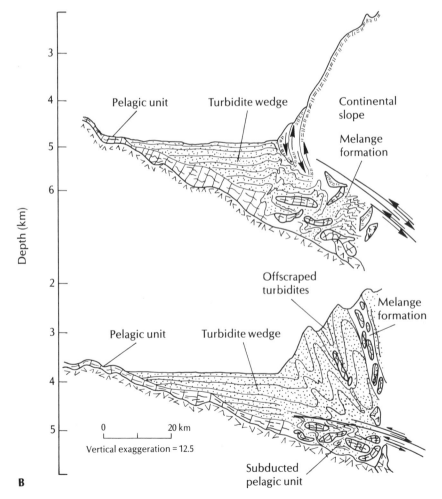

B

Figure 14.12

(A) Partially rounded exotic blocks in Franciscan melange, near San Simeon, Coast Ranges of California. Largest blocks are 1 meter across. Photo by the author. (B) Mode of formation of melange as scrapings from a subducting plate (terrigenous and pelagic sedimentary blocks and oceanic ophiolites) accumulate and are intensely folded, sheared, and metamorphosed. From Scholl and Marlow (1974).

Table 14.2

The Differences between Melange and Olistostrome

Criterion	Melange	Olistostrome
Clast character	Angular, fractured, sheared, may be deformed into boudins or smeared phacoids	Angular to rounded, may be fractured but not sheared
Clast source	Overlying and/ or underlying unit	Overlying unit only
Matrix	Sheared, plastic intrusion into cracks	Not necessarily sheared
Contacts with other units	Sheared	May have sedimentary contacts with "normal" slope or trench sediments

From Hsü (1974).

because their presence implies completely different tectonic and sedimentary settings.

The metamorphism of the accretionary prism is unique as well. Because the downgoing slab is cold relative to the mantle, metamorphism takes place at high pressure but at relatively low temperature. This is called *blueschist-grade metamorphism* because the dominant mineral is the blue amphibole *glaucophane*. The presence of glaucophane blueschist is diagnostic of a subduction zone.

On top of the slabs of ophiolite, melange, blueschists, and deformed oceanic shales and cherts of the accretionary wedge is a thin mantle of pelagic sedimentary cover. In some places, there even are basins on top of the accreted material (see Fig. 14.11). The sediment trapped in these basins has a good chance of becoming entrained in the deformation process and eventually becoming deformed and metamorphosed.

Above the trench-slope break is a basin in the arc-trench gap, the *forearc basin* (see Fig. 14.10). The forearc basin can be shallow marine or subaerial, but most of its sediments are volcanic debris shed from the arc, from the accretionary wedge, or transported by longitudinal marine currents. Many forearc basins are filled with coarse, shallow marine clastics and volcaniclastics, and with deltaic and fluvial deposits as well. In other instances, the forearc basin is very deep and accumulates turbidites. The classic ancient example is the Great Valley sequence, which underlies the Central Valley of California. It is a basin almost 70 kilometers wide and

12 kilometers deep, filled with a mixture of volcaniclastics, turbidites, and submarine-fan deposits. On the margin, shallower marine deposits occur. Because it is a deep, elongate basin in a compressional region, some synclinal folding has developed.

The *arc* itself is generally made of stratified volcanics and volcaniclastic debris. Many arcs have small fault graben basins at their crests, forming *intra-arc basins* (see Fig. 14.10). These are best known from the Altiplano of the Andes of Peru and Bolivia. One such basin is filled by Lake Titicaca, the world's highest lake. As one would predict, intra-arc basins fill with coarse, immature alluvial detritus and volcaniclastics, and occasionally with lacustrine clays.

The two backarc settings (see Fig. 14.10) produce different sedimentary products. *Interarc basins* are made of oceanic or thinned, "oceanized" continental crust and are completely marine. Their main sediments are turbidites and volcaniclastics from the arc, montmorillonitic clays from weathered volcanics, biogenic ooze, and small amounts of eolian dust from the continent. The continent sheds debris into the proximal part of the basin. *Retroarc basins* form in response to the compression of the intracontinental arc, which forms the foreland fold-thrust belt (see Fig. 14.10B). This thrusting and basement uplift typically brings up *metamorphic infrastructure*, which consists of high-grade metamorphic terranes from deep in the continental basement. The sediments that fill the retroarc basin are shed not only from the arc and foreland fold-thrust belt but also from epicontinental seaways as well. There can be a complete mixture of shallow marine, deltaic, and fluvial sands and shales, which are relatively mature except near a faulted upland source.

Transform Margins

In transform margins, the dominant motion is neither compressional nor extensional but strike-slip. Theoretically, there should be simple translation of crust with no opportunity for basins to form. In the real world, however, transform margins are seldom straight; instead they curve and are composed of numerous parallel shear zones. This is well demonstrated by the San Andreas fault zone in California. Because of irregularities in the sliding plates, there is local compression and extension, forming deep, narrow, fault-bounded troughs called *pull-apart basins*. The pull-apart basin has steep, fault-bounded walls and can drop precipitously to great depths. For example, the Mio-Pliocene Ridge Basin of California (discussed in Box 9.1) accumulated over 13,500 meters of breccias, alluvial fan deposits, lacustrine silts, marine turbidites, and evaporites. Other such basins are starved of clastics and accumulate deep pelagic silts and oozes, such as the Miocene Monterey Formation, a cherty shale in California. Such a thick, narrow, and immature sedimentary accumulation can happen only in pull-apart basins on transform margins.

Geosynclines and Plate Tectonics

The elements of the classical geosyncline can now be reinterpreted in the light of plate tectonics. The classic eugeosyncline-miogeosyncline association is the normal product of passive margin sinking. The miogeosynclinal sequence is produced by continental shelf sedimentation, and the eugeosyncline accumulates on the continental slope and rise (see Figs. 14.4, 14.13A).

To complicate matters, some elements of the eugeosyncline are due to active-margin tectonics. The abundant volcanics are due to the approach of an island arc during the closing of an ocean (Fig. 14.13B,C). The de-

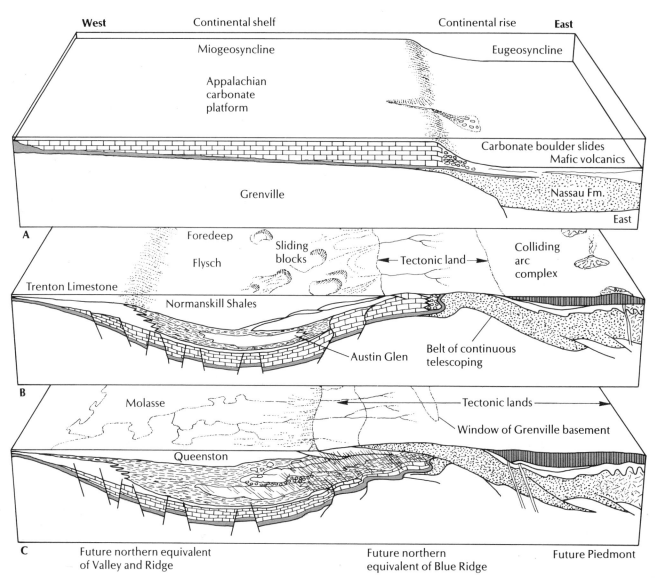

Figure 14.13

The Cambro-Ordovician "geosyncline" in western New England. (A) During the Cambrian, passive margin conditions persisted, with a thick shelf-platform sequence (miogeosyncline) and slope-rise package (eugeosyncline). (B) In the mid-Ordovician, the passive margin changed to an active margin when an island arc complex from the proto-Atlantic began to collide with the continent. The former miogeosyncline was downbuckled into a eugeosynclinal trough, trapping flysch and huge submarine slides shed from the upwarped "tectonic land" to the west. (C) in the late Ordovician, the collision stopped and the basins filled with nonmarine molasse from the remnant "tectonic land." After Dewey and Bird (1970).

bris shed from an upland on the seaward side of the trough was once thought to be derived from a mysterious "tectonic land," but this upland was the arc or continent that finally collided during last phases of closing. For example, much of the debris that filled the Appalachian eugeosyncline during the Ordovician Taconic Orogeny and the Devonian Acadian Orogeny was shed from Europe or Africa, now no longer in the vicinity. The characteristic compressive deformation and overthrusting of the eugeosyncline onto the miogeosyncline are also due to the collision of plates in the last phases of ocean closure. Sometimes this collisional suture zone includes elements of the accretionary prism, such as ophiolites and melanges.

Thus, the complex association of geologic features that puzzled the geosyncline theorists for so long is not due to a single cause but to a combination of features formed by active-margin tectonics superimposed on rocks formed by passive margin sedimentation.

Tectonics and Sandstone Petrology

We have seen how plate tectonic models can be used to predict what types of sediments will be found in various types of basin. The final step is to reverse the process and infer ancient plate configurations from the types of sediments. The overall geometry and lithology of a stratigraphic sequence is the most diagnostic feature. For example, a complex of coarse fanglomerates and deep marine turbidites, thousands of meters thick, could be formed only in a strike-slip pull-apart basin. Rock associations such as melanges and ophiolites are also diagnostic of their tectonic settings. In a few instances, a single mineral, such as glaucophane, is diagnostic of a particular tectonic regime, namely blueschist metamorphism and a subduction zone. For the vast majority of sedimentary rocks, however, the clues are more subtle.

In a series of classic papers, Dickinson and colleagues (Dickinson and Suczek, 1979; Dickinson and Valloni, 1980; Dickinson, 1982; Dickinson et al., 1983) have used triangle diagrams of quartz, feldspar, and lithic fragments (QFL plots) to interpret the tectonic provenance of sandstones (Fig. 14.14). A slightly different form of the classic QFL plot is the QmFLt plot, which has poles at monocrystalline quartz (Qm), feldspar, and all polycrystalline lithic fragments, including polycrystalline quartz (Lt). In these plots, samples from known tectonic regimes were plotted with respect to the three end-member components. A clear pattern of fields representing distinct tectonic terranes shows up on the plots (Fig. 14.15). Craton interior sandstones are usually very mature, pure monocrystalline quartz, so they plot very close to the Q or

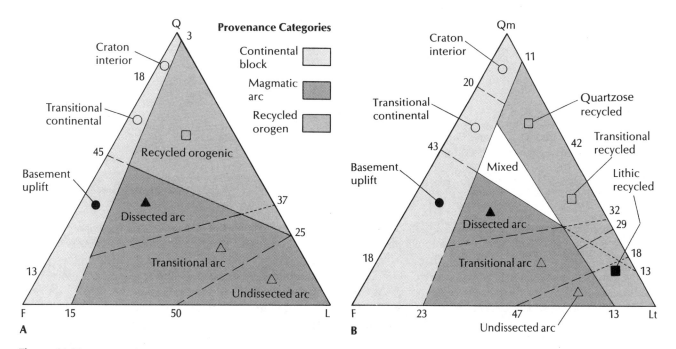

Figure 14.14

Provenance of sandstones is strongly influenced by the type of tectonic sources that were present. On (A) a QFL (quartz, feldspar, lithics) plot and (B) a QmFLt (monocrystalline quartz, feldspar, lithics including polycrystalline quartz) plot, pure quartz sandstones typically come from the cratonic interior, feldspathic sandstones from basement uplifts, and lithic sandstones from eroded arcs. Finer subdivisions are also shown. From Dickinson et al. (1983).

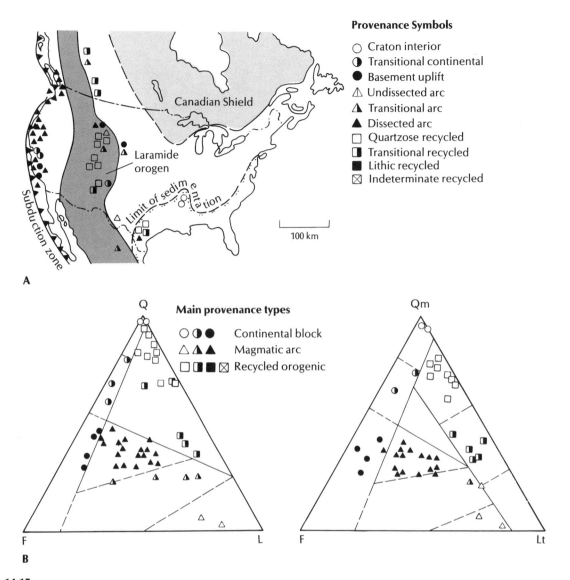

Figure 14.15

(A) Paleotectonic map, showing the tectonic source of selected Cretaceous-Paleogene sandstones (indicated by circles, triangles, and squares). (B) QFL and QmFLt plots for these sandstones showing how each of the tectonic sources clusters into discrete fields based on their composition. From Dickinson et al. (1983).

Qm pole. Basement uplifts are typically eroded, producing considerable feldspar from plutonic basement rocks, resulting in arkosic sandstones that cluster near the F pole. Undissected arc terranes generate much lithic debris, so they cluster near the L or Lt poles. As arc terranes are progressively dissected, they move toward the center of the plot. Orogenic debris that has been recycled also plots near the middle of the triangles, though it is richer in quartz than in the other two components.

Although tectonic sources influence sandstone composition, they are not the only influence. Suttner, Basu, and Mack (1981) have shown that climate can influence its composition as well (Fig. 14.16). For example, metamorphic-source terranes in humid climates produce a weathered quartz-rich sandstone whereas plutonic igneous-source terranes in humid climates are more feldspar-rich. Arid climates preserve the lithic fragments produced from metamorphic terranes, and both the feldspar and the lithics produced from plutonic terranes. Thus caution must be used when interpreting sandstones for tectonic influences. Ideally, the petrologic data should be complemented by data from other rock types and total field relations, so the effects of tectonic source and climate can be separated.

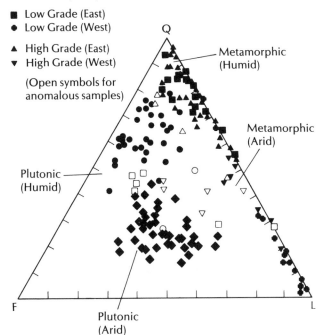

Plutonic (East)
Plutonic (West)

Low Grade (East)
Low Grade (West)

High Grade (East)
High Grade (West)

(Open symbols for anomalous samples)

Q

Metamorphic (Humid)

Metamorphic (Arid)

Plutonic (Humid)

F   L

Plutonic (Arid)

Figure 14.16

Climate can also strongly influence the composition of sandstones. Here first-cycle sand from a modern river with known sources is plotted on a QFL diagram. Metamorphic and plutonic sources gave very different clusters on the plot, depending on whether they came from a humid region with active chemical weathering or an arid region with limited chemical weathering and breakdown of feldspars and lithic fragments. From Suttner et al. (1981).

FOR FURTHER READING

Dickinson, W. R., (ed.), 1974. *Tectonics and Sedimentation.* Soc. Econ. Paleo. Mineral. Spec. Publ. 22. Based on a 1973 symposium, contains many classic papers on tectonic models for sedimentary rocks and sequences. The opening paper by Dickinson is the best short overview of the subject in print.

Dott, R. H., Jr., and R. H. Shaver (eds.), 1974. *Modern and Ancient Geosynclinal Sedimentation.* Soc. Econ. Paleo. Mineral. Spec. Publ. 19. Based on a 1972 symposium similar to that of the preceding book but also published in 1974. Many of the papers are classics that anticipated the plate tectonics revolution, but others are dated by their old-fashioned geosynclinal approach. This symposium caught geology in transition between the geosynclinal and plate tectonics paradigms. What a difference a year can make!

Ingersoll, R. V., 1988. "Tectonics of Sedimentary Basins," *Geol. Soc. Amer. Bull.* 100: 1704–1719. Excellent review article which brings sedimentary tectonics up to date.

Kay, M., 1951. *North American Geosynclines.* Geol. Soc. Amer. Memoir 48. Kay's classic monograph on geosyncline theory; the epitome of this school of thought. Although dated, it is still worth reading for its great description and insights.

In addition to the preceding works, the books by Reading (*Sedimentary Environments and Facies*) and Miall (*Principles of Sedimentary Basin Analysis*) have excellent, well-illustrated chapters on tectonics and sedimentation, giving numerous examples. The full references for these books are given in preceding chapters.

APPENDIXES

APPENDIX A

North American Stratigraphic Code[1]
North American Commission on
Stratigraphic Nomenclature

FOREWORD

This code of recommended procedures for classifying and naming stratigraphic and related units has been prepared during a four-year period, by and for North American earth scientists, under the auspices of the North American Commission on Stratigraphic Nomenclature. It represents the thought and work of scores of persons, and thousands of hours of writing and editing. Opportunities to participate in and review the work have been provided throughout its development, as cited in the Preamble, to a degree unprecedented during preparation of earlier codes.

Publication of the International Stratigraphic Guide in 1976 made evident some insufficiencies of the American Stratigraphic Codes of 1961 and 1970. The Commission considered whether to discard our codes, patch them over, or rewrite them fully, and chose the last. We believe it desirable to sponsor a code of stratigraphic practice for use in North America, for we can adapt to new methods and points of view more rapidly than a worldwide body. A timely example was the recognized need to develop modes of establishing formal nonstratiform (igneous and high-grade metamorphic) rock units, an objective which is met in this Code, but not yet in the Guide.

The ways in which this Code differs from earlier American codes are evident from the Contents. Some categories have disappeared and others are new, but this Code has evolved from earlier codes and from the International Stratigraphic Guide. Some new units have not yet stood the test of long practice, and conceivably may not, but they are introduced toward meeting recognized and defined needs of the profession. Take this Code, use it, but do not condemn it because it contains something new or not of direct interest to you. Innovations that prove unacceptable to the profession will expire without damage to other concepts and procedures, just as did the geologic-climate units of the 1961 Code.

This Code is necessarily somewhat innovative because of: (1) the decision to write a new code, rather than to revise the old; (2) the open invitation to members of the geologic profession to offer suggestions and ideas, both in writing and orally; and (3)

[1] Reprinted by permission from American Association of Petroleum Geologists Bulletin, v. 67, no. 5 (May, 1983), pp. 841–875.
Copies are available at $1.00 per copy prepaid. Order from American Association of Petroleum Geologists, Box 979, Tulsa, Oklahoma 74101.

the progress in the earth sciences since completion of previous codes. This report strives to incorporate the strength and acceptance of established practice, with suggestions for meeting future needs perceived by our colleagues; its authors have attempted to bring together the good from the past, the lessons of the Guide, and carefully reasoned provisions for the immediate future.

Participants in preparation of this Code are listed in Appendix I, but many others helped with their suggestions and comments. Major contributions were made by the members, and especially the chairmen, of the named subcommittees and advisory groups under the guidance of the Code Committee, chaired by Steven S. Oriel, who also served as principal, but not sole, editor. Amidst the noteworthy contributions by many, those of James D. Aitken have been outstanding. The work was performed for and supported by the Commission, chaired by Malcolm P. Weiss from 1978 to 1982.

This Code is the product of a truly North American effort. Many former and current commissioners representing not only the ten organizational members of the North American Commission on Stratigraphic Nomenclature (Appendix II), but other institutions as well, generated the product. Endorsement by constituent organizations is anticipated, and scientific communication will be fostered if Canadian, United States, and Mexican scientists, editors, and administrators consult Code recommendations for guidance in scientific reports. The Commission will appreciate reports of formal adoption or endorsement of the Code, and asks that they be transmitted to the Chairman of the Commission (c/o American Association of Petroleum Geologists, Box 979, Tulsa, Oklahoma 74101, U.S.A.).

Any code necessarily represents but a stage in the evolution of scientific communication. Suggestions for future changes of, or additions to, the North American Stratigraphic Code are welcome. Suggested and adopted modifications will be announced to the profession, as in the past, by serial Notes and Reports published in the *Bulletin* of the American Association of Petroleum Geologists. Suggestions may be made to representatives of your association or agency who are current commissioners, or directly to the Commission itself. The Commission meets annually, during the national meetings of the Geological Society of America.

1982 NORTH AMERICAN COMMISSION
ON STRATIGRAPHIC NOMENCLATURE

CONTENTS

PART I. PREAMBLE

BACKGROUND

PERSPECTIVE

Codes of Stratigraphic Nomenclature prepared by the American Commission on Stratigraphic Nomenclature (ACSN, 1961) and its predecessor (Committee on Stratigraphic Nomenclature, 1933) have been used widely as a basis for stratigraphic terminology. Their formulation was a response to needs recognized during the past century by government surveys (both national and local) and by editors of scientific journals for uniform standards and common procedures in defining and classifying formal rock bodies, their fossils, and the time spans represented by them. The most recent Code (ACSN, 1970) is a slightly revised version of that published in 1961, incorporating some minor amendments adopted by the Commission between 1962 and 1969. The Codes have served the profession admirably and have been drawn upon heavily for codes and guides prepared in other parts of the world (ISSC, 1976, p. 104-106). The principles embodied by any code, however, reflect the state of knowledge at the time of its preparation, and even the most recent code is now in need of revision.

New concepts and techniques developed during the past two decades have revolutionized the earth sciences. Moreover, increasingly evident have been the limitations of previous codes in meeting some needs of Precambrian and Quaternary geology and in classification of plutonic, high-grade metamorphic, volcanic, and intensely deformed rock assemblages. In addition, the important contributions of numerous international stratigraphic organizations associated with both the International Union of Geological Sciences (IUGS) and UNESCO, including working groups of the International Geological Correlation Program (IGCP), merit recognition and incorporation into a North American code.

For these and other reasons, revision of the American Code has been undertaken by committees appointed by the North American Commission on Stratigraphic Nomenclature (NACSN). The Commission, founded as the American Commission on Stratigraphic Nomenclature in 1946 (ACSN, 1947), was renamed the NACSN in 1978 (Weiss, 1979b) to emphasize that delegates from ten organizations in Canada, the United States, and Mexico represent the geological profession throughout North America (Appendix II).

Although many past and current members of the Commission helped prepare this revision of the Code, the participation of all interested geologists has been sought (for example, Weiss, 1979a). Open forums were held at the national meetings of both the Geological Society of America at San Diego in November, 1979, and the American Association of Petroleum Geologists at Denver in June, 1980, at which comments and suggestions were offered by more than 150 geologists. The resulting draft of this report was printed, through the courtesy of the Canadian Society of Petroleum Geologists, on October 1, 1981, and additional comments were invited from the profession for a period of one year before submittal of this report to the Commission for adoption. More than 50 responses were received with sufficient suggestions for improvement to prompt moderate revision of the printed draft (NACSN, 1981). We are particularly indebted to Hollis D. Hedberg and Amos Salvador for their exhaustive and perceptive reviews of early drafts of this Code, as well as to those who responded to the request for comments. Participants in the preparation and revisions of this report, and conferees, are listed in Appendix I.

Some of the expenses incurred in the course of this work were defrayed by National Science Foundation Grant EAR 7919845, for which we express appreciation. Institutions represented by the participants have been especially generous in their support.

SCOPE

The North American Stratigraphic Code seeks to describe explicit practices for classifying and naming all formally defined geologic units. *Stratigraphic procedures* and principles, although developed initially to bring order to strata and the events recorded therein, are applicable to all earth materials, not solely to strata. They promote systematic and rigorous study of the composition, geometry, sequence, history, and genesis of rocks and unconsolidated materials. They provide the framework within which time and space relations among rock bodies that constitute the Earth are ordered systematically. Stratigraphic procedures are used not only to reconstruct the history of the Earth and of extra-terrestrial bodies, but also to define the distribution and geometry of some commodities needed by society. *Stratigraphic classification* systematically arranges and partitions bodies of rock or unconsolidated materials of the Earth's crust into units based on their inherent properties or attributes.

A *stratigraphic code* or guide is a formulation of current views on stratigraphic principles and procedures designed to promote standardized classification and formal nomenclature of rock materials. It provides the basis for formalization of the language used to denote rock units and their spatial and temporal relations. To be effective, a code must be widely accepted and used; geologic organizations and journals may adopt its recommendations for nomenclatural procedure. Because any code embodies only current concepts and principles, it should have the flexibility to provide for both changes and additions to improve its relevance to new scientific problems.

Any system of nomenclature must be sufficiently explicit to enable users to distinguish objects that are embraced in a class from those that are not. This stratigraphic code makes no attempt to systematize structural, petrographic, paleontologic, or physiographic terms. Terms from these other fields that are used as part of formal stratigraphic names should be sufficiently general as to be unaffected by revisions of precise petrographic or other classifications.

The objective of a system of classification is to promote unambiguous communication in a manner not so restrictive as to inhibit scientific progress. To minimize ambiguity, a code must promote recognition of the distinction between observable features (reproducible data) and inferences or interpretations. Moreover, it should be sufficiently adaptable and flexible to promote the further development of science.

Stratigraphic classification promotes understanding of the *geometry* and *sequence* of rock bodies. The development of stratigraphy as a science required formulation of the Law of Superposition to explain sequential stratal relations. Although superposition is not applicable to many igneous, metamorphic, and tectonic rock assemblages, other criteria (such as cross-cutting relations and isotopic dating) can be used to determine sequential arrangements among rock bodies.

The term *stratigraphic unit* may be defined in several ways. Etymological emphasis requires that it be a stratum or assemblage of adjacent strata distinguished by any or several of the many properties that rocks may possess (ISSC, 1976, p. 13). The scope of stratigraphic classification and procedures, however, suggests a broader definition: a naturally occurring body of rock or rock material distinguished from adjoining rock on the basis of some stated property or properties. Commonly used properties include composition, texture, included fossils, magnetic signature, radioactivity, seismic velocity, and age. Sufficient care is required in defining the boundaries of a unit to enable others to distinguish the material body from those adjoining it. Units based on one property commonly do not coincide with those based on another and, therefore, distinctive terms are needed to identify the property used in defining each unit.

The adjective *stratigraphic* is used in two ways in the remainder of this report. In discussions of lithic (used here as synonymous with "lithologic") units, a conscious attempt is made to restrict the term to lithostratigraphic or layered rocks and sequences that obey the Law of Superposition. For nonstratiform rocks (of plutonic or tectonic origin, for example), the term *lithodemic* (see Article 27) is used. The adjective *stratigraphic* is also used in a broader sense to refer to those procedures derived from stratigraphy which are now applied to all classes of earth materials.

An assumption made in the material that follows is that the reader has some degree of familiarity with basic principles of stratigraphy as outlined, for example, by Dunbar and Rodgers (1957), Weller (1960), Shaw (1964), Matthews (1974), or the International Stratigraphic Guide (ISSC, 1976).

RELATION OF CODES TO INTERNATIONAL GUIDE

Publication of the International Stratigraphic Guide by the International Subcommission on Stratigraphic Classification (ISSC, 1976), which is being endorsed and adopted throughout the world, played a part in prompting examination of the American Stratigraphic Code and the decision to revise it.

The International Guide embodies principles and procedures that had been adopted by several national and regional stratigraphic committees and commissions. More than two decades of effort by H. D. Hedberg and other members of the Subcommission (ISSC, 1976, p. VI, 1, 3) developed the consensus required for preparation of the Guide. Although the Guide attempts to cover all kinds of rocks and the diverse ways of investigating them, it is necessarily incomplete. Mechanisms are needed to stimulate individual innovations toward promulgating new concepts, principles, and practices which subsequently may be found worthy of inclusion in later editions of the Guide. The flexibility of national and regional committees or commissions enables them to perform this function more readily than an international subcommission, even while they adopt the Guide as the international standard of stratigraphic classification.

A guiding principle in preparing this Code has been to make it as consistent as possible with the International Guide, which was endorsed by the ACSN in 1976, and at the same time to foster further innovations to meet the expanding and changing needs of earth scientists on the North American continent.

OVERVIEW

CATEGORIES RECOGNIZED

An attempt is made in this Code to strike a balance between serving the needs of those in evolving specialties and resisting the proliferation of categories of units. Consequently, more formal categories are recognized here than in previous codes or in the International Guide (ISSC, 1976). On the other hand, no special provision is made for formalizing certain kinds of units (deep oceanic, for example) which may be accommodated by available categories.

Four principal categories of units have previously been used widely in traditional stratigraphic work; these have been termed lithostratigraphic, biostratigraphic, chronostratigraphic, and geochronologic and are distinguished as follows:

1. A *lithostratigraphic unit* is a stratum or body of strata, generally but not invariably layered, generally but not invariably tabular, which conforms to the Law of Superposition and is distinguished and delimited on the basis of lithic characteristics and stratigraphic position. Example: Navajo Sandstone.

2. A *biostratigraphic unit* is a body of rock defined and characterized by its fossil content. Example: *Discoaster multiradiatus* Interval Zone.

3. A *chronostratigraphic unit* is a body of rock established to serve as the material reference for all rocks formed during the same span of time. Example: Devonian System. Each boundary of a chronostratigraphic unit is synchronous. Chronostratigraphy provides a means of organizing strata into units based on their age relations. A chronostratigraphic body also serves as the basis for defining the specific interval of geologic time, or geochronologic unit, represented by the referent.

4. A *geochronologic unit* is a division of time distinguished on the basis of the rock record preserved in a chronostratigraphic

unit. Example: Devonian Period.

The first two categories are comparable in that they consist of material units defined on the basis of content. The third category differs from the first two in that it serves primarily as the standard for recognizing and isolating materials of a specific age. The fourth, in contrast, is not a material, but rather a conceptual, unit; it is a division of time. Although a geochronologic unit is not a stratigraphic body, it is so intimately tied to chronostratigraphy that the two are discussed properly together.

Properties and procedures that may be used in distinguishing geologic units are both diverse and numerous (ISSC, 1976, p. 1, 96; Harland, 1977, p. 230), but all may be assigned to the following principal classes of categories used in stratigraphic classification (Table 1), which are discussed below:

I. Material categories based on content, inherent attributes, or physical limits,

II. Categories distinguished by geologic age:
   A. Material categories used to define temporal spans, and
   B. Temporal categories.

Table 1. Categories of Units Defined*

MATERIAL CATEGORIES BASED ON CONTENT OR PHYSICAL LIMITS
Lithostratigraphic (22)
Lithodemic (31)**
Magnetopolarity (44)
Biostratigraphic (48)
Pedostratigraphic (55)
Allostratigraphic (58)

CATEGORIES EXPRESSING OR RELATED TO GEOLOGIC AGE
Material Categories Used to Define Temporal Spans
Chronostratigraphic (66)
Polarity-Chronostratigraphic (83)
Temporal (Non-Material) Categories
Geochronologic (80)
Polarity-Chronologic (88)
Diachronic (91)
Geochronometric (96)

*Numbers in parentheses are the numbers of the Articles where units are defined.

** Italicized categories are those introduced or developed since publication of the previous code (ACSN, 1970).

Material Categories Based on Content or Physical Limits

The basic building blocks for most geologic work are rock bodies defined on the basis of composition and related lithic characteristics, or on their physical, chemical, or biologic content or properties. Emphasis is placed on the relative objectivity and reproducibility of data used in defining units within each category.

Foremost properties of rocks are composition, texture, fabric, structure, and color, which together are designated *lithic characteristics*. These serve as the basis for distinguishing and defining the most fundamental of all formal units. Such units based primarily on composition are divided into two categories (Henderson and others, 1980): lithostratigraphic (Article 22) and lithodemic (defined here in Article 31). A lithostratigraphic unit obeys the Law of Superposition, whereas a lithodemic unit does not. A *lithodemic unit* is a defined body of predominantly intrusive, highly metamorphosed, or intensely deformed rock that, because it is intrusive or has lost primary structure through metamorphism or tectonism, generally does not conform to the Law of Superposition.

Recognition during the past several decades that remanent magnetism in rocks records the Earth's past magnetic characteristics (Cox, Doell, and Dalrymple, 1963) provides a powerful new tool encompassed by magnetostratigraphy (McDougall, 1977; McElhinny, 1978). *Magnetostratigraphy* (Article 43) is the study of remanent magnetism in rocks; it is the record of the Earth's magnetic polarity (or field reversals), dipole-field-pole position (including apparent polar wander), the non-dipole component (secular variation), and field intensity. Polarity is of particular utility and is used to define a *magnetopolarity unit* (Article 44) as a body of rock identified by its remanent magnetic polarity (ACSN, 1976; ISSC, 1979). Empirical demonstration of uniform polarity does not necessarily have direct temporal connotations because the remanent magnetism need not be related to rock deposition or crystallization. Nevertheless, polarity is a physical attribute that may characterize a body of rock.

Biologic remains contained in, or forming, strata are uniquely important in stratigraphic practice. First, they provide the means of defining and recognizing material units based on fossil content (biostratigraphic units). Second, the irreversibility of organic evolution makes it possible to partition enclosing strata temporally. Third, biologic remains provide important data for the reconstruction of ancient environments of deposition.

Composition also is important in distinguishing pedostratigraphic units. A *pedostratigraphic unit* is a body of rock that consists of one or more pedologic horizons developed in one or more lithic units now buried by a formally defined lithostratigraphic or allostratigraphic unit or units. A pedostratigraphic unit is the part of a buried soil characterized by one or more clearly defined soil horizons containing pedogenically formed minerals and organic compounds. Pedostratigraphic terminology is discussed below and in Article 55.

Many upper Cenozoic, especially Quaternary, deposits are distinguished and delineated on the basis of content, for which lithostratigraphic classification is appropriate. However, others are delineated on the basis of criteria other than content. To facilitate the reconstruction of geologic history, some compositionally similar deposits in vertical sequence merit distinction as separate stratigraphic units because they are the products of different processes; others merit distinction because they are of demonstrably different ages. Lithostratigraphic classification of these units is impractical and a new approach, allostratigraphic classification, is introduced and may prove applicable to older deposits as well. An *allostratigraphic unit* is a mappable stratiform body of sedimentary rock defined and identified on the basis of bounding discontinuities (Article 58 and related Remarks).

Geologic-Climate units, defined in the previous Code (ACSN, 1970, p. 31), are abandoned here because they proved to be of dubious utility. Inferences regarding climate are subjective and too tenuous a basis for the definition of formal geologic units. Such inferences commonly are based on deposits assigned more appropriately to lithostratigraphic or allostratigraphic units and may be expressed in terms of diachronic units (defined below).

Categories Expressing or Related to Geologic Age

Time is a single, irreversible continuum. Nevertheless, various categories of units are used to define intervals of geologic time, just as terms having different bases, such as Paleolithic, Renaissance, and Elizabethan, are used to designate specific periods of human history. Different temporal categories are established to express intervals of time distinguished in different ways.

Major objectives of stratigraphic classification are to provide a basis for systematic ordering of the time and space relations of rock bodies and to establish a time framework for the discussion of geologic history. For such purposes, units of geologic time traditionally have been named to represent the span of time during which a well-described sequence of rock, or a chronostratigraphic unit, was deposited ("time units based on material referents," Fig. 1). This procedure continues, to the exclusion of other possible approaches, to be standard practice in studies of Phanerozoic rocks. Despite admonitions in previous American codes and the International Stratigraphic Guide (ISSC, 1976, p. 81) that similar procedures should be applied to the Precambrian, no comparable chronostratigraphic units, or geochronologic units derived therefrom, proposed for the Precambrian have yet been accepted worldwide. Instead, the IUGS Subcommission on Precambrian Stratigraphy (Sims, 1979) and its Working Groups (Harrison and Peterman, 1980) recommend division of Precambrian time into *geochronometric units* having no material referents.

A distinction is made throughout this report between *isochronous* and *synchronous*, as urged by Cumming, Fuller, and Porter (1959, p. 730), although the terms have been used synonymously by many. *Isochronous* means of equal duration; *synchronous* means simultaneous, or occurring at the same time. Although two rock bodies of very different ages may be formed during equal durations of time, the term *isochronous* is not applied to them in the earth sciences. Rather, isochronous bodies are those bounded by synchronous surfaces and formed during the same span of time. *Isochron*, in contrast, is used for a line connecting points of equal age on a graph representing physical or chemical phenomena; the line represents the same or equal time. The adjective *diachronous* is applied either to a rock unit with one or two bounding surfaces which are not synchronous, or to a boundary which is not synchronous (which "transgresses time").

Two classes of time units based on material referents, or stratotypes, are recognized (Fig. 1). The first is that of the traditional and conceptually isochronous units, and includes *geochronologic units*, which are based on *chronostratigraphic units*, and *polarity-geochronologic units*. These isochronous units have worldwide applicability and may be used even in areas lacking a material record of the named span of time. The second class of time units, newly defined in this Code, consists of *diachronic units* (Article 91), which are based on rock bodies known to be diachronous. In contrast to isochronous units, a diachronic term is used only where a material referent is present; a diachronic unit is coextensive with the material body or bodies on which it is based.

A *chronostratigraphic unit*, as defined above and in Article 66, is a body of rock established to serve as the material reference for all rocks formed during the same span of time; its boundaries are synchronous. It is the referent for a *geochronologic unit*, as defined above and in Article 80. Internationally accepted and traditional chronostratigraphic units were based initially on the time spans of lithostratigraphic units, biostratigraphic units, or other features of the rock record that have specific durations. In sum, they form the Standard Global Chronostratigraphic Scale (ISSC, 1976, p. 76-81; Harland, 1978), consisting of established systems and series.

A *polarity-chronostratigraphic unit* is a body of rock that contains a primary magnetopolarity record imposed when the rock was deposited or crystallized (Article 83). It serves as a material standard or referent for a part of geologic time during which the Earth's magnetic field had a characteristic polarity or sequence of polarities; that is, for a *polarity-chronologic unit* (Article 88).

A *diachronic unit* comprises the unequal spans of time represented by one or more specific diachronous rock bodies (Article 91). Such bodies may be lithostratigraphic, biostratigraphic, pedostratigraphic, allostratigraphic, or an assemblage of such units. A diachronic unit is applicable only where its material referent is present.

A *geochronometric* (or chronometric) *unit* is an isochronous direct division of geologic time expressed in years (Article 96). It has no material referent.

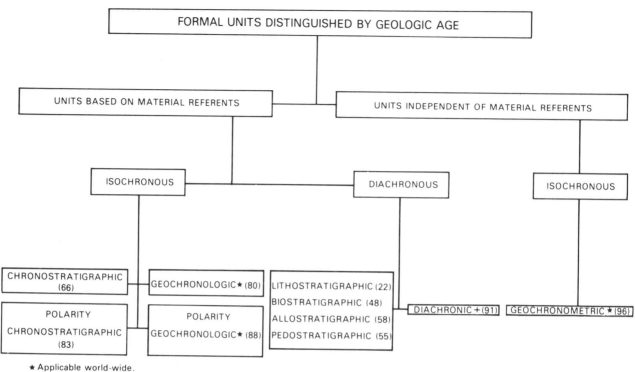

FORMAL UNITS DISTINGUISHED BY GEOLOGIC AGE

UNITS BASED ON MATERIAL REFERENTS

UNITS INDEPENDENT OF MATERIAL REFERENTS

ISOCHRONOUS

DIACHRONOUS

ISOCHRONOUS

CHRONOSTRATIGRAPHIC (66)

GEOCHRONOLOGIC ★ (80)

LITHOSTRATIGRAPHIC (22)
BIOSTRATIGRAPHIC (48)
ALLOSTRATIGRAPHIC (58)
PEDOSTRATIGRAPHIC (55)

DIACHRONIC + (91)

GEOCHRONOMETRIC ★ (96)

POLARITY CHRONOSTRATIGRAPHIC (83)

POLARITY GEOCHRONOLOGIC ★ (88)

★ Applicable world-wide.
+Applicable only where material referents are present.
( )Number of article in which defined.

FIG. 1.—Relation of geologic time units to the kinds of rock-unit referents on which most are based.

Pedostratigraphic Terms

The definition and nomenclature for pedostratigraphic[2] units in this Code differ from those for soil-stratigraphic units in the previous Code (ACSN, 1970, Article 18), by being more specific with regard to content, boundaries, and the basis for determining stratigraphic position.

The term "soil" has different meanings to the geologist, the soil scientist, the engineer, and the layman, and commonly has no stratigraphic significance. The term *paleosol* is currently used in North America for any soil that formed on a landscape of the past; it may be a buried soil, a relict soil, or an exhumed soil (Ruhe, 1965; Valentine and Dalrymple, 1976).

A *pedologic soil* is composed of one or more soil horizons.[3] A *soil horizon* is a layer within a pedologic soil that (1) is approximately parallel to the soil surface, (2) has distinctive physical, chemical, biological, and morphological properties that differ from those of adjacent, genetically related, soil horizons, and (3) is distinguished from other soil horizons by objective compositional properties that can be observed or measured in the field. The physical boundaries of buried pedologic horizons are objective traceable boundaries with stratigraphic significance. A buried pedologic soil provides the material basis for definition of a stratigraphic unit in pedostratigraphic classification (Article 55), but a buried pedologic soil may be somewhat more inclusive than a pedostratigraphic unit. A pedologic soil may contain both an 0-horizon and the entire C-horizon (Fig. 6), whereas the former is excluded and the latter need not be included in a pedostratigraphic unit.

The definition and nomenclature for pedostratigraphic units

in this Code differ from those of soil stratigraphic units proposed by the International Union for Quaternary Research and International Society of Soil Science (Parsons, 1981). The pedostratigraphic unit, geosol, also differs from the proposed INQUA-ISSS soil-stratigraphic unit, pedoderm, in several ways, the most important of which are: (1) a geosol may be in any part of the geologic column, whereas a pedoderm is a surficial soil; (2) a geosol is a buried soil, whereas a pedoderm may be a buried, relict, or exhumed soil; (3) the boundaries and stratigraphic position of a geosol are defined and delineated by criteria that differ from those for a pedoderm; and (4) a geosol may be either all or only a part of a buried soil, whereas a pedoderm is the entire soil.

The term *geosol*, as defined by Morrison (1967, p. 3), is a laterally traceable, mappable, geologic weathering profile that has a consistent stratigraphic position. The term is adopted and redefined here as the fundamental and only unit in formal pedostratigraphic classification (Article 56).

FORMAL AND INFORMAL UNITS

Although the emphasis in this Code is necessarily on formal categories of geologic units, informal nomenclature is highly useful in stratigraphic work.

Formally named units are those that are named in accordance with an established scheme of classification; the fact of formality is conveyed by capitalization of the initial letter of the *rank* or *unit* term (for example, Morrison Formation). Informal units, whose unit terms are ordinary nouns, are not protected by the stability provided by proper formalization and recommended classification procedures. Informal terms are devised for both economic and scientific reasons. Formalization is appropriate for those units requiring stability of nomenclature, particularly those likely to be extended far beyond the locality in which they were first recognized. Informal terms are appropriate for casually mentioned, innovative, and most economic units, those

---

[2]From Greek, *pedon*, ground or soil.
[3]As used in a geological sense, a *horizon* is a surface or line. In pedology, however, it is a body of material, and such usage is continued here.

defined by unconventional criteria, and those that may be too thin to map at usual scales.

Casually mentioned geologic units not defined in accordance with this Code are informal. For many of these, there may be insufficient need or information, or perhaps an inappropriate basis, for formal designations. Informal designations as beds or lithozones (the pebbly beds, the shaly zone, third coal) are appropriate for many such units.

Most economic units, such as aquifers, oil sands, coal beds, quarry layers, and ore-bearing "reefs," are informal, even though they may be named. Some such units, however, are so significant scientifically and economically that they merit formal recognition as beds, members, or formations.

Innovative approaches in regional stratigraphic studies have resulted in the recognition and definition of units best left as informal, at least for the time being. Units bounded by major regional unconformities on the North American craton were designated "sequences" (example: Sauk sequence) by Sloss (1963). Major unconformity-bounded units also were designated "synthems" by Chang (1975), who recommended that they be treated formally. Marker-defined units that are continuous from one lithofacies to another were designated "formats" by Forgotson (1957). The term "chronosome" was proposed by Schultz (1982) for rocks of diverse facies corresponding to geographic variations in sedimentation during an interval of deposition identified on the basis of bounding stratigraphic markers. Successions of faunal zones containing evolutionarily related forms, but bounded by non-evolutionary biotic discontinuities, were termed "biomeres" (Palmer, 1965). The foregoing are only a few selected examples to demonstrate how informality provides a continuing avenue for innovation.

The terms *magnafacies* and *parvafacies*, coined by Caster (1934) to emphasize the distinction between lithostratigraphic and chronostratigraphic units in sequences displaying marked facies variation, have remained informal despite their impact on clarifying the concepts involved.

Tephrochronologic studies provide examples of informal units too thin to map at conventional scales but yet invaluable for dating important geologic events. Although some such units are named for physiographic features and places where first recognized (e.g., Guaje pumice bed, where it is not mapped as the Guaje Member of the Bandelier Tuff), others bear the same name as the volcanic vent (e.g., Huckleberry Ridge ash bed of Izett and Wilcox, 1981).

Informal geologic units are designated by ordinary nouns, adjectives or geographic terms and lithic or unit-terms that are not capitalized (chalky formation or beds, St. Francis coal).

No geologic unit should be established and defined, whether formally or informally, unless its recognition serves a clear purpose.

CORRELATION

Correlation is a procedure for demonstrating correspondence between geographically separated parts of a geologic unit. The term is a general one having diverse meanings in different disciplines. Demonstration of temporal correspondence is one of the most important objectives of stratigraphy. The term "correlation" frequently is misused to express the idea that a unit has been identified or recognized.

Correlation is used in this Code as the demonstration of correspondence between two geologic units in both some defined property and relative stratigraphic position. Because correspondence may be based on various properties, three kinds of correlation are best distinguished by more specific terms. *Lithocorrelation* links units of similar lithology and stratigraphic position (or sequential or geometric relation, for lithodemic

units). *Biocorrelation* expresses similarity of fossil content and biostratigraphic position. *Chronocorrelation* expresses correspondence in age and in chronostratigraphic position.

Other terms that have been used for the similarity of content and stratal succession are homotaxy and chronotaxy. *Homotaxy* is the similarity in separate regions of the serial arrangement or succession of strata of comparable compositions or of included fossils. The term is derived from *homotaxis*, proposed by Huxley (1862, p. xlvi) to emphasize that similarity in succession does not prove age equivalence of comparable units. The term *chronotaxy* has been applied to similar stratigraphic sequences composed of units which are of equivalent age (Henbest, 1952, p. 310).

Criteria used for ascertaining temporal and other types of correspondence are diverse (ISSC, 1976, p. 86-93) and new criteria will emerge in the future. Evolving statistical tests, as well as isotopic and paleomagnetic techniques, complement the traditional paleontologic and lithologic procedures. Boundaries defined by one set of criteria need not correspond to those defined by others.

PART II. ARTICLES

INTRODUCTION

Article 1.—**Purpose.** This Code describes explicit stratigraphic procedures for classifying and naming geologic units accorded formal status. Such procedures, if widely adopted, assure consistent and uniform usage in classification and terminology and therefore promote unambiguous communication.

Article 2.—**Categories.** Categories of formal stratigraphic units, though diverse, are of three classes (Table 1). The first class is of rock-material categories based on inherent attributes or content and stratigraphic position, and includes lithostratigraphic, lithodemic, magnetopolarity, biostratigraphic, pedostratigraphic, and allostratigraphic units. The second class is of material categories used as standards for defining spans of geologic time, and includes chronostratigraphic and polarity-chronostratigraphic units. The third class is of non-material temporal categories, and includes geochronologic, polarity-chronologic, geochronometric, and diachronic units.

GENERAL PROCEDURES

DEFINITION OF FORMAL UNITS

Article 3.—**Requirements for Formally Named Geologic Units.** Naming, establishing, revising, redefining, and abandoning formal geologic units require publication in a recognized scientific medium of a comprehensive statement which includes: (i) intent to designate or modify a formal unit; (ii) designation of category and rank of unit; (iii) selection and derivation of name; (iv) specification of stratotype (where applicable); (v) description of unit; (vi) definition of boundaries; (vii) historical background; (viii) dimensions, shape, and other regional aspects; (ix) geologic age; (x) correlations; and possibly (xi) genesis (where applicable). These requirements apply to subsurface and offshore, as well as exposed, units.

Article 4.—**Publication.**[4] "Publication in a recognized scientific medium" in conformance with this Code means that a work, when first issued, must (1) be reproduced in ink on paper or by some method that assures numerous identical copies and wide distribution; (2) be issued for the purpose of scientific, public, permanent record; and (3) be readily obtainable by purchase or free distribution.

Remarks. (a) **Inadequate publication.**—The following do not constitute publication within the meaning of the Code: (1) distribution of microfilms, microcards, or matter reproduced by similar methods; (2)

---

[4]This article is modified slightly from a statement by the International Commission of Zoological Nomenclature (1964, p. 7-9).

Table 2. Categories and Ranks of Units Defined in This Code*

A. Material Units

LITHOSTRATIGRAPHIC	LITHODEMIC	MAGNETOPOLARITY	BIOSTRATIGRAPHIC	PEDOSTRATIGRAPHIC	ALLOSTRATIGRAPHIC
Supergroup	Supersuite				
Group	Suite (Complex)	Polarity Superzone			Allogroup
Formation	Lithodeme	Polarity zone	Biozone (Interval, Assemblage or Abundance)	Geosol	Alloformation
Member (or Lens, or Tongue)		Polarity Subzone	Subbiozone		Allomember
Bed(s) or Flow(s)					

B. Temporal and Related Chronostratigraphic Units

CHRONO-STRATIGRAPHIC	GEOCHRONOLOGIC GEOCHRONOMETRIC	POLARITY CHRONO-STRATIGRAPHIC	POLARITY CHRONOLOGIC	DIACHRONIC
Eonothem	Eon	Polarity Superchronozone	Polarity Superchron	
Erathem (Supersystem)	Era (Superperiod)			
System (Subsystem)	Period (Subperiod)	Polarity Chronozone	Polarity Chron	Episode
Series	Epoch			Phase
Stage (Substage)	Age (Subage)	Polarity Subchronozone	Polarity Subchron	Span
Chronozone	Chron			Cline

*Fundamental units are italicized.

distribution to colleagues or students of a note, even if printed, in explanation of an accompanying illustration; (3) distribution of proof sheets; (4) open-file release; (5) theses, dissertations, and dissertation abstracts; (6) mention at a scientific or other meeting; (7) mention in an abstract, map explanation, or figure caption; (8) labeling of a rock specimen in a collection; (9) mere deposit of a document in a library; (10) anonymous publication; or (11) mention in the popular press or in a legal document.

(b). **Guidebooks.**—A guidebook with distribution limited to participants of a field excursion does not meet the test of availability. Some organizations publish and distribute widely large editions of serial guidebooks that include refereed regional papers; although these do meet the tests of scientific purpose and availability, and therefore constitute valid publication, other media are preferable.

Article 5.—Intent and Utility. To be valid, a new unit must serve a clear purpose and be duly proposed and duly described, and the intent to establish it must be specified. Casual mention of a unit, such as "the granite exposed near the Middleville schoolhouse," does not establish a new formal unit, nor does mere use in a table, columnar section, or map.

Remark. (a) **Demonstration of purpose served.**—The initial definition or revision of a named geologic unit constitutes, in essence, a proposal. As such, it lacks status until use by others demonstrates that a clear purpose has been served. A unit becomes established through repeated demonstration of its utility. The decision not to use a newly proposed or a newly revised term requires a full discussion of its unsuitability.

Article 6.—Category and Rank. The category and rank of a new or revised unit must be specified.

Remark. (a) **Need for specification.**—Many stratigraphic controversies have arisen from confusion or misinterpretation of the

category of a unit (for example, lithostratigraphic vs. chronostratigraphic). Specification and unambiguous description of the category is of paramount importance. Selection and designation of an appropriate rank from the distinctive terminology developed for each category help serve this function (Table 2).

Article 7.—Name. The name of a formal geologic unit is compound. For most categories, the name of a unit should consist of a geographic name combined with an appropriate rank (Wasatch Formation) or descriptive term (Viola Limestone). Biostratigraphic units are designated by appropriate biologic forms (*Exus albus* Assemblage Biozone). Worldwide chronostratigraphic units bear long established and generally accepted names of diverse origins (Triassic System). The first letters of all words used in the names of formal geologic units are capitalized (except for the trivial species and subspecies terms in the name of a biostratigraphic unit).

Remarks. (a) **Appropriate geographic terms.**—Geographic names derived from permanent natural or artificial features at or near which the unit is present are preferable to those derived from impermanent features such as farms, schools, stores, churches, crossroads, and small communities. Appropriate names may be selected from those shown on topographic, state, provincial, county, forest service, hydrographic, or comparable maps, particularly those showing names approved by a national board for geographic names. The generic part of a geographic name, e.g., river, lake, village, should be omitted from new terms, unless required to distinguish between two otherwise identical names (e.g., Redstone Formation and Redstone River Formation). Two names should not be derived from the same geographic feature. A unit should not be named for the source of its components; for example, a deposit inferred to have been derived from the Keewatin glaciation center should not be designated the "Keewatin Till."

(b) **Duplication of names.**—Responsibility for avoiding duplication, either in use of the same name for different units (homonymy) or in use of different names for the same unit (synonymy), rests with the proposer. Although the same geographic term has been applied to different categories of units (example: the lithostratigraphic Word Formation and the chronostratigraphic Wordian Stage) now entrenched in the literature, the practice is undesirable. The extensive geologic nomenclature of North America, including not only names but also nomenclatural history of formal units, is recorded in compendia maintained by the Committee on Stratigraphic Nomenclature of the Geological Survey of Canada, Ottawa, Ontario; by the Geologic Names Committee of the United States Geological Survey, Reston, Virginia; by the Instituto de Geología, Ciudad Universitaria, México, D.F.; and by many state and provincial geological surveys. These organizations respond to inquiries regarding the availability of names, and some are prepared to reserve names for units likely to be defined in the next year or two.

(c) **Priority and preservation of established names.**—Stability of nomenclature is maintained by use of the rule of priority and by preservation of well-established names. Names should not be modified without explaining the need. Priority in publication is to be respected, but priority alone does not justify displacing a well-established name by one neither well-known nor commonly used; nor should an inadequately established name be preserved merely on the basis of priority. Redefinitions in precise terms are preferable to abandonment of the names of well-established units which may have been defined imprecisely but nonetheless in conformance with older and less stringent standards.

(d) **Differences of spelling and changes in name.**—The geographic component of a well-established stratigraphic name is not changed due to differences in spelling or changes in the name of a geographic feature. The name Bennett Shale, for example, used for more than half a century, need not be altered because the town is named Bennet. Nor should the Mauch Chunk Formation be changed because the town has been renamed Jim Thorpe. Disappearance of an impermanent geographic feature, such as a town, does not affect the name of an established geologic unit.

(e) **Names in different countries and different languages.**—For geologic units that cross local and international boundaries, a single name for each is preferable to several. Spelling of a geographic name commonly conforms to the usage of the country and linguistic group involved. Although geographic names are not translated (Cuchillo is not translated to Knife), lithologic or rank terms are (Edwards Limestone, Caliza Edwards; Formación La Casita, La Casita Formation).

Article 8.—**Stratotypes.** The designation of a unit or boundary stratotype (type section or type locality) is essential in the definition of most formal geologic units. Many kinds of units are best defined by reference to an accessible and specific sequence of rock that may be examined and studied by others. A stratotype is the standard (original or subsequently designated) for a named geologic unit or boundary and constitutes the basis for definition or recognition of that unit or boundary; therefore, it must be illustrative and representative of the concept of the unit or boundary being defined.

Remarks. (a) **Unit stratotypes.**—A unit stratotype is the type section for a stratiform deposit or the type area for a nonstratiform body that serves as the standard for definition and recognition of a geologic unit. The upper and lower limits of a unit stratotype are designated points in a specific sequence or locality and serve as the standards for definition and recognition of a stratigraphic unit's boundaries.

(b) **Boundary stratotype.**—A boundary stratotype is the type locality for the boundary reference point for a stratigraphic unit. Both boundary stratotypes for any unit need not be in the same section or region. Each boundary stratotype serves as the standard for definition and recognition of the base of a stratigraphic unit. The top of a unit may be defined by the boundary stratotype of the next higher stratigraphic unit.

(c) **Type locality.**—A type locality is the specified geographic locality where the stratotype of a formal unit or unit boundary was originally defined and named. A type area is the geographic territory encompassing the type locality. Before the concept of a stratotype was developed, only type localities and areas were designated for many geologic units which are now long- and well-established. Stratotypes, though now mandatory in defining most stratiform units, are impractical in definitions of many large nonstratiform rock bodies whose diverse major components may be best displayed at several reference localities.

(d) **Composite-stratotype.**—A composite-stratotype consists of several reference sections (which may include a type section) required to demonstrate the range or totality of a stratigraphic unit.

(e) **Reference sections.**—Reference sections may serve as invaluable standards in definitions or revisions of formal geologic units. For those well-established stratigraphic units for which a type section never was specified, a principal reference section (lectostratotype of ISSC, 1976, p. 26) may be designated. A principal reference section (neostratotype of ISSC, 1976, p. 26) also may be designated for those units or boundaries whose stratotypes have been destroyed, covered, or otherwise made inaccessible. Supplementary reference sections often are designated to illustrate the diversity or heterogeneity of a defined unit or some critical feature not evident or exposed in the stratotype. Once a unit or boundary stratotype section is designated, it is never abandoned or changed; however, if a stratotype proves inadequate, it may be supplemented by a principal reference section or by several reference sections that may constitute a composite-stratotype.

(f) **Stratotype descriptions.**—Stratotypes should be described both geographically and geologically. Sufficient geographic detail must be included to enable others to find the stratotype in the field, and may consist of maps and/or aerial photographs showing location and access, as well as appropriate coordinates or bearings. Geologic information should include thickness, descriptive criteria appropriate to the recognition of the unit and its boundaries, and discussion of the relation of the unit to other geologic units of the area. A carefully measured and described section provides the best foundation for definition of stratiform units. Graphic profiles, columnar sections, structure-sections, and photographs are useful supplements to a description; a geologic map of the area including the type locality is essential.

Article 9.—**Unit Description.** A unit proposed for formal status should be described and defined so clearly that any subsequent investigator can recognize that unit unequivocally. Distinguishing features that characterize a unit may include any or several of the following: composition, texture, primary structures, structural attitudes, biologic remains, readily apparent mineral composition (e.g., calcite vs. dolomite), geochemistry, geophysical properties (including magnetic signatures), geomorphic expression, unconformable or cross-cutting relations, and age. Although all distinguishing features pertinent to the unit category should be described sufficiently to characterize the unit, those not pertinent to the category (such as age and inferred genesis for lithostratigraphic units, or lithology for biostratigraphic units) should not be made part of the definition.

Article 10.—**Boundaries.** The criteria specified for the recognition of boundaries between adjoining geologic units are of paramount importance because they provide the basis for scientific reproducibility of results. Care is required in describing the criteria, which must be appropriate to the category of unit involved.

Remarks. (a) **Boundaries between intergradational units.**—Contacts between rocks of markedly contrasting composition are appropriate boundaries of lithic units, but some rocks grade into, or intertongue with, others of different lithology. Consequently, some boundaries are necessarily arbitrary as, for example, the top of the uppermost limestone in a sequence of interbedded limestone and shale. Such arbitrary boundaries commonly are diachronous.

(b) **Overlaps and gaps.**—The problem of overlaps and gaps between long-established adjacent chronostratigraphic units is being addressed by international IUGS and IGCP working groups appointed to deal with various parts of the geologic column. The procedure recommended by the Geological Society of London (George and others, 1969; Holland and others, 1978), of defining only the basal boundaries of chronostratigraphic units, has been widely adopted (e.g., McLaren, 1977) to resolve the problem. Such boundaries are defined by a carefully selected and agreed-upon boundary-stratotype (marker-point type section or "golden spike") which becomes the standard for the base of a chronostratigraphic unit. The concept of the mutual-boundary stratotype (ISSC, 1976, p. 84-86), based on the assumption of continuous deposition in selected sequences, also has been used to define chronostratigraphic units.

Although international chronostratigraphic units of series and higher rank are being redefined by IUGS and IGCP working groups, there may be a continuing need for some provincial series. Adoption of the basal boundary-stratotype concept is urged.

Article 11.—**Historical Background.** A proposal for a new name must include a nomenclatorial history of rocks assigned to the proposed unit, describing how they were treated previously and by whom (references), as well as such matters as priorities, possible synonymy, and other pertinent considerations. Consideration of the historical background of an older unit commonly provides the basis for justifying definition of a new unit.

Article 12.—**Dimensions and Regional Relations.** A perspective on the magnitude of a unit should be provided by such information as may be available on the geographic extent of a unit; observed ranges in thickness, composition, and geomorphic expression; relations to other kinds and ranks of stratigraphic units; correlations with other nearby sequences; and the bases for recognizing and extending the unit beyond the type locality. If the unit is not known anywhere but in an area of limited extent, informal designation is recommended.

Article 13.—**Age.** For most formal material geologic units, other than chronostratigraphic and polarity-chronostratigraphic, inferences regarding geologic age play no proper role in their definition. Nevertheless, the age, as well as the basis for its assignment, are important features of the unit and should be stated. For many lithodemic units, the age of the protolith should be distinguished from that of the metamorphism or deformation. If the basis for assigning an age is tenuous, a doubt should be expressed.

Remarks. (a) **Dating.**—The geochronologic ordering of the rock record, whether in terms of radioactive-decay rates or other processes, is generally called "dating." However, the use of the noun "date" to mean "isotopic age" is not recommended. Similarly, the term "absolute age" should be suppressed in favor of "isotopic age" for an age determined on the basis of isotopic ratios. The more inclusive term "numerical age" is recommended for all ages determined from isotopic ratios, fission tracks, and other quantifiable age-related phenomena.

(b) **Calibration**—The dating of chronostratigraphic boundaries in terms of numerical ages is a special form of dating for which the word "calibration" should be used. The geochronologic time-scale now in use has been developed mainly through such calibration of chronostratigraphic sequences.

(c) **Convention and abbreviations.**—The age of a stratigraphic unit or the time of a geologic event, as commonly determined by numerical dating or by reference to a calibrated time-scale, may be expressed in years before the present. The unit of time is the modern year as presently recognized worldwide. Recommended (but not mandatory) abbreviations for such ages are SI (International System of Units) multipliers coupled with "a" for annum: ka, Ma, and Ga[5] for kilo-annum (10^3 years), Mega-annum (10^6 years), and Giga-annum (10^9 years), respectively. Use of these terms after the age value follows the convention established in the field of C-14 dating. The "present" refers to 1950 AD, and such qualifiers as "ago" or "before the present" are omitted after the value because measurement of the duration from the present to the past is implicit in the designation. In contrast, the duration of a remote interval of geologic time, as a number of years, should not be expressed by the same symbols. Abbreviations for numbers of years, without reference to the present, are informal (e.g., y or yr for years; my, m.y., or m.yr. for millions of years; and so forth, as preference dictates). For example, boundaries of the Late Cretaceous Epoch currently are calibrated at 63 Ma and 96 Ma, but the interval of time represented by this epoch is 33 m.y.

(d) **Expression of "age" of lithodemic units.**—The adjectives "early," "middle," and "late" should be used with the appropriate

geochronologic term to designate the age of lithodemic units. For example, a granite dated isotopically at 510 Ma should be referred to using the geochronologic term "Late Cambrian granite" rather than either the chronostratigraphic term "Upper Cambrian granite" or the more cumbersome designation "granite of Late Cambrian age."

Article 14.—**Correlation.** Information regarding spatial and temporal counterparts of a newly defined unit beyond the type area provides readers with an enlarged perspective. Discussions of criteria used in correlating a unit with those in other areas should make clear the distinction between data and inferences.

Article 15.—**Genesis.** Objective data are used to define and classify geologic units and to express their spatial and temporal relations. Although many of the categories defined in this Code (e.g., lithostratigraphic group, plutonic suite) have genetic connotations, inferences regarding geologic history or specific environments of formation may play no proper role in the definition of a unit. However, observations, as well as inferences, that bear on genesis are of great interest to readers and should be discussed.

Article 16.—**Subsurface and Subsea Units.** The foregoing procedures for establishing formal geologic units apply also to subsurface and offshore or subsea units. Complete lithologic and paleontologic descriptions or logs of the samples or cores are required in written or graphic form, or both. Boundaries and divisions, if any, of the unit should be indicated clearly with their depths from an established datum.

Remarks. (a) **Naming subsurface units.**—A subsurface unit may be named for the borehole (Eagle Mills Formation), oil field (Smackover Limestone), or mine which is intended to serve as the stratotype, or for a nearby geographic feature. The hole or mine should be located precisely, both with map and exact geographic coordinates, and identified fully (operator or company, farm or lease block, dates drilled or mined, surface elevation and total depth, etc).

(b) **Additional recommendations.**—Inclusion of appropriate borehole geophysical logs is urged. Moreover, rock and fossil samples and cores and all pertinent accompanying materials should be stored, and available for examination, at appropriate federal, state, provincial, university, or museum depositories. For offshore or subsea units (Clipperton Formation of Tracey and others, 1971, p. 22; Argo Salt of McIver, 1972, p. 57), the names of the project and vessel, depth of sea floor, and pertinent regional sampling and geophysical data should be added.

(c) **Seismostratigraphic units.**—High-resolution seismic methods now can delineate stratal geometry and continuity at a level of confidence not previously attainable. Accordingly, seismic surveys have come to be the principal adjunct of the drill in subsurface exploration. On the other hand, the method identifies rock types only broadly and by inference. Thus, formalization of units known only from seismic profiles is inappropriate. Once the stratigraphy is calibrated by drilling, the seismic method may provide objective well-to-well correlations.

REVISION AND ABANDONMENT OF FORMAL UNITS

Article 17.—**Requirements for Major Changes.** Formally defined and named geologic units may be redefined, revised, or abandoned, but revision and abandonment require as much justification as establishment of a new unit.

Remark. (a) **Distinction between redefinition and revision.**—Redefinition of a unit involves changing the view or emphasis on the content of the unit without changing the boundaries or rank, and differs only slightly from redescription. Neither redefinition nor redescription is considered revision. A redescription corrects an inadequate or inaccurate description, whereas a redefinition may change a descriptive (for example, lithologic) designation. Revision involves either minor changes in the definition of one or both boundaries or in the rank of a unit (normally, elevation to a higher rank). Correction of a misidentification of a unit outside its type area is neither redefinition nor revision.

---

[5]Note that the initial letters of Mega- and Giga- are capitalized, but that of kilo- is not, by SI convention.

Article 18.—**Redefinition.** A correction or change in the descriptive term applied to a stratigraphic or lithodemic unit is a redefinition which does not require a new geographic term.

Remarks. (a) **Change in lithic designation.**—Priority should not prevent more exact lithic designation if the original designation is not everywhere applicable; for example, the Niobrara Chalk changes gradually westward to a unit in which shale is prominent, for which the designation "Niobrara Shale" or "Formation" is more appropriate. Many carbonate formations originally designated "limestone" or "dolomite" are found to be geographically inconsistent as to prevailing rock type. The appropriate lithic term or "formation" is again preferable for such units.

(b) **Original lithic designation inappropriate.**—Restudy of some long-established lithostratigraphic units has shown that the original lithic designation was incorrect according to modern criteria; for example, some "shales" have the chemical and mineralogical composition of limestone, and some rocks described as felsic lavas now are understood to be welded tuffs. Such new knowledge is recognized by changing the lithic designation of the unit, while retaining the original geographic term. Similarly, changes in the classification of igneous rocks have resulted in recognition that rocks originally described as quartz monzonite now are more appropriately termed granite. Such lithic designations may be modernized when the new classification is widely adopted. If heterogeneous bodies of plutonic rock have been misleadingly identified with a single compositional term, such as "gabbro," the adoption of a neutral term, such as "intrusion" or "pluton," may be advisable.

Article 19.—**Revision.** Revision involves either minor changes in the definition of one or both boundaries of a unit, or in the unit's rank.

Remarks. (a) **Boundary change.**—Revision is justifiable if a minor change in boundary or content will make a unit more natural and useful. If revision modifies only a minor part of the content of a previously established unit, the original name may be retained.

(b) **Change in rank.**—Change in rank of a stratigraphic or temporal unit requires neither redefinition of its boundaries nor alteration of the geographic part of its name. A member may become a formation or vice versa, a formation may become a group or vice versa, and a lithodeme may become a suite or vice versa.

(c) **Examples of changes from area to area.**—The Conasauga Shale is recognized as a formation in Georgia and as a group in eastern Tennessee; the Osgood Formation, Laurel Limestone, and Waldron Shale in Indiana are classed as members of the Wayne Formation in a part of Tennessee; the Virgelle Sandstone is a formation in western Montana and a member of the Eagle Sandstone in central Montana; the Skull Creek Shale and the Newcastle Sandstone in North Dakota are members of the Ashville Formation in Manitoba.

(d) **Example of change in single area.**—The rank of a unit may be changed without changing its content. For example, the Madison Limestone of early work in Montana later became the Madison Group, containing several formations.

(e) **Retention of type section.**—When the rank of a geologic unit is changed, the original type section or type locality is retained for the newly ranked unit (see Article 22c).

(f) **Different geographic name for a unit and its parts.**—In changing the rank of a unit, the same name may not be applied both to the unit as a whole and to a part of it. For example, the Astoria Group should not contain an Astoria Sandstone, nor the Washington Formation, a Washington Sandstone Member.

(g) **Undesirable restriction.**—When a unit is divided into two or more of the same rank as the original, the original name should not be used for any of the divisions. Retention of the old name for one of the units precludes use of the name in a term of higher rank. Furthermore, in order to understand an author's meaning, a later reader would have to know about the modification and its date, and whether the author is following the original or the modified usage. For these reasons, the normal practice is to raise the rank of an established unit when units of the same rank are recognized and mapped within it.

Article 20.—**Abandonment.** An improperly defined or obsolete stratigraphic, lithodemic, or temporal unit may be formally abandoned, provided that (a) sufficient justification is presented to demonstrate a concern for nomenclatural stability, and (b) recommendations are made for the classification and nomenclature to be used in its place.

Remarks. (a) **Reasons for abandonment.**—A formally defined unit may be abandoned by the demonstration of synonymy or homonymy, of assignment to an improper category (for example, definition of a lithostratigraphic unit in a chronostratigraphic sense), or of other direct violations of a stratigraphic code or procedures prevailing at the time of the original definition. Disuse, or the lack of need or useful purpose for a unit, may be a basis for abandonment; so, too, may widespread misuse in diverse ways which compound confusion. A unit also may be abandoned if it proves impracticable, neither recognizable nor mappable elsewhere.

(b) **Abandoned names.**—A name for a lithostratigraphic or lithodemic unit, once applied and then abandoned, is available for some other unit only if the name was introduced casually, or if it has been published only once in the last several decades and is not in current usage, and if its reintroduction will cause no confusion. An explanation of the history of the name and of the new usage should be a part of the designation.

(c) **Obsolete names.**—Authors may refer to national and provincial records of stratigraphic names to determine whether a name is obsolete (see Article 7b).

(d) **Reference to abandoned names.**—When it is useful to refer to an obsolete or abandoned formal name, its status is made clear by some such term as "abandoned" or "obsolete," and by using a phrase such as "La Plata Sandstone of Cross (1898)". (The same phrase also is used to convey that a named unit has not yet been adopted for usage by the organization involved.)

(e) **Reinstatement.**—A name abandoned for reasons that seem valid at the time, but which subsequently are found to be erroneous, may be reinstated. Example: the Washakie Formation, defined in 1869, was abandoned in 1918 and reinstated in 1973.

CODE AMENDMENT

Article 21.—**Procedure for Amendment.** Additions to, or changes of, this Code may be proposed in writing to the Commission by any geoscientist at any time. If accepted for consideration by a majority vote of the Commission, they may be adopted by a two-thirds vote of the Commission at an annual meeting not less than a year after publication of the proposal.

FORMAL UNITS DISTINGUISHED BY CONTENT, PROPERTIES, OR PHYSICAL LIMITS

LITHOSTRATIGRAPHIC UNITS

Nature and Boundaries

Article 22.—**Nature of Lithostratigraphic Units.** A lithostratigraphic unit is a defined body of sedimentary, extrusive igneous, metasedimentary, or metavolcanic strata which is distinguished and delimited on the basis of lithic characteristics and stratigraphic position. A lithostratigraphic unit generally conforms to the Law of Superposition and commonly is stratified and tabular in form.

Remarks. (a) **Basic units.**—Lithostratigraphic units are the basic units of general geologic work and serve as the foundation for delineating strata, local and regional structure, economic resources, and geologic history in regions of stratified rocks. They are recognized and defined by observable rock characteristics; boundaries may be placed at clearly distinguished contacts or drawn arbitrarily within a zone of gradation. Lithification or cementation is not a necessary property; clay, gravel, till, and other unconsolidated deposits may constitute valid lithostratigraphic units.

(b) **Type section and locality.**—The definition of a lithostratigraphic unit should be based, if possible, on a stratotype consisting of readily accessible rocks in place, e.g., in outcrops, excavations, and mines, or of rocks accessible only to remote sampling devices, such as those in drill holes and underwater. Even where remote methods are used, definitions must be based on lithic criteria and not on the geophysical characteristics of the rocks, nor the implied age of their contained fossils. Definitions must be based on descriptions of actual rock material. Regional validity must be demonstrated for all such units. In regions where the stratigraphy has been established through studies of surface exposures, the naming of new units in the subsurface is justified only where the subsurface section differs materially from the surface section, or where there is doubt as to the equivalence of a subsurface and a surface unit. The establishment of subsurface reference sections for units originally defined in outcrop is encouraged.

(c) **Type section never changed.**—The definition and name of a lithostratigraphic unit are established at a type section (or locality) that, once specified, must not be changed. If the type section is poorly designated or delimited, it may be redefined subsequently. If the originally specified stratotype is incomplete, poorly exposed, structurally complicated, or unrepresentative of the unit, a principal reference section or several reference sections may be designated to supplement, but not to supplant, the type section (Article 8e).

(d) **Independence from inferred geologic history.**—Inferred geologic history, depositional environment, and biological sequence have no place in the definition of a lithostratigraphic unit, which must be based on composition and other lithic characteristics; nevertheless, considerations of well-documented geologic history properly may influence the choice of vertical and lateral boundaries of a new unit. Fossils may be valuable during mapping in distinguishing between two lithologically similar, noncontiguous lithostratigraphic units. The fossil content of a lithostratigraphic unit is a legitimate lithic characteristic; for example, oyster-rich sandstone, coquina, coral reef, or graptolitic shale. Moreover, otherwise similar units, such as the Formación Mendez and Formación Velasco mudstones, may be distinguished on the basis of coarseness of contained fossils (foraminifera).

(e) **Independence from time concepts.**—The boundaries of most lithostratigraphic units may transgress time horizons, but some may be approximately synchronous. Inferred time-spans, however measured, play no part in differentiating or determining the boundaries of any lithostratigraphic unit. Either relatively short or relatively long intervals of time may be represented by a single unit. The accumulation of material assigned to a particular unit may have begun or ended earlier in some localities than in others; also, removal of rock by erosion, either within the time-span of deposition of the unit or later, may reduce the time-span represented by the unit locally. The body in some places may be entirely younger than in other places. On the other hand, the establishment of formal units that straddle known, identifiable, regional disconformities is to be avoided, if at all possible. Although concepts of time or age play no part in defining lithostratigraphic units nor in determining their boundaries, evidence of age may aid recognition of similar lithostratigraphic units at localities far removed from the type sections or areas.

(f) **Surface form.**—Erosional morphology or secondary surface form may be a factor in the recognition of a lithostratigraphic unit, but properly should play a minor part at most in the definition of such units. Because the surface expression of lithostratigraphic units is an important aid in mapping, it is commonly advisable, where other factors do not countervail, to define lithostratigraphic boundaries so as to coincide with lithic changes that are expressed in topography.

(g) **Economically exploited units.**—Aquifers, oil sands, coal beds, and quarry layers are, in general, informal units even though named. Some such units, however, may be recognized formally as beds, members, or formations because they are important in the elucidation of regional stratigraphy.

(h) **Instrumentally defined units.**—In subsurface investigations, certain bodies of rock and their boundaries are widely recognized on borehole geophysical logs showing their electrical resistivity, radioactivity, density, or other physical properties. Such bodies and their boundaries may or may not correspond to formal lithostratigraphic units and their boundaries. Where other considerations do not countervail, the boundaries of subsurface units should be defined so as to correspond to useful geophysical markers; nevertheless, units defined exclusively on the basis of remotely sensed physical properties, although commonly useful in stratigraphic analysis, stand completely apart from the hierarchy of formal lithostratigraphic units and are considered informal.

(i) **Zone.**—As applied to the designation of lithostratigraphic units, the term "zone" is informal. Examples are "producing zone," "mineralized zone," "metamorphic zone," and "heavy-mineral zone." A zone may include all or parts of a bed, a member, a formation, or even a group.

(j) **Cyclothems.**—Cyclic or rhythmic sequences of sedimentary rocks, whose repetitive divisions have been named cyclothems, have been recognized in sedimentary basins around the world. Some cyclothems have

been identified by geographic names, but such names are considered informal. A clear distinction must be maintained between the division of a stratigraphic column into cyclothems and its division into groups, formations, and members. Where a cyclothem is identified by a geographic name, the word *cyclothem* should be part of the name, and the geographic term should not be the same as that of any formal unit embraced by the cyclothem.

(k) **Soils and paleosols.**—Soils and paleosols are layers composed of the in-situ products of weathering of older rocks which may be of diverse composition and age. Soils and paleosols differ in several respects from lithostratigraphic units, and should not be treated as such (see "Pedostratigraphic Units," Articles 55 et seq).

(l) **Depositional facies.**—Depositional facies are informal units, whether objective (conglomeratic, black shale, graptolitic) or genetic and environmental (platform, turbiditic, fluvial), even when a geographic term has been applied, e.g., Lantz Mills facies. Descriptive designations convey more information than geographic terms and are preferable.

Article 23.—Boundaries. Boundaries of lithostratigraphic units are placed at positions of lithic change. Boundaries are placed at distinct contacts or may be fixed arbitrarily within zones of gradation (Fig. 2a). Both vertical and lateral boundaries are based on the lithic criteria that provide the greatest unity and utility.

Remarks. (a) **Boundary in a vertically gradational sequence.**—A named lithostratigraphic unit is preferably bounded by a single lower and a single upper surface so that the name does not recur in a normal stratigraphic succession (see Remark b). Where a rock unit passes vertically into another by intergrading or interfingering of two or more kinds of rock, unless the gradational strata are sufficiently thick to warrant designation of a third, independent unit, the boundary is necessarily arbitrary and should be selected on the basis of practicality (Fig. 2b). For example, where a shale unit overlies a unit of interbedded limestone and shale, the boundary commonly is placed at the top of the highest readily traceable limestone bed. Where a sandstone unit grades upward into shale, the boundary may be so gradational as to be difficult to place even arbitrarily; ideally it should be drawn at the level where the rock is composed of one-half of each component. Because of creep in outcrops and caving in boreholes, it is generally best to define such arbitrary boundaries by the highest occurrence of a particular rock type, rather than the lowest.

(b) **Boundaries in lateral lithologic change.**—Where a unit changes laterally through abrupt gradation into, or intertongues with, a markedly different kind of rock, a new unit should be proposed for the different rock type. An arbitrary lateral boundary may be placed between the two equivalent units. Where the area of lateral intergradation or intertonguing is sufficiently extensive, a transitional interval of interbedded rocks may constitute a third independent unit (Fig. 2c). Where tongues (Article 25b) of formations are mapped separately or otherwise set apart without being formally named, the unmodified formation name should not be repeated in a normal stratigraphic sequence, although the modified name may be repeated in such phrases as "lower tongue of Mancos Shale" and "upper tongue of Mancos Shale." To show the order of superposition on maps and cross sections, the unnamed tongues may be distinguished informally (Fig. 2d) by number, letter, or other means. Such relationships may also be dealt with informally through the recognition of depositional facies (Article 22-1).

(c) **Key beds used for boundaries.**—Key beds (Article 26b) may be used as boundaries for a formal lithostratigraphic unit where the internal lithic characteristics of the unit remain relatively constant. Even though bounding key beds may be traceable beyond the area of the diagnostic overall rock type, geographic extension of the lithostratigraphic unit bounded thereby is not necessarily justified. Where the rock between key beds becomes drastically different from that of the type locality, a new name should be applied (Fig. 2e), even though the key beds are continuous (Article 26b). Stratigraphic and sedimentologic studies of stratigraphic units (usually informal) bounded by key beds may be very informative and useful, especially in subsurface work where the key beds may be recognized by their geophysical signatures. Such units, however, may be a kind of chronostratigraphic, rather than lithostratigraphic, unit (Article 75, 75c), although others are diachronous because one, or both, of the key beds are also diachronous.

(d) **Unconformities as boundaries.**—Unconformities, where recognizable objectively on lithic criteria, are ideal boundaries for lithostratigraphic units. However, a sequence of similar rocks may include an

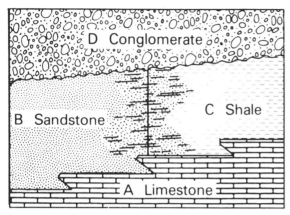

A.--Boundaries at sharp lithologic contacts and in laterally gradational sequence.

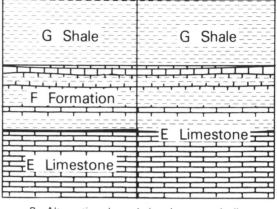

B.--Alternative boundaries in a vertically gradational or interlayered sequence.

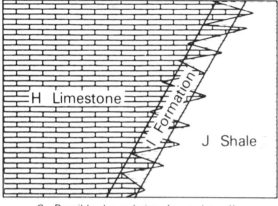

C.--Possible boundaries for a laterally intertonguing sequence.

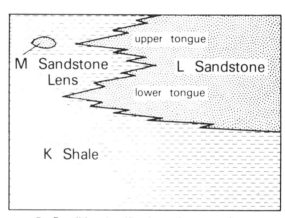

D.--Possible classification of parts of an intertonguing sequence.

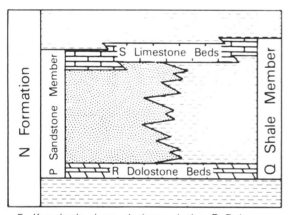

E.--Key beds, here designated the R Dolostone Beds and the S Limestone Beds, are used as boundaries to distinguish the Q Shale Member from the other parts of the N Formation. A lateral change in composition between the key beds requires that another name, P Sandstone Member, be applied. The key beds are part of each member.

EXPLANATION

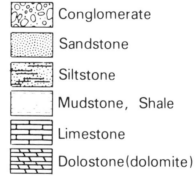

FIG. 2.—**Diagrammatic examples of lithostratigraphic boundaries and classification.**

obscure unconformity so that separation into two units may be desirable but impracticable. If no lithic distinction adequate to define a widely recognizable boundary can be made, only one unit should be recognized, even though it may include rock that accumulated in different epochs, periods, or eras.

(e) **Correspondence with genetic units.**—The boundaries of lithostratigraphic units should be chosen on the basis of lithic changes and, where feasible, to correspond with the boundaries of genetic units, so that subsequent studies of genesis will not have to deal with units that straddle formal boundaries.

Ranks of Lithostratigraphic Units

Article 24.—Formation. The formation is the fundamental unit in lithostratigraphic classification. A formation is a body of rock identified by lithic characteristics and stratigraphic position; it is prevailingly but not necessarily tabular and is mappable at the Earth's surface or traceable in the subsurface.

Remarks. (a) **Fundamental unit.**—Formations are the basic lithostratigraphic units used in describing and interpreting the geology of a region. The limits of a formation normally are those surfaces of lithic change that give it the greatest practicable unity of constitution. A formation may represent a long or short time interval, may be composed of materials from one or several sources, and may include breaks in deposition (see Article 23d).

(b) **Content.**—A formation should possess some degree of internal lithic homogeneity or distinctive lithic features. It may contain between its upper and lower limits (i) rock of one lithic type, (ii) repetitions of two or more lithic types, or (iii) extreme lithic heterogeneity which in itself may constitute a form of unity when compared to the adjacent rock units.

(c) **Lithic characteristics.**—Distinctive lithic characteristics include chemical and mineralogical composition, texture, and such supplementary features as color, primary sedimentary or volcanic structures, fossils (viewed as rock-forming particles), or other organic content (coal, oil-shale). A unit distinguishable only by the taxonomy of its fossils is not a lithostratigraphic but a biostratigraphic unit (Article 48). Rock type may be distinctively represented by electrical, radioactive, seismic, or other properties (Article 22h), but these properties by themselves do not describe adequately the lithic character of the unit.

(d) **Mappability and thickness.**—The proposal of a new formation must be based on tested mappability. Well-established formations commonly are divisible into several widely recognizable lithostratigraphic units; where formal recognition of these smaller units serves a useful purpose, they may be established as members and beds, for which the requirement of mappability is not mandatory. A unit formally recognized as a formation in one area may be treated elsewhere as a group, or as a member of another formation, without change of name. Example: the Niobrara is mapped at different places as a member of the Mancos Shale, of the Cody Shale, or of the Colorado Shale, and also as the Niobrara Formation, as the Niobrara Limestone, and as the Niobrara Shale.

Thickness is not a determining parameter in dividing a rock succession into formations; the thickness of a formation may range from a feather edge at its depositional or erosional limit to thousands of meters elsewhere. No formation is considered valid that cannot be delineated at the scale of geologic mapping practiced in the region when the formation is proposed. Although representation of a formation on maps and cross sections by a labeled line may be justified, proliferation of such exceptionally thin units is undesirable. The methods of subsurface mapping permit delineation of units much thinner than those usually practicable for surface studies; before such thin units are formalized, consideration should be given to the effect on subsequent surface and subsurface studies.

(e) **Organic reefs and carbonate mounds.**—Organic reefs and carbonate mounds ("buildups") may be distinguished formally, if desirable, as formations distinct from their surrounding, thinner, temporal equivalents. For the requirements of formalization, see Article 30f.

(f) **Interbedded volcanic and sedimentary rock.**—Sedimentary rock and volcanic rock that are interbedded may be assembled into a formation under one name which should indicate the predominant or distinguishing lithology, such as Mindego Basalt.

(g) **Volcanic rock.**—Mappable distinguishable sequences of stratified volcanic rock should be treated as formations or lithostratigraphic units of higher or lower rank. A small intrusive component of a dominantly stratiform volcanic assemblage may be treated informally.

(h) **Metamorphic rock.**—Formations composed of low-grade metamorphic rock (defined for this purpose as rock in which primary structures are clearly recognizable) are, like sedimentary formations, distinguished mainly by lithic characteristics. The mineral facies may differ from place to place, but these variations do not require definition of a new formation. High-grade metamorphic rocks whose relation to established formations is uncertain are treated as lithodemic units (see Articles 31 et seq).

Article 25.—Member. A member is the formal lithostratigraphic unit next in rank below a formation and is always a part of some formation. It is recognized as a named entity within a formation because it possesses characteristics distinguishing it from adjacent parts of the formation. A formation need not be divided into members unless a useful purpose is served by doing so. Some formations may be divided completely into members; others may have only certain parts designated as members; still others may have no members. A member may extend laterally from one formation to another.

Remarks. (a) **Mapping of members.**—A member is established when it is advantageous to recognize a particular part of a heterogeneous formation. A member, whether formally or informally designated, need not be mappable at the scale required for formations. Even if all members of a formation are locally mappable, it does not follow that they should be raised to formational rank, because proliferation of formation names may obscure rather than clarify relations with other areas.

(b) **Lens and tongue.**—A geographically restricted member that terminates on all sides within a formation may be called a lens (lentil). A wedging member that extends outward beyond a formation or wedges ("pinches") out within another formation may be called a tongue.

(c) **Organic reefs and carbonate mounds.**—Organic reefs and carbonate mounds may be distinguished formally, if desirable, as members within a formation. For the requirements of formalization, see Article 30f.

(d) **Division of members.**—A formally or informally recognized division of a member is called a bed or beds, except for volcanic flow-rocks, for which the smallest formal unit is a flow. Members may contain beds or flows, but may never contain other members.

(e) **Laterally equivalent members.**—Although members normally are in vertical sequence, laterally equivalent parts of a formation that differ recognizably may also be considered members.

Article 26.—Bed(s). A bed, or beds, is the smallest formal lithostratigraphic unit of sedimentary rocks.

Remarks. (a) **Limitations.**—The designation of a bed or a unit of beds as a formally named lithostratigraphic unit generally should be limited to certain distinctive beds whose recognition is particularly useful. Coal beds, oil sands, and other beds of economic importance commonly are named, but such units and their names usually are not a part of formal stratigraphic nomenclature (Articles 22g and 30g).

(b) **Key or marker beds.**—A key or marker bed is a thin bed of distinctive rock that is widely distributed. Such beds may be named, but usually are considered informal units. Individual key beds may be traced beyond the lateral limits of a particular formal unit (Article 23c).

Article 27.—Flow. A flow is the smallest formal lithostratigraphic unit of volcanic flow rocks. A flow is a discrete, extrusive, volcanic body distinguishable by texture, composition, order of superposition, paleomagnetism, or other objective criteria. It is part of a member and thus is equivalent in rank to a bed or beds of sedimentary-rock classification. Many flows are informal units. The designation and naming of flows as formal rock-stratigraphic units should be limited to those that are distinctive and widespread.

Article 28.—Group. A group is the lithostratigraphic unit next higher in rank to formation; a group may consist entirely of named formations, or alternatively, need not be composed entirely of named formations.

Remarks. (a) **Use and content.**—Groups are defined to express the natural relationships of associated formations. They are useful in small-scale mapping and regional stratigraphic analysis. In some reconnaissance work, the term "group" has been applied to lithostratigraphic units that appear to be divisible into formations, but have not yet been so divided. In such cases, formations may be erected subsequently for one or all of the practical divisions of the group.

(b) **Change in component formations.**—The formations making up a group need not necessarily be everywhere the same. The Rundle Group, for example, is widespread in western Canada and undergoes several changes in formational content. In southwestern Alberta, it comprises the Livingstone, Mount Head, and Etherington Formations in the Front Ranges, whereas in the foothills and subsurface of the adjacent plains, it comprises the Pekisko, Shunda, Turner Valley, and Mount Head Formations. However, a formation or its parts may not be assigned to two vertically adjacent groups.

(c) **Change in rank.**—The wedge-out of a component formation or formations may justify the reduction of a group to formation rank, retaining the same name. When a group is extended laterally beyond where it is divided into formations, it becomes in effect a formation, even if it is still called a group. When a previously established formation is divided into two or more component units that are given formal formation rank, the old formation, with its old geographic name, should be raised to group status. Raising the rank of the unit is preferable to restricting the old name to a part of its former content, because a change in rank leaves the sense of a well-established unit unchanged (Articles 19b, 19g).

Article 29.—Supergroup. A supergroup is a formal assemblage of related or superposed groups, or of groups and formations. Such units have proved useful in regional and provincial syntheses. Supergroups should be named only where their recognition serves a clear purpose.

Remark. (a) **Misuse of "series" for group or supergroup.**—Although "series" is a useful general term, it is applied formally only to a chronostratigraphic unit and should not be used for a lithostratigraphic unit. The term "series" should no longer be employed for an assemblage of formations or an assemblage of formations and groups, as it has been, especially in studies of the Precambrian. These assemblages are groups or supergroups.

Lithostratigraphic Nomenclature

Article 30.—Compound Character. The formal name of a lithostratigraphic unit is compound. It consists of a geographic name combined with a descriptive lithic term or with the appropriate rank term, or both. Initial letters of all words used in forming the names of formal rock-stratigraphic units are capitalized.

Remarks. (a) **Omission of part of a name.**—Where frequent repetition would be cumbersome, the geographic name, the lithic term, or the rank term may be used alone, once the full name has been introduced; as "the Burlington," "the limestone," or "the formation," for the Burlington Limestone.

(b) **Use of simple lithic terms.**—The lithic part of the name should indicate the predominant or diagnostic lithology, even if subordinate lithologies are included. Where a lithic term is used in the name of a lithostratigraphic unit, the simplest generally acceptable term is recommended (for example, limestone, sandstone, shale, tuff, quartzite). Compound terms (for example, clay shale) and terms that are not in common usage (for example, calcirudite, orthoquartzite) should be avoided. Combined terms, such as "sand and clay," should not be used for the lithic part of the names of lithostratigraphic units, nor should an adjective be used between the geographic and the lithic terms, as "Chattanooga Black Shale" and "Biwabik Iron-Bearing Formation."

(c) **Group names.**—A group name combines a geographic name with the term "group," and no lithic designation is included; for example, San Rafael Group.

(d) **Formation names.**—A formation name consists of a geographic name followed by a lithic designation or by the word "formation." Examples: Dakota Sandstone, Mitchell Mesa Rhyolite, Monmouth Formation, Halton Till.

(e) **Member names.**—All member names include a geographic term and the word "member;" some have an intervening lithic designation, if useful; for example, Wedington Sandstone Member of the Fayetteville Shale. Members designated solely by lithic character (for example, siliceous shale member), by position (upper, lower), or by letter or number, are informal.

(f) **Names of reefs.**—Organic reefs identified as formations or members are formal units only where the name combines a geographic name with the appropriate rank term, e.g., Leduc Formation (a name applied to the several reefs enveloped by the Ireton Formation), Rainbow Reef Member.

(g) **Bed and flow names.**—The names of beds or flows combine a geographic term, a lithic term, and the term "bed" or "flow;" for example, Knee Hills Tuff Bed, Ardmore Bentonite Beds, Negus Variolitic Flows.

(h) **Informal units.**—When geographic names are applied to such informal units as oil sands, coal beds, mineralized zones, and informal members (see Articles 22g and 26a), the unit term should not be capitalized. A name is not necessarily formal because it is capitalized, nor does failure to capitalize a name render it informal. Geographic names should be combined with the terms "formation" or "group" only in formal nomenclature.

(i) **Informal usage of identical geographic names.**—The application of identical geographic names to several minor units in one vertical sequence is considered informal nomenclature (lower Mount Savage coal, Mount Savage fireclay, upper Mount Savage coal, Mount Savage rider coal, and Mount Savage sandstone). The application of identical geographic names to the several lithologic units constituting a cyclothem likewise is considered informal.

(j) **Metamorphic rock.**—Metamorphic rock recognized as a normal stratified sequence, commonly low-grade metavolcanic or metasedimentary rocks, should be assigned to named groups, formations, and members, such as the Deception Rhyolite, a formation of the Ash Creek Group, or the Bonner Quartzite, a formation of the Missoula Group. High-grade metamorphic and metasomatic rocks are treated as lithodemes and suites (see Articles 31, 33, 35).

(k) **Misuse of well-known name.**—A name that suggests some well-known locality, region, or political division should not be applied to a unit typically developed in another less well-known locality of the same name. For example, it would be inadvisable to use the name "Chicago Formation" for a unit in California.

LITHODEMIC UNITS

Nature and Boundaries

Article 31.—Nature of Lithodemic Units. A lithodemic[6] unit is a defined body of predominantly intrusive, highly deformed, and/or highly metamorphosed rock, distinguished and delimited on the basis of rock characteristics. In contrast to lithostratigraphic units, a lithodemic unit generally does not conform to the Law of Superposition. Its contacts with other rock units may be sedimentary, extrusive, intrusive, tectonic, or metamorphic (Fig. 3).

Remarks. (a) **Recognition and definition.**—Lithodemic units are defined and recognized by observable rock characteristics. They are the practical units of general geological work in terranes in which rocks generally lack primary stratification; in such terranes they serve as the foundation for studying, describing, and delineating lithology, local and regional structure, economic resources, and geologic history.

(b) **Type and reference localities.**—The definition of a lithodemic unit should be based on as full a knowledge as possible of its lateral and vertical variations and its contact relationships. For purposes of nomenclatural stability, a type locality and, wherever appropriate, reference localities should be designated.

(c) **Independence from inferred geologic history.**—Concepts based on inferred geologic history properly play no part in the definition of a lithodemic unit. Nevertheless, where two rock masses are lithically similar but display objective structural relations that preclude the possibility of their being even broadly of the same age, they should be assigned to different lithodemic units.

(d) **Use of "zone."**—As applied to the designation of lithodemic units, the term "zone" is informal. Examples are: "mineralized zone," "contact zone," and "pegmatitic zone."

---

[6]From the Greek *demas, -os:* "living body, frame".

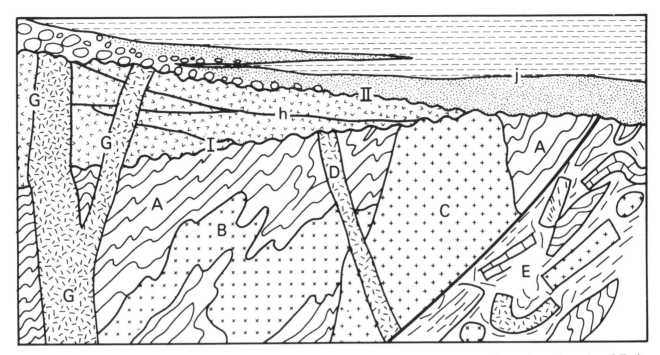

FIG. 3.—Lithodemic (upper case) and lithostratigraphic (lower case) units. A *lithodeme* of *gneiss* (A) contains an *intrusion* of diorite
(B) that was deformed with the gneiss. A and B may be treated jointly as a *complex*. A younger *granite* (C) is cut by a dike of *syenite*
(D), that is cut in turn by unconformity I. All the foregoing are in fault contact with a *structural complex* (E). A *volcanic complex* (G)
is built upon unconformity I, and its feeder dikes cut the unconformity. Laterally equivalent volcanic strata in orderly, mappable succession (h) are treated as lithostratigraphic units. A *gabbro* feeder (G′), to the volcanic complex, where surrounded by gneiss is readily
distinguished as a separate lithodeme and named as a *gabbro* or an *intrusion*. All the foregoing are overlain, at unconformity II, by
sedimentary rocks (j) divided into formations and members.

Article 32.—**Boundaries.** Boundaries of lithodemic units are
placed at positions of lithic change. They may be placed at clearly
distinguished contacts or within zones of gradation. Boundaries,
both vertical and lateral, are based on the lithic criteria that provide the greatest unity and practical utility. Contacts with other
lithodemic and lithostratigraphic units may be depositional,
intrusive, metamorphic, or tectonic.

Remark. (a) **Boundaries within gradational zones.**—Where a lithodemic unit changes through gradation into, or intertongues with, a rock-
mass with markedly different characteristics, it is usually desirable to
propose a new unit. It may be necessary to draw an arbitrary boundary
within the zone of gradation. Where the area of intergradation or intertonguing is sufficiently extensive, the rocks of mixed character may constitute a third unit.

Ranks of Lithodemic Units

Article 33.—**Lithodeme.** The lithodeme is the fundamental
unit in lithodemic classification. A lithodeme is a body of intrusive, pervasively deformed, or highly metamorphosed rock, generally non-tabular and lacking primary depositional structures,
and characterized by lithic homogeneity. It is mappable at the
Earth's surface and traceable in the subsurface. For cartographic
and hierarchical purposes, it is comparable to a formation (see
Table 2).

Remarks. (a) **Content.**—A lithodeme should possess distinctive
lithic features and some degree of internal lithic homogeneity. It may consist of (i) rock of one type, (ii) a mixture of rocks of two or more types, or
(iii) extreme heterogeneity of composition, which may constitute in itself
a form of unity when compared to adjoining rock-masses (see also "complex," Article 37).
(b) **Lithic characteristics.**—Distinctive lithic characteristics may
include mineralogy, textural features such as grain size, and structural
features such as schistose or gneissic structure. A unit distinguishable

from its neighbors only by means of chemical analysis is informal.
(c) **Mappability.**—Practicability of surface or subsurface mapping is
an essential characteristic of a lithodeme (see Article 24d).

Article 34.—**Division of Lithodemes.** Units below the rank of
lithodeme are informal.

Article 35.—**Suite.** A *suite* (metamorphic suite, intrusive
suite, plutonic suite) is the lithodemic unit next higher in rank to
lithodeme. It comprises two or more associated lithodemes of the
same class (e.g., plutonic, metamorphic). For cartographic and
hierarchical purposes, suite is comparable to group (see Table 2).

Remarks. (a) **Purpose.**—Suites are recognized for the purpose of
expressing the natural relations of associated lithodemes having significant lithic features in common, and of depicting geology at compilation
scales too small to allow delineation of individual lithodemes. Ideally, a
suite consists entirely of named lithodemes, but may contain both named
and unnamed units.
(b) **Change in component units.**—The named and unnamed units
constituting a suite may change from place to place, so long as the original
sense of natural relations and of common lithic features is not violated.
(c) **Change in rank.**—Traced laterally, a suite may lose all of its formally named divisions but remain a recognizable, mappable entity. Under
such circumstances, it may be treated as a lithodeme but retain the same
name. Conversely, when a previously established lithodeme is divided
into two or more mappable divisions, it may be desirable to raise its rank
to suite, retaining the original geographic component of the name. To
avoid confusion, the original name should not be retained for one of the
divisions of the original unit (see Article 19g).

Article 36.—**Supersuite.** A supersuite is the unit next higher in
rank to a suite. It comprises two or more suites or complexes having a degree of natural relationship to one another, either in the
vertical or the lateral sense. For cartographic and hierarchical
purposes, supersuite is similar in rank to supergroup.

Article 37.—**Complex.** An assemblage or mixture of rocks of *two or more genetic classes*, i.e., igneous, sedimentary, or metamorphic, with or without highly complicated structure, may be named a *complex*. The term "complex" takes the place of the lithic or rank term (for example, Boil Mountain Complex, Franciscan Complex) and, although unranked, commonly is comparable to suite or supersuite and is named in the same manner (Articles 41, 42).

Remarks (a) **Use of "complex."**—Identification of an assemblage of diverse rocks as a complex is useful where the mapping of each separate lithic component is impractical at ordinary mapping scales. "Complex" is unranked but commonly comparable to suite or supersuite; therefore, the term may be retained if subsequent, detailed mapping distinguishes some or all of the component lithodemes or lithostratigraphic units.

(b) **Volcanic complex.**—Sites of persistent volcanic activity commonly are characterized by a diverse assemblage of extrusive volcanic rocks, related intrusions, and their weathering products. Such an assemblage may be designated a *volcanic complex*.

(c) **Structural complex.**—In some terranes, tectonic processes (e.g., shearing, faulting) have produced heterogeneous mixtures or disrupted bodies of rock in which some individual components are too small to be mapped. *Where there is no doubt that the mixing or disruption is due to tectonic processes,* such a mixture may be designated as a structural complex, whether it consists of two or more classes of rock, or a single class only. A simpler solution for some mapping purposes is to indicate intense deformation by an overprinted pattern.

(d) **Misuse of "complex".**—Where the rock assemblage to be united under a single, formal name consists of diverse types of a *single class* of rock, as in many terranes that expose a variety of either intrusive igneous or high-grade metamorphic rocks, the term "intrusive suite," "plutonic suite," or "metamorphic suite" should be used, rather than the unmodified term "complex." Exceptions to this rule are the terms *structural complex* and *volcanic complex* (see Remarks c and b, above).

Article 38.—**Misuse of "Series" for Suite, Complex, or Supersuite.** The term "series" has been employed for an assemblage of lithodemes or an assemblage of lithodemes and suites, especially in studies of the Precambrian. This practice now is regarded as improper; these assemblages are suites, complexes, or supersuites. The term "series" also has been applied to a sequence of rocks resulting from a succession of eruptions or intrusions. In these cases a different term should be used; "group" should replace "series" for volcanic and low-grade metamorphic rocks, and "intrusive suite" or "plutonic suite" should replace "series" for intrusive rocks of group rank.

Lithodemic Nomenclature

Article 39.—**General Provisions.** The formal name of a lithodemic unit is compound. It consists of a geographic name combined with a descriptive or appropriate rank term. The principles for the selection of the geographic term, concerning suitability, availability, priority, etc, follow those established in Article 7, where the rules for capitalization are also specified.

Article 40.—**Lithodeme Names.** The name of a lithodeme combines a geographic term with a lithic or descriptive term, e.g., Killarney Granite, Adamant Pluton, Manhattan Schist, Skaergaard Intrusion, Duluth Gabbro. The term *formation* should not be used.

Remarks. (a) **Lithic term.**—The lithic term should be a common and familiar term, such as schist, gneiss, gabbro. Specialized terms and terms not widely used, such as websterite and jacupirangite, and compound terms, such as graphitic schist and augen gneiss, should be avoided.

(b) **Intrusive and plutonic rocks.**—Because many bodies of intrusive rock range in composition from place to place and are difficult to characterize with a single lithic term, and because many bodies of plutonic rock

are considered not to be intrusions, latitude is allowed in the choice of a lithic or descriptive term. Thus, the descriptive term should preferably be compositional (e.g., gabbro, granodiorite), but may, if necessary, denote form (e.g., dike, sill), or be neutral (e.g., intrusion, pluton[7]). In any event, specialized compositional terms not widely used are to be avoided, as are form terms that are not widely used, such as bysmalith and chonolith. Terms implying genesis should be avoided as much as possible, because interpretations of genesis may change.

Article 41.—**Suite Names.** The name of a suite combines a geographic term, the term "suite," and an adjective denoting the fundamental character of the suite; for example, Idaho Springs Metamorphic Suite, Tuolumne Intrusive Suite, Cassiar Plutonic Suite. The geographic name of a suite may not be the same as that of a component lithodeme (see Article 19f). Intrusive assemblages, however, may share the same geographic name if an intrusive lithodeme is representative of the suite.

Article 42.—**Supersuite Names.** The name of a supersuite combines a geographic term with the term "supersuite."

MAGNETOSTRATIGRAPHIC UNITS

Nature and Boundaries

Article 43.—**Nature of Magnetostratigraphic Units.** A magnetostratigraphic unit is a body of rock unified by specified remanent-magnetic properties and is distinct from underlying and overlying magnetostratigraphic units having different magnetic properties.

Remarks. (a) **Definition.**—Magnetostratigraphy is defined here as all aspects of stratigraphy based on remanent magnetism (paleomagnetic signatures). Four basic paleomagnetic phenomena can be determined or inferred from remanent magnetism: polarity, dipole-field-pole position (including apparent polar wander), the non-dipole component (secular variation), and field intensity.

(b) **Contemporaneity of rock and remanent magnetism.**—Many paleomagnetic signatures reflect earth magnetism at the time the rock formed. Nevertheless, some rocks have been subjected subsequently to physical and/or chemical processes which altered the magnetic properties. For example, a body of rock may be heated above the blocking temperature or Curie point for one or more minerals, or a ferromagnetic mineral may be produced by low-temperature alteration long after the enclosing rock formed, thus acquiring a component of remanent magnetism reflecting the field at the time of alteration, rather than the time of original rock deposition or crystallization.

(c) **Designations and scope.**—The prefix *magneto* is used with an appropriate term to designate the aspect of remanent magnetism used to define a unit. The terms "magnetointensity" or "magnetosecular-variation" are possible examples. This Code considers only polarity reversals, which now are recognized widely as a stratigraphic tool. However, apparent-polar-wander paths offer increasing promise for correlations within Precambrian rocks.

Article 44.—**Definition of Magnetopolarity Unit.** A magnetopolarity unit is a body of rock unified by its remanent magnetic polarity and distinguished from adjacent rock that has different polarity.

Remarks. (a) **Nature.**—Magnetopolarity is the record in rocks of the polarity history of the Earth's magnetic-dipole field. Frequent past reversals of the polarity of the Earth's magnetic field provide a basis for magnetopolarity stratigraphy.

(b) **Stratotype.**—A stratotype for a magnetopolarity unit should be designated and the boundaries defined in terms of recognized lithostratigraphic and/or biostratigraphic units in the stratotype. The formal definition of a magnetopolarity unit should meet the applicable specific requirements of Articles 3 to 16.

(c) **Independence from inferred history.**—Definition of a magnetopolarity unit does not require knowledge of the time at which the unit acquired its remanent magnetism; its magnetism may be primary or secondary. Nevertheless, the unit's present polarity is a property that may be

---

[7]Pluton—a mappable body of plutonic rock.

ascertained and confirmed by others.

(d) **Relation to lithostratigraphic and biostratigraphic units.**—Magnetopolarity units resemble lithostratigraphic and biostratigraphic units in that they are defined on the basis of an objective recognizable property, but differ fundamentally in that most magnetopolarity unit boundaries are thought not to be time transgressive. Their boundaries may coincide with those of lithostratigraphic or biostratigraphic units, or be parallel to but displaced from those of such units, or be crossed by them.

(e) **Relation of magnetopolarity units to chronostratigraphic units.**—Although transitions between polarity reversals are of global extent, a magnetopolarity unit does not contain within itself evidence that the polarity is primary, or criteria that permit its unequivocal recognition in chronocorrelative strata of other areas. Other criteria, such as paleontologic or numerical age, are required for both correlation and dating. Although polarity reversals are useful in recognizing chronostratigraphic units, magnetopolarity alone is insufficient for their definition.

Article 45.—Boundaries. The upper and lower limits of a magnetopolarity unit are defined by boundaries marking a change of polarity. Such boundaries may represent either a depositional discontinuity or a magnetic-field transition. The boundaries are either polarity-reversal horizons or polarity transition-zones, respectively.

Remark. (a) **Polarity-reversal horizons and transition-zones.**—A polarity-reversal horizon is either a single, clearly definable surface or a thin body of strata constituting a transitional interval across which a change in magnetic polarity is recorded. Polarity-reversal horizons describe transitional intervals of 1 m or less; where the change in polarity takes place over a stratigraphic interval greater than 1 m, the term "polarity transition-zone" should be used. Polarity-reversal horizons and polarity transition-zones provide the boundaries for polarity zones, although they may also be contained within a polarity zone where they mark an internal change subsidiary in rank to those at its boundaries.

Ranks of Magnetopolarity Units

Article 46.—Fundamental Unit. A polarity zone is the fundamental unit of magnetopolarity classification. A polarity zone is a unit of rock characterized by the polarity of its magnetic signature. Magnetopolarity zone, rather than polarity zone, should be used where there is risk of confusion with other kinds of polarity.

Remarks. (a) **Content.**—A polarity zone should possess some degree of internal homogeneity. It may contain rocks of (1) entirely or predominantly one polarity, or (2) mixed polarity.

(b) **Thickness and duration.**—The thickness of rock of a polarity zone or the amount of time represented should play no part in the definition of the zone. The polarity signature is the essential property for definition.

(c) **Ranks.**—When continued work at the stratotype for a polarity zone, or new work in correlative rocks elsewhere, reveals smaller polarity units, these may be recognized formally as polarity subzones. If it should prove necessary or desirable to group polarity zones, these should be termed polarity superzones. The rank of a polarity unit may be changed when deemed appropriate.

Magnetopolarity Nomenclature

Article 47.—Compound Name. The formal name of a magnetopolarity zone should consist of a geographic name and the term *Polarity Zone.* The term may be modified by *Normal, Reversed*, or *Mixed* (example: Deer Park Reversed Polarity Zone). In naming or revising magnetopolarity units, appropriate parts of Articles 7 and 19 apply. The use of informal designations, e.g., numbers or letters, is not precluded.

BIOSTRATIGRAPHIC UNITS

Nature and Boundaries

Article 48.—Nature of Biostratigraphic Units. A biostratigraphic unit is a body of rock defined or characterized by its fossil content. The basic unit in biostratigraphic classification is the biozone, of which there are several kinds.

Remarks. (a) **Enclosing strata.**—Fossils that define or characterize a biostratigraphic unit commonly are contemporaneous with the body of rock that contains them. Some biostratigraphic units, however, may be represented only by their fossils, preserved in normal stratigraphic succession (e.g., on hardgrounds, in lag deposits, in certain types of remanié accumulations), which alone represent the rock of the biostratigraphic unit. In addition, some strata contain fossils derived from older or younger rocks or from essentially coeval materials of different facies; such fossils should not be used to define a biostratigraphic unit.

(b) **Independence from lithostratigraphic units.**—Biostratigraphic units are based on criteria which differ fundamentally from those for lithostratigraphic units. Their boundaries may or may not coincide with the boundaries of lithostratigraphic units, but they bear no inherent relation to them.

(c) **Independence from chronostratigraphic units.**—The boundaries of most biostratigraphic units, unlike the boundaries of chronostratigraphic units, are both characteristically and conceptually diachronous. An exception is an abundance biozone boundary that reflects a mass-mortality event. The vertical and lateral limits of the rock body that constitutes the biostratigraphic unit represent the limits in distribution of the defining biotic elements. The lateral limits never represent, and the vertical limits rarely represent, regionally synchronous events. Nevertheless, biostratigraphic units are effective for interpreting chronostratigraphic relations.

Article 49.—Kinds of Biostratigraphic Units. Three principal kinds of biostratigraphic units are recognized: *interval, assemblage,* and *abundance* biozones.

Remark: (a) **Boundary definitions.**—Boundaries of interval zones are defined by lowest and/or highest occurrences of single taxa; boundaries of some kinds of assemblage zones (Oppel or concurrent range zones) are defined by lowest and/or highest occurrences of more than one taxon; and boundaries of abundance zones are defined by marked changes in relative abundances of preserved taxa.

Article 50.—Definition of Interval Zone. An interval zone (or subzone) is the body of strata between two specified, documented lowest and/or highest occurrences of single taxa.

Remarks. (a) **Interval zone types.**—Three basic types of interval zones are recognized (Fig. 4). These include the range zones and interval zones of the International Stratigraphic Guide (ISSC, 1976, p. 53, 60) and are:

1. The interval between the documented lowest and highest occurrences of a single taxon (Fig. 4A). This is the *taxon range zone* of ISSC (1976, p. 53).

2. The interval included between the documented lowest occurrence of one taxon and the documented highest occurrence of another taxon (Fig. 4B). When such occurrences result in stratigraphic overlap of the taxa (Fig. 4B-1), the interval zone is the *concurrent range zone* of ISSC (1976, p. 55), that involves only two taxa. When such occurrences do not result in stratigraphic overlap (Fig. 4B-2), but are used to partition the range of a third taxon, the interval is the *partial range zone* of George and others (1969).

3. The interval between documented successive lowest occurrences or successive highest occurrences of two taxa (Fig. 4C). When the interval is between successive documented lowest occurrences within an evolutionary lineage (Fig. 4C-1), it is the *lineage zone* of ISSC (1976, p. 58). When the interval is between successive lowest occurrences of unrelated taxa or between successive highest occurrences of either related or unrelated taxa (Fig. 4C-2), it is a kind of *interval zone* of ISSC (1976, p. 60).

(b) **Unfossiliferous intervals.**—Unfossiliferous intervals between or within biozones are the *barren interzones* and *intrazones* of ISSC (1976, p. 49).

Article 51.—Definition of Assemblage Zone. An assemblage zone is a biozone characterized by the association of three or more taxa. It may be based on all kinds of fossils present, or restricted to only certain kinds of fossils.

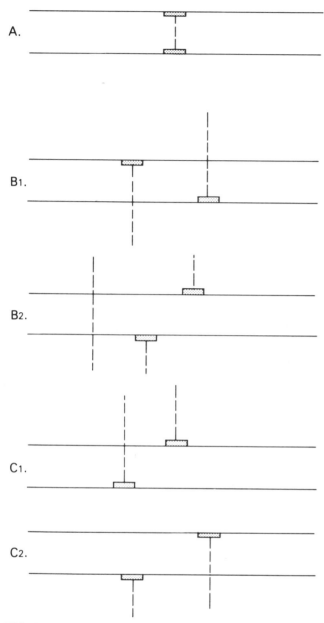

A.

B₁.

B₂.

C₁.

C₂.

FIG. 4.—Examples of biostratigraphic interval zones.
Vertical broken lines indicate ranges of taxa; bars indicate lowest or highest documented occurrences.

Remarks. (a) **Assemblage zone contents.**—An assemblage zone may consist of a geographically or stratigraphically restricted assemblage, or may incorporate two or more contemporaneous assemblages with shared characterizing taxa (*composite assemblage zones* of Kauffman, 1969) (Fig. 5c).

(b) **Assemblage zone types.**—In practice, two assemblage zone concepts are used:

1. The *assemblage zone* (or cenozone) of ISSC (1976, p. 50), which is characterized by taxa without regard to their range limits (Fig. 5a). Recognition of this type of assemblage zone can be aided by using techniques of multivariate analysis. Careful designation of the characterizing taxa is especially important.

2. The *Oppel zone,* or the *concurrent range zone* of ISSC (1976, p. 55, 57), a type of zone characterized by more than two taxa and having boundaries based on two or more documented first and/or last occurrences of the included characterizing taxa (Fig. 5b).

Article 52.—**Definition of Abundance Zone.** An abundance

zone is a biozone characterized by quantitatively distinctive maxima of relative abundance of one or more taxa. This is the *acme zone* of ISSC (1976, p. 59).

Remark. (a) **Ecologic controls.**—The distribution of biotic assemblages used to characterize some assemblage and abundance biozones may reflect strong local ecological control. Biozones based on such assemblages are included within the concept of ecozones (Vella, 1964), and are informal.

Ranks of Biostratigraphic Units

Article 53.—**Fundamental Unit.** The fundamental unit of biostratigraphic classification is a biozone.

Remarks. (a) **Scope.**—A single body of rock may be divided into various kinds and scales of biozones or subzones, as discussed in the International Stratigraphic Guide (ISSC, 1976, p. 62). Such usage is recommended if it will promote clarity, but only the unmodified term *biozone* is accorded formal status.

(b) **Divisions.**—A biozone may be completely or partly divided into formally designated sub-biozones (subzones), if such divisions serve a useful purpose.

Biostratigraphic Nomenclature

Article 54.—**Establishing Formal Units.** Formal establishment of a biozone or subzone must meet the requirements of Article 3 and requires a unique name, a description of its content and its boundaries, reference to a stratigraphic sequence in which the zone is characteristically developed, and a discussion of its spatial extent.

Remarks. (a) **Name.**—The name, which is compound and designates the kind of biozone, may be based on:

1. One or two characteristic and common taxa that are restricted to the biozone, reach peak relative abundance within the biozone, or have their total stratigraphic overlap within the biozone. These names most commonly are those of genera or subgenera, binomial designations of species, or trinomial designations of subspecies. If names of the nominate taxa change, names of the zones should be changed accordingly. Generic or subgeneric names may be abbreviated. Trivial species or subspecies names should not be used alone because they may not be unique.

2. Combinations of letters derived from taxa which characterize the biozone. However, alpha-numeric code designations (e.g., N1, N2, N3...) are informal and not recommended because they do not lend themselves readily to subsequent insertions, combinations, or eliminations. Biozonal systems based *only* on simple progressions of letters or numbers (e.g., A, B, C, or 1, 2, 3) are also not recommended.

(b) **Revision.**—Biozones and subzones are established empirically and may be modified on the basis of new evidence. Positions of established biozone or subzone boundaries may be stratigraphically refined, new characterizing taxa may be recognized, or original characterizing taxa may be superseded. If the concept of a particular biozone or subzone is substantially modified, a new unique designation is required to avoid ambiguity in subsequent citations.

(c) **Specifying kind of zone.**—Initial designation of a formally proposed biozone or subzone as an abundance zone, or as one of the types of interval zones, or assemblage zones (Articles 49-52), is strongly recommended. Once the type of biozone is clearly identified, the designation may be dropped in the remainder of a text (e.g., *Exus albus* taxon range zone to *Exus albus* biozone).

(d) **Defining taxa.**—Initial description or subsequent emendation of a biozone or subzone requires designation of the defining and characteristic taxa, and/or the documented first and last occurrences which mark the biozone or subzone boundaries.

(e) **Stratotypes.**—The geographic and stratigraphic position and boundaries of a formally proposed biozone or subzone should be defined precisely or characterized in one or more designated reference sections. Designation of a stratotype for each new biostratigraphic unit and of reference sections for emended biostratigraphic units is required.

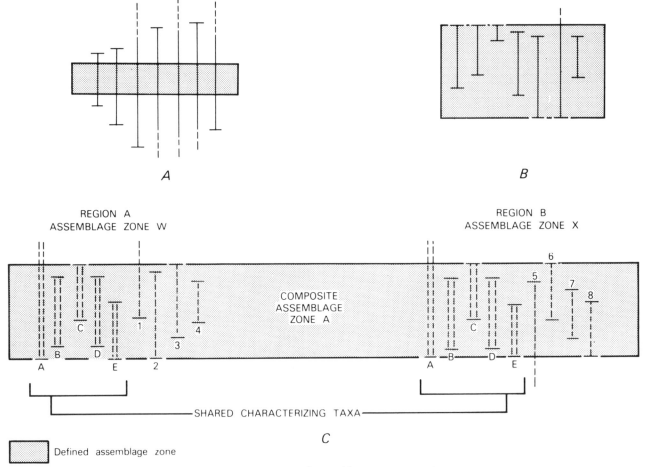

FIG. 5.—Examples of assemblage zone concepts.

PEDOSTRATIGRAPHIC UNITS

Nature and Boundaries

Article 55.—**Nature of Pedostratigraphic Units.** A pedostratigraphic unit is a body of rock that consists of one or more pedologic horizons developed in one or more lithostratigraphic, allostratigraphic, or lithodemic units (Fig. 6) and is overlain by one or more formally defined lithostratigraphic or allostratigraphic units.

Remarks. (a) **Definition.**—A pedostratigraphic[8] unit is a buried, traceable, three-dimensional body of rock that consists of one or more differentiated pedologic horizons.

(b) **Recognition.**—The distinguishing property of a pedostratigraphic unit is the presence of one or more distinct, differentiated, pedologic horizons. Pedologic horizons are products of soil development (pedogenesis) which occurred subsequent to formation of the lithostratigraphic, allostratigraphic, or lithodemic unit or units on which the buried soil was formed; these units are the parent materials in which pedogenesis occurred. Pedologic horizons are recognized in the field by diagnostic features such as color, soil structure, organic-matter accumulation, texture, clay coatings, stains, or concretions. Micromorphology, particle size, clay mineralogy, and other properties determined in the laboratory also may be used to identify and distinguish pedostratigraphic units.

(c) **Boundaries and stratigraphic position.**—The upper boundary of a pedostratigraphic unit is the top of the uppermost pedologic horizon formed by pedogenesis in a buried soil profile. The lower boundary of a pedostratigraphic unit is the lowest *definite* physical boundary of a pedologic horizon within a buried soil profile. The stratigraphic position of a pedostratigraphic unit is determined by its relation to overlying and underlying stratigraphic units (see Remark d).

(d) **Traceability.**—Practicability of subsurface tracing of the upper boundary of a buried soil is essential in establishing a pedostratigraphic unit because (1) few buried soils are exposed continuously for great distances, (2) the physical and chemical properties of a specific pedostratigraphic unit may vary greatly, both vertically and laterally, from place to place, and (3) pedostratigraphic units of different stratigraphic significance in the same region generally do not have unique identifying physical and chemical characteristics. Consequently, extension of a pedostratigraphic unit is accomplished by lateral tracing of the contact between a buried soil and an overlying, formally defined lithostratigraphic or allostratigraphic unit, or between a soil and two or more demonstrably correlative stratigraphic units.

(e) **Distinction from pedologic soils.**—Pedologic soils may include organic deposits (e.g., litter zones, peat deposits, or swamp deposits) that overlie or grade laterally into differentiated buried soils. The organic deposits are not products of pedogenesis, and O horizons are not included in a pedostratigraphic unit (Fig. 6); they may be classified as biostratigraphic or lithostratigraphic units. Pedologic soils also include the entire C horizon of a soil. The C horizon in pedology is not rigidly defined; it is merely the part of a soil profile that underlies the B horizon. The base of the C horizon in many soil profiles is gradational or unidentifiable; commonly it is placed arbitrarily. The need for clearly defined and easily recognized physical boundaries for a stratigraphic unit requires that the lower boundary of a pedostratigraphic unit be defined as the lowest *definite* physical boundary of a pedologic horizon in a buried soil profile, and part or all of the C horizon may be excluded from a pedostratigraphic unit.

---

[8]Terminology related to pedostratigraphic classification is summarized on page 850.

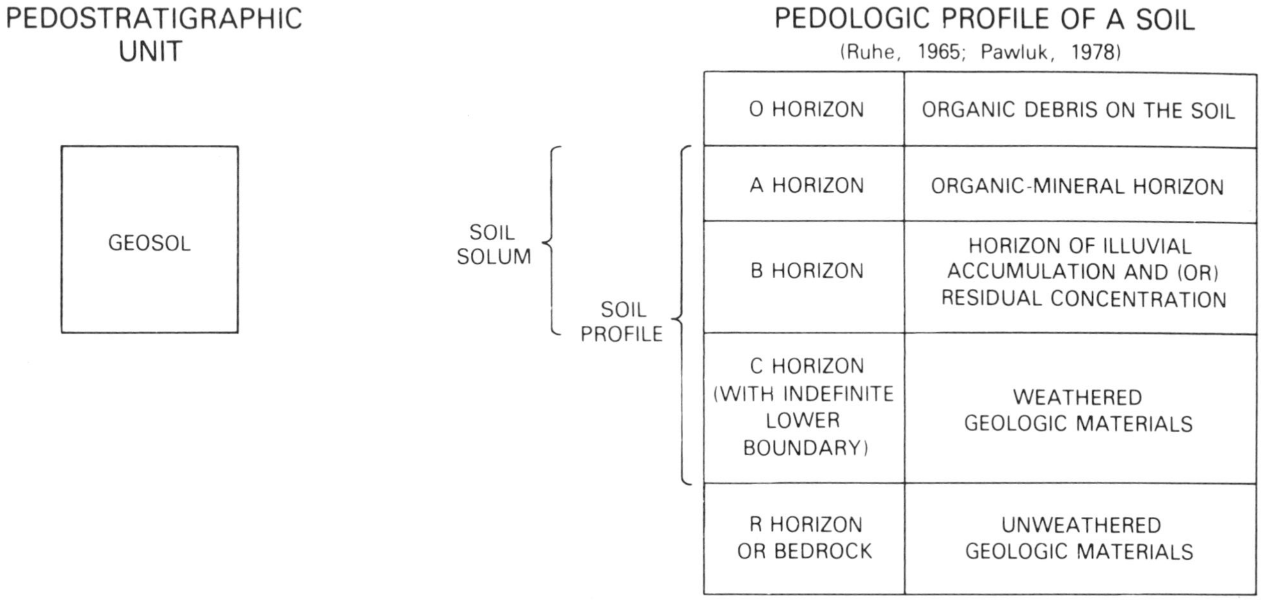

PEDOSTRATIGRAPHIC UNIT

PEDOLOGIC PROFILE OF A SOIL
(Ruhe, 1965; Pawluk, 1978)

O HORIZON	ORGANIC DEBRIS ON THE SOIL
A HORIZON	ORGANIC-MINERAL HORIZON
B HORIZON	HORIZON OF ILLUVIAL ACCUMULATION AND (OR) RESIDUAL CONCENTRATION
C HORIZON (WITH INDEFINITE LOWER BOUNDARY)	WEATHERED GEOLOGIC MATERIALS
R HORIZON OR BEDROCK	UNWEATHERED GEOLOGIC MATERIALS

GEOSOL

SOIL SOLUM

SOIL PROFILE

FIG. 6.—Relationship between pedostratigraphic units and pedologic profiles.
The base of a geosol is the lowest clearly defined physical boundary of a pedologic horizon in a buried soil profile. In this example it is the lower boundary of the B horizon because the base of the C horizon is not a clearly defined physical boundary. In other profiles the base may be the lower boundary of a C horizon.

(f) **Relation to saprolite and other weathered materials.**—A material derived by in situ weathering of lithostratigraphic, allostratigraphic, and(or) lithodemic units (e.g., saprolite, bauxite, residuum) may be the parent material in which pedologic horizons form, but is not a pedologic soil. A pedostratigraphic unit may be based on the pedologic horizons of a buried soil developed in the product of in-situ weathering, such as saprolite. The parents of such a pedostratigraphic unit are both the saprolite and, indirectly, the rock from which it formed.

(g) **Distinction from other stratigraphic units.**—A pedostratigraphic unit differs from other stratigraphic units in that (1) it is a product of surface alteration of one or more older material units by specific processes (pedogenesis), (2) its lithology and other properties differ markedly from those of the parent material(s), and (3) a single pedostratigraphic unit may be formed in situ in parent material units of diverse compositions and ages.

(h) **Independence from time concepts.**—The boundaries of a pedostratigraphic unit are time-transgressive. Concepts of time spans, however measured, play no part in defining the boundaries of a pedostratigraphic unit. Nonetheless, evidence of age, whether based on fossils, numerical ages, or geometrical or other relationships, may play an important role in distinguishing and identifying non-contiguous pedostratigraphic units at localities away from the type areas. The name of a pedostratigraphic unit should be chosen from a geographic feature in the type area, and not from a time span.

Pedostratigraphic Nomenclature and Unit

Article 56.—**Fundamental Unit.** The fundamental and only unit in pedostratigraphic classification is a geosol.

Article 57.—**Nomenclature.**—The formal name of a pedostratigraphic unit consists of a geographic name combined with the term "geosol." Capitalization of the initial letter in each word serves to identify formal usage. The geographic name should be selected in accordance with recommendations in Article 7 and should not duplicate the name of another formal geologic unit. Names based on subjacent and superjacent rock units, for example the super-Wilcox–sub-Claiborne soil, are informal, as are

those with time connotations (post-Wilcox–pre-Claiborne soil).

Remarks. (a) **Composite geosols.**—Where the horizons of two or more merged or "welded" buried soils can be distinguished, formal names of pedostratigraphic units based on the horizon boundaries can be retained. Where the horizon boundaries of the respective merged or "welded" soils cannot be distinguished, formal pedostratigraphic classification is abandoned and a combined name such as Hallettville-Jamesville geosol may be used informally.

(b) **Characterization.**—The physical and chemical properties of a pedostratigraphic unit commonly vary vertically and laterally throughout the geographic extent of the unit. A pedostratigraphic unit is characterized by the *range* of physical and chemical properties of the unit in the type area, rather than by "typical" properties exhibited in a type section. Consequently, a pedostratigraphic unit is characterized on the basis of a composite stratotype (Article 8d).

(c) **Procedures for establishing formal pedostratigraphic units.**—A formal pedostratigraphic unit may be established in accordance with the applicable requirements of Article 3, and additionally by describing major soil horizons in each soil facies.

ALLOSTRATIGRAPHIC UNITS

Nature and Boundaries

Article 58.—**Nature of Allostratigraphic Units.** An allostratigraphic[9] unit is a mappable stratiform body of sedimentary rock that is defined and identified on the basis of its bounding discontinuities.

Remarks. (a) **Purpose.**—Formal allostratigraphic units may be defined to distinguish between different (1) superposed discontinuity-bounded deposits of similar lithology (Figs. 7, 9), (2) contiguous discontinuity-bounded deposits of similar lithology (Fig. 8), or (3) geographically separated discontinuity-bounded units of similar lithology (Fig. 9), or to distinguish as single units discontinuity-bounded deposits characterized by lithic heterogeneity (Fig. 8).

(b) **Internal characteristics.**—Internal characteristics (physical, chemical, and paleontological) may vary laterally and vertically throughout the unit.

(c) **Boundaries.**—Boundaries of allostratigraphic units are laterally traceable discontinuities (Figs. 7, 8, and 9).

(d) **Mappability.**—A formal allostratigraphic unit must be mappable

---

[9]From the Greek *allo*: "other, different."

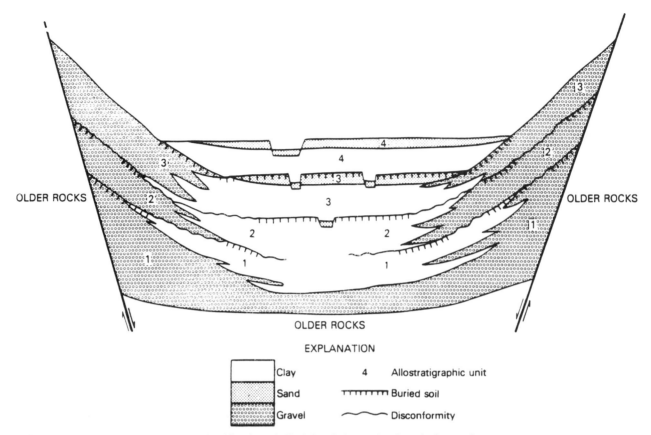

EXPLANATION

▢ Clay	**4** Allostratigraphic unit
▢ Sand	⊤⊤⊤⊤⊤⊤ Buried soil
▢ Gravel	∿∿∿ Disconformity

FIG. 7.—Example of allostratigraphic classification of alluvial and lacustrine deposits in a graben.

The alluvial and lacustrine deposits may be included in a single formation, or may be separated laterally into formations distinguished on the basis of contrasting texture (gravel, clay). Textural changes are abrupt and sharp, both vertically and laterally. The gravel deposits and clay deposits, respectively, are lithologically similar and thus cannot be distinguished as members of a formation. Four allostratigraphic units, each including two or three textural facies, may be defined on the basis of laterally traceable discontinuities (buried soils and disconformities).

at the scale practiced in the region where the unit is defined.

(e) **Type locality and extent.**—A type locality and type area must be designated; a composite stratotype or a type section and several reference sections are desirable. An allostratigraphic unit may be laterally contiguous with a formally defined lithostratigraphic unit; a vertical cut-off between such units is placed where the units meet.

(f) **Relation to genesis.**—Genetic interpretation is an inappropriate basis for defining an allostratigraphic unit. However, genetic interpretation may influence the choice of its boundaries.

(g) **Relation to geomorphic surfaces.**—A geomorphic surface may be used as a boundary of an allostratigraphic unit, but the unit should not be given the geographic name of the surface.

(h) **Relation to soils and paleosols.**—Soils and paleosols are composed of products of weathering and pedogenesis and differ in many respects from allostratigraphic units, which are depositional units (see "Pedostratigraphic Units," Article 55). The upper boundary of a surface or buried soil may be used as a boundary of an allostratigraphic unit.

(i) **Relation to inferred geologic history.**—Inferred geologic history is not used to define an allostratigraphic unit. However, well-documented geologic history may influence the choice of the unit's boundaries.

(j) **Relation to time concepts.**—Inferred time spans, however measured, are not used to define an allostratigraphic unit. However, age relationships may influence the choice of the unit's boundaries.

(k) **Extension of allostratigraphic units.**—An allostratigraphic unit is extended from its type area by tracing the boundary discontinuities or by tracing or matching the deposits between the discontinuities.

Ranks of Allostratigraphic Units

Article 59.—Hierarchy. The hierarchy of allostratigraphic units, in order of decreasing rank, is allogroup, alloformation, and allomember.

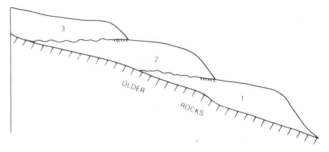

FIG. 8.—Example of allostratigraphic classification of contiguous deposits of similar lithology.

Allostratigraphic units 1, 2, and 3 are physical records of three glaciations. They are lithologically similar, reflecting derivation from the same bedrock, and constitute a single lithostratigraphic unit.

Remarks. (a) **Alloformation.**—The alloformation is the fundamental unit in allostratigraphic classification. An alloformation may be completely or only partly divided into allomembers, if some useful purpose is served, or it may have no allomembers.

(b) **Allomember.**—An allomember is the formal allostratigraphic unit next in rank below an alloformation.

(c) **Allogroup.**—An allogroup is the allostratigraphic unit next in rank above an alloformation. An allogroup is established only if a unit of that rank is essential to elucidation of geologic history. An allogroup may consist entirely of named alloformations or, alternatively, may contain one or more named alloformations which jointly do not comprise the entire allogroup.

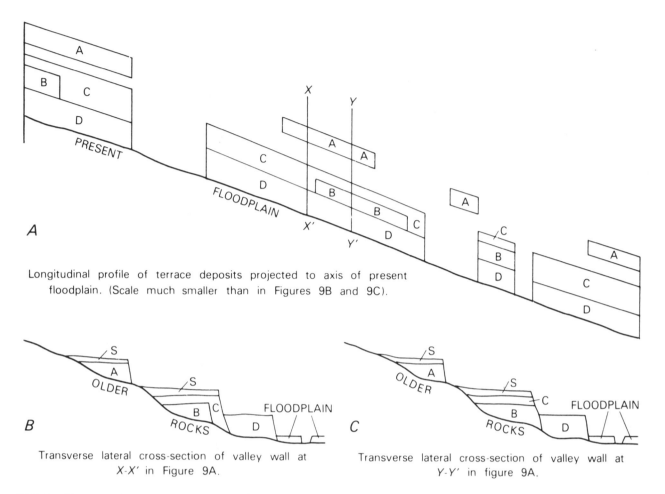

Longitudinal profile of terrace deposits projected to axis of present floodplain. (Scale much smaller than in Figures 9B and 9C).

Transverse lateral cross-section of valley wall at *X-X'* in Figure 9A.

Transverse lateral cross-section of valley wall at *Y-Y'* in figure 9A.

FIG. 9.—**Example of allostratigraphic classification of lithologically similar, discontinuous terrace deposits.**

A, B, C, and D are terrace gravel units of similar lithology at different topographic positions on a valley wall. The deposits may be defined as separate formal allostratigraphic units if such units are useful and if bounding discontinuities can be traced laterally. Terrace gravels of the same age commonly are separated geographically by exposures of older rocks. Where the bounding discontinuities cannot be traced continuously, they may be extended geographically on the basis of objective correlation of internal properites of the deposits other than lithology (e.g., fossil content, included tephras), topographic position, numerical ages, or relative-age criteria (e.g., soils or other weathering phenomena). The criteria for such extension should be documented. Slope deposits and eolian deposits (S) that mantle terrace surfaces may be of diverse ages and are not included in a terrace-gravel allostratigraphic unit. A single terrace surface may be underlain by more than one allostratigraphic unit (units B and C in sections b and c).

(d) **Changes in rank.**—The principles and procedures for elevation and reduction in rank of formal allostratigraphic units are the same as those in Articles 19b, 19g, and 28.

Allostratigraphic Nomenclature

Article 60.—**Nomenclature.** The principles and procedures for naming allostratigraphic units are the same as those for naming of lithostratigraphic units (see Articles 7, 30).

Remark. (a) **Revision.**—Allostratigraphic units may be revised or otherwise modified in accordance with the recommendations in Articles 17 to 20.

FORMAL UNITS DISTINGUISHED BY AGE

GEOLOGIC-TIME UNITS

Nature and Types

Article 61.—**Types.** Geologic-time units are conceptual, rather than material, in nature. Two types are recognized: those based on material standards or referents (specific rock sequences or bodies), and those independent of material referents (Fig. 1).

Units Based on Material Referents

Article 62.—**Types Based on Referents.** Two types of formal geologic-time units based on material referents are recognized: they are isochronous and diachronous units.

Article 63.—**Isochronous Categories.** Isochronous time units and the material bodies from which they are derived are twofold: geochronologic units (Article 80), which are based on corresponding material chronostratigraphic units (Article 66), and polarity-geochronologic units (Article 88), based on corresponding material polarity-chronostratigraphic units (Article 83).

Remark. (a) **Extent.**—Isochronous units are applicable worldwide; they may be referred to even in areas lacking a material record of the named span of time. The duration of the time may be represented by a unit-stratotype referent. The beginning and end of the time are represented by point-boundary-stratotypes either in a single stratigraphic sequence or in separate stratotype sections (Articles 8b, 10b).

Article 64.—**Diachronous Categories.** Diachronic units (Article 91) are time units corresponding to diachronous material allostratigraphic units (Article 58), pedostratigraphic units (Article 55), and most lithostratigraphic (Article 22) and biostratigraphic (Article 48) units.

Remarks. (a) **Diachroneity.**—Some lithostratigraphic and biostratigraphic units are clearly diachronous, whereas others have boundaries which are not demonstrably diachronous within the resolving power of available dating methods. The latter commonly are treated as isochronous and are used for purposes of chronocorrelation (see biochronozone, Article 75). However, the assumption of isochroneity must be tested continually.

(b) **Extent.**—Diachronic units are coextensive with the diachronous material stratigraphic units on which they are based and are not used beyond the extent of their material referents.

Units Independent of Material Referents

Article 65.—**Numerical Divisions of Time.** Isochronous geologic-time units based on numerical divisions of time in years are geochronometric units (Article 96) and have no material referents.

CHRONOSTRATIGRAPHIC UNITS

Nature and Boundaries

Article 66.—**Definition.** A chronostratigraphic unit is a body of rock established to serve as the material reference for all rocks formed during the same span of time. Each of its boundaries is synchronous. The body also serves as the basis for defining the specific interval of time, or geochronologic unit (Article 80), represented by the referent.

Remarks. (a) **Purposes.**—Chronostratigraphic classification provides a means of establishing the temporally sequential order of rock bodies. Principal purposes are to provide a framework for (1) temporal correlation of the rocks in one area with those in another, (2) placing the rocks of the Earth's crust in a systematic sequence and indicating their relative position and age with respect to earth history as a whole, and (3) constructing an internationally recognized Standard Global Chronostratigraphic Scale.

(b) **Nature.**—A chronostratigraphic unit is a material unit and consists of a body of strata formed during a specific time span. Such a unit represents all rocks, and only those rocks, formed during that time span.

(c) **Content.**—A chronostratigraphic unit may be based upon the time span of a biostratigraphic unit, a lithic unit, a magnetopolarity unit, or any other feature of the rock record that has a time range. Or it may be any arbitrary but specified sequence of rocks, provided it has properties allowing chronocorrelation with rock sequences elsewhere.

Article 67.—**Boundaries.** Boundaries of chronostratigraphic units should be defined in a designated stratotype on the basis of observable paleontological or physical features of the rocks.

Remark. (a) **Emphasis on lower boundaries of chronostratigraphic units.**—Designation of point boundaries for both base and top of chronostratigraphic units is not recommended, because subsequent information on relations between successive units may identify overlaps or gaps. One means of minimizing or eliminating problems of duplication or gaps in chronostratigraphic successions is to define formally as a point-boundary stratotype only the base of the unit. Thus, a chronostratigraphic unit with its base defined at one locality, will have its top defined by the base of an overlying unit at the same, but more commonly another, locality (Article 8b).

Article 68.—**Correlation.** Demonstration of time equivalence is required for geographic extension of a chronostratigraphic unit from its type section or area. Boundaries of chronostratigraphic units can be extended only within the limits of resolution of available means of chronocorrelation, which currently include paleontology, numerical dating, remanent magnetism, thermoluminescence, relative-age criteria (examples are superposition and cross-cutting relations), and such indirect and inferential physical criteria as climatic changes, degree of weathering, and relations to unconformities. Ideally, the boundaries of chronostratigraphic units are independent of lithology, fossil content, or other material bases of stratigraphic division, but, in practice, the correlation or geographic extension of these boundaries relies at least in part on such features. Boundaries of chronostratigraphic units commonly are intersected by boundaries of most other kinds of material units.

Ranks of Chronostratigraphic Units

Article 69.—**Hierarchy.** The hierarchy of chronostratigraphic units, in order of decreasing rank, is eonothem, erathem, system, series, and stage. Of these, system is the primary unit of worldwide major rank; its primacy derives from the history of development of stratigraphic classification. All systems and units of higher rank are divided completely into units of the next lower rank. Chronozones are non-hierarchical and commonly lower-rank chronostratigraphic units. Stages and chronozones in sum do not necessarily equal the units of next higher rank and need not be contiguous. The rank and magnitude of chronostratigraphic units are related to the time interval represented by the units, rather than to the thickness or areal extent of the rocks on which the units are based.

Article 70.—**Eonothem.** The unit highest in rank is eonothem. The Phanerozoic Eonothem encompasses the Paleozoic, Mesozoic, and Cenozoic Erathems. Although older rocks have been assigned heretofore to the Precambrian Eonothem, they also have been assigned recently to other (Archean and Proterozoic) eonothems by the IUGS Precambrian Subcommission. The span of time corresponding to an eonothem is an *eon*.

Article 71.—**Erathem.** An erathem is the formal chronostratigraphic unit of rank next lower to eonothem and consists of several adjacent systems. The span of time corresponding to an erathem is an *era*.

Remark. (a) **Names.**—Names given to traditional Phanerozoic erathems were based upon major stages in the development of life on Earth: Paleozoic (old), Mesozoic (intermediate), and Cenozoic (recent) life. Although somewhat comparable terms have been applied to Precambrian units, the names and ranks of Precambrian divisions are not yet universally agreed upon and are under consideration by the IUGS Subcommission on Precambrian Stratigraphy.

Article 72.—**System.** The unit of rank next lower to erathem is the system. Rocks encompassed by a system represent a time-span and an episode of Earth history sufficiently great to serve as a worldwide chronostratigraphic reference unit. The temporal equivalent of a system is a *period*.

Remark. (a) **Subsystem and supersystem.**—Some systems initially established in Europe later were divided or grouped elsewhere into units ranked as systems. *Subsystems* (Mississippian Subsystem of the Carboniferous System) and *supersystems* (Karoo Supersystem) are more appropriate.

Article 73.—**Series.** Series is a conventional chronostratigraphic unit that ranks below a system and always is a division of a system. A series commonly constitutes a major unit of chronostratigraphic correlation within a province, between provinces, or between continents. Although many European series are being adopted increasingly for dividing systems on other continents, provincial series of regional scope continue to be useful. The temporal equivalent of a series is an *epoch*.

Article 74.—**Stage.** A stage is a chronostratigraphic unit of smaller scope and rank than a series. It is most commonly of greatest use in intra-continental classification and correlation, although it has the potential for worldwide recognition. The geochronologic equivalent of stage is *age*.

Remark. (a) **Substage.**—Stages may be, but need not be, divided completely into substages.

Article 75.—**Chronozone.** A chronozone is a non-hierarchical, but commonly small, formal chronostratigraphic unit, and its boundaries may be independent of those of ranked units. Although a chronozone is an isochronous unit, it may be based on a biostratigraphic unit (example: *Cardioceras cordatum* Biochronozone), a lithostratigraphic unit (Woodbend Lithochronozone), or a magnetopolarity unit (Gilbert Reversed-Polarity Chronozone). Modifiers (litho-, bio-, polarity) used in formal names of the units need not be repeated in general discussions where the meaning is evident from the context, e.g., *Exus albus* Chronozone.

Remarks. (a) **Boundaries of chronozones.**—The base and top of a *chronozone* correspond in the unit's stratotype to the observed, defining, physical and paleontological features, but they are extended to other areas by any means available for recognition of synchroneity. The temporal equivalent of a chronozone is a chron.

(b) **Scope.**—The scope of the non-hierarchical chronozone may range markedly, depending upon the purpose for which it is defined either formally or informally. The informal "biochronozone of the ammonites," for example, represents a duration of time which is enormous and exceeds that of a system. In contrast, a biochronozone defined by a species of limited range, such as the *Exus albus* Chronozone, may represent a duration equal to or briefer than that of a stage.

(c) **Practical utility.**—Chronozones, especially thin and informal biochronozones and lithochronozones bounded by key beds or other "markers," are the units used most commonly in industry investigations of selected parts of the stratigraphy of economically favorable basins. Such units are useful to define geographic distributions of lithofacies or biofacies, which provide a basis for genetic interpretations and the selection of targets to drill.

Chronostratigraphic Nomenclature

Article 76.—**Requirements.** Requirements for establishing a formal chronostratigraphic unit include: (i) statement of intention to designate such a unit; (ii) selection of name; (iii) statement of kind and rank of unit; (iv) statement of general concept of unit including historical background, synonymy, previous treatment, and reasons for proposed establishment; (v) description of characterizing physical and/or biological features; (vi) designation and description of boundary type sections, stratotypes, or other kinds of units on which it is based; (vii) correlation and age relations; and (viii) publication in a recognized scientific medium as specified in Article 4.

Article 77.—**Nomenclature.** A formal chronostratigraphic unit is given a compound name, and the initial letter of all words, except for trivial taxonomic terms, is capitalized. Except for chronozones (Article 75), names proposed for new chronostratigraphic units should not duplicate those for other stratigraphic units. For example, naming a new chronostratigraphic unit simply by adding "-an" or "-ian" to the name of a lithostratigraphic unit is improper.

Remarks. (a) **Systems and units of higher rank.**—Names that are generally accepted for systems and units of higher rank have diverse origins, and they also have different kinds of endings (Paleozoic, Cambrian, Cretaceous, Jurassic, Quaternary).

(b) **Series and units of lower rank.**—Series and units of lower rank are commonly known either by geographic names (Virgilian Series, Ochoan Series) or by names of their encompassing units modified by the capitalized adjectives Upper, Middle, and Lower (Lower Ordovician). Names of chronozones are derived from the unit on which they are based (Article 75). For series and stage, a geographic name is preferable because it may be related to a type area. For geographic names, the adjectival endings -an or -ian are recommended (Cincinnatian Series), but it is permissible to use the geographic name without any special ending, if more euphonious. Many series and stage names already in use have been based on lithic units (groups, formations, and members) and bear the names of these units

(Wolfcampian Series, Claibornian Stage). Nevertheless, a stage preferably should have a geographic name not previously used in stratigraphic nomenclature. Use of internationally accepted (mainly European) stage names is preferable to the proliferation of others.

Article 78.—**Stratotypes.** An ideal stratotype for a chronostratigraphic unit is a completely exposed unbroken and continuous sequence of fossiliferous stratified rocks extending from a well-defined lower boundary to the base of the next higher unit. Unfortunately, few available sequences are sufficiently complete to define stages and units of higher rank, which therefore are best defined by boundary-stratotypes (Article 8b).

Boundary-stratotypes for major chronostratigraphic units ideally should be based on complete sequences of either fossiliferous monofacial marine strata or rocks with other criteria for chronocorrelation to permit widespread tracing of synchronous horizons. Extension of synchronous surfaces should be based on as many indicators of age as possible.

Article 79.—**Revision of units.** Revision of a chronostratigraphic unit without changing its name is allowable but requires as much justification as the establishment of a new unit (Articles 17, 19, and 76). Revision or redefinition of a unit of system or higher rank requires international agreement. If the definition of a chronostratigraphic unit is inadequate, it may be clarified by establishment of boundary stratotypes in a principal reference section.

GEOCHRONOLOGIC UNITS

Nature and Boundaries

Article 80.—**Definition and Basis.** Geochronologic units are divisions of time traditionally distinguished on the basis of the rock record as expressed by chronostratigraphic units. A geochronologic unit is not a stratigraphic unit (i.e., it is not a material unit), but it corresponds to the time span of an established chronostratigraphic unit (Articles 65 and 66), and its beginning and ending corresponds to the base and top of the referent.

Ranks and Nomenclature of Geochronologic Units

Article 81.—**Hierarchy.** The hierarchy of geochronologic units in order of decreasing rank is *eon, era, period, epoch,* and *age.* Chron is a non-hierarchical, but commonly brief, geochronologic unit. Ages in sum do not necessarily equal epochs and need not form a continuum. An eon is the time represented by the rocks constituting an eonothem; era by an erathem; period by a system; epoch by a series; age by a stage; and chron by a chronozone.

Article 82.—**Nomenclature.** Names for periods and units of lower rank are identical with those of the corresponding chronostratigraphic units; the names of some eras and eons are independently formed. Rules of capitalization for chronostratigraphic units (Article 77) apply to geochronologic units. The adjectives Early, Middle, and Late are used for the geochronologic epochs equivalent to the corresponding chronostratigraphic Lower, Middle, and Upper series, where these are formally established.

POLARITY-CHRONOSTRATIGRAPHIC UNITS

Nature and Boundaries

Article 83.—**Definition.** A polarity-chronostratigraphic unit is a body of rock that contains the primary magnetic-polarity record imposed when the rock was deposited, or crystallized, during a specific interval of geologic time.

Remarks. (a) **Nature.**—Polarity-chronostratigraphic units depend fundamentally for definition on actual sections or sequences, or measurements on individual rock units, and without these standards they are meaningless. They are based on material units, the polarity zones of magnetopolarity classification. Each polarity-chronostratigraphic unit is the record of the time during which the rock formed and the Earth's magnetic field had a designated polarity. Care should be taken to define polarity-chronologic units in terms of polarity-chronostratigraphic units, and not vice versa.

(b) **Principal purposes.**—Two principal purposes are served by polarity-chronostratigraphic classification: (1) correlation of rocks at one place with those of the same age and polarity at other places; and (2) delineation of the polarity history of the Earth's magnetic field.

(c) **Recognition.**—A polarity-chronostratigraphic unit may be extended geographically from its type locality only with the support of physical and/or paleontologic criteria used to confirm its age.

Article 84.—**Boundaries.** The boundaries of a polarity chronozone are placed at polarity-reversal horizons or polarity transition-zones (see Article 45).

Ranks and Nomenclature of Polarity-Chronostratigraphic Units

Article 85.—**Fundamental Unit.** The polarity chronozone consists of rocks of a specified primary polarity and is the fundamental unit of worldwide polarity-chronostratigraphic classification.

Remarks. (a) **Meaning of term.**—A polarity chronozone is the worldwide body of rock strata that is collectively defined as a polarity-chronostratigraphic unit.

(b) **Scope.**—Individual polarity zones are the basic building blocks of polarity chronozones. Recognition and definition of polarity chronozones may thus involve step-by-step assembly of carefully dated or correlated individual polarity zones, especially in work with rocks older than the oldest ocean-floor magnetic anomalies. This procedure is the method by which the Brunhes, Matuyama, Gauss, and Gilbert Chronozones were recognized (Cox, Doell, and Dalrymple, 1963) and defined originally (Cox, Doell, and Dalrymple, 1964).

(c) **Ranks.**—Divisions of polarity chronozones are designated polarity subchronozones. Assemblages of polarity chronozones may be termed polarity superchronozones.

Article 86.—**Establishing Formal Units.** Requirements for establishing a polarity-chronostratigraphic unit include those specified in Articles 3 and 4, and also (1) definition of boundaries of the unit, with specific references to designated sections and data; (2) distinguishing polarity characteristics, lithologic descriptions, and included fossils; and (3) correlation and age relations.

Article 87.—**Name.** A formal polarity-chronostratigraphic unit is given a compound name beginning with that for a named geographic feature; the second component indicates the normal, reversed, or mixed polarity of the unit, and the third component is *chronozone*. The initial letter of each term is capitalized. If the same geographic name is used for both a magnetopolarity zone and a polarity-chronostratigraphic unit, the latter should be distinguished by an -an or -ian ending. Example: Tetonian Reversed-Polarity Chronozone.

Remarks: (a) **Preservation of established name.**—A particularly well-established name should not be displaced, either on the basis of priority, as described in Article 7c, or because it was not taken from a geographic feature. Continued use of Brunhes, Matuyama, Gauss, and Gilbert, for example, is endorsed so long as they remain valid units.

(b) **Expression of doubt.**—Doubt in the assignment of polarity zones to polarity-chronostratigraphic units should be made explicit if criteria of time equivalence are inconclusive.

POLARITY-CHRONOLOGIC UNITS

Nature and Boundaries

Article 88.—**Definition.** Polarity-chronologic units are divisions of geologic time distinguished on the basis of the record of magnetopolarity as embodied in polarity-chronostratigraphic units. No special kind of magnetic time is implied; the designations used are meant to convey the parts of geologic time during which the Earth's magnetic field had a characteristic polarity or sequence of polarities. These units correspond to the time spans represented by polarity chronozones, e.g., Gauss Normal Polarity Chronozone. They are not material units.

Ranks and Nomenclature of Polarity-Chronologic Units

Article 89.—**Fundamental Unit.** The polarity chron is the fundamental unit of geologic time designating the time span of a polarity chronozone.

Remark. (a) **Hierarchy.**—Polarity-chronologic units of decreasing hierarchical ranks are polarity superchron, polarity chron, and polarity subchron.

Article 90.—**Nomenclature.** Names for polarity chronologic units are identical with those of corresponding polarity-chronostratigraphic units, except that the term chron (or superchron, etc) is substituted for chronozone (or superchronozone, etc).

DIACHRONIC UNITS

Nature and Boundaries

Article 91.—**Definition.** A diachronic unit comprises the unequal spans of time represented either by a specific lithostratigraphic, allostratigraphic, biostratigraphic, or pedostratigraphic unit, or by an assemblage of such units.

Remarks. (a) **Purposes.**—Diachronic classification provides (1) a means of comparing the spans of time represented by stratigraphic units with diachronous boundaries at different localities, (2) a basis for broadly establishing in time the beginning and ending of deposition of diachronous stratigraphic units at different sites, (3) a basis for inferring the rate of change in areal extent of depositional processes, (4) a means of determining and comparing rates and durations of deposition at different localities, and (5) a means of comparing temporal and spatial relations of diachronous stratigraphic units (Watson and Wright, 1980).

(b) **Scope.**—The scope of a diachronic unit is related to (1) the relative magnitude of the transgressive division of time represented by the stratigraphic unit or units on which it is based and (2) the areal extent of those units. A diachronic unit is not extended beyond the geographic limits of the stratigraphic unit or units on which it is based.

(c) **Basis.**—The basis for a diachronic unit is the diachronous referent.

(d) **Duration.**—A diachronic unit may be of equal duration at different places despite differences in the times at which it began and ended at those places.

Article 92.—**Boundaries.** The boundaries of a diachronic unit are the times recorded by the beginning and end of deposition of the material referent at the point under consideration (Figs. 10, 11).

Remark. (a) **Temporal relations.**—One or both of the boundaries of a diachronic unit are demonstrably time-transgressive. The varying time significance of the boundaries is defined by a series of boundary reference sections (Article 8b, 8e). The duration and age of a diachronic unit differ from place to place (Figs. 10, 11).

Ranks and Nomenclature of Diachronic Units

Article 93.—**Ranks.** A diachron is the fundamental and non-hierarchical diachronic unit. If a hierarchy of diachronic units is needed, the terms episode, phase, span, and cline, in order of decreasing rank, are recommended. The rank of a hierarchical

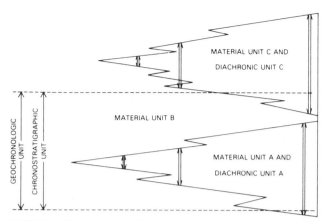

FIG. 10.—Comparison of geochronologic, chronostratigraphic, and diachronic units.

unit is determined by the scope of the unit (Article 91 b), and not by the time span represented by the unit at a particular place.

Remarks. (a) **Diachron.**—Diachrons may differ greatly in magnitude because they are the spans of time represented by individual or grouped lithostratigraphic, allostratigraphic, biostratigraphic, and(or) pedostratigraphic units.

(b) **Hierarchical ordering permissible.**—A hierarchy of diachronic units may be defined if the resolution of spatial and temporal relations of diachronous stratigraphic units is sufficiently precise to make the hierarchy useful (Watson and Wright, 1980). Although all hierarchical units of rank lower than episode are part of a unit next higher in rank, not all parts of an episode, phase, or span need be represented by a unit of lower rank.

(c) **Episode.**—An episode is the unit of highest rank and greatest scope in hierarchical classification. If the "Wisconsinan Age" were to be redefined as a diachronic unit, it would have the rank of episode.

Article 94.—**Name.** The name for a diachronic unit should be compound, consisting of a geographic name followed by the term diachron or a hierarchical rank term. Both parts of the compound name are capitalized to indicate formal status. If the diachronic unit is defined by a single stratigraphic unit, the geographic name of the unit may be applied to the diachronic unit. Otherwise, the geographic name of the diachronic unit should not duplicate that of another formal stratigraphic unit. Genetic terms (e.g., alluvial, marine) or climatic terms (e.g., gla-

cial, interglacial) are not included in the names of diachronic units.

Remarks. (a) **Formal designation of units.**—Diachronic units should be formally defined and named only if such definition is useful.

(b) **Inter-regional extension of geographic names.**—The geographic name of a diachronic unit may be extended from one region to another if the stratigraphic units on which the diachronic unit is based extend across the regions. If different diachronic units in contiguous regions eventually prove to be based on laterally continuous stratigraphic units, one name should be applied to the unit in both regions. If two names have been applied, one name should be abandoned and the other formally extended. Rules of priority (Article 7d) apply. Priority in publication is to be respected, but priority alone does not justify displacing a well-established name by one not well-known or commonly used.

(c) **Change from geochronologic to diachronic classification.**—Lithostratigraphic units have served as the material basis for widely accepted chronostratigraphic and geochronologic classifications of Quaternary nonmarine deposits, such as the classifications of Frye et al (1968), Willman and Frye (1970), and Dreimanis and Karrow (1972). In practice, time-parallel horizons have been extended from the stratotypes on the basis of markedly time-transgressive lithostratigraphic and pedostratigraphic unit boundaries. The time ("geochronologic") units, defined on the basis of the stratotype sections but extended on the basis of diachronous stratigraphic boundaries, are diachronic units. Geographic names established for such "geochronologic" units may be used in diachronic classification if (1) the chronostratigraphic and geochronologic classifications are formally abandoned and diachronic classifications are proposed to replace the former "geochronologic" classifications, and (2) the units are redefined as formal diachronic units. Preservation of well-established names in these specific circumstances retains the intent and purpose of the names and the units, retains the practical significance of the units, enhances communication, and avoids proliferation of nomenclature.

Article 95.—**Establishing Formal Units.** Requirements for establishing a formal diachronic unit, in addition to those in Article 3, include (1) specification of the nature, stratigraphic relations, and geographic or areal relations of the stratigraphic unit or units that serve as a basis for definition of the unit, and (2) specific designation and description of multiple reference sections that illustrate the temporal and spatial relations of the defining stratigraphic unit or units and the boundaries of the unit or units.

Remark. (a) **Revision or abandonment.**—Revision or abandonment of the stratigraphic unit or units that serve as the material basis for defini-

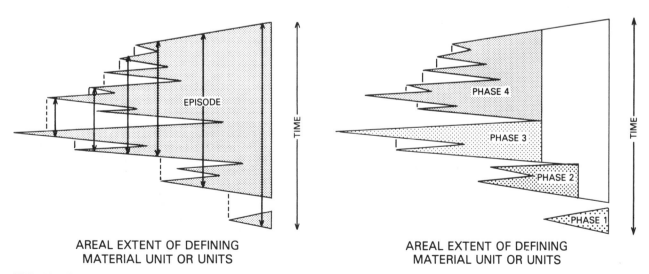

AREAL EXTENT OF DEFINING
MATERIAL UNIT OR UNITS

AREAL EXTENT OF DEFINING
MATERIAL UNIT OR UNITS

FIG. 11.—Schematic relation of phases to an episode.
Parts of a phase similarly may be divided into spans, and spans into clines. Formal definition of spans and clines is unnecessary in most diachronic unit hierarchies.

tion of a diachronic unit may require revision or abandonment of the diachronic unit. Procedure for revision must follow the requirements for establishing a new diachronic unit.

GEOCHRONOMETRIC UNITS

Nature and Boundaries

Article 96.—**Definition**. Geochronometric units are units established through the direct division of geologic time, expressed in years. Like geochronologic units (Article 80), geochronometric units are abstractions, i.e., they are not material units. Unlike geochronologic units, geochronometric units are not based on the time span of designated chronostratigraphic units (stratotypes), but are simply time divisions of convenient magnitude for the purpose for which they are established, such as the development of a time scale for the Precambrian. Their boundaries are arbitrarily chosen or agreed-upon ages in years.

Ranks and Nomenclature of Geochronometric Units

Article 97.—**Nomenclature**. Geochronologic rank terms (eon, era, period, epoch, age, and chron) may be used for geochronometric units when such terms are formalized. For example, Archean Eon and Proterozoic Eon, as recognized by the IUGS Subcommission on Precambrian Stratigraphy, are formal geochronometric units in the sense of Article 96, distinguished on the basis of an arbitrarily chosen boundary at 2.5 Ga. Geochronometric units are not defined by, but may have, corresponding chronostratigraphic units (eonothem, erathem, system, series, stage, and chronozone).

PART III: ADDENDA

REFERENCES[10]

American Commission on Stratigraphic Nomenclature, 1947, Note 1—Organization and objectives of the Stratigraphic Commission: American Association of Petroleum Geologists Bulletin, v. 31, no. 3, p. 513-518.

———— ,1961, Code of Stratigraphic Nomenclature: American Association of Petroleum Geologists Bulletin, v. 45, no. 5, p. 645-665.

———— ,1970, Code of Stratigraphic Nomenclature (2d ed.): American Association of Petroleum Geologists, Tulsa, Okla., 45 p.

———— ,1976, Note 44—Application for addition to code concerning magnetostratigraphic units: American Association of Petroleum Geologists Bulletin, v. 60, no. 2, p. 273-277.

Caster, K. E., 1934, The stratigraphy and paleontology of northwestern Pennsylvania, Part 1, Stratigraphy: Bulletins of American Paleontology, v. 21, 185 p.

Chang, K. H., 1975, Unconformity-bounded stratigraphic units: Geological Society of America Bulletin, v. 86, no. 11, p. 1544-1552.

Committee on Stratigraphic Nomenclature, 1933, Classification and nomenclature of rock units: Geological Society of America Bulletin, v. 44, no. 2, p. 423-459, and American Association of Petroleum Geologists Bulletin, v. 17, no. 7, p. 843-868.

Cox, A. V., R. R. Doell, and G. B. Dalrymple, 1963, Geomagnetic polarity epochs and Pleistocene geochronometry: Nature, v. 198, p. 1049-1051.

———— ,1964, Reversals of the Earth's magnetic field: Science, v. 144, no. 3626, p. 1537-1543.

Cross, C. W., 1898, Geology of the Telluride area: U.S. Geological Survey 18th Annual Report, pt. 3, p. 759.

Cumming, A. D., J. G. C. M. Fuller, and J. W. Porter, 1959, Separation of strata: Paleozoic limestones of the Williston basin: American Journal of Science, v. 257, no. 10, p. 722-733.

Dreimanis, Aleksis, and P. F. Karrow, 1972, Glacial history of the Great Lakes–St. Lawrence region, the classification of the Wisconsin(an) Stage, and its correlatives: International Geologic Congress, 24th Session, Montreal, 1972, Section 12, Quaternary Geology, p. 5-15.

Dunbar, C. O., and John Rodgers, 1957, Principles of stratigraphy: Wiley, New York, 356 p.

Forgotson, J. M., Jr., 1957, Nature, usage and definition of marker-defined vertically segregated rock units: American Association of Petroleum Geologists Bulletin, v. 41, no. 9, p. 2108-2113.

Frye, J. C., H. B. Willman, Meyer Rubin, and R. F. Black, 1968, Definition of Wisconsinan Stage: U.S. Geological Survey Bulletin 1274-E, 22 p.

George, T. N., and others, 1969, Recommendations on stratigraphical usage: Geological Society of London, Proceedings no. 1656, p. 139-166.

Harland, W. B., 1977, Essay review [of] International Stratigraphic Guide, 1976: Geology Magazine, v. 114, no. 3, p. 229-235.

———— ,1978, Geochronologic scales, in G. V. Cohee et al, eds., Contributions to the Geologic Time Scale: American Association of Petroleum Geologists, Studies in Geology, no. 6, p. 9-32.

Harrison, J. E., and Z. E. Peterman, 1980, North American Commission on Stratigraphic Nomenclature Note 52—A preliminary proposal for a chronometric time scale for the Precambrian of the United States and Mexico: Geological Society of America Bulletin, v. 91, no. 6, p. 377-380.

Henbest, L. G., 1952, Significance of evolutionary explosions for diastrophic division of Earth history: Journal of Paleontology, v. 26, p. 299-318.

Henderson, J. B., W. G. E. Caldwell, and J. E. Harrison, 1980, North American Commission on Stratigraphic Nomenclature, Report 8—Amendment of code concerning terminology for igneous and high-grade metamorphic rocks: Geological Society of America Bulletin, v. 91, no. 6, p. 374-376.

Holland, C. H., and others, 1978, A guide to stratigraphical procedure: Geological Society of London, Special Report 10, p. 1-18.

Huxley, T. H., 1862, The anniversary address: Geological Society of London, Quarterly Journal, v. 18, p. xl-liv.

International Commission on Zoological Nomenclature, 1964: International Code of Zoological Nomenclature adopted by the XV International Congress of Zoology: International Trust for Zoological Nomenclature, London, 176 p.

International Subcommission on Stratigraphic Classification (ISSC), 1976, International Stratigraphic Guide (H. D. Hedberg, ed.): John Wiley and Sons, New York, 200 p.

International Subcommission on Stratigraphic Classification, 1979, Magnetostratigraphy polarity units—a supplementary chapter of the ISSC International Stratigraphic Guide: Geology, v. 7, p. 578-583.

Izett, G. A., and R. E. Wilcox, 1981, Map showing the distribution of the Huckleberry Ridge, Mesa Falls, and Lava Creek volcanic ash beds (Pearlette family ash beds) of Pliocene and Pleistocene age in the western United States and southern Canada: U. S. Geological Survey Miscellaneous Geological Investigations Map I-1325.

Kauffman, E. G., 1969, Cretaceous marine cycles of the Western Interior: Mountain Geologist: Rocky Mountain Association of Geologists, v. 6, no. 4, p. 227-245.

Matthews, R. K., 1974, Dynamic stratigraphy—an introduction to sedimentation and stratigraphy: Prentice-Hall, New Jersey, 370 p.

McDougall, Ian, 1977, The present status of the geomagnetic polarity time scale: Research School of Earth Sciences, Australian National University, Publication no. 1288, 34 p.

McElhinny, M. W., 1978, The magnetic polarity time scale; prospects and possibilities in magnetostratigraphy, in G. V. Cohee et al, eds., Contributions to the Geologic Time Scale, American Association of Petroleum Geologists, Studies in Geology, no. 6, p. 57-65.

McIver, N. L., 1972, Cenozoic and Mesozoic stratigraphy of the Nova Scotia shelf: Canadian Journal of Earth Science, v. 9, p. 54-70.

McLaren, D. J., 1977, The Silurian-Devonian Boundary Committee. A final report, in A. Martinsson, ed., The Silurian-Devonian boundary: IUGS Series A, no. 5, p. 1-34.

Morrison, R. B., 1967, Principles of Quaternary soil stratigraphy, in R. B. Morrison and H. E. Wright, Jr., eds., Quaternary soils: Reno, Nevada, Center for Water Resources Research, Desert Research Institute, Univ. Nevada, p. 1-69.

North American Commission on Stratigraphic Nomenclature, 1981, Draft North American Stratigraphic Code: Canadian Society of

---

[10]Readers are reminded of the extensive and noteworthy bibliography of contributions to stratigraphic principles, classification, and terminology cited by the International Stratigraphic Guide (ISSC, 1976, p. 111-187).

Petroleum Geologists, Calgary, 63 p.

Palmer, A. R., 1965, Biomere-a new kind of biostratigraphic unit: Journal of Paleontology, v. 39, no. 1, p. 149-153.

Parsons, R. B., 1981, Proposed soil-stratigraphic guide, *in* International Union for Quaternary Research and International Society of Soil Science: INQUA Commission 6 and ISSS Commission 5 Working Group, Pedology, Report, p. 6-12.

Pawluk, S., 1978, The pedogenic profile in the stratigraphic section, *in* W. C. Mahaney, ed., Quaternary soils: Norwich, England, GeoAbstracts, Ltd., p. 61-75.

Ruhe, R. V., 1965, Quaternary paleopedology, *in* H. E. Wright, Jr., and D. G. Frey, eds., The Quaternary of the United States: Princeton, N.J., Princeton University Press, p. 755-764.

Schultz, E. H., 1982, The chronosome and supersome--terms proposed for low-rank chronostratigraphic units: Canadian Petroleum Geology, v. 30, no. 1, p. 29-33.

Shaw, A. B., 1964, Time in stratigraphy: McGraw-Hill, New York, 365 p.

Sims, P. K., 1979, Precambrian subdivided: Geotimes, v. 24, no. 12, p. 15.

Sloss, L. L., 1963, Sequences in the cratonic interior of North America: Geological Society of America Bulletin, v. 74, no. 2, p. 94-114.

Tracey, J. I., Jr., and others, 1971, Initial reports of the Deep Sea Drilling Project, v. 8: U.S. Government Printing Office, Washington, 1037 p.

Valentine, K. W. G., and J. B. Dalrymple, 1976, Quaternary buried paleosols: A critical review: Quaternary Research, v. 6, p. 209-222.

Vella, P., 1964, Biostratigraphic units: New Zealand Journal of Geology and Geophysics, v. 7, no. 3, p. 615-625.

Watson, R. A., and H. E. Wright, Jr., 1980, The end of the Pleistocene: A general critique of chronostratigraphic classification: Boreas, v. 9, p. 153-163.

Weiss, M. P., 1979a, Comments and suggestions invited for revision of American Stratigraphic Code: Geological Society of America, News and Information, v. 1, no. 7, p. 97-99.

———— ,1979b, Stratigraphic Commission Note 50--Proposal to change name of Commission: American Association of Petroleum Geologists Bulletin, v. 63, no. 10, p. 1986.

Weller, J. M., 1960, Stratigraphic principles and practice: Harper and Brothers, New York, 725 p.

Willman, H. B., and J. C. Frye, 1970, Pleistocene stratigraphy of Illinois: Illinois State Geological Survey Bulletin 94, 204 p.

APPENDIX I: PARTICIPANTS AND CONFEREES IN CODE REVISION

Code Committee

Steven S. Oriel (U.S. Geological Survey), chairman, Hubert Gabrielse (Geological Survey of Canada), William W. Hay (Joint Oceanographic Institutions), Frank E. Kottlowski (New Mexico Bureau of Mines), John B. Patton (Indiana Geological Survey).

Lithostratigraphic Subcommittee

James D. Aitken (Geological Survey of Canada), chairman, Monti Lerand (Gulf Canada Resources, Ltd.), Mitchell W. Reynolds (U.S. Geological Survey), Robert J. Weimer (Colorado School of Mines), Malcolm P. Weiss (Northern Illinois University).

Biostratigraphic Subcommittee

Allison R. (Pete) Palmer (Geological Society of America), chairman, Ismael Ferrusquia (University of Mexico), Joseph E. Hazel (U.S. Geological Survey), Erle G. Kauffman (University of Colorado), Colin McGregor (Geological Survey of Canada), Michael A. Murphy (University of California, Riverside), Walter C. Sweet (Ohio State University).

Chronostratigraphic Subcommittee

Zell E. Peterman (U.S. Geological Survey), chairman, Zoltan de Cserna (Sociedad Geológica Mexicana), Edward H. Schultz (Suncor, Inc., Calgary), Norman F. Sohl (U.S. Geological Survey), John A. Van Couvering (American Museum of Natural History).

Plutonic-Metamorphic Advisory Group

Jack E. Harrison (U.S. Geological Survey), chairman, John B. Henderson (Geological Survey of Canada), Harold L. James (retired), Leon T. Silver (California Institute of Technology), Paul C. Bateman (U.S. Geological Survey).

Magnetostratigraphic Advisory Group

Roger W. Macqueen (University of Waterloo), chairman, G. Brent Dalrymple (U.S. Geological Survey), Walter F. Fahrig (Geological Survey of Canada), J. M. Hall (Dalhousie University).

Volcanic Advisory Group

Richard V. Fisher (University of California, Santa Barbara), chairman, Thomas A. Steven (U.S. Geological Survey), Donald A. Swanson (U.S. Geological Survey).

Tectonostratigraphic Advisory Group

Darrel S. Cowan (University of Washington), chairman, Thomas W. Donnelly (State University of New York at Binghamton), Michael W. Higgins and David L. Jones (U.S. Geological Survey), Harold Williams (Memorial University, Newfoundland).

Quaternary Advisory Group

Norman P. Lasca (University of Wisconsin-Milwaukee), chairman, Mark M. Fenton (Alberta Research Council), David S. Fullerton (U.S. Geological Survey), Robert J. Fulton (Geological Survey of Canada), W. Hilton Johnson (University of Illinois), Paul F. Karrow (University of Waterloo), Gerald M. Richmond (U.S. Geological Survey).

Conferees

W. G. E. Caldwell (University of Saskatchewan), Lucy E. Edwards (U.S. Geological Survey), Henry H. Gray (Indiana Geological Survey), Hollis D. Hedberg (Princeton University), Lewis H. King (Geological Survey of Canada), Rudolph W. Kopf (U.S. Geological Survey), Jerry A. Lineback (Robertson Research U.S.), Marjorie E. MacLachlan (U.S. Geological Survey), Amos Salvador (University of Texas, Austin), Brian R. Shaw (Samson Resources, Inc.), Ogden Tweto (U.S. Geological Survey).

APPENDIX II: 1977-1982 COMPOSITION OF THE NORTH AMERICAN COMMISSION ON STRATIGRAPHIC NOMENCLATURE

Each Commissioner is appointed, with few exceptions, to serve a 3-year term (shown by such numerals as 80-82 for 1980-1982) and a few are reappointed.

American Association of Petroleum Geologists

Timothy A. Anderson (Gulf Oil Co.) 77-83, Orlo E. Childs (Texas Tech University) 76-79, Kenneth J. Englund (U.S. Geological Survey) 74-77, Susan Longacre (Getty Oil Co.) 78-84, Donald E. Owen (Cities Service Co.) 79-82, Grant Steele (Gulf Oil Co.) 75-78.

Association of American State Geologists

Larry D. Fellows (Arizona Bureau of Geology) 81-82, Lee C. Gerhard (North Dakota Geological Survey) 79-81, Donald C. Haney (Kentucky Geological Survey) 80-83, Wallace B. Howe (Missouri Division of Geology) 74-77, Robert R. Jordan (Delaware Geological Survey) 78-84, vice-chairman, Frank E. Kottlowski (New Mexico Bureau of Mines) 76-79, Meredith E. Ostrom (Wisconsin Geological Survey) 77-80, John B. Patton (Indiana Geological Survey) 75-78.

Geological Society of America

Clarence A. Hall, Jr. (University of California, Los Angeles) 78-81, Jack E. Harrison (U.S. Geological Survey) 74-77, William W. Hay (University of Miami) 75-78, Robert S. Houston (University of Wyoming) 77-80, Michael A. Murphy (University of California, Riverside) 81-84, Allison R. Palmer (Geological Society of America) 80-83, Malcolm P. Weiss (Northern Illinois University) 76-82, chairman.

United States Geological Survey

Earl E. Brabb (Menlo Park) 78-82, David S. Fullerton (Denver) 78-84, E. Dale Jackson (Menlo Park) 76-78, Kenneth L. Pierce (Denver) 75-78, Norman F. Sohl (Washington) 74-83.

Geological Survey of Canada

James D. Aitken (Calgary) 75-78, Kenneth D. Card (Kanata) 80-83, Donald G. Cook (Calgary) 78-81, Robert J. Fulton (Ottawa) 81-84, John B. Henderson (Ottawa) 74-77, Lewis H. King (Dartmouth) 79-82, Maurice B. Lambert (Ottawa) 77-80, Christopher J. Yorath (Sydney) 76-79.

Canadian Society of Petroleum Geologists

Roland F. deCaen (Union Oil Co. of Canada) 79-82, J. Ross McWhae (Petro Canada Exploration) 77-80, Edward H. Schultz (Suncor, Inc.) 74-77, 80-83, Ulrich Wissner (Union Oil Co. of Canada) 76-79.

Geological Association of Canada

W. G. E. Caldwell (University of Saskatchewan) 76-79, R. K. Jull (University of Windsor) 78-79, Paul S. Karrow (University of Waterloo) 81-84, Alfred C. Lenz (University of Western Ontario) 79-81, David E. Pearson (British Columbia Mines and Petroleum Resources) 79-81, Paul E. Schenk (Dalhousie University) 75-78.

Asociación Mexicana de Geólogos Petróleros

Jose Carillo Bravo (Petróleos Mexicanos) 78-81, Baldomerro Carrasco V., 75-78.

Sociedad Geólogica Mexicana

Zoltan de Cserna (Universidad Nacional Autónoma de México) 76-82.

Instituto de Geologia de la Universidad Nacional Autónoma de México

Ismael Ferrusquia Villafranca (Universidad Nacional Autónoma de México) 76-81, Fernando Ortega Gutiérrez (Universidad Nacional Autónoma de México) 81-84.

APPENDIX III: REPORTS AND NOTES OF THE AMERICAN COMMISSION ON STRATIGRAPHIC NOMENCLATURE

Reports (formal declarations, opinions, and recommendations)
1. Moore, Raymond C., Declaration on naming of subsurface stratigraphic units: AAPG Bulletin, v. 33, no. 7, p. 1280-1282, 1949.
2. Hedberg, Hollis D., Nature, usage, and nomenclature of time-stratigraphic and geologic-time units: AAPG Bulletin, v. 36, no. 8, p. 1627-1638, 1952.
3. Harrison, J. M., Nature, usage, and nomenclature of time-stratigraphic and geologic-time units as applied to the Precambrian: AAPG Bulletin, v. 39, no. 9, p. 1859-1861, 1955.
4. Cohee, George V., and others, Nature, usage, and nomenclature of rock-stratigraphic units: AAPG Bulletin, v. 40, no. 8, p. 2003-2014, 1956.
5. McKee, Edwin D., Nature, usage and nomenclature of biostratigraphic units: AAPG Bulletin, v. 41, no. 8, p. 1877-1889, 1957.
6. Richmond, Gerald M., Application of stratigraphic classification and nomenclature to the Quaternary: AAPG Bulletin, v. 43, no. 3, pt. I, p. 663-675, 1959.
7. Lohman, Kenneth E., Function and jurisdictional scope of the American Commission on Stratigraphic Nomenclature: AAPG Bulletin, v. 47, no. 5, p. 853-855, 1963.
8. Henderson, John B., W. G. E. Caldwell, and Jack E. Harrison, Amendment of code concerning terminology for igneous and high-grade metamorphic rocks: GSA Bulletin, pt. I, v. 91, no. 6, p. 374-376, 1980.
9. Harrison, Jack E., and Zell E. Peterman, Adoption of geochronometric units for divisions of Precambrian time: AAPG Bulletin, v. 66, no. 6, p. 801-802, 1982.

Notes (informal statements, discussions, and outlines of problems)

1. Organization and objectives of the Stratigraphic Commission: AAPG Bulletin, v. 31, no. 3, p. 513-518, 1947.
2. Nature and classes of stratigraphic units: AAPG Bulletin, v. 31, no. 3, p. 519-528, 1947.
3. Moore, Raymond C., Rules of geologic nomenclature of the Geological Survey of Canada: AAPG Bulletin, v. 32, no. 3, p. 366-367, 1948.
4. Jones, Wayne V., and Raymond C. Moore, Naming of subsurface stratigraphic units: AAPG Bulletin, v. 32, no. 3, p. 367-371, 1948.
5. Flint, Richard Foster, and Raymond C. Moore, Definition and adoption of the terms stage and age: AAPG Bulletin, v. 32, no. 3, p. 372-376, 1948.
6. Moore, Raymond C., Discussion of nature and classes of stratigraphic units: AAPG Bulletin, v. 21, no. 3, p. 376-381, 1948.
7. Records of the Stratigraphic Commission for 1947-1948: AAPG Bulletin, v. 33, no. 7, p. 1271-1273, 1949.
8. Australian Code of Stratigraphical Nomenclature: AAPG Bulletin, v. 33, no. 7, p. 1273-1276, 1949.
9. The Pliocene-Pleistocene boundary: AAPG Bulletin, v. 33, no. 7, p. 1276-1280, 1949.
10. Moore, Raymond C., Should additional categories of stratigraphic units be recognized?: AAPG Bulletin, v. 34, no. 12, p. 2360-2361, 1950.
11. Moore, Raymond C., Records of the Stratigraphic Commission for 1949-1950: AAPG Bulletin, v. 35, no. 5, p. 1074-1076, 1951.
12. Moore, Raymond C., Divisions of rocks and time: AAPG Bulletin, v. 35, no. 5, p. 1076, 1951.
13. Williams, James Steele, and Aureal T. Cross, Third Congress of Carboniferous Stratigraphy and Geology: AAPG Bulletin, v. 36, no. 1, p. 169-172, 1952.
14. Official report of round table conference on stratigraphic nomenclature at Third Congress of Carboniferous Stratigraphy and Geology, Heerlen, Netherlands, June 26-28, 1951: AAPG Bulletin, v. 36, no. 10, p. 2044-2048, 1952.
15. Records of the Stratigraphic Commission for 1951-1952: AAPG Bulletin, v. 37, no. 5, p. 1078-1080, 1953.
16. Records of the Stratigraphic Commission for 1953-1954: AAPG Bulletin, v. 39, no. 9, p. 1861-1863, 1955.
17. Suppression of homonymous and obsolete stratigraphic names: AAPG Bulletin, v. 40, no. 12, p. 2953-2954, 1956.
18. Gilluly, James, Records of the Stratigraphic Commission for 1955-1956: AAPG Bulletin, v. 41, no. 1, p. 130-133, 1957.
19. Richmond, Gerald M., and John C. Frye, Status of soils in stratigraphic nomenclature: AAPG Bulletin, v. 31, no. 4, p. 758-763, 1957.
20. Frye, John C., and Gerald M. Richmond, Problems in applying standard stratigraphic practice in nonmarine Quaternary deposits: AAPG Bulletin, v. 42, no. 8, p. 1979-1983, 1958.
21. Frye, John C., Preparation of new stratigraphic code by American Commission on Stratigraphic Nomenclature: AAPG Bulletin, v. 42, no. 8, p. 1984-1986, 1958.
22. Records of the Stratigraphic Commission for 1957-1958: AAPG Bulletin, v. 43, no. 8, p. 1967-1971, 1959.
23. Rodgers, John, and Richard B. McConnell, Need for rock-stratigraphic units larger than group: AAPG Bulletin, v. 43, no. 8, p. 1971-1975, 1959.
24. Wheeler, Harry E., Unconformity-bounded units in stratigraphy: AAPG Bulletin, v. 43, no. 8, p. 1975-1977, 1959.
25. Bell, W. Charles, and others, Geochronologic and chronostratigraphic units: AAPG Bulletin, v. 45, no. 5, p. 666-670, 1961.
26. Records of the Stratigraphic Commission for 1959-1960: AAPG Bul-

letin, v. 45, no. 5, p. 670-673, 1961.

27. Frye, John C., and H. B. Willman, Morphostratigraphic units in Pleistocene stratigraphy: AAPG Bulletin, v. 46, no. 1, p. 112-113, 1962.

28. Shaver, Robert H., Application to American Commission on Stratigraphic Nomenclature for an amendment of Article 4f of the Code of Stratigraphic Nomenclature on informal status of named aquifers, oil sands, coal beds, and quarry layers: AAPG Bulletin, v. 46, no. 10, p. 1935, 1962.

29. Patton, John B., Records of the Stratigraphic Commission for 1961-1962: AAPG Bulletin, v. 47, no. 11, p. 1987-1991, 1963.

30. Richmond, Gerald M., and John G. Fyles, Application to American Commission on Stratigraphic Nomenclature for an amendment of Article 31, Remark (b) of the Code of Stratigraphic Nomenclature on misuse of the term "stage": AAPG Bulletin, v. 48, no. 5, p. 710-711, 1964.

31. Cohee, George V., Records of the Stratigraphic Commission for 1963-1964: AAPG Bulletin, v. 49, no. 3, pt. I of II, p. 296-300, 1965.

32. International Subcommission on Stratigraphic Terminology, Hollis D. Hedberg, ed., Definition of geologic systems: AAPG Bulletin, v. 49, no. 10, p. 1694-1703, 1965.

33. Hedberg, Hollis D., Application to American Commission on Stratigraphic Nomenclature for amendments to Articles 29, 31, and 37 to provide for recognition of erathem, substage, and chronozone as time-stratigraphic terms in the Code of Stratigraphic Nomenclature: AAPG Bulletin, v. 50, no. 3, p. 560-561, 1966.

34. Harker, Peter, Records of the Stratigraphic Commission for 1964-1966: AAPG Bulletin, v. 51, no. 9, p. 1862-1869, 1967.

35. DeFord, Ronald K., John A. Wilson, and Frederick M. Swain, Application to American Commission on Stratigraphic Nomenclature for an amendment of Article 3 and Article 13, Remarks (c) and (e), of the Code of Stratigraphic Nomenclature to disallow recognition of new stratigraphic names that appear only in abstracts, guidebooks, microfilms, newspapers, or in commercial or trade journals: AAPG Bulletin, v. 51, no. 9, p. 1868-1869, 1967.

36. Cohee, George V., Ronald K. DeFord, and H. B. Willman, Amendment of Article 5, Remarks (a) and (e) of the Code of Stratigraphic Nomenclature for treatment of geologic names in a gradational or interfingering relationship of rock-stratigraphic units: AAPG Bulletin, v. 53, no. 9, p. 2005-2006, 1969.

37. Kottlowski, Frank E., Records of the Stratigraphic Commission for 1966-1968: AAPG Bulletin, v. 53, no. 10, p. 2179-2186, 1969.

38. Andrews, J., and K. Jinghwa Hsü, A recommendation to the American Commission on Stratigraphic Nomenclature concerning nomenclatural problems of submarine formations: AAPG Bulletin, v. 54, no. 9, p. 1746-1747, 1970.

39. Wilson, John Andrew, Records of the Stratigraphic Commission for 1968-1970: AAPG Bulletin, v. 55, no. 10, p. 1866-1872, 1971.

40. James, Harold L., Subdivision of Precambrian: An interim scheme to be used by U.S. Geological Survey: AAPG Bulletin, v. 56, no. 6, p. 1128-1133, 1972.

41. Oriel, Steven S., Application for amendment of Article 8 of code, concerning smallest formal rock-stratigraphic unit: AAPG Bulletin, v. 59, no. 1, p. 134-135, 1975.

42. Oriel, Steven S., Records of Stratigraphic Commission for 1970-1972: AAPG Bulletin, v. 59, no. 1, p. 135-139, 1975.

43. Oriel, Steven S., and Virgil E. Barnes, Records of Stratigraphic Commission for 1972-1974: AAPG Bulletin, v. 59, no. 10, p. 2031-2036, 1975.

44. Oriel, Steven S., Roger W. Macqueen, John A. Wilson, and G. Brent Dalrymple, Application for addition to code concerning magnetostratigraphic units: AAPG Bulletin, v. 60, no. 2, p. 273-277, 1976.

45. Sohl, Norman F., Application for amendment concerning terminology for igneous and high-grade metamorphic rocks: AAPG Bulletin, v. 61, no. 2, p. 248-251, 1977.

46. Sohl, Norman F., Application for amendment of Articles 8 and 10 of code, concerning smallest formal rock-stratigraphic unit: AAPG Bulletin, v. 61, no. 2, p. 252, 1977.

47. Macqueen, Roger W., and Steven S. Oriel, Application for amendment of Articles 27 and 34 of stratigraphic code to introduce point-boundary stratotype concept: AAPG Bulletin, v. 61, no. 7, p. 1083-1085, 1977.

48. Sohl, Norman F., Application for amendment of Code of Stratigraphic Nomenclature to provide guidelines concerning formal terminology for oceanic rocks: AAPG Bulletin, v. 62, no. 7, p. 1185-1186, 1978.

49. Caldwell, W.G.E., and N. F. Sohl, Records of Stratigraphic Commission for 1974-1976: AAPG Bulletin, v. 62, no. 7, p. 1187-1192, 1978.

50. Weiss, Malcolm P., Proposal to change name of commission: AAPG Bulletin, v. 63, no. 10, p. 1986, 1979.

51. Weiss, Malcolm P., and James D. Aitken, Records of Stratigraphic Commission, 1976-1978: AAPG Bulletin, v. 64, no. 1, p. 136-137, 1980.

52. Harrison, Jack E., and Zell E. Peterman, A preliminary proposal for a chronometric time scale for the Precambrian of the United States and Mexico: GSA Bulletin, pt. I, v. 91, no. 6, p. 377-380, 1980.

Geologic Time Scales

Cenozoic Time Scale (After Haq et al., 1987)

| Series | | | Stage | Relative Change of Coastal Onlap | Eustatic Curves (meters) | Magnetostratigraphy | | |

[a] As noted in Chapter 13, the Eocene/Oligocene portion of the time scale is still undergoing revision.

[b] ▓, normal polarity; □, reversed polarity.

Mesozoic Time Scale (After Harland et al., 1982)

Age (Ma)	Eon	Era	Subera	Period	Epoch	Age	Best Age Estimate (Ma)	Relative Change of Coastal Onlap	Magnetostratigraphy Polarity[a]
55	Phanerozoic	Cenozoic	Tertiary	Paleogene	Paleocene	Thanetian (L)	54.9	1.0 ← Landward / Seaward → 0.5 0	24r, 25, 25r, 26
60						Danian (E)	60.2		26r, 27, 27r, 28, 28r
66.5				Pg			66.5		30, 31, 31r, 32, 32r, 33
				Cretaceous	Late	Maastrichtian	73		33r
						Campanian	83		34
						Santonian	87.5		
						Coniacian	88.5		
						Turonian	91		
						Cenomanian	97.5		
100		Mesozoic			Early	Albian	113		
						Aptian	119		M0, M1n, M1
						Barremian	125		M3
					Neocomiam	Hauterivian	131		M5, M9, M10Nn, M11
						Valanginian	138		M12, M14, M16n, M17
				K		Berriasian	144		M19n, M20
150				Jurassic	Late	Tithonian (Malm)	150		M22n, M24, M25An, M26n
						Kimmeridgian	156		M29
						Oxfordian	163		
					Middle	Callovian (Dogger)	169		
						Bathonian	175		
						Bajocian	181		
						Aalenian	188		
					Early	Toarcian (Lias)	194		
200						Pliensbachian	200		
						Sinemurian	206		
				J		Hettangian	213		
				Triassic	Late	Rhaetian	219		
						Norian	225		
						Carnian	231		
					Middle	Ladinian	238		
						Anisian	243		
248				Tr	Early	Scythian	248		
250		Paleozoic		Permian	Late	Tatarian	253		
						Kazanian	253		
						Ufimian	258		
						Kungurian	263		
					Early	Artinskian	268		
						Sakmarian	268		
300				P		Asselian	286		

[a] ▒, normal polarity; ☐, reversed polarity.

Paleozoic Time Scale (After Harland et al., 1982)

Age (Ma)	Eon	Era	Period	Epoch	Age	Best Age Estimate (Ma)	Relative Change of Coastal Onlap
	Phanerozoic	Mesozoic	Triassic	Late	Rhaetian	219	
					Norian	225	
					Carnian	231	
				Middle	Ladinian		
					Anisian	238	
					Scythian	243	
248			Tr	Early	Tatarian	248	
250		Paleozoic	Permian	Late	Kazanian	253	
					Ufimian	258	
				Early	Kungurian	263	
					Artinskian	268	
			P	Gzelian	Sakmarian		
				Kasimovian	Asselian	286	
300			Carboniferous	Pennsylvanian / Pen	Moscovian	296	
					Bashkirian	315	
					Serpukhovian	320	
				Mississippian / Mis	Visean	333	
350			C		Tournaisian	352	
			Devonian	Late	Famennian	360 / 367	
					Frasnian	374	
				Middle	Givetian	380	
					Eifelian	387	
400				Early	Emsian	394	
					Siegenian	401	
			D		Gedinnian	408	
			Silurian	Pridoli		414	
				Ludlow		421	
				Wenlock		428	
			S	Llandovery		438	
450			Ordovician	Ashgill		448	
				Caradoc		458	
				Llandeilo		468	
				Llanvirn		478	
				Arenig		488	
500			O	Tremadoc		505	
			Cambrian	Merioneth		525	
				St Davids		540	
550				Caerfai			
					—?—	570	
			€		Tommotian		
590			Vendian	Ediacaran		590	
600		Sinian	V	Varangian		630 / 670	
800			Sturtian U			800 / 900	

Relative Change of Coastal Onlap scale: 1.0 — 0.5 — 0 — Landward ← → Seaward

Precambrian Time Scale (After Harland et al., 1982)

Age (Ma)	Eon	Era	Period	Epoch	Age	Best Age Estimate (Ma)
				Merioneth		525
				St Davids		540
550	Phanerozoic	Paleozoic	Cambrian	Caerfai		570
					Tommotian	
			€		– – – ? – – –	590
590				Ediacaran		630
600		Sinian	Vendian V			670
			Sturtian U	Varangian		800
						900
1000						1050
		Riphean	Yurmatin Y			1350
	Proterozoic		Burzyan B			1650
2000					Precambrian subera and period names have no international status	2100
			Huronian H			2400
						2500
			Randian Ran			2630
3000						2800
	Archean	Swazian				3000
						3500
						3750
			Isuan I			3900
4000						4000
	Priscoan		Hadean			
			Hde			
5000						

Bibliography

Ager, D. V., 1964. "The British Mesozoic Committee." *Nature,* 203: 1059.

Ager, D. V., 1973. *The Nature of the Stratigraphical Record,* 1st ed. Macmillian, London.

Ager, D. V., 1981. *The Nature of the Stratigraphical Record,* 2d ed. John Wiley, New York.

Aitken, J. D., 1978. "Revised Model for Depositional Grand Cycles, Cambrian of the Southern Rocky Mountains, Canada." *Bull. Can. Petrol. Geol.,* 26: 515–542.

Allen, J. R. L., 1970. "Sediments of the modern Niger Delta; A Summary and Review." *Soc. Econ. Paleont. Mineral. Spec. Publ.,* 15: 138–151.

Allen, J. R. L., 1985. *Principles of Physical Sedimentology.* George Allen and Unwin, London.

Alvarez, L. W., W. Alvarez, F. Asaro, and H. V. Michel, 1980. "Extraterrestrial Cause for the Cretaceous-Tertiary Extinction." *Science,* 208: 1095–1108.

Alvarez, W., M. A. Arthur, A. G. Fischer, W. Lowrie, G. Napoleone, I. Premoli-Silva, and W. R. Roggenthen, 1977. "Upper Cretaceous-Paleocene Magnetic Stratigraphy at Gubbio, Italy: V. Type Section for the Late Cretaceous-Paleocene Geomagnetic Reversal Time Scale." *Geol. Soc. Amer. Amer. Bull.,* 88: 383–389.

Alvaro, M., R. Capote, and R. Vegas, 1979. "Un Modelo de Evolución Geotectónica para la Cadena Celtibérica." *Acta Geol. Hispánica,* 14: 172–177.

Anders, M. H., S. W. Krueger, and P. M. Sadler, 1987. "A New Look at Sedimentation Rates and the Completeness of the Stratigraphic Record." *J. Geol.,* 95: 1–14.

Anderson, E. J., P. W. Goodwin, and T. H. Sobieski, 1984. "Episodic Accumulation and the Origin of Formation Boundaries in the Helderberg Group of New York State." *Geology,* 12: 120–123.

Anstey, N. A., 1982. *Simple Seismics.* International Human Resources Development Corporation, Boston.

Arthur, M. A., W. E. Dean, R. M. Pollastro, G. E. Claypool, and P. A. Scholle, 1985. "Comparative Geochemical and Mineralogical Studies of Two Cyclic Transgressive Pelagic Limestone Units, Cretaceous Western Interior Basin, U.S." In L. M. Pratt,

E. G. Kauffman, and F. B. Zelt (eds.), *Fine-Grained Deposits and Biofacies of the Cretaceous Western Interior Seaway: Evidence of Cyclic Sedimentary Processes*, 16–27. Soc. Econ. Paleont, Mineral. Field Trip Guidebook 4.

Ashley, G. H., et al., 1933. "Classification and Nomenclature of Rock Units." *Geol. Soc. Amer. Bull.,* 44: 423–459.

Asquith, G. B., 1982. *Basic Well Log Analysis for Geologists.* Amer. Assoc. Petrol. Geol. Methods in Exploration Series, Tulsa, Okla.

Aubry, M.-P., W. A. Berggren, D. V. Kent, J. J. Flynn, K. D. Klitgord, J. D. Obradovich, and D. R. Prothero, 1988. "Paleogene Geochronology: A Critique and Response." *Paleoceanog.,* 3: 707–742.

Baars, D. L., 1962. "Permian System of the Colorado Plateau." *Amer. Assoc. Petrol. Geol. Bull.,* 46: 149–218.

Baganz, B. P., J. C. Horne, and J. C. Ferm, 1975. "Carboniferous and Recent Mississippi Lower Delta Plains: A Comparison." *Trans. Gulf Coast Assoc. Geol. Socs.,* 25: 183–191.

Bachhuber, F. W., S. Rowland, and P. W. Huntoon, 1987. "Geology of the Lower Grand Canyon and Upper Lake Mead by Boat—An Overview." *Ariz. Bur. Geol. Min. Tech., Geol. Surv. Branch Spec. Paper* 5: 39–51.

Bally, A. W., 1980. "Basins and Subsidence—A Summary." In A. W. Bally, P. L. Bender, T. R. McGetchin, and R. I. Walcott (eds.), *Dynamics of Plate Interiors.* Amer. Geophys. Union, Geodynamics Series 1: 5–20.

Barrande, J., 1846. *Notice Préliminaire sur le Système Silurien et les Trilobites de la Bohême.* Leipzig.

Barrell, J., 1917. "Rhythms and the Measurement of Geologic Time." *Geol. Soc. Amer. Bull.,* 28: 745–904.

Barron, E. J., M. A. Arthur, and E. G. Kauffman, 1985. "Cretaceous Rhythmic Bedding Sequences: A Plausible Link between Orbital Variations and Climate." *Earth Planet. Sci. Lett.,* 73: 327–340.

Barron, E. J., and W. M. Washington, 1982. "Cretaceous Climate: A Comparison of Atmospheric Simulations with the Geologic Record." *Palaeogeogr., Palaeoclimat., Palaeoecol.,* 40: 103–133.

Barron, E. J., and W. M. Washington, 1985. "Warm Cretaceous Climates: High Atmospheric CO_2 As a Plausible Mechanism." *Geophys. Monogr. Series,* 32: 546–553.

Barwis, J. H., and M. O. Hayes, 1979. "Regional Patterns of Barrier Island and Tidal-Inlet Deposition Applied to Hydrocarbon Exploration." In R. S. Saxena (ed.), *Stratigraphic Concepts in Hydrocarbon Exploration,* 113–158. Short course notes, 78th Gulf

Coast Geol. Soc.–Soc. Econ. Paleont. Mineral. Meeting.

Bassett, M. G., 1985. "Towards a Common Language in Stratigraphy." *Episodes,* 8(2): 87–92.

Bateman, P. C., 1965. "Geology of Tungsten Mineralization of the Bishop District, California." *U.S. Geol. Surv. Prof. Paper,* 470.

Bathurst, R. G. C., 1975. *Carbonate Sediments and Their Diagenesis,* 2d. ed. Developments in Sedimentology 12. Elsevier, New York.

Berger, W. H., 1974. "Deep-sea Sedimentation." In C. A. Burk and C. L. Drake (eds.), *The Geology of Continental Margins,* 213–241. Springer-Verlag, New York.

Berger, W. H., A. W. H. Bé, and W. V. Sliter, 1975. *Dissolution of Deep-sea Carbonates: An Introduction.* Spec. Publ. Cushman Foundation 13.

Berggren, W. A., 1971. "Tertiary Boundaries and Correlations." In B. M Funnell and W. R. Riedel (eds.), *The Micropaleontology of Oceans,* 693–809. Cambridge Univ. Press, Cambridge.

Berggren, W. A., 1986. "Geochronology of the Eocene-Oligocene Boundary." In C. Pomerol and I. Premoli-Silva (eds.), *Terminal Eocene Events,* 349–356. Elsevier, Amsterdam.

Berggren, W. A., and M.-P. Aubry, 1983. "Rb-Sr Isochron of the Eocene Castle Hayne Limestone, North Carolina—Further Discussion." *Geol. Soc. Amer. Bull.,* 94: 364–370.

Berggren, W. A., and J. A. Van Couvering, 1978. "Biochronology." *Amer. Assoc. Petrol. Geol. Mem.,* 6: 39–55.

Berggren, W. A., D. V. Kent, and J. J. Flynn, 1985. "Paleogene Geochronology and Chronostratigraphy." In N. J. Snelling (ed.), *The Chronology of the Geological Record,* 141–195 Geol. Soc. Lond. Mem., 10.

Berggren, W. A., M. C. McKenna, J. Hardenbol, and J. D. Obradovich, 1978. "Revised Paleogene Polarity Time-scale." *J. Geol.,* 86: 67–81.

Bergman, K. M., and R. G. Walker, 1987. "The Importance of Sea-Level Fluctuations in the Formation of Linear Conglomerate Bodies: Carrot Creek Member of the Cardium Formation, Cretaceous Western Interior Seaway, Alberta, Canada." *J. Sed. Petrol.,* 57(4): 651–665.

Berry, W. B. N., 1987. *Growth of a Prehistoric Time Scale Based on Organic Evolution,* 2d ed. Blackwell Scientific Publ., Palo Alto, Calif.

Bertrand-Sarfati, J., and M. R. Walter, 1981. "Stromatolite Biostratigraphy." *Precamb. Res.,* 15: 353–371

Beyrich, H. E. von, 1854. "Über die Stellung die hessischen Tertiar-bildungen." *K. Preuss. Akad. Wiss. Berlin Monatsber,* Nov. 1854: 664–666.

Billings, M. P., 1950. "Stratigraphy and the Study of Metamorphic Rocks." *Geol. Soc. Amer. Bull.*, 61: 435–448.

Blatt, H., 1982. *Sedimentary Petrology*. W. H. Freeman and Co., San Francisco.

Blatt, H., G. V. Middleton, and R. C. Murray, 1980. *Origin of Sedimentary Rocks*, 2d ed. Prentice-Hall, Englewood Cliffs, N.J.

Boggs, S., Jr., 1987. *Principles of Sedimentology and Stratigraphy*. Merrill Publishing Co., Columbus, Ohio.

Bouma, A. H., 1962. *Sedimentology of Some Flysch Deposits—A Graphic Approach to Facies Interpretation*. Elsevier, New York.

Bouma, A. H., W. R. Normark, and N. E. Barnes (eds.), 1985. *Submarine Fans and Related Turbidite Systems*. Springer-Verlag, New York.

Bourgeois, J., 1980. "A Transgressive Shelf Sequence Exhibiting Hummocky Cross-stratification: The Cape Sebastian Sandstone (Upper Cretaceous), Southwest Oregon." *J. Sed. Petrol.*, 50: 681–702.

Bowles, F. A., R. N. Jack, and I. S. E. Carmichael, 1973. "Investigations of Deep-sea Volcanic Ash Layers from Equatorial Pacific Cores." *Geol. Soc. Amer. Bull.*, 84: 237–238.

Brett, C. E., and G. C. Baird, 1986. "Symmetrical and Upward Shallowing Cycles in the Middle Devonian of New York State and their Implications for Punctuated Aggradational Cycles." *Paleoceanog.*, 1: 431–445.

Brice, W. R., 1982. "Bishop Ussher, John Lightfoot, and the Age of Creation." *J. Geol. Educ.*, 30: 18–24.

Brock, A., and G. L. Isaac, 1974. "Paleomagnetic Stratigraphy and Chronology of Hominid-Bearing Sediments East of Lake Rudolf, Kenya." *Nature*, 247: 344–348.

Broecker, W. S., 1972. *Chemical Oceanography*. Harcourt Brace Jovanovich, New York.

Brunhes, B., 1906. "Recherches sur le direction d'aimantation des roches volcaniques." *J. Physique*, 5: 705–724.

Buchheim, H. P., and R. Biaggi, 1988. "Laminae Counts Within a Synchronous Oil Shale Unit: A Challenge to the Varve Concept." *Geol. Soc. Amer. Abst. with Prog.*, 20(7): A217.

Bull, W. B., 1972. "Recognition of Alluvial Fan Deposits in the Stratigraphic Record." In J. K. Rigby and W. K. Hamblin (eds.), *Recognition of Ancient Sedimentary Environments*, 63–83. Soc. Econ. Paleont. Mineral. Spec. Publ. 16.

Bull, W. B., 1977. "The Alluvial Fan Environment." *Prog. Phys. Geog.*, 1: 222–270.

Burk, C. A., and C. L Drake (eds.), 1974. *The Geology of Continental Margins*. Springer-Verlag, New York.

Burke, K. C. A., and J. F. Dewey, 1973. "Plume-Gen-erated Triple Junctions: Key Indicators in Applying Plate Tectonics to Old Rocks. *J. Geol.*, 81: 406–433.

Busch, R. M., and H. T. Rollins, 1984. "Correlation of Carboniferous Strata Using a Hierarchy of Transgressive-Regressive Units." *Geology*, 12: 471–474.

Busch, R. M., and R. R. West, 1987. "Hierarchical Genetic Stratigraphy: A Framework For Paleoceanography." *Paleoceanog.*, 2: 141–164.

Byers, C. A., 1982. "Stratigraphy—The Fall of Continuity." *J. Geol. Educ.*, 30: 215–221.

Carter, R. M., 1988. "Plate Boundary Tectonics, Global Sea-level Changes, and the Development of the Eastern South Island Continental Margin, New Zealand, Southwest Pacific." *Marine and Petrol. Geol.*, 4: 1–80.

Cavelier, C., J.-J. Chateauneuf, C. Pomerol, D. Rabussier, M. Renard, and C. Vergnaud-Grazzini, 1981. "The Geological Events at the Eocene-Oligocene Boundary." *Palaeogeogr., Palaeoclimat., Palaeoecol.*, 36: 223–248.

Cerling, T. E., and F. H. Brown, 1982. "Tuffaceous Marker Horizons in the Koobi Fora Region and the Lower Omo Valley." *Nature*, 299: 216–221.

Chamberlain, T. C., and R. D. Salisbury, 1906. *Geology*. Holt, New York.

Chang, K. H., 1975. "Unconformity-Bounded Stratigraphic Units." *Geol. Soc. Amer. Bull.*, 86: 1544–1552.

Chlupác, I., H. Jaeger, and J. Zikmundova, 1972. "The Silurian-Devonian boundary in the Barrandian." *Bull. Can. Petrol. Geol.*, 20: 104–174.

Cluff, R. M., M. L. Reinbold, and J. A. Lineback, 1981. "The New Albany Shale Group of Illinois." *Ill. State Geol. Surv. Circ.* 518: 1–83.

Cobban, W. A., and J. T. Reeside, 1952. "Correlation of the Cretaceous Formations of the Western Interior of the United States." *Amer. Assoc. Petrol. Geol. Bull.*, 63: 1011–1044.

Cohee, G. V., M. F. Glaessner, and H. D. Hedberg (eds.), 1978. *"Contributions to the Geologic Time Scale*. Amer. Assoc. Petrol. Geol. Stud. Geol. 6: 1–388.

Colbert, E. H., 1935. "Siwalik Mammals in the American Museum of Natural History." *Trans. Amer. Phil. Soc.*, 26: 1–401.

Coleman, J. M., 1976. *Deltas: Processes of Deposition and Models for Exploration*. Continuing Education Publishing Co., Champaign, Ill.

Collinson, J. D., and D. B. Thompson, 1982. *Sedimentary Structures*. George Allen and Unwin, London.

Conkin, J. E., B. F. Conkin, and L. Z. Lipschutz, 1980. "Devonian Black Shales in the Eastern United States." *Univ. Louisville Stud. Paleont. Strat.*, 12: 1–65.

Conybeare, C. E. B., 1979. *Lithostratigraphic Analysis of Sedimentary Basins.* Academic Press, New York.

Cook, F. A., L. D. Brown, and J. E. Oliver, 1980. "The Southern Appalachians and the Growth of Continents." *Sci. Amer.,* 243(4): 156–168.

Cooke, B., 1976. "Suidae from the Plio-Pleistocene Strata of the Rudolf Basin." In Y. Coppens, F. C. Howell, G. L. Isaac, and R. E. F. Leakey (eds.), *Earliest Man and Environments in the Lake Rudolf Basin,* 251–263. Univ. Chicago Press, Chicago.

Cooper, T., 1560. *Chronicle.* London.

Cowie, J. A., and M. R. W. Johnson, 1985. "Late Precambrian and Cambrian Geological Time-Scale." In N. J. Snelling (ed.), *The Chronology of the Geological Record,* 47–64. Geol. Soc. Lond. Mem. 10.

Cox, A. V., 1982. "Magnetostratigraphic Time Scale." In W. B. Harland (ed.), *A Geologic Time Scale,* 63–84. Cambridge Univ. Press, Cambridge.

Crimes, T. P., 1975. "The Stratigraphical Significance of Trace Fossils." In R. W. Frey (ed.), *The Study of Trace Fossils,* 109–130. Springer-Verlag, New York.

Crowell, J. C., 1978. "Gondwana Glaciation, Cyclothems, Continental Positioning, and Climate Change." *Amer. J. Sci.,* 278: 1345–1372.

Crowell, J. C., 1982. "The Tectonics of the Ridge Basin, Southern California." In J. C. Crowell and M. H. Link (eds.), *Geologic History of Ridge Basin, Southern California,* 25–42. Pacific Sec. Soc. Econ. Paleont. Mineral., 22.

Crowell, J. C., and M. H. Link (eds.), 1982. *Geologic History of the Ridge Basin, Southern California.* Soc. Econ. Paleont. Mineral, Pacific Sect., Spec. Publ. 22 (Field Trip Guidebook).

Crowell, J. C., and L. A. Frakes, 1970. "Phanerozoic Glaciation and the Causes of Ice Ages." *Amer. J. Sci.,* 268: 193–224.

Cubitt, J. M., and R. A. Reyment (eds.), 1982. *Quantitative Stratigraphic Correlation.* John Wiley, New York.

Curry, D., 1985. "Oceanic Magnetic Lineaments and the Calibration of the Late Mesozoic-Cenozoic Time-Scale." In N. J. Snelling (ed.), *The Chronology of the Geological Record,* 269–272. Geol. Soc. Lond. Mem. 10.

Curry, D., and G. S. Odin, 1982. "Dating of the Palaeogene." In G. S. Odin (ed.), *Numerical Dating in Stratigraphy,* 607–630. John Wiley, New York.

Curtis, G. H., R. E. Drake, T. Cerling, and J. Hampel, 1975. "Age of the KBS Tuff in Koobi Fora Formation, East Rudolf, Kenya." *Nature,* 258: 395–398.

Dalrymple, G. B., 1972. "Potassium-Argon Dating of Geomagnetic Reversals and North American Glaciations." In W. W. Bishop and J. A. Miller (eds.), *Calibration of Hominid Evolution,* 107–134. Scottish Academic Press, Edinburgh.

Dalrymple, G. B., 1979. "Critical Tables for Conversion of K-Ar Ages from Old to New Constants." *Geology,* 7: 558–560.

Dalrymple, G. B., A. Cox, and R. R. Doell, 1965. "Potassium-Argon Age and Paleomagnetism of the Bishop Tuff, California." *Geol. Soc. Amer. Bull.,* 75: 665–674.

Dalrymple, G. B., and M. A. Lanphere, 1969. *Potassium-Argon Dating.* W. H. Freeman and Co., San Francisco.

Dana, J. D., 1873. "On Some Results of the Earth's Contraction from Cooling, Including a Discussion of the Origin of Mountains, and the Nature of the Earth's Interior." *Amer. J. Sci.,* 5: 423–443.

Davaud, E., and J. Guex, 1978. "Traitement analytique 'manuel' et algorithmique de problemes complexes de corrélations biochronologiques." *Eclogae Geol. Helv.,* 71: 581–610.

Davis, R. A., Jr. (ed), 1978. *Coastal Sedimentary Environments.* Springer-Verlag, New York.

Davis, R. A., Jr., 1983. *Depositional Systems—A Genetic Approach to Sedimentary Geology.* Prentice-Hall, New York.

Davis, R. A., Jr., and R. L. Ethington (eds.), 1976. *Beach and Nearshore Sedimentation.* Soc. Econ. Paleont. Mineral. Spec. Publ. 24.

Denton, G. H., and T. J. Hughes, 1981. *The Last Great Ice Sheets.* John Wiley, New York.

DePaolo, D. J., and B. L. Ingram, 1985. "High-Resolution Stratigraphy with Strontium Isotopes." *Science,* 227: 938–941.

Desborough, G. A., 1978. "A Biogenic Chemical Stratified Lake Model for the Origin of Oil Shale of the Green River Formation: An Alternative to the Playa-Lake Model." *Geol. Soc. Amer. Bull.,* 89: 961–971.

Dewey, J. F., and J. M. Bird, 1970. "Mountain Belts and the New Global Tectonics." *J. Geophys. Res.,* 75: 2625–2647.

Dewey, J. F., and K. C. A. Burke, 1974. "Hotspots and Continental Breakup: Implications for Collisional Orogeny." *Geology,* 2: 57–60.

Dickinson, W. R. (ed.), 1974. *Tectonics and Sedimentation.* Soc. Econ. Paleont. Mineral. Spec. Publ. 22.

Dickinson, W. R., 1982. "Composition of Sandstones in Circum-pacific Subduction Complexes and Fore-arc Basins." *Amer. Assoc. Petrol. Geol. Bull.,* 66: 121–137.

Dickinson, W. R., L. S. Beard, G. R. Brakenridge, J. L. Erjavec, R. C. Ferguson, K. F. Inman, R. A. Knepp, F. A. Lindberg, and P. T. Ryberg, 1983. "Provenance of North American Phanerozoic Sandstones in Relation to Tectonic Setting." *Geol. Soc. Amer. Bull.,* 94: 222–235.

Dickinson, W. R., and C. A. Suczek, 1979. "Plate Tectonics and Sandstone Composition." *Amer. Assoc. Petrol. Geol. Bull.,* 63: 2164–2182.

Dickinson, W. R., and R. Valloni, 1980. "Plate Settings and Provenance of Sands in Modern Ocean Basins." *Geology,* 8: 82–86.

Dickinson, W. R., and H. Yarborough. "Plate Tectonics and Hydrocarbon Accumulations." *Amer. Assoc. Petrol. Geol. Educ. Course Note Series* 1: 1–55.

Dietz, R. S., and J. C. Holden, 1974. "Collapsing Continental Rises: Actualistic Concept of Geosynclines—A Review." *Soc. Econ. Paleont. Mineral. Spec. Publ.,* 19: 14–25.

Dineley, D. L., 1984. *Aspects of a Stratigraphic System: The Devonian.* John Wiley, New York.

Dobrin, M. B., 1976. *Introduction to Geophysical Prospecting,* 3d. ed. McGraw-Hill, New York.

Donnelly, A. T., 1976. "The Refugian Stage of the California Tertiary: Foraminifera, Zonation, Geologic History, and Correlations with the Pacific Northwest. Univ. Calif. Berkeley, unpubl. doct. dissert.

Donovan, D. T., and E. J. W. Jones, 1979. "Causes of Worldwide Changes in Sea Level." *J. Geol. Soc. London,* 136: 187–192.

Dott, R. H., Jr., 1983. "Episodic Sedimentation—How Normal is Average? How Rare is Rare? Does It Matter?" *J. Sed. Petrol.,* 53: 5–23.

Dott, R. H., Jr., and R. L. Batten, 1988. *Evolution of the Earth,* 3d ed. McGraw-Hill, New York.

Dott, R. H., Jr., and R. H. Shaver (eds.), 1974. *Modern and Ancient Geosynclinal Sedimentation.* Soc. Econ. Paleont. Mineral. Spec. Publ. 19: 1–380.

Doyle, L. J., and O. H. Pilkey (eds.), 1979. *Geology of Continental Slopes,* Soc. Econ. Paleont. Mineral. Spec. Publ. 27: 1–374.

Drake, R. E., G. H. Curtis, T. E. Cerling, B. W. Cerling, and J. Hampel, 1980. "KBS Tuff Dating and Geochronology of Tuffaceous Sediments in the Koobi Fora and Shungura Formations, East Africa." *Nature,* 283: 368–372.

Dunbar, C. O., and J. Rodgers, 1957. *Principles of Stratigraphy.* John Wiley, New York.

Dunham, R. J., 1970. "Stratigraphic Reef versus Ecologic Reefs." *Amer. Assoc. Petrol. Geol. Bull.,* 54: 1931–1932.

Dutton, C. E., 1882. "Tertiary History of the Grand Cañon District." *U.S. Geol. Surv. Monogr.,* 2: 1–264.

Dzulynski, S., and E. K. Walton, 1965. *Sedimentary Features of Flysch and Greywackes.* Developments in Sedimentology ·7. Elsevier, New York.

Edwards, L. E., 1982a. "Quantitative Biostratigraphy: The Methods Should Suit the Data." In J. M. Cubitt and R. A. Reyment (eds.), *Quantitative Stratigraphic Correlation,* 45–60. John Wiley, New York.

Edwards, L. E., 1982b. "Numerical and Semi-objective Biostratigraphy: Review and Predictions." *Proc. Third. N. Amer. Paleont. Conv.,* 1: 47–52.

Ehlers, E. G., and H. Blatt, 1982. *Petrology: Igneous, Metamorphic, and Sedimentary.* W. H. Freeman and Co., San Francisco.

Eicher, D. L., 1976. *Geologic Time,* 2d ed. Prentice-Hall, Englewood Cliffs, N.J.

Elderfield, H., 1986. "Strontium Isotope Stratigraphy." *Palaeogeogr., Palaeoclimat., Palaeoecol.,* 57: 71–90.

Elderfield, H., J. M. Gieskes, P. A. Baker, R. K. Oldfield, C. J. Hawkesworth, and R. Miller, 1982. "^{87}Sr/Sr^{86} and ^{18}O/^{16}O Ratios, Interstitial Water Chemistry, and Diagenesis in Deep-Sea Carbonate Sediments of the Ontong Java Plateau." *Geochim. Cosmochim. Acta,* 45: 2201–2212.

Eldredge, N., and S. J. Gould, 1972. "Punctuated Equilibria: An Alternative to Phyletic Gradualism." In T. J. M. Schopf (ed.), *Models in Paleobiology,* 82–115. Freeman, Cooper, San Francisco.

Eldredge, N., and S. J. Gould, 1977. "Evolutionary Models and Biostratigraphic Strategies." In E. G. Kauffman and J. E. Hazel (eds.), *Concepts and Methods of Biostratigraphy,* 25–40. Dowden, Hutchison, and Ross, Stroudsburg, Penn.

Embry, A. F., and J. E. Klovan, 1971. "A Late Devonian Reef Tract on the Northeastern Banks Island, N. W. T." *Bull. Can. Petrol. Geol.,* 19: 730–781.

Emery, K. O., 1968. "Relict Sediments on Continental Shelves of the World." *Amer. Assoc. Petrol. Geol. Bull.,* 52: 445–464

Emiliani, C., 1955. "Pleistocene Temperatures." *J. Geol.,* 63: 538–578.

Emiliani, C., 1966. "Isotopic Paleotemperatures." *Science,* 154(3751): 851–857.

Emry, R. J., 1973. "Stratigraphy and Preliminary Biostratigraphy of the Flagstaff Rim Area, Natrona County, Wyoming." *Smithsonian Contrib. Paleobiol.,* 25: 1–20.

Ensley, R. A., and K. L. Verosub, 1982. "Biostratigraphy and Magnetostratigraphy of Southern Ridge Basin, Central Transverse Ranges, California." in J. C. Crowell and M. H. Link (eds.), *Geologic History of Ridge Basin, Southern California,* 13–24. Pacific Sec. Soc. Econ. Paleont. Mineral. 22.

Eskola, P., 1915. "Om sambandet mellan kemisk och mineralogisk sammansättning hos Orijärvitraktens metamorfa bergarter." *Bull. Comm. Géol. Finlande,* 44: 1–145.

Ethridge, F. G., T. J. Jackson, and A. D. Youngberg, 1981. "Floodbasin Sequences of a Fine-Grained Meander Belt Subsystem: The Coal-Bearing Lower Wasatch and Upper Fort Union Formations, Southern Powder River Basin, Wyoming." *Soc. Econ. Paleont. Mineral. Spec. Publ.,* 31: 191–212.

Ethridge, F. G., and R. M. Flores (eds.), 1981. *Recent and Ancient Nonmarine Depositional Environments: Models for Exploration.* Soc. Econ. Paleont. Mineral. Spec. Publ. 31.

Eugster, H. P., 1970. "Chemistry and Origin of the Brines of Lake Magadi, Kenya." *Mineral. Soc. Amer. Spec. Publ.,* 3: 215–235.

Eugster, H. P., and L. A. Hardie, 1975. "Sedimentation in an Ancient Playa-Lake Complex: The Wilkins Peak Member of the Green River Formation of Wyoming." *Geol. Soc. Amer. Bull.,* 86: 319–334.

Evernden, J. F., and G. H. Curtis, 1965. "The Present Status of Potassium-Argon Dating of Tertiary and Quaternary Rocks." *Int. Assoc. Quat. Res., 6th Cong., Warsaw,* 1: 643–651.

Evernden, J. F., D. E. Savage, G. H. Curtis, and G. T. James, 1964. "Potassium-Argon Dates and the Cenozoic Mammalian Chronology of North America. *Amer. J. Sci.,* 262: 145–198.

Eyles, J. M., 1985. "William Smith, Sir Joseph Banks, and the French Geologists." In A. Wheeler and J. H. Price (eds.), *From Linnaeus to Darwin,* 37–50. Society for the History of Natural History, London.

Faure, G., 1986. *Principles of Isotope Geology,* 2d ed. John Wiley, New York.

Ferm, J. C., 1974. "Carboniferous Environmental Models in Eastern United States and Their Significance." *Geol. Soc. Amer. Spec. Paper* 148: 79–95.

Findlater, I. C., 1978. "Isochronous Surfaces within the Plio-Pleistocene Sediments East of Lake Turkana." In W. W. Bishop (ed.), *Geological Background to Fossil Man,* 415–420. Scottish Academic Press, Edinburgh.

Fischer, A. G., 1961. "Stratigraphic Record of Transgressing Seas in Light of Sedimentation on the Atlantic Coast of New Jersey." *Amer. Assoc. Petrol. Geol. Bull.,* 45: 1656–1666.

Fischer, A. G., 1964. "The Lofer Cyclothems of the Alpine Triassic." *Kans. State Geol. Surv. Bull.,* 169: 107–149.

Fischer, A. G., 1981. "Climatic Oscillations in the Biosphere." in M. H. Nitecki (ed.), *Biotic Crises in Ecological and Evolutionary Time,* 102–131. Academic Press, New York.

Fischer, A. G., 1982. "Long-Term Climatic Oscillations Recorded in Stratigraphy." In W. H. Berger and J. C. Crowell (eds.), *Climate in Earth History,* 97–104. Nat. Acad. Press, Washington, D. C.

Fischer, A. G., 1984. "The two Phanerozoic supercycles." In W. A. Berggren and J. A. Van Couvering (eds.), *Catastrophes in Earth History,* 129–150. Princeton Univ. Press, Princeton, N.J.

Fischer, A. G., 1986. "Climatic Rhythms Recorded in Strata." *Ann. Rev. Earth Planet. Sci.,* 14: 351–376.

Fischer, A. G., T. Herbert, and I. Premoli-Silva, 1985. "Carbonate Bedding Cycles in Cretaceous Pelagic and Hemipelagic Sequences." In L. M. Pratt E. G. Kauffman, and F. B. Zelt (eds.), *Fine-Grained Deposits and Biofacies of the Cretaceous Western Interior Seaway: Evidence of Cyclic Sedimentary Processes,* 1–10.

Soc. Econ. Paleont. Mineral. Field Trip Guidebook 4.

Fisher, R. V., and J. M. Rensberger, 1972. "Physical Stratigraphy of the John Day Formation." *Univ. Calif. Publ. Geol. Sci.,* 101: 1–45.

Fitch, F. J., and J. A. Miller, 1970. "Radioisotopic Age Determinations of Lake Rudolf Artefact Site." *Nature,* 226: 226–228.

Fitch, F. J., P. J. Hooker, and J. A. Miller, 1976. "^{40}Ar/^{39}Ar Dating of the KBS Tuff in Koobi Fora Formation, East Rudolf, Kenya." *Nature,* 263: 740–744.

Flynn, J. J., 1986. "Correlation and Geochronology of Middle Eocene Strata from the Western United States." *Palaeogeogr., Palaeoclimat., Palaeoecol.,* 55: 335–406.

Flynn, J. J., B. J. MacFadden, and M. C. McKenna, 1984. "Land Mammal Ages, Faunal Heterochrony, and Temporal Resolution in Cenozoic Terrestrial Sequences." *J. Geol.,* 92: 687–705.

Folk, R. L., 1974. *Petrology of Sedimentary Rocks.* Hemphill's, Austin, Texas.

Frey, R. W., 1972. "Trace Fossils of the Fort Hays Limestone Member of Niobrara Chalk (Upper Cretaceous), West-Central Kansas." *Univ. Kans. Paleont. Contrib.,* 58: 1–72.

Friedman, G. M. (ed.), 1969. *Depositional Environments in Carbonate Rocks.* Soc. Econ. Paleont. Mineral. Spec. Publ. 14.

Friedman, G. M., and J. E. Sanders, 1978. *Principles of Sedimentology.* John Wiley, New York.

Frost, S. H., M. P. Weiss, and J. B. Saunders (eds.), 1977. *Reefs and Related Carbonates—Ecology and Sedimentology.* Amer. Assoc. Petrol. Geol. Stud. Geol. 4: 1–421.

Fryberger, S. G., T. S. Ahlbrandt, and S. Andrews, 1979. "Origin, Sedimentary Features, and Significance of Low-Angle Eolian 'Sand Sheet' Deposits, Great Sand Dunes National Monument and Vicinity, Colorado." *J. Sed. Petrol.,* 49: 733–746.

Gale, N. H., 1985. "Numerical Calibration of the Paleozoic Timescale: Ordovician, Silurian, and Devonian Periods." In N. J. Snelling, (ed.), *The Chronology of the Geological Record,* 81–88. Geol. Soc. Lond. Mem. 10.

Galloway, W. A., 1978. *Exploration for Stratigraphic Traps in Terrigenous Clastic Depositional Systems.* Amer. Assoc. Petrol. Geol. Short Course Notes 3.

Galloway. W. A., and D. K. Hobday, 1983. *Terrigenous Clastic Depositional Systems.* Springer-Verlag, New York.

Gast, P. W., 1962. "The Isotopic Composition of Strontium and the Age of Stone Meteorites, I." *Geochim. Cosmochim. Acta,* 26: 927–943.

Gilbert, C. M., 1938. "Welded Tuff in Eastern California." *Geol. Soc. Amer. Bull.,* 49: 1829–1862.

Gilbert, G. K., "The Transportation of Debris by Running Water." *U.S. Geol. Surv. Prof. Paper* 86: 1–243.

Gill, J. R., and W. A. Cobban, 1973. "Stratigraphy and Geologic History of the Montana Group and Equivalent Rocks, Montana, Wyoming, and North and South Dakota." *U.S. Geol. Surv. Prof. Paper* 776: 1–36.

Gilluly, J., 1977. "American Geology Since 1910—A Personal Appraisal." *Ann. Rev. Earth Planet. Sci.*, 5: 1–12.

Gilreath, J. A., J. S. Healy, and J. N. Yelverton, 1969. "Depositional Environments Defined by Dipmeter Interpretation." *Gulf Coast Assoc. Geol. Soc. Trans.*, 19: 101–109.

Ginsburg, R. N., 1971. "Landward Movement of Carbonate Mud: New Model for Regressive Cycles in Carbonates." *Bull. Amer. Assoc. Petro. Geol.*, 55: 340 (abs).

Ginsburg, R. N., (ed.), 1975. *Tidal Deposits, A Casebook of Recent Examples and Fossil Counterparts.* Springer-Verlag, New York.

Glass, B. P., and J. R. Crosbie, 1982. "Age of Eocene/Oligocene Boundary Based on Extrapolation from North American Microtektite Layer." *Amer. Assoc. Petrol. Geol. Bull.*, 66: 471–476.

Glass, B. P., C. M. Hall, and D. York, 1986. "^{40}Ar/^{39}Ar Laser-Probe Dating of North American Tektite Fragments from Barbados and the Age of the Eocene-Oligocene Boundary." *Chemical Geology*, 69: 181–186.

Glass, B. P., M. B. Swincki, and P. A. Swort, 1979. "Australasian, Ivory Coast, and North American Tektite Strewn Fields: Size, Mass, and Correlation with Geomagnetic Reversal and other Earth Events." *Proc. Lunar Planet. Sci. Conf.*, 10: 2535–2545.

Gleadow, A. J. W., 1980. "Fission Track Age of the KBS Tuff and Associated, Hominid Remains in Northern Kenya." *Nature*, 284: 228–230.

Gloppen, T. G., and R. J. Steel, 1981. "The Deposits, Internal Structure and Geometry in Six Alluvial Fan–Fan Delta Bodies (Devonian-Norway)—A Study in the Significance of Bedding Sequence in Conglomerates." *Soc. Econ. Paleont. Mineral. Spec. Publ.*, 31: 49–69.

Goddard, E. N., et al., 1948. *Rock Color Chart.* Geol. Soc. Amer., Boulder, Colo.

Goodwin, P. W., and E. J., Anderson, 1980. *Helderberg PACs.* Eastern Sect. Soc. Econ. Paleont. Mineral. Guidebook.

Goodwin, P. W., and E. J. Anderson, 1985. "Punctuated Aggradational Cycles: A General Hypothesis of Episodic Stratigraphic Accumulation." *J. Geol.*, 93: 515–553.

Goodwin, P. W., E. J. Anderson, W. M. Goodman, and L. J. Saraka, 1986. "Punctuated Aggradational

Cycles: Implications for Stratigraphic Analysis." *Paleoceanog.*, 1:417–429.

Gould, H. R., 1970. "The Mississippi Delta Complex." *Soc. Econ. Paleont. Mineral. Spec. Publ.*, 15: 3–30.

Gould, S. J., 1987. *Time's Arrow, Time's Cycle.* Harvard Univ. Press, Cambridge.

Gould, S. J., and N. Eldredge, 1977. "Punctuated Equilibria: The Tempo and Mode of Evolution Reconsidered." *Paleobiology*, 3: 115–151.

Gradstein, F. M., F. P. Agterberg, M.-P. Aubry, W. A. Berggren, J. J. Flynn, R. Hewitt, D. V. Kent, K. D. Klitgord, K. G. Miller, J. D. Obradovich, J. G. Ogg, D. R. Prothero, and G. E. G. Westermann, 1988. "Sea Level History." *Science*, 241: 599–601.

Gradstein, F. M., F. P. Agterberg, J. C. Brower, and W. J. Schwarzacher (eds.), 1985. *Quantitative Stratigraphy.* D. Reidel Publ. Co., Dordrecht, Netherlands.

Gressly, A., 1838. "Observations géologiques dur le Jura Soleurois." *Neue Denkschr. Allg. Schweizerische Gesellsch. ges. Naturw.*, 2: 1–112.

Guex, J., 1977. "Une nouvelle méthode d'analyse biochronologique, note préliminaire." *Bull. Soc. Vaud. Sci. Nat.*, 351: 73.

Hall, J., Jr., 1859. *Paleontology.* Vol. III. Geological Survey of New York. Van Benthuysen, Albany.

Hallam, A., 1981. *Facies Interpretation and the Stratigraphic Record.* W. H. Freeman and Co., San Francisco.

Hallam, A., 1988. "A Reevaluation of Jurassic Eustasy in the Light of New Data and the Revised Exxon Curve." *Soc. Econ. Paleont. Mineral. Spec. Publ.* 42: 261–274.

Hamblin, W. K., 1965. "Internal Structures of Homogenous Sandstones." *Kansas Geol. Surv. Bull.*, 175(pt. 1): 1–37.

Hancock, J. M., 1975. "The Sequence of Facies in the Upper Cretaceous of Northern Europe Compared with That in the Western Interior." In W. G. E. Caldwell (ed.), *The Cretaceous System in the Western Interior of North America*, 83–118. Geol. Assoc. Canada Spec. Paper 13.

Hancock, J. M., 1977. "The Historic Development of Biostratigraphic Correlation." In E. G. Kauffman and J. E. Hazel (eds.), *Concepts and Methods of Biostratigraphy* 3–22. Dowden, Hutchinson and Ross, Stroudsburg, Penn.

Haq, B. U., J. Hardenbol, and P. R. Vail, 1987. "Chronology of Fluctuating Sea-Levels Since the Triassic." *Science*, 235: 1156–1167.

Haq, B. U., J. Hardenbol, and P. R. Vail, 1988. "Mesozoic and Cenozoic Chronostratigraphy and Cycles of Sea-level Change." *Soc. Econ. Paleont. Mineral. Spec. Publ.*, 42: 71–108.

Haq, B. U., T. R. Worsley, L. H. Burckle, K. G. Douglas, L. D. Keigwin, Jr., N. D. Opdyke, S.

M. Savin, M. H. Sommer, E. Vincent, and F. Woodruff, 1980. "Late Miocene Marine Carbon-Isotope Shift and Synchroneity of Some Phytoplanktonic Biostratigraphic events." *Geology*, 8: 427–431.

Hardenbol, J., and W. A. Berggren, 1978. "A New Paleogene Numerical Time Scale." *Amer. Assoc. Petrol. Geol. Mem.*, 6: 213–234.

Hardie, L. A., 1986. "Carbonate Tidal Flat Deposition." *Quart. J. Colo. Sch. Mines.*, 81: 3–74.

Harland, W. B., A. V. Cox, P. G. Llewellyn, C. A. G. Pickton, A. G. Smith, and R. Walters, 1982. *A Geologic Time Scale.* Cambridge Univ. Press, Cambridge.

Harms, J. C., 1979. "Primary Sedimentary Structures." *Ann. Rev. Earth Planet. Sci.*, 7: 227–248.

Harris, W. B., and P. D. Fullagar, 1989. "Comparison of Rb-Sr and K-Ar Dates of Middle Eocene Bentonite and Glauconite, Southeastern Atlantic Coastal Plain." *Geol. Soc. Amer. Bull.*, 101: 573–577. 577.

Harris, W. B., P. D. Fullagar, and J. A. Winters, 1984. "Rb-Sr Glauconite Ages, Sabinian, Claibornian and Jacksonian Units, Southeastern Atlantic Coastal Plain, U.S.A." *Palaeogeogr., Palaeoclimat., Palaeoecol.*, 47: 53–76.

Harris, W. B., and V. A. Zullo, 1980. "Rb-Sr Glauconite Isochron of the Eocene Castle Hayne Limestone, North Carolina." *Geol. Soc. Amer. Bull.*, 93: 587–592.

Harrison, C. G. A., I. McDougall, and N. D. Watkins, 1979. "A Geomagnetic Field Reversal Time Scale Back to 13.0 Million Years before Present." *Earth Planet. Sci. Lett.*, 42: 143–152.

Harshbarger, J. W., C. A. Repenning, and J. H. Irwin, 1957. "Stratigraphy of the Uppermost Triassic and the Jurassic Rocks of the Navajo Country (Colorado Plateau)." *U.S. Geol. Surv. Prof. Paper* 291: 1–74.

Hattin, D. E., 1975. "Stratigraphic Study of the Carlile-Niobrara (Upper Cretaceous) Unconformity in Kansas and Northeastern Nebraska." *Geol. Assoc. Canada Spec. Paper* 13: 31–54.

Hattin, D. E., 1982. "Stratigraphy and Depositional Environments of the Smoky Hill Chalk Member, Niobrara Chalk (Upper Cretaceous) of the Type Area, Western Kansas." *Kans. Geol. Surv. Bull.*, 255: 1–108.

Hattin, D. E., 1985. "Distribution and Significance of Widespread, Time-parallel Pelagic Limestone Beds in Greenhorn Limestone (Upper Cretaceous) of the Central Great Plains and Southern Rocky Mountains." *Soc. Econ. Paleont. Mineral. Field Trip Guidebook*, 4: 28–37.

Haug, E., 1907. *Traité de Géologie.* 1: 1–536. Libr. Armand Cohn, Paris.

Hay, R. L., 1975. *Geology of Olduvai Gorge.* Univ. Calif. Press, Berkeley, Calif.

Hay, W. W., 1972. "Probabilistic Stratigraphy." *Eclogae Geol. Helv.*, 65: 255–266.

Hay, W. W., and J. R. Southam, 1978. "Quantifying Biostratigraphic Correlation." *Ann. Rev. Earth Planet. Sci.*, 6: 353–375.

Hays, J. D., J. Imbrie, and N. J. Shackleton, 1976. "Variations in the Earth's Orbit—Pacemaker of the Ice Ages." *Science*, 194: 1121–1132.

Hays, J. D., and N. D. Opdyke, 1967. "Antarctic Radiolaria, Magnetic Reversals, and Climatic Change." *Science*, 158: 1001–1011.

Hays, J. D., and W. L. Pitman, 1973. "Lithospheric Plate Motion, Sea-Level Changes and Climatic and Ecological Consequences." *Nature*, 246: 18–21.

Hays, J. D., and N. J. Shackleton, 1976. "Globally Synchronous Extinction of the Radiolarian *Stylatractus universus.*" *Geology*, 4: 649–652.

Hazel, J. E., 1977. "Use of Certain Multivariate and Other Techniques in Assemblage Zonal Biostratigraphy: Examples Utilizing Cambrian, Cretaceous, and Tertiary Benthic Invertebrates." In E. G. Kauffman and J. Hazel (eds.), *Concepts and Methods in Biostratigraphy*, 187–212. Dowden, Hutchinson and Ross, Stroudsburg, Penn.

Heckel P. H., 1974. "Carbonate Buildups in the Geological Record: A Review." *Soc. Econ. Paleont. Mineral. Spec. Publ.*, 18: 90–154.

Heckel, P. H., 1977. "Origin of Phosphatic Black Shale Facies in Pennsylvanian Cyclothems of Midcontinent North America." *Amer. Assoc. Petrol. Geol. Bull.*, 61: 1045–1068.

Heckel, P. H., 1980. "Paleogeography of Eustatic Model for Deposition of Midcontinent Upper Pennsylvanian Cyclothems." In T. D. Fouch and E. R. Magathan (eds.), *Paleozoic Paleogeography of the West-Central United States*, 197–215. Soc. Econ. Paleont. Mineral., Rocky Mtn. Sect., Paleogeography Symposium 1.

Heckel, P. H., 1986. "Sea-level Curves for Pennsylvanian Eustatic Marine Transgressive-Regressive Depositional Cycles along the Midcontinent Outcrop Belt, North America." *Geology*, 14: 330–334.

Hedberg, H. D., 1976. *International Stratigraphic Guide. A Guide to Stratigraphical Classification, Terminology, and Procedure.* John Wiley, New York.

Heezen, B. C., and C. D. Hollister, 1971. *The Face of the Deep.* Oxford Univ. Press, New York.

Heirtzler, J. R., G. O. Dickson, E. M. Herron, W. C. Pittman, III, and X. Le Pichon, 1968. "Marine Magnetic Anomalies, Geomagnetic Field Reversals, and Motions of the Ocean Floor and Continents." *J. Geophys. Res.*, 73: 2119–2136.

Hess, J., M. L. Bender, and J. G. Schilling, 1986. "Seawater $^{87}Sr/^{86}Sr$ Evolution from Cretaceous to

Present—Applications to Paleoceanography." *Science*, 231: 979–984.

Hillhouse, J. W., T. E. Cerling, and F. H. Brown, 1986. "Magnetostratigraphy of the Koobi Fora Formation, Lake Turkana, Kenya." *J. Geophys. Res.,* 9(B11): 11581–11595.

Hillhouse, J. W., J. W. M. Ndombi, A. Cox, and A. Brock, 1977. "Additional Results on Paleomagnetic Stratigraphy of the Koobi Fora Formation, East of Lake Turkana (Lake Rudolf), Kenya." *Nature,* 256: 411–415.

Hobday, D. K., and A. J. Tankard, 1978. "Transgressive-Barrier and Shallow-Shelf Interpretation of the Lower Paleozoic Peninsula Formation, South Africa." *Geol. Soc. Amer. Bull.,* 89: 1733–1744.

Hohn, M. E., 1978. "Stratigraphic Correlation by Principal Components: Effect of Missing Data." *J. Geol.,* 86: 524–532.

Hohn, M. E., 1982. "Properties of Composite Sections Constructed by Least Squares." In J. M. Cubitt and R. A. Reyment (eds.), *Quantitative Stratigraphic Correlation,* 107–117. John Wiley, New York.

Holmes, A., 1911. "The Association of Lead with Uranium in Rock Minerals, and its Application to the Measurement of Geologic Time." *Proc. R. Soc. London A,* 85: 248–256.

Holmes, A., 1913. *The Age of the Earth.* Harper, New York.

Hook, R. W., and J. C. Ferm, 1985. "A Depositional Model for the Linton Tetrapod Assemblage (Westphalian D, Upper Carboniferous) and Its Palaeoenvironmental Significance," *Phil. Trans. R. Soc. London B,* 311 (1148): 101–109.

Hooke, R. L., 1967. "Processes on Arid-Region Alluvial Fans." *J. Geol.,* 75: 438–460.

Horne, J. C., and J. C. Ferm, 1976. *Carboniferous Depositional Environments in the Pocahontas Basin, Eastern Kentucky and Southern West Virginia: A Field Guide.* Department of Geology, University of South Carolina.

Horne, J. C., J. C. Ferm, F. T. Caruccio, and B. P. Baganz, 1978. "Depositional Models in Coal Exploration and Mine Planning in the Appalachian Region." *Amer. Assoc. Petrol. Geol. Bull.,* 62: 2379–2411.

House, M. R., 1985. "The Ammonoid Time-Scale and the Ammonoid Evolution." In N. J. Snelling (ed.), *The Chronology of the Geological Record,* 273–283. Geol. Soc. Lond. Mem. 10.

Howell, D. J., and J. C. Ferm, 1980. "Exploration Model for Pennsylvanian Upper Delta Plain Coals, Southwest West Virginia." *Amer. Assoc. Petrol. Geol. Bull.,* 64: 938–941.

Hsü, K. J., 1974. "Melanges and Their Distinction from Olistostromes." *Soc. Econ. Paleont. Mineral. Spec. Publ.,* 19: 321–333.

Hubbard, R. J., 1988. "Age and Significance of Sequence Boundaries on Jurassic and Early Cretaceous Rifted Continental Margins." *Amer. Assoc. Petrol Geol. Bull.,* 72: 49–72.

Hughes, J. D., 1984. "Geology and Depositional Setting of the Late Cretaceous Upper Bearpaw and Lower Horseshoe Canyon Formation in the Dodds-Round Hill Coalfield of Central Alberta—A Computer-Based Study of Closely Spaced Exploration Data." *Geol. Surv. Canada Bull.,* 361: 1–81.

Huntoon, P. W., 1977. "Cambrian Stratigraphic Nomenclature and Ground-Water Prospecting Failures on the Hualapai Plateau, Arizona." *Ground Water,* 15: 426–433.

Hurford, A. J., A. J. W. Gleadow, and C. W. Naeser, 1976. "Fission-Track Dating of Pumice from the KBS Tuff, East Rudolf, Kenya." *Nature,* 263: 738–740.

Hutchinson, R. W., and G. G. Engels, 1970. "Tectonic Significance of Regional Geology and Evaporite Lithofacies in Northeastern Ethiopia." *Phil. Trans. R. Soc. A,* 267: 313–329.

Hutton, J., 1788. "Theory of the Earth; Or an Investigation of the Laws Observable in the Composition, Dissolution, and Restoration of Land upon the Globe." *Trans. R. Soc. Edinburgh,* 1: 209–304.

Hutton, J., 1795. *Theory of the Earth with Proofs and Illustrations,* 2 vols. Wm. Creech, Edinburgh.

Huxley, T. H., 1862. "The Anniversary Address." *Quart. J. Geol. Soc. London,* 18: xl–liv.

Hyne, N. J., 1984. *Geology for Petroleum Exploration, Drilling and Production.* McGraw-Hill, New York.

Imbrie, J., and K. P. Imbrie, 1979. *Ice Ages: Solving the Mystery.* Enslow Publ., Short Hills, N.J.,

Ingersoll, R. V., 1988. "Tectonics of Sedimentary Basins." *Geol. Soc. Amer. Bull.,* 100: 1704–1719.

Israelsky, M. C., 1949. "Oscillation chart." *Amer. Assoc. Petrol. Geol. Bull.,* 33: 92–98.

Izett, G. A., 1981. "Volcanic Ash Beds: Records of Upper Cenozoic Silicic Pyroclastic Volcanism in the Western United States." *J. Geophys. Res.,* 86(B11): 10,200–10,222.

Izett, G. A., and C. W. Naeser, 1976. "Age of the Bishop Tuff of Eastern California As Determined by Fission-Track Method." *Geology,* 4: 587–590.

Izett, G. A., R. E. Wilcox, H. A. Powers, and G. A. Desborough, 1970. "The Bishop Ash Bed, A Pleistocene Marker Bed in the Western United States." *Quat. Res.,* 1: 121–132.

Jaeger, J.-J. and J.-L. Hartenberger, 1975. "Pour utilization systématique de niveaux-repéres en biochronologie mammalienne." *3e Réunion Annuel Sci. Terre,* p. 201.

James, N. P., 1984. "Introduction to Carbonate Facies Models; Shallowing-Upward Sequences in Carbonates; Reefs." In R. G. Walker (ed.), *Facies models,* 2d ed., 209–244. Geoscience Canada Reprint ser. 1.

Johanson, D., and M. Edey, 1981. *Lucy, the Beginnings of Humankind*. Simon and Schuster, New York.

Johnson, N. M., N. D. Opdyke, G. D. Johnson, E. H. Lindsay, and R. A. K. Tahirkeli, 1982. "Magnetic Polarity Stratigraphy and Ages of Siwalik Group Rocks of the Potwar Plateau, Pakistan." *Palaeogeogr., Palaeoclimat., Palaeoecol.*, 37: 17–42.

Jones, B. C., P. F. Carr, and A. J. Wright, 1981. "Silurian and Early Devonian Geochronology—A Reappraisal, with New Evidence from the Bungonian Limestone."*Alcheringa*, 5: 197–207.

Jopling, A. V., 1967. "Origin of Laminae Deposited by the Movement of Ripples along a Stream Bed: Laboratory Study. *J. Geol.*, 75: 287–305.

Karig, D. E., and G. F. Sharman, III, 1975. "Subduction and Accretion in Trenches." *Geol. Soc. Amer. Bull.*, 86: 377–389.

Kauffman, E. G., 1967. "Coloradoan Macroinvertebrate Assemblages, Central Western Interior United States." *Colo. School. Mines Publ.*, 67–143.

Kauffman, E. G., 1969. "Cretaceous Marine Cycles of the Western Interior." *Mountain Geol.*, 6: 227–245.

Kauffman, E. G., 1970. "Population Systematics, Radiometrics, and Zonation—A New Biostratigraphy." *Proc. N. Amer. Paleont. Conv*, F: 612–666.

Kauffman, E. G., 1985. "Cretaceous Evolution of the Western Interior Basin of the United States." In L. M. Pratt, E. G. Kauffman, and F. B. Zelt (eds.), *Fine-Grained Deposits and Biofacies of the Cretaceous Western Interior Seaway: Evidence of Cyclic Sedimentary Processes*, iv–xiii. Soc. Econ. Paleont. Mineral. Field Trip Guidebook 4.

Kauffman, E. G., 1988. "Concepts and Methods of High-Resolution Event Stratigraphy." *Ann. Rev. Earth Planet. Sci.*, 16: 605–654.

Kauffman, E. G., and J. E. Hazel (eds.), 1977. *Concepts and Methods of Biostratigraphy*. Dowden, Hutchison, and Ross, Stroudsburg, Penn.

Kauffman, E. G., and L. M. Pratt, 1985. "Field Reference Section." In L. M. Pratt, E. G. Kauffman, and F. B. Zelt (eds.), *Fine-Grained Deposits and Biofacies of the Cretaceous Western Interior Seaway: Evidence of Cyclic Sedimentary Processes*, 1–26. Soc. Econ. Paleont. Mineral. Field Trip Guidebook 4.

Kay, M., 1951. "North American Geosynclines." *Geol. Soc. Amer. Mem.* 48.

Keller, J., W. B. F. Ryan, and D. Ninkovitch, 1978. "Explosive Volcanic Activity in the Mediterranean over the Past 200,000 Years As Recorded in Deep-Sea Sediments." *Geol. Soc. Amer. Bull.*, 84: 591–604.

Kendall, C. G. St. C., and P. A. d'E. Skipwith, 1969. "Holocene and Shallow-Water Carbonate and Evaporite Sediments of Khor al Bazam, Abu Dhabi, Southwest Persian Gulf." *Amer. Assoc. Petrol. Geol. Bull.*, 53: 841–869.

Kennett, J. P. (ed.), 1980. *Magnetic Stratigraphy of Sediments*. Benchmark Papers in Geology 54. Dowden, Hutchison, and Ross, Stroudsburg, Penn.

Kennett, J. P., 1982. *Marine Geology*. Prentice-Hall, Englewood Cliffs, N.J.

King, P. B., 1977. *The Evolution of North America*, 2. ed. Princeton Univ. Press, Princeton, N.J.

Klein, G. deV., 1970. "Depositional and Dispersal Dynamics of Intertidal Sand Bars." *J. Sed. Petrol.*, 40: 1095–1127.

Klein, G. deV., 1972. "Determination of Paleotidal Range in Clastic Sedimentary Rocks." *24th Int. Geol. Cong., Comptes Rendus*, 6: 397–405.

Klein, G. deV., 1977. *Clastic Tidal Facies*. Cont. Edu. Publ. Co., Champaign, Ill.

Kocurek, G., and R. H. Dott, 1983. "Jurassic Paleogeography and Paleoclimate of the Central and Southern Rocky Mountains Region." In M. W. Reynolds, and E. D. Dolly (eds.), *Mesozoic Paleogeography of the West-Central United States*, 101–116. Rocky Mtn. Sect. Soc. Econ. Paleont. Mineral.

Koepnick, R. B., R. E. Denison, and D. A. Dahl, 1988. "The Cenozoic Seawater $^{87}Sr/^{86}Sr$ Curve: Data Review and Implications for Correlation of Marine Strata." *Paleoceanography*, 3: 743–756.

Kottlowski, F. E., 1965. *Measuring Stratigraphic Sections*. Holt, Rinehart and Winston, New York.

Kradyna, J., and C. Mehrtens, 1984. "Episodic Accumulation and the Origin of Formation Boundaries in the Helderberg Group of New York State: Comment." *Geology*, 12: 637.

Krumbein, W. C., 1934. "Size Frequency Distribution of Sediments." *J. Sed. Petrol.*, 4: 65–77.

Krumbein, W. C., and L. L. Sloss, 1963. *Stratigraphy and Sedimentation*. W. H. Freeman and Co., San Francisco.

Kuenen, P. H., and C. I. Migliorini, 1950. "Turbidity Currents As a Cause of Graded Bedding. *J. Geol.*, 58: 91–127.

Kumar, N., 1973. "Modern and Ancient Barrier Sediments: New Interpretation Based on Stratal Sequence in Inlet-Filling Sands and on Recognition of Nearshore Storm Deposits." *Ann. N.Y. Acad. Sci.*, 220(5): 245–340.

La Fon, N. A., 1981. "Offshore Bar Deposits of Semilla Sandstone Member of Mancos Shale (Upper Cretaceous), San Juan Basin, New Mexico." *Amer. Assoc. Petrol. Geol. Bull.*, 65: 706–721.

LaBrecque, J. L., D. V. Kent, and S. C. Cande, 1977. "Revised Magnetic Polarity Time Scale for Late Cretaceous and Cenozoic time." *Geology*, 5: 330–335.

LaBrecque, J. L., and others, 1983. "Contributions to the Paleogene Stratigraphy in Nomenclature, Chronology, and Sedimentation Rates." *Palaeogeogr., Palaeoclimat., Palaeoecol.*, 42: 91–125.

Laffitte, R., W. B. Harland, H. K. Erben, W. H. Blow, W. Haas, N. F. Hughes, W. H. C. Ramsbottom, P. Rat, H. Tintant, and W. Ziegler, 1972. "Some International Agreement on Essentials of Stratigraphy." *Geol. Mag.*, 109: 1–15.

Laming, D. J. C., 1966. "Imbrication, Paleocurrents, and Other Sedimentary Features in the Lower New Red Sandstone, Devonshire, England." *J. Sed. Petrol.*, 36: 940–959.

Langford, R., and M. A. Chan, 1988. "Flood Surfaces and Deflation Surfaces within the Cutler Formation and Cedar Mesa Sandstone (Permian), Southeastern Utah." *Geol. Soc. Amer. Bull.*, 100: 1541–1549.

Langstaff, C. S., and D. Morrill, 1981. *Geologic Cross Sections.* International Human Resources Development Corporation, Boston.

Laporte, L. F., 1967. "Carbonate Deposition Near Mean Sea-Level and Resultant Facies Mosaic: Manlius Formation (Lower Devonian) of New York State." *Amer. Assoc. Petrol. Geol. Bull.*, 51: 73–101.

Laporte, L. F., 1969. "Recognition of a Transgressive Carbonate Sequence within an Epeiric Sea: Helderberg Group (Lower Devonian) of New York State." *Soc. Econ. Paleont. Mineral. Spec. Publ.* 14: 98–119.

Laporte, L. F., 1975. "Carbonate Tidal Deposits of the Early Devonian Manlius Formation of New York State." In R. N. Ginsburg (ed.), *Tidal Deposits: A Casebook of Recent Examples and Fossil Counterparts,* 243–250. Springer-Verlag, Berlin.

Laporte, L. F. (ed.), 1974. *Reefs in Time and Space.* Soc. Econ. Paleont. Mineral. Spec. Publ. 18.

Lapworth, C., 1879. "On the Tripartite Classification of the Lower Palaeozoic Rocks." *Geol. Mag.*, 6: 1–15.

Lavoisier, A., 1789. "Observations générales sur les couches horizontales, qui ont été deposées par la mer, et sur les consequences, qu'on peut tirer de leurs dispositions, relativement à l'ancienneté du globe terrestre." *Acad. Sci. Mem.*, 351–371.

Lazarus, D. B., and D. R. Prothero, 1984. "The Role of Stratigraphic and Morphologic Data in Phylogeny Reconstruction." *J. Paleont.*, 58: 163–172.

Leakey, L. S. B., J. F. Evernden, and G. H. Curtis, 1961. "Age of Bed I, Olduvai." *Nature*, 191: 478–479.

Leakey, M. D., 1978. "Olduvai Gorge, 1911–1975: A History of the Investigation." In W. W. Bishop (ed.), *Geological Background to Fossil Man,* 151–156. Scottish Academic Press, Edinburgh.

Leeder, M. R., 1982. *Sedimentology: Process and Product.* George Allen and Unwin, London.

LeFournier, J., and G. M. Friedman, 1974. "Rate of Lateral Migration of Adjoining Sea-Marginal Sedimentary Environments Shown by Historical Records, Authie Bay, France." *Geology*, 2: 497–498.

LeRoy, L. W., and J. W. Low, 1954. *Graphic Problems in Petroleum Geology.* Harper and Brothers, New York.

Levorsen, A. I., 1967. *Geology of Petroleum.* W. H. Freeman and Co., San Francisco.

Lewin, R., 1987. *Bones of Contention: Controversies in the Search for Human Origins.* Simon and Schuster, New York.

Lewis, D. W., 1984. *Practical Sedimentology.* Hutchison Ross Publ., Stroudsburg, Penn.

Lineback, J. A., 1968. "Turbidites and Other Sandstone Bodies of the Borden Siltstone (Mississippian) in Illinois." *Ill. Geol. Surv. Circ.*, 425: 1–29.

Lineback, J. A., 1970. "Stratigraphy of the New Albany Shale in Indiana." *Ind. Geol. Surv. Bull.*, 44: 1–73.

Lisitzin, A. P., 1972. *Sedimentation in the World Ocean.* Soc. Econ. Paleont. Mineral. Spec. Publ. 17.

Lofton, C. L., and W. M. Adams, 1971. "Possible Future Petroleum Provinces of the Eocene and Paleocene, Western Gulf Basin." *Amer. Assoc. Petrol. Geol. Mem.*, 15: 855–886.

Loope, D. B., 1985. "Episodic Deposition and Preservation of Eolian Sands, a Late Paleozoic Example from South-eastern Utah." *Geology*, 13: 73–76.

Lowrie, W., and W. Alvarez, 1981. "One Hundred Million Years of Geomagnetic Polarity History." *Geology*, 9: 392–397.

Lowrie, W., W. Alvarez, G. Napoleone, K. Perch-Nielsen, I. Premoli-Silva, and M. Toumarkine, 1982. "Paleogene Magnetic Stratigraphy in Umbrian Pelagic Carbonate Rocks. The Contessa Sections, Gubbio." *Geol. Soc. Amer. Bull.*, 93: 414–432.

Lowrie, W., W. Alvarez, and I. Premoli-Silva, 1980. "Lower Cretaceous Magnetic Stratigraphy in Umbrian Pelagic Carbonate Rocks." *Geophys. J. R. Astron. Soc.*, 60: 263–281.

Lupe, R., and T. S. Ahlbrandt, 1979. "Sediments of the Ancient Eolian Environment—Reservoir Inhomogeneity." *U.S. Geol. Surv. Prof. Paper* 1052: 241–252.

Lyell, Sir Charles, 1830–3. *Principles of Geology*, 3 vols. John Murray, London.

Maglio, V. J., 1972. "Vertebrate Faunas and Chronology of Hominid-Bearing Sediments East of Lake Rudolf, Kenya." *Nature*, 239: 379–385.

Maglio, V. J., and H. B. S. Cooke (eds.), 1972. *Evolution of African Mammals.* Harvard Univ. Press, Cambridge.

Mallory, V. S., 1959. *Lower Tertiary Biostratigraphy of the California Coast Ranges.* Amer. Assoc. Petrol. Geol., Tulsa, Okla.

Mankinen, E. A., and G. B. Dalrymple, 1979. "Revised Geomagnetic Polarity Time Scale for the Interval 0–5 m.y. B.P." *J. Geophys. Res.*, 84: 615–626.

Markevich, V. P., 1960. "The Concept of Facies." *Int. Geol. Rev.*, 2: 376–379, 498–507, 582–604.

Matthews, R. K., 1984. *Dynamic Stratigraphy*, 2d ed. Prentice-Hall, Englewood Cliffs, N.J.

McBride, E. F., 1962. "Flysch and Associated Beds of the Martinsburg Formation (Ordovician), Central Appalachians." *J. Sed. Petrol.*, 32: 39–91.

McDougall, I., 1985. "K-Ar and ^{40}Ar/^{39}Ar Dating of the Hominid-Bearing Pliocene-Pleistocene Sequence at Koobi Fora, Lake Turkana, Northern Kenya." *Geol. Soc. Amer. Bull.*, 96: 159–175.

McDougall, I., R. Maiaer, P. Sutherland-Hawkes, and A. J. W. Gleadow, 1980. "K-Ar Age Estimates for the KBS Tuff, East Turkana, Kenya." *Nature*, 284: 230–234.

McDougall, K., 1980. "Paleoecological Evaluation of Late Eocene Biostratigraphic Zonations of the Pacific Coast of North America." *J. Paleont.*, 54(4): 1–75 (supplement).

McElhinney, M. W., 1973. *Palaeomagnetism and Plate Tectonics.* Cambridge Univ. Press, Cambridge, England.

McKee, E. D., 1982. "The Supai Group of the Grand Canyon." *U.S. Geol. Surv. Prof. Paper* 1173: 1–504.

McKee, E. D., and R. C. Gutschick, 1969. "History of Redwall Limestone of Northern Arizona." *Geol. Soc. Amer. Mem.*, 114: 1–726.

McKee, E. D., and C. E. Resser, 1945. "Cambrian History of the Grand Canyon region." *Carnegie Inst. Wash. Publ.*, 563: 1–168.

McKerrow, W. S., R. St. J. Lambert, and V. E. Chamberlain, 1980. "The Ordovician, Silurian, and Devonian Time Scales." *Earth Planet. Sci. Lett.*, 51: 1–8.

McKerrow, W. S., R. St. J. Lambert, and L. R. M. Cocks, 1985. "The Ordovician, Silurian, and Devonian periods." In N. J. Snelling, (ed.), *The Chronology of the Geological Record*, 73–80. Geol. Soc. Lond. Mem. 10.

McLaren, D. J., 1973. "The Silurian-Devonian Boundary." *Geol. Mag.*, 110: 302–303.

McLaren, D. J., 1977. "The Silurian-Devonian Boundary Committee: A Final Report." In the *Silurian-Devonian Boundary*, 1–34. IUGS Series A, 5.

McPhee, J., 1980. *Basin and Range.* Farrar, Straus and Giroux, New York.

McPhee, J., 1986. *Rising from the Plains.* Farrar, Straus and Giroux, New York.

Meckel, L. D., 1967. "Origin of Pottsville Conglomerates (Pennsylvanian) in the Central Appalachians." *Geol. Soc. Amer. Bull.*, 78: 223–258.

Miall, A. D., 1984. *Principles of Sedimentary Basin Analysis.* Springer-Verlag, New York.

Miall, A. D., 1986. "Eustatic Sea-level Changes Interpreted from Seismic Stratigraphy: A Critique of the Methodology with Particular Reference to the North Sea Jurassic Record." *Amer. Assoc. Petrol. Geol. Bull.*, 70: 131–137.

Middleton, G. V., and A. H. Bouma (eds.), 1973. *Turbidites and Deep Water Sedimentation.* Pacific Section Soc. Econ. Paleont. Mineral. Short Course, Anaheim, Calif.

Middleton, L. T., and R. C. Blakey, 1983. "Processes and Controls on the Intertonguing of the Kayenta and Navajo Formations, Northern Arizona: Eolian-fluvial Interactions." In M. E. Brookfield, and T. S. Ahlbrandt (eds.), *Eolian Sediments and Processes*, 613–624. Elsevier, Amsterdam.

Miller, F. X., 1977. "The Graphic Correlation Method in Biostratigraphy." In E. G. Kauffman and J. Hazel (eds.), *Concepts and Methods in Biostratigraphy*, 165–186. Dowden, Hutchinson and Ross, Stroudsburg, Penn.

Mitchum, R. M., Jr., P. R. Vail, and J. B. Sangree, 1977. "Stratigraphic Interpretation of Seismic Reflection Patterns in Depositional Sequences." *Amer. Assoc. Petrol. Geol. Mem.*, 26: 135–143.

Molostovsky, E. A., M. A. Peuzner, D. M. Petchersky, V. D. Rodionov, and A. N. Khramov, 1976. "Phanerozoic Magnetostratigraphic Scale and Geomagnetic Field Inversion Regime." *Geomagn. Res.*, 17: 45–52.

Montanari, A., R. Drake, D. M. Bice, W. Alvarez, G. H. Curtis, B. Turrin, and D. J. DePaolo, 1985. "Radiometric Time Scale for the Upper Eocene and Oligocene Based on K/Ar and Rb/Sr Dating of Volcanic Biotites." *Geology*, 13: 596–599.

Montanari, A., A. Deino, R. Drake, B. D. Turrin, D. J. DePaolo, G. S. Odin, G. H. Curtis, W. Alvarez, and D. M. Bice, 1988. "Radioisotopic Dating of the Eocene-Oligocene Boundary in the Pelagic Sequence of the Northern Apennines." In I. Premoli-Silva, R. Coccioni, and A. Montanari (eds.), *The Eocene-Oligocene Boundary in the Marche-Umbria Basin (Italy).* IUGS Commission on Stratigraphy, Ancona, Italy.

Moore, C. H., Jr., E. H. Graham, and L. S. Land, 1976. "Sediment Transport and Dispersal across the Deep Fore-reef and Island Slope (−55 m to −305 m), Discovery Bay, Jamaica." *J. Sed. Petrol.*, 46: 174–187.

Moore, G. F., and D. E. Karig, 1976. "Development of Sedimentary Basins on the Lower Trench Slope." *Geology*, 4: 693–697.

Moore, R. C., 1949. "Meaning of Facies." *Geol. Soc. Amer. Mem.*, 39: 1–34.

Moore, R. C., 1955. "Invertebrates and the Geological Time Scale." *Geol. Soc. Amer. Spec. Paper* 62: 547–573.

Moore, R. C., et al., 1944. "Correlation of Pennsylvanian Formations of North America." *Geol. Soc. Amer. Bull.*, 55: 657–706.

Morgan, J. P., and R. H. Shaver (eds.), 1970. *Deltaic Sedimentation: Modern and Ancient*. Soc. Econ. Paleont. Mineral. Spec. Publ. 15.

Murphy, M. A., 1977. "On Chronostratigraphic Units." *J. Paleont.,* 52: 123–219.

Murray, J., and A. F. Renard, 1891. *Report on Deep-Sea Deposits Based on Specimens Collected during the Voyage of H.M.S. Challenger in the Years 1873–1876.* H.M.S.O., Edinburgh.

Mutti, E., 1985. "Turbidite Systems and Their Relations to Depositional Sequences." *NATO Adv. Study Inst. Ser. C,* 148: 65–73.

Mutti, E., and F. Ricci Lucchi, 1972. "Turbidites of the Northern Apennines: Introduction to Facies Analysis." *Geology Rev.,* 20(2): 125–166, and translated 1978 by T. H. Nilsen; reprinted by the American Geological Institute, Falls Church, Va.

Naeser, C., 1979. "Thermal History of Sedimentary Basins: Fission-track Dating of Subsurface Rocks." *Soc. Econ. Paleont. Mineral. Spec. Publ.,* 26: 109–112.

Ness, G., S. Levi, and G. Couch, 1980. "Marine Magnetic Anomaly Timescales for the Cenozoic and Late Cretaceous: A Precis, Critique, and Synthesis." *Rev. Geophys. Space Phys.,* 18: 753–770.

Nier, A. O., 1940. "A Mass Spectrometer for Routine Isotope Abundance Measurements." *Rev. Sci. Instrum.,* 11: 212–216.

Ninkovitch, D., and N. J. Shackleton, 1975. "Distribution, Stratigraphic Position, and Age of Ash Layer 'L' in the Panama Basin Region." *Earth Planet. Sci. Lett.,* 27: 20–34.

North, F. K., 1985. *Petroleum Geology.* Allen and Unwin, Boston.

North American Commission on Stratigraphic Nomenclature, 1983. "North American Stratigraphic Code." *Amer. Assoc. Petrol. Geol. Bull.,* 67: 841–875.

Obradovich, J. D., and W. A. Cobban, 1975. "A Time-Scale for the Late Cretaceous of the Western Interior of North America." *Geol. Assoc. Canada Spec. Paper,* 13: 31–54.

Obradovich, J. D., 1988. "A Different Perspective on Glauconite as a Chronometer for Geologic Time Scale Studies." *Paleoceanography,* 3: 757–770.

Odin, G. S., 1978. "Isotopic Dates for the Paleogene Time Scale." *Amer. Assoc. Petrol. Geol. Stud. Geol.,* 6: 247–257.

Odin, G. S. (ed.), 1982. *Numerical Dating in Stratigraphy,* 2 vols. John Wiley, New York.

Odin, G. S., 1985. "Remarks on the Numerical Scale of Ordovician to Devonian Time." In N. J. Snelling (ed)., *The Chronology of the Geological Record,* 93–98. Geol. Soc. London Mem. 10.

Odin, G. S., and D. Curry, 1981. L'échelle numerique des temps paléogènes en 1981. *C. R. Acad. Sci. Paris.,* 293(II): 1003–1006.

Odin, G. S., and D. Curry, 1985. "The Palaeogene Time-Scale: Radiometric Dating versus Magnetostratigraphic Approach." *J. Geol. Soc. London,* 142: 1179–1188.

Olsen, P. E., 1984. "Periodicities of Lake-Level Cycles in the Late Triassic Lockatong Formation of the Newark Basin (Newark Supergroup), New Jersey and Pennsylvania." In A. Berger, J. Imbrie, J. Hays, G. Kukla, and B. Saltzman (eds.), *Milankovitch and Climate, Part 1:* 129–146. The Hague, Netherlands.

Opdyke, N. D., L. H. Burckle, and A. Todd, 1974. "The Extension of the Magnetic Time Scale in Sediments of the Central Pacific Ocean." *Earth Planet. Sci. Lett.,* 22: 300–306.

Opdyke, N. D., B. Glass, J. D. Hays, and J. Foster, 1966. "Paleomagnetic Study of Antarctic Deep-Sea Cores." *Science,* 154: 349–357.

Oppel, A., 1856–1858. *Die Juraformation Englands, Frankreichs, und des südwestlichen Deutschlands.* Stuttgart.

Orbigny, A. D. d', 1842. *Paléontologie Francaise, Terraines Jurassiques. Pt. 1, Cephalopodes.* Masson, Paris.

Osborn, H. F., and W. D. Matthew, 1909. "Cenozoic Mammal Horizons of Western North America." *U.S. Geol. Surv. Bull.,* 361: 1–138.

Owen, D. E., 1987. "Commentary: Usage of Stratigraphic Terminology in Papers, Illustrations, and Talks." *J. Sed., Petrol.,* 57: 363–372.

Palmer, M. R., and H. Elderfield, 1985. "Sr Isotope Composition of Sea Water over the Past 75 Myr." *Nature,* 314: 526–528.

Parkinson, N., and C. Summerhayes, 1985. "Synchronous Global Sequence Boundaries." *Amer. Assoc. Petrol. Geol. Bull,* 69: 685–687.

Parrish, R., and J. C. Roddick, 1985. *Geochronology and Isotope Geology for the Geologist and Explorationist.* Geol. Assoc. Canada, Cordilleran Section, Short Course 4.

Payton, C. E. (ed.), 1977. *Seismic Stratigraphy—Applications to Hydrocarbon Exploration.* Amer. Assoc. Petrol. Geol. Memoir 26.

Pettijohn, F. J., and P. E. Potter, 1964. *Atlas and Glossary of Primary Sedimentary Structures.* Springer-Verlag, New York.

Phillips, J., 1840. "Palaeozoic Series." In G. Long, (ed.), *The Penny Cyclopaedia of the Society for the Diffusion of Useful Knowledge,* 17: 153–154. Charles Knight, London.

Pitman, W. C., III., 1978. "Relationship between Eustacy and Stratigraphic Sequences of Passive Margins." *Geol. Soc. Amer. Bull.,* 89: 1389–1403.

Playfair, J., 1802. *Illustrations of the Huttonian Theory of the Earth.* Wm. Creech, Edinburgh.

Playfair, J., 1805. "Biographical Account of the Late James Hutton, F.R.S., Edinburgh." *Trans. R. Soc. Edinburgh*, 5(3): 39–99.

Playford, P. E., 1980. "Devonian 'Great Barrier Reef' of Canning Basin, Western Australia." *Amer. Assoc. Petrol. Geol. Bull.*, 64: 814–840.

Playford, P. E., 1981. *Devonian Reef Complexes of the Canning Basin, Western Australia.* Geol. Soc. Aust. Fifth Aust. Geol. Conv. Field Excursion Guidebook.

Playford, P. E., 1984. "Platform-Margin and Marginal-Slope Relationships in Devonian Reef Complexes of the Canning Basin." In P. G. Purcell (ed.), *The Canning Basin*, 189–214. Proc. Geol. Soc. Aust., Petrol. Explor. Soc. Aust. Symposium, Perth.

Playford, P. E., and D. C. Lowry, 1966. "Devonian Reef Complexes of the Canning Basin, Western Australia." *Geol. Surv. West. Aust. Bull.*, 118: 1–150.

Poag, C. W., 1977. "Biostratigraphy in Gulf Coast Petroleum Exploration." In E. G. Kauffman and J. Hazel (eds.), *Concepts and Methods in Biostratigraphy*, 213–233. Dowden, Hutchinson and Ross, Stroudsburg, Penn.

Poag, C. W., and J. S. Schlee, 1984. "Depositional Sequences and Stratigraphic Gaps on Submerged United States Atlantic Margin." *Amer. Assoc. Petrol. Geol. Mem.*, 36: 165–182.

Poore, R. Z., L. Tauxe, S. F. Percival, Jr., and J. L. LaBrecque, 1982. "Late Eocene-Oligocene Magnetostratigraphy and Biostratigraphy at South Atlantic DSDP Site 527." *Geology*, 10: 508–511.

Porter, J. W., R. A. Price, and R. G. McCrossan, 1982. "The Western Canada Sedimentary Basin." *Phil. Trans. R. Soc. A*, 305(1489): 169–173.

Posamentier, H. W., M. T. Jervey, and P. R. Vail, 1988. "Eustatic Controls on Clastic Deposition I—Conceptual Framework." *Soc. Econ. Paleont. Mineral. Spec. Publ.* 42: 109–124.

Posamentier, H. W., and P. R. Vail, 1988. "Eustatic Controls on Clastic Deposition II—Sequence and Systems Tract Models." *Soc. Econ. Paleont. Mineral. Spec. Publ.* 42: 125–154.

Potter, P. E., and F. J. Pettijohn, 1977. *Paleocurrents and Basin Analysis*, 2d ed. Springer-Verlag, New York.

Powell, J. W., 1875. *Exploration of the Colorado River of the West and Its Tributaries.* Government Printing Office, Washington, D.C.

Pratt, L. M., 1985. "Isotopic Studies of Organic Matter and Carbonate in Rocks of the Greenhorn Marine Cycle." In L. M. Pratt, E. G. Kauffman, and F. B. Zelt (eds.), *Finegrained Deposits and Biofacies of the Cretaceous Western Interior Seaway: Evidence of Cyclic Sedimentary Processes*, 38–48. Soc. Econ. Paleont. Mineral. Field. Trip Guidebook 4.

Pratt, L. M., E. G. Kauffman, and F. B. Zelt (eds.), 1985. *Fine-Grained Deposits and Biofacies of the Cretaceous Western Interior Seaway: Evidence of Cyclic Sedimentary Processes.* Soc. Econ. Paleont. Mineral. Field Trip Guidebook 4.

Prothero, D. R., 1982. "How Isochronous Are Mammalian Biostratigraphic Events? *Proc. Third N. Amer. Paleont. Conv.*, 2: 405–409.

Prothero, D. R., 1985a. "Chadronian (Early Oligocene) Magnetostratigraphy of Eastern Wyoming: Implications for the Eocene-Oligocene Boundary." *J. Geol.*, 93: 555–565.

Prothero, D. R., 1985b. "North American Mammalian Diversity and Eocene-Oligocene Extinctions." *Paleobiol.*, 11: 389–405.

Prothero, D. R., C. R. Denham, and H. G. Farmer, 1982. "Oligocene Calibration of the Magnetic Polarity Timescale." *Geology*, 10: 650–653.

Prothero, D. R., C. R. Denham, and H. G. Farmer, 1983. "Magnetostratigraphy of the White River Group, and Its Implications for Oligocene Geochronology." *Palaeogeogr., Palaeoclimat., Palaeoecol.*, 42: 151–166.

Quenstedt, F. A., 1856–1858. *Der Jura.* H. Laupp, Tübingen.

Raaf, J. F. M. de, J. R. Boersma, and A. van Gelder, 1977. "Wave Generated Structures and Sequences from a Shallow Marine Succession, Lower Carboniferous, County Cork, Ireland." *Sedimentology*, 4: 1–52.

Ralph, E. K., 1971. "Carbon-14 Dating." In H. N. Michael and E. K. Ralph (eds.), *Dating Techniques for the Archeologist*, 1–48, M.I.T. Press, Cambridge, Mass.

Ramos, A., and A. Sopeña, 1983. "Gravel Bars in Low Sinuosity Streams (Permian and Triassic, Central Spain)." In J. D. Collinson and J. Lewin (eds.), *Modern and Ancient Fluvial Systems*, 301–302. Spec. Publ. Int. Ass. Sediment., 6.

Ramos, A., A. Sopeña, and M. Perez-Arlucea, 1986. "Evolution of Buntsandstein Fluvial Sedimentation in the Northwest Iberian Ranges (Central Spain)." *J. Sed. Petrol.*, 56: 862–875.

Ramsbottom, W. H. C., 1979. "Rates of Transgression and Regression in the Carboniferous of Northwest Europe." *J. Geol. Soc. London*, 136: 147–153.

Rau, W. W., 1958. "Stratigraphy and Foraminiferal Zonation in Some Tertiary Rocks of Southwestern Washington." *U.S. Geol. Surv. Oil and Gas. Invest. Chart OC 57.*

Rau, W. W., 1966. "Stratigraphy and Foraminifera of the Satsop River Area, Southern Olympic Peninsula, Washington." *Wash. Div. Mines and Geol. Bull.*, 53: 1–66.

Reading, H. G. (ed.), 1986. *Sedimentary Environments and Facies,* 2d ed. Blackwell Scientific Publications, Oxford.

Reineck, H. E., and I. B. Singh, 1973. *Depositional Sedimentary Environments.* Springer-Verlag, New York.

Reineck, H.-E., and F. Wunderlich, 1968a. "Classification and Origin of Flaser and Lenticular Bedding." *Sedimentology,* 11: 99–104.

Reineck, H.-E., and F. Wunderlich, 1968b. "Zeitmessungen und Bezeitenschichten." *Natur und Museum,* 97: 193–197.

Repenning, C. A., 1967. "Palearctic-Nearctic Mammalian Dispersal in the Late Cenozoic." In D. M. Hopkins (ed.), *The Bering Land Bridge,* 288–311. Stanford Univ. Press Stanford, Calif.

Ricci Lucchi, F., and E. Valmori, 1980. "Basin-Wide Turbidites in a Miocene Oversupplied Deep-Sea Plain." *Sedimentol.,* 27: 241–270.

Rickard, L. V., 1962. "Late Cayugan (Upper Silurian) and Helderbergian (Lower Devonian) Stratigraphy in New York." *N.Y. State Mus. Sci. Serv. Bull.,* 386: 1–157.

Rigby, J. K., and W. K. Hamblin (eds.), 1972. *Recognition of Ancient Sedimentary Environments.* Soc. Econ. Paleont. Mineral. Spec. Publ. 16.

Riley, J. P., and R. L. Chester (eds.), 1976. *Chemical Oceanography,* vol. 5, 2d ed. Academic Press, New York.

Ronov, A. B., V. E. Khain, A. N. Balukhovsky, and K. B. Seslavinsky, 1980. "Quantitative Analysis of Phanerozoic Sedimentation." *Sed. Geol.,* 25: 311–325.

Ross, C. A., and J. R. P. Ross, 1985. "Late Paleozoic Depositional Sequences Are Synchronous and Worldwide." *Geology,* 13: 194–197.

Rudwick, M. J. S., 1978. "Charles Lyell's Dream of a Statistical Palaeontology." *Palaeontol.,* 21: 225–244.

Rust, I. C., 1973. "The Evolution of the Paleozoic Cape Basin, Southern Margin of Africa." In A. E. M. Nairn and F. H. Stehli (eds.), *The Ocean Basins and Margins,* 247–276. New York.

Rust, I. C., 1977. "Evidence of Shallow Marine and Tidal Sedimentation in the Ordovician Graafwater Formation, Cape Province, South Africa." *Sed. Geol.,* 18: 123–133.

Ryder, R. T., T. D. Fouch, and J. H. Elison, 1976. "Early Tertiary Sedimentation in the Western Uinta Basin, Utah." *Geol. Soc. Amer. Bull.,* 87: 496–512.

Ryer, T. A., 1977. "Patterns of Cretaceous Shallow Marine Sedimentation, Coalville and Rockport Areas, Utah." *Geol. Soc. Amer. Bull.,* 88: 177–188.

Sadler, P. M., 1981. "Sediment Accumulation Rates and the Completeness of Stratigraphic Sections." *J. Geol.,* 89: 569–584.

Sadler, P. M., and L. W. Dingus, 1982. "Expected Completeness of Sedimentary Sections: Estimating a Time-Scale Dependent, Limiting Factor in the Resolution of the Fossil Record." *Proc. Third. N. Amer. Paleont. Conv.,* 2: 461–464.

Sanders, J. E., 1970. "Coastal-zone Geology and Its Relationship to Water Pollution Problems." In A. A. Johnson (ed.), *Water Pollution in the Greater New York Area,* 23–35. Gordon & Breach, New York.

Sarna-Wojcicki, A. M., S. D. Morrison, C. E. Meyer, and J. W. Hillhouse, 1987. "Correlation of Upper Cenozoic Tephra Layers between Sediments of the Western United States and the Eastern Pacific Ocean, and Comparison with Biostratigraphic and Magnetostratigraphic Age Data." *Geol. Soc. Amer. Bull.,* 98: 207–223.

Scheltema, R. S., 1977. "Dispersal of Marine Invertebrate Organisms: Paleobiogeographic and Biostratigraphic Implications." In E. G. Kauffman and J. E. Hazel (eds.), *Concepts and Methods of Biostratigraphy,* 73–108. Dowden, Hutchison and Ross, Stroudsburg, Penn.

Schenck, H. G., and R. M. Kleinpell, 1936. "Refugian Stage of the Pacific Coast Tertiary." *Amer. Assoc. Petrol. Geol. Bull.,* 20: 215–255.

Schenck, H. G., and S. W. Muller, 1941. "Stratigraphic Terminology." *Geol. Soc. Amer. Bull.,* 52: 1419–1426.

Schlee, J. S., and L. F. Jansa, 1981. "The Paleoenvironment and Development of the Eastern North American Continental Margin." *Colloque C3: Géologie des marges continentales. 26th IGC, Oceanologica Acta* 4: 71–80.

Scholl, D. W., and M. S. Marlow, 1974. "Sedimentary Sequence in Modern Pacific Trenches and the Deformed Circum-Pacific Eugeosyncline." *Soc. Econ. Paleont. Mineral. Spec. Publ.,* 19: 193–211.

Scholle, P. A., and D. Spearing (eds.), 1982. *Sandstone Depositional Environments.* American Association of Petroleum Geologists Memoir 31.

Scholle, P. A., D. G. Bebout, and C. H. Moore (eds.), 1983. *Carbonate Depositional Environments.* Amer. Assoc. Petrol. Geol. Mem. 33.

Schwartz, R. K., 1975. "Nature and Genesis of Some Storm Washover Deposits." *U.S. Army Corps Eng. Coastal Eng. Res. Center Tech. Mem.,* 61: 69.

Scoffin, T. P., 1987. *An Introduction to Carbonate Sediments and Rocks.* Blackie and Son, Ltd., London.

Secord, J. A., 1986. *Controversy in Victorian Geology: The Cambrian-Silurian Dispute.* Princeton Univ. Press, Princeton, N.J.

Sedgwick, A., and R. I. Murchison, 1839. "On the Older Rocks of Devonshire and Cornwall." *Proc. Geol. Soc. London,* 3(63): 121–123.

Selley, R. C., 1978. *Ancient Sedimentary Environments,* 2d ed. Cornell University Press, Ithica, New York.

Sellwood, B. W., 1975. "Lower Jurassic Tidal Flat Deposits, Bornholm, Denmark." In R. N. Ginsburg (ed.), *Tidal Deposits,* 93–101. Springer-Verlag, New York.

Semikhatov, M. A., 1980. "On the Upper Precambrian Stromatolite Standard of Northern Eurasia." *Precamb. Res.,* 16: 235–247.

Sepkoski, J. J., and A. H. Knoll, 1983. "Precambrian-Cambrian Boundary: The Spike is Driven and the Monolith Crumbles." *Paleobiology,* 9: 199–206.

Shackleton, N. J., and N. D. Opdyke, 1973. "Oxygen Isotope and Paleomagnetic Stratigraphy of Equatorial Pacific Core V28-238: Oxygen Isotope Temperatures and Ice Volumes on a 10^5 Year and 10^6 Year Scale." *Quat. Res.,* 3: 39–55.

Shackleton, N. J., and N. D. Opdyke, 1976. "Oxygen Isotope and Paleomagnetic Stratigraphy of Equatorial Pacific Core V28-239, Late Pliocene to Latest Pleistocene." *Geol. Soc. Amer. Mem.,* 145: 449–464.

Shanmugam, G., and R. J. Moiola, 1982. "Eustatic Control of Turbidites and Winnowed Turbidites." *Geology,* 10: 131–135.

Shanmugam, G., R. J. Moiola, and J. E. Damuth, 1985. "Eustatic Control of Submarine Fan Development." In A. H. Bouma, W. R. Normark, and N. E. Barnes (eds.), *Submarine Fans and Related Turbidite Systems,* 23–28. Springer-Verlag, New York.

Shaw, A. B., 1964. *Time in Stratigraphy.* McGraw-Hill, New York.

Shelton, J. S., 1966. *Geology Illustrated.* W. H. Freeman and Co., San Francisco.

Shepard, F. P., F. B. Phleger, and T. J. van Andel (eds.), *Recent Sediments of the Northwest Gulf of Mexico.* Amer. Assoc. Petrol. Geol., Tulsa, Okla.

Sheridan, R. E., 1974. "Atlantic Continental Margin of North America." In C. A. Burk, and C. L. Drake (eds.), *The Geology of Continental Margins,* 391–407. Springer-Verlag, New York.

Sheridan, R. E., 1983. "Phenomenon of Pulsation Tectonics Related to the Breakup of the Eastern North American Continental Margin." *Init. Rept. Deep-Sea Drill. Proj.,* 76: 897–909.

Sheridan, R. E., 1987a. "Pulsation Tectonics as the Control of Continental Breakup." *Tectonophys.,* 143: 59–73.

Sheridan, R. E., 1987b. "Pulsation Tectonics as the Control of Long-term Stratigraphic Cycles." *Paleoceanography,* 2: 97–118.

Sheriff, R. E., 1978. *A First Course in Geophysical Exploration and Interpretation.* International Human Resource Development Corporation, Boston.

Sheriff, R. E., 1980. *Seismic Stratigraphy.* International Human Resource Development Corporation, Boston.

Shimer, H. W., and R. R. Shrock, 1944. *Index Fossils of North America.* MIT Press, Cambridge, Mass.

Shinn, E. A., and R. N. Ginsburg, 1964. "Formation of Recent Dolomite in Florida and the Bahamas (abs.)." *Amer. Assoc. Petrol. Geol. Bull.,* 48: 547.

Shinn, E. A., R. N. Ginsburg, and R. M. Lloyd, 1965. "Recent Supratidal Dolomite from Andros Island, Bahamas." *Soc. Econ. Paleont. Mineral. Spec. Publ.,* 13: 112–123.

Shinn, E. A., R. M. Lloyd, and R. N. Ginsburg, 1969. "Anatomy of a Modern Carbonate Tidal-Flat, Andros Island, Bahamas." *J. Sed. Petrol.,* 39: 1202–1228.

Siever, R., 1988. *Sand.* W. H. Freeman and Company, New York.

Skipp, B. 1979. "Great Basin Region." *U.S. Geol. Surv. Prof. Paper* 1010: 273–328.

Sloss, L. L., 1963. "Sequences in the Cratonic Interior of North America." *Geol. Soc. Amer. Bull.,* 74: 93–114.

Sloss, L. L., 1972. "Synchrony of Phanerozoic Sedimentary-Tectonic Events of the North American Craton and the Russian Platform." *24th Int. Geol. Congr.,* 6: 24–32.

Sloss, L. L., 1979. "Global Sea Level Change: A View from the Craton." *Amer. Assoc. Petrol. Geol. Mem.,* 29: 461–467.

Sloss, L. L., 1982. "The Midcontinent Province: United States." *Geol. Soc. Amer., DNAG Spec. Publ.,* 1: 27–39.

Sloss, L. L., 1984. "Comparative Anatomy of Cratonic Unconformities." *Amer. Assoc. Petrol. Geol. Mem.,* 36: 1–6.

Sloss, L. L., W. C. Krumbein, and E. C. Dapples, 1949. "Integrated Facies Analysis." *Geol. Soc. Amer. Mem.,* 39: 94–124.

Smith, W., 1815. *A Delineation of the Strata of England and Wales with Part of Scotland.* Privately published, London.

Snelling, N. J., 1985. "An Interim Time-scale." In N. J. Snelling (ed.), *The Chronology of the Geological Record.* Geol. Soc. Lond. Mem., 10: 261–265.

Soares, P. C., P. M. B. Landim, and W. S. Fulfair, 1978. "Tectonic Cycles and Sedimentary Sequences in the Brazilian Intracratonic Basins." *Geol. Soc. Amer. Bull.,* 89: 181–191.

Spearing, D., 1974. *Summary Sheets of Sedimentary Deposits.* Geol. Soc. Amer. Maps and Charts series MC-8.

Srinivasan, M. S., and J. P. Kennett, 1981. "A Review of Neogene Planktic Foraminiferal Biostratigraphy: Applications in the Equatorial and South Pacific." *Soc. Econ. Paleont. Mineral. Spec. Publ.,* 32: 395–432.

Stanley, D. J., and D. J. P. Swift (eds.), 1976. *Marine Sediment Transport and Environmental Management.* John Wiley, New York.

Stanley, S. M., 1989. *Earth and Life through Time*, 2d ed. W. H. Freeman and Co., New York.

Stanley, S. M., W. O. Addicott, and K. Chinzei, 1980. "Lyellian Curves in Paleontology: Possibilities and Limitations." *Geology*, 8: 422–426.

Steel, R. J., and T. G. Gloppen, 1980. "Late Caledonian (Devonian) Basin Formation, Western Norway—Signs of Strike-slip Tectonics During Infilling." *Spec. Publ. Int. Assoc. Sedimentol.*, 4: 79–103.

Steel, R. J., S. Mæhle, H. Nilsen. S. L. Røe, and Å. Spinnangr, 1977. "Coarsening-Upward Cycles in the Alluvium of Hornelen Basin (Devonian), Norway: Sedimentary Response to Tectonic Events." *Geol. Soc. Amer. Bull.*, 88: 1124–1134.

Steiger, R. H., and E. Jäger, 1977. "Subcommission on Geochronology: Convention on the Use of Decay Constants in Geo- and Cosmochronology." *Earth Planet Sci. Lett.*, 36: 359–362.

Steno, N., 1669. *De solido intra solidum naturaliter contento dissertationis prodromus*. The Star, Florence.

Stille, H., 1936. *Wege und Ergebnisse der geologisch-tektonischen Forschung*. Kaiser Wilhelm Ges., Berlin.

Summerhayes, C. P., 1986. "Sea Level Curves Based on Seismic Stratigraphy: Their Chronostratigraphic Significance." *Palaeogeogr., Palaeoclimat., Palaeoecol.*, 57: 27–42.

Surdam, R. C., and C. A. Wolfbauer, 1975. "Green River Formation, Wyoming: A Playa-Lake Complex." *Geol. Soc. Amer. Bull.*, 86: 335–345.

Suttner, L. J., A. Basu, and G. H. Mack, 1981. "Climate and Origin of Quartz Arenites." *J. Sed. Petrol.*, 51: 1235–1246.

Swift, D. J. P., 1968. "Coastal Erosion and Transgressive Stratigraphy." *J. Geol.*, 76: 444–456.

Swift, D. J. P., 1974. "Continental Shelf Sedimentation. In C. A. Burk and C. L. Drake (eds.), *The Geology of Continental Margins*, 117–135. Springer-Verlag, Berlin.

Swift, D. J. P., D. B. Duane, and T. F. McKinney, 1973. "Ridge and Swale Topography of the Middle Atlantic Bight, North America: Secular Response to the Holocene Hydraulic Regime." *Marine Geol.*, 15: 227–247.

Swift, D. J. P., D. B. Duane, and O. H. Pilkey (eds.), 1972. *Shelf Sediment Transport: Process and Product*. Dowden, Hutchinson and Ross, Stroudsburg, Penn.

Swift D. J. P., D. J. Stanley and J. R. Curray, 1971. "Relict Sediments on Continental Shelves: A Reconsideration." *J. Geol.*, 79: 322–346.

Tankard, A. B., and D. K. Hobday, 1977. "Tide-Dominated Back-barrier Sedimentation, Early Ordovician Cape Basin, Cape Peninsula, South Africa." *Sed. Geol.*, 18: 135–159.

Tarling, D. H., 1983. *Palaeomagnetism: Principles and Applications in Geology, Geophysics and Archaeology.* Chapman and Hall, London.

Tauxe, L., R. Butler, and J. C. Herguera, 1987. "Magnetostratigraphy: In Pursuit of Missing Links." *Rev. Geophys.*, 25(5): 939–950.

Tazieff, H., 1970. "The Afar Triangle. "*Sci. Amer.*, 222(2): 32–51.

Tedford, R. H., 1970. "Principles and Practices of Mammalian Geochronology in North America." *Proc. North Amer. Paleont. Conv.*, F: 666–703.

Teichert, C., 1958. "Concepts of Facies." *Amer. Assoc. Petrol. Geol. Bull.*, 42: 2718–2744.

Thaler, L., 1972. "Datation, Zonation et Mammifères." *Bur. Rech. Géol. Minières Mém.*, 77: 411–424.

Thompson, G. R., and J. Hower., 1973. "An Explanation for Low Radiometric Ages from Glauconite." *Geochim. Cosmochim. Acta*, 37: 1473–1491.

Thorne, J., and A. B. Watts, 1984. "Seismic Reflectors and Unconformities at Passive Continental Margins." *Nature*, 311: 365–368.

Tillman, R. W., and C. W. Siemars (eds.), 1984. *Siliciclastic Shelf Sediments*. Soc. Econ. Paleont. Mineral. Spec. Publ. 34.

Tillman, R. W., D. J. P Swift, and R. G. Walker, 1985. *Shelf Sands and Sandstone Reservoirs*. Soc. Econ. Paleont. Mineral., Short Course 13.

Tucker, M. E., 1982. *The Field Description of Sedimentary Rocks*. Open University Press, Milton Keynes, England.

Twain, M., 1883. *Life on the Mississippi*. J. R. Osgood, Boston.

Ulrich, E. O., 1916. "Correlation by Displacement of the Strandline and the Function and Proper Use of Fossils in Correlation. *Geol. Soc. Amer. Bull.*, 27: 451–490.

Vail, P. R., J. Hardenbol, and R. G. Todd, 1984. "Jurassic Unconformities, Chronostratigraphy and Sea Level Changes from Seismic Stratigraphy and Biostratigraphy." *Amer. Assoc. Petrol. Geol. Mem.*, 129–144.

Vail, P. R., R. M. Mitchum, Jr., and S. Thompson, III, 1977. "Global Cycles of Relative Changes in Sea Level." *Amer. Assoc. Petrol. Geol. Mem.*, 26: 83–98.

Vail, P. R., R. G. Todd, and J. B. Sangree, 1977. "Chronostratigraphic Significance of Seismic Reflections." *Amer. Assoc. Petrol. Geol. Mem.*, 16: 99–116.

Van Andel, T. H., G. R. Heath, and T. C. Moore, 1975. "Cenozoic History and Paleoceanography of the Central Equatorial Pacific." *Geol. Soc. Amer. Mem.*, 143: 1–134.

Vidal, G., and V. Zoubek (eds.), 1981. "Biostratigraphic Schemes." *Precamb. Res.*, 15: 95–96.

Vogt, P. R., 1975. "Changes in Geomagnetic Reversal Frequency at Time of Tectonic Change: Evidence for Coupling between Core and Upper Mantle Processes." *Earth Planet. Sci. Lett.*, 25: 313–321.

Walker, R. G., (ed.), 1984. *Facies Models*, 2d ed. Geoscience Canada Reprint Series 1, Geol. Assoc. Canada, Toronto.

Walker, T. R., and J. C. Harms, 1972. "Eolian Origin of Flagstone Beds, Lyons Sandstone (Permian), Type Area, Boulder County, Colorado." *Mountain Geol.*, 9: 279–288.

Walther, J., 1894. *Einleitung in die Geologie als historische Wissenschaft, Bd. 3, Lithogenesis der Gegenwart*, pp. 534–1055. G. Fischer, Jena.

Wanless, H. R., and J. M. Weller, 1932. "Correlation and Extent of Pennsylvanian Cyclothems." *Geol. Soc. Amer. Bull.*, 43: 1003–1016.

Warme, J. E., R. G. Douglas, and E. L. Winterer (eds.), 1981. *The Deep Sea Drilling Project: A Decade of Progress*. Soc. Econ. Paleont. Mineral Spec. Publ. 32.

Watts, A. B., 1982. "Tectonic Subsidence, Flexure, and Global Changes in Sea Level." *Nature*, 297: 469–474.

Watts, A. B., and J. Thorne, 1984. "Tectonics, Global Changes in Sea Level and Their Relationship to Stratigraphical Sequences in the U.S. Atlantic Continental Margin." *Mar. Petrol. Geol.*, 1: 319–339.

Weller, J. M., 1947. "Relations of the Invertebrate Paleontologist to Geology." *J. Paleont.*, 21: 570–575.

Weller, J. M., 1958. "Stratigraphic Facies Differentiation and Nomenclature. "*Amer. Assoc. Petrol. Geol. Bull.*, 42: 609–639.

Weller, J. M., 1960. *Stratigraphic Principles and Practice.* Harper and Brothers, New York.

Wheeler, H. E., 1964. "Baselevel, Lithospheric Surface, and Time-Stratigraphy." *Geol. Soc. Amer. Bull.*, 75: 599–610.

White, T. D., and J. M. Harris, 1977. "Suid Evolution and Correlation of African Hominid Localities." *Science*, 198(4312): 13–21.

Wilgus, C. K., B. S. Hastings, C. G. St. C. Kendall, H. W. Posamentier, C. A. Ross, and J. C. Van Wagoner (eds.), 1988. *Sea-level Changes: An Integrated Approach.* Soc. Econ. Paleont. Mineral. Spec. Publ. 42.

Wilkinson, B. H., 1982. "Cyclic Cratonic Carbonates and Phanerozoic Calcite Seas." *J. Geol. Educ.*, 30: 189–203.

Wilkinson, B. H., J. M. Budai, and R. K. Given, 1984. "Episodic Accumulation and the Origin of Formation Boundaries in the Helderberg Group of New York State: Comment." *Geology*, 12: 572–573.

Williams, D. F., 1988. "Evidence For and Against Sea-level Changes from the Stable Isotope Record of the Cenozoic." *Soc. Econ. Paleont. Mineral. Spec. Publ.* 42: 31–37.

Williams, E. C., J. C. Ferm, A. L. Guber, and R. E. Bergenback, 1964. *Cyclic Sedimentation in the Carboniferous of Western Pennsylvania.* Guidebook 29th Ann. Field Conf. Penn. Geol., Penn. State Univ., University Park, Penn.

Williams, H. S., 1901. "The Discrimination of Time-values in Geology." *J. Geol.*, 9: 570–585.

Wilson, J. L., 1975. *Carbonate Facies in Geologic History.* Springer-Verlag, New York.

Wood, H. E., II, R. W. Chaney, J. Clark, E. H. Colbert, G. L. Jepsen, J. B. Reeside, and C. Stock, 1941. "Nomenclature and Correlation of the North American Continental Tertiary." *Geol. Soc. Amer. Bull.*, 52: 1–48.

Woodburne, M. O., 1977. "Definition and Characterization in Mammalian Chronostratigraphy." *J. Paleont.*, 51: 220–234.

Woodburne, M. O. (ed.), 1987. *Cenozoic Mammals of North America: Geochronology and Biostratigraphy.* Univ. Calif. Press, Berkeley.

Woodburne, M. O., 1989. "Hipparion Horses: A Pattern of Worldwide Dispersal and Endemic Evolution." In D. R. Prothero and R. M. Schoch (eds.), *The Evolution of Perissodactyls*, 197–233. Oxford Univ. Press, Oxford.

Worsley, T. R., D. Nance, and J. B. Moody, 1984. Global Tectonics and Eustasy for the Past 2 Billion years. *Marine Geol.*, 58: 373–400.

Wunderlich, F., 1970. "Genesis and Environment of the Nellenköpfchenschichten (Lower Emsian, Rheinian Devon) at Locus Typicus in Comparison with Modern Coastal Environment of the German Bay." *J. Sed. Petrol.*, 40: 102–130.

York, D., and R. M. Farquhar, 1972. *The Earth's Age and Geochronology.* Pergamon Press, Oxford.

Illustration Credits

Fig. **1.5** from J.M. Weller, *Stratigraphic Principles and Practice*, pp. 34–35, Fig. 7; copyright © 1960 by the author; reprinted by permission of Harper & Row Publishers, Inc.

Fig. **1.7** from J.M. Weller, *Stratigraphic Principles and Practice*, p. 36, Fig. 8; copyright © 1960 by the author; reprinted by permission of Harper & Row Publishers, Inc.

Fig. **2.6A** from Blatt, Middleton, and Murray, *Origin of Sedimentary Rocks*, 2d ed., p. 137; copyright © 1980; reprinted by permission of Prentice-Hall, Inc., Englewood Cliffs, N.J.

Fig. **2.7** from A. V. Jopling, "Origin of laminae deposited by the movement of ripples along a streambed: A laboratory study," *J. Geol.*, v. 75, p. 298, Fig. 19; copyright © 1967, by permission of the University of Chicago Press.

Fig. **2.9** reproduced with permission from the *Annual Review of Earth Sciences*, v. 7, pp. 236–237, Fig. 8; copyright © 1979 by Annual Reviews, Inc.

Fig. **3.2** from Spearing, *Geol. Soc. Amer. Map Sheet* MC-8, Sheet 1, Fig. 1C; reprinted with permission of the Geological Society of America.

Fig. **3.3** from R. L. Hooke, "Processes on arid-region alluvial fans," *J. Geol.*, v. 75, p. 450, Fig. 8; copyright © 1967, by permission of the University of Chicago Press.

Fig. **3.14A** from Blatt, Middleton, and Murray, *Origin of Sedimentary Rocks*, 2d ed. p. 637; copyright © 1980; reprinted by permission of Prentice-Hall, Inc., Englewood Cliffs, N.J.

Fig. **3.21** from Surdam and Wolfbauer, *Geol. Soc. Amer. Bull.*, v. 86, p. 340, Fig. 8; reprinted with permission of the Geological Society of America.

Fig. **3.22** reproduced with permission from the *Annual Review of Earth Sciences*, v. 7, p. 242, Fig. 11; copyright © 1979 by Annual Reviews, Inc.

Fig. **4.2** reprinted by permission of the American Association of Petroleum Geologists.

Fig. **4.3** from H. G. Reading, *Sedimentary Environments and Facies*, 1986, p. 117, Fig. 6.3; by permission of Blackwell Scientific Publishers.

Fig. **4.7** from Ferm, *Geol. Soc. Amer. Spec. Paper* 148, p. 87, Fig. 6; reprinted with permission of the Geological Society of America.

Fig. **4.9** reprinted by permission of the American Association of Petroleum Geologists.

Fig. **4.11A** reprinted by permission of the American Association of Petroleum Geologists.

Fig. **4.14** from Klein, *Geol. Soc. Amer. Bull.*, v. 83, p. 540,

Fig. 2; reprinted with permission of the Geological Society of America.

Fig. 4.19 reproduced with permission from the *Annual Review of Earth Sciences*, v. 7, p. 239, Fig. 9, copyright © 1979 by Annual Reviews, Inc.

Fig. 4.23B, C reprinted by permission of the American Association of Petroleum Geologists.

Fig. 4.26A from *Sedimentary Geology*, v. 18, p. 149, Fig. 18; reprinted by permission of Elsevier Science Publishers.

Fig. 4.27 from *Sedimentary Geology*, v. 18, p. 138, Fig. 2; reprinted by permission of Elsevier Science Publishers.

Fig. 5.3 from *Marine Geology*, v. 18, p. 113, Fig. 6; reprinted by permission of Elseview Science Publishers.

Fig. 5.5 from H. G. Reading, *Sedimentary Environments and Facies*, 1986, p. 246, Fig. 9.17; by permission of Blackwell Scientific Publishers.

Fig. 5.8 from *Sedimentology*, v. 24, p. 459, Fig. 7; by permission of Blackwell Scientific Publishers.

Fig. 5.10 from Moore, *Geol. Soc. Amer. Paper* 107, p. 79, Fig. 22; reprinted with permission of the Geological Society of America.

Fig. 5.11 from H. G. Reading, *Sedimentary Environments and Facies*, 1986, p. 402, Fig. 12.3; by permission of Blackwell Scientific Publishers.

Fig. 5.14 from A. H. Bouma, *Sedimentology of Some Flysch Deposits*, p. 49, Fig. 8; reprinted by permission of Elsevier Science Publishers.

Fig. 5.15B from Shanmugam and Moiola, *Geology*, v. 13, p. 234, Fig. 1; reprinted with permission of the Geological Society of America.

Fig. 5.18 from J. P. Kennett, *Marine Geology*, p. 98; copyright © 1982; reprinted by permission of Prentice-Hall, Inc., Englewood Cliffs, N.J.

Fig. 5.21 from van Andel, Heath, and Moore, *Geol. Soc. Amer. Memoir* 143, p. 40, Fig. 24; reprinted with permission of the Geological Society of America.

Fig. 5.27 from H. G. Reading, *Sedimentary Environments and Facies*, 1986, p. 276, Fig. 9.55; by permission of Blackwell Scientific Publishers.

Fig. 5.29B, C reprinted by permission of the American Association of Petroleum Geologists.

Fig. 6.3 reprinted by permission of the American Association of Petroleum Geologists.

Fig. 6.10B from Blatt, Middleton, and Murray, *Origin of Sedimentary Rocks*, 2d ed., p. 706; copyright © 1980; reprinted by permission of Prentice-Hall, Inc., Englewood Cliffs, N.J.

Fig. 6.10B from E. G. Purdy, "Recent calcium carbonate facies of the Grand Bahama Banks," *J. Geol.*, v. 71, p. 473, Fig. 1; copyright © 1963, by permission of the University of Chicago Press.

Fig. 6.14 reprinted by permission of the American Association of Petroleum Geologists.

Fig. 6.16A reprinted by permission of the American Association of Petroleum Geologists.

Fig. 6.21 from Scoffin, *Introduction to Carbonate Sediments and Rocks*, Fig. 14.36, p. 191; reproduced by permission of Blackie & Sons, Ltd.

Fig. 6.22 reprinted by permission of the American Association of Petroleum Geologists.

Fig. 6.23A reprinted by permission of the American Association of Petroleum Geologists.

Fig. 6.24 reprinted by permission of the American Association of Petroleum Geologists.

Fig. 7.5 from Ryer, *Geol. Soc. Amer. Bull.*, v. 88, p. 175, Fig. 5, p. 185, Fig. 14, and p. 186, Fig. 15; reprinted with permission of the Geological Society of America.

Fig. 7.8 from *Sedimentary Geology*, v. 25, p. 319; reprinted by permission of Elsevier Science Publishers.

Fig. 7.9 from Barrell, *Geol. Soc. Amer. Bull.*, v. 28, p. 796, Fig. 5; reprinted with permission of the Geological Society of America.

Fig. 7.10 from Loope, *Geology*, v. 13, p. 75, Fig. 2; reprinted with permission of the Geological Society of America.

Fig. 7.14 from Wheeler, *Geol. Soc. Amer. Bull.*, v. 75, p. 606, Fig. 2, reprinted with permission of the Geological Society of America.

Page 184 from *Sedimentology*, v. 27, p. 255, Fig. 10; by permission of Blackwell Scientific Publishers.

Fig. 8.8 from Anderson et al., *Geology*, v. 12, pp. 121, 123; reprinted with permission of the Geological Society of America.

Fig. 8.9 from Sloss, *Geol. Soc. Amer. Bull.*, v. 74, p. 110, Fig. 6; reprinted with permission of the Geological Society of America.

Fig. 8.10 from Ross and Ross, *Geology*, v. 13, p. 156, Fig. 2; reprinted with permission of the Geological Society of America.

Fig. 8.11 from W. A. Berggren and J. A. Van Couvering (eds.), *Catastrophes and Earth History*, p. 132, Fig. 7-1; copyright © 1984 by Princeton University Press; reprinted with permission of Princeton University Press.

Fig. 8.12 from Pitman, *Geol. Soc. Amer. Bull.*, v. 89, p. 1391, Fig. 3; reprinted with permission of the Geological Society of America.

Fig. 8.13 from Sheridan, *Tectonophysics*, v. 143, pp. 64–67, 1987; copyright by the American Geophysical Union.

Fig. 8.15 from Busch and Rollins, *Paleoceanograph*, v. 2, p. 160, Fig. 18; copyright by the American Geophysical Union.

Fig. 8.17B from Izett and Naeser, *Geology*, v. 4, p. 587, Fig. 1; reprinted with permission of the Geological Society of America

Fig. 8.19 reprinted with permission from *Proc. 10th*

Lunar Planet. Sci. Conf., p. 2539, Fig. 2, by B. P. Glass, M. B. Swincki, and P. A. Zwort, "Australasian Ivory Coast, and North American Tektite Strewn Fields: Size, Mass, and Correlation with Geomagnetic Reversals and Other Earth Events;" copyright © 1979, Pergamon Press.

Page 208 from the Geological Survey of Canada; reprinted with the permission of the Minister of Services Canada, 1989.

Fig. 9.5 reprinted by permission of the American Association of Petroleum Geologists.

Fig. 9.7 from *Phil. Trans. Royal Soc.* A, v. 305, pp. 169–192, by permission of the Royal Society.

Fig. 9.10 from the Geological Survey of Canada; reprinted with the permission of the Minister of Supply and Services Canada, 1989.

Fig. 9.11B from Meckel, *Geol. Soc. Amer. Bull.*, v. 78, p. 237, Fig. 6; reprinted with permission of the Geological Society of America.

Fig. 10.1 reprinted by permission of the American Association of Petroleum Geologists.

Page 248 reprinted by permission of the American Association of Petroleum Geologists.

Fig. 11.5 from *Illinois Geol. Surv. Circular* 425, p. 15, Fig. 12; reprinted with permission of the Illinois State Geological Survey.

Fig. 11.6 reprinted by permission of the American Association of Petroleum Geologists.

Fig. 11.14 reprinted by permission of the American Association of Petroleum Geologists.

Fig. 11.15 reprinted by permission of the American Association of Petroleum Geologists.

Fig. 11.16 reprinted by permission of the American Association of Petroleum Geologists.

Fig. 11.18 reprinted by permission of the American Association of Petroleum Geologists.

Fig. 11.19 reprinted by permission of the American Association of Petroleum Geologists.

Fig. 11.22 from W. W. Bishop and J. A. Miller (eds.), *Calibration of Hominid Evolution*; copyright © 1972; by permission of Scottish Academic Press.

Fig. 11.23B from D. H. Tarling, *Palaeomagnetism*, p. 211, Fig. 9.6; by permission of Chapman & Hall Publishers.

Fig. 11.24 from *Science*, v. 154, pp. 349–357, Fig. 1, by Opdyke et al.; reprinted by permission of the author and *Science*; copyright © 1966 by the A.A.A.S.

Fig. 11.25 from *Palaeogeogr., Palaeoclimat., Palaeoecol.*, v. 37, pp. 34, 38; reprinted by permission of Elsevier Science Publishers.

Fig. 11.27 from Shackleton and Opdyke, *Geol. Soc. Amer. Memoir* 145, pp. 449–464; reprinted with permission of the Geological Society of America.

Fig. 11.28 from *Earth Planet. Sci. Lett.*, v. 27, p. 32, Fig. 12; reprinted by permission of Elsevier Science Publishers.

Fig. 11.29 from J. P. Kennett, *Marine Geology*, p. 86; copyright © 1982; reprinted by permission of Prentice-Hall, Inc., Englewood Cliffs, N.J.

Fig. 11.30 from *Science*, v. 227, pp. 938–941, Fig. 1, by DePaolo and Ingram; reprinted by permission of the author and *Science*; copyright © 1985 by the A.A.A.S.

Fig. 12.6 from P. W. Gast, "The isotopic composition of strontium and the age of stone meteorites, I," *Geochim. Cosmochim. Acta*, v. 26, p. 932, Fig. 1; copyright © 1962, Pergamon Press, Inc.

Table 12.1 from Dalrymple, *Geology*, v. 7, p. 558, Table 1; reprinted with permission of the Geological Society of America.

Fig. 12.7 from *The Chronology of the Geological Record*, N.J. Snelling (ed.), 1985, p. 86, Fig. 2; by permission of Blackwell Scientific Publishers.

Fig. 12.10 from McDougall, *Geol. Soc. Amer. Bull.*, v. 96, pp. 165,171; reprinted with permission of the Geological Society of America.

Fig. 12.12 from *Science*, v. 166, p. 743, Fig. 2 by Rankin et al.; reprinted by permission of the author and *Science*; copyright © 1969 by the A.A.A.S.

Fig. 12.13 from *Science*, v. 173, p. 386, Fig. 4, by Wetherill; reprinted by permission of the author and *Science*; copyright © 1971 by the A.A.A.S.

Fig. 13.2 from J. J. Flynn, B. J. MacFadden, and M. C. McKenna, "Land mammal ages, faunal heterochrony, and temporal resolution in Cenozoic terrestrial sequences," *J. Geol.*, v. 92, p. 689, Fig. 1; copyright © 1984, by permission of the University of Chicago Press.

Fig. 13.3 from *The Chronology of the Geological Record*, N.J. Snelling (ed.), 1985, p. 275, Fig. 2; by permission of Blackwell Scientific Publishers.

Fig. 13.4 reprinted by permission of the American Association of Petroleum Geologists.

Fig. 13.5 from McDougall, *Geol. Soc. Amer. Bull.*, v. 96, p. 173; reprinted with permission of the Geological Society of America.

Fig. 13.6 from Hillhouse et al., *J. Geophys. Res.*, v. 91, no. B11, p. 11594, copyright © by the American Geophysical Union.

Fig. 13.10 reproduced with permission from the *Annual Review of Earth Sciences*, v. 16, p. 626, Fig. 8; copyright © 1988 by Annual Reviews, Inc.

Fig. 13.11 reproduced with permission from the *Annual Review of Earth Sciences*, v. 16, p. 646, Fig. 16; copyright © 1988 by Annual Reviews, Inc.

Fig. 13.12 reprinted by permission of the American Association of Petroleum Geologists.

Fig. 13.14 from *The Chronology of the Geological Record*, N.J. Snelling (ed.), 1985, p. 144, Fig. 1; by permission of Blackwell Scientific Publishers.

Fig. 13.15 from *The Chronology of the Geological Record*, N.J. Snelling (ed.), 1985, p. 147, Fig. 2; by permission of Blackwell Scientific Publishers.

Fig. 14.1 from Sloss, *Decade of North American Geology Spec. Publ.* 1, p. 36, Fig. 3; reprinted with permission of the Geological Society of America.

Fig. 14.2 from P. B. King, *The Evolution of North America*, p. 28. Fig. 15 and p. 35, Fig. 22; copyright © 1959, 1977 by Princeton University Press; reprinted with permission of Princeton University Press.

Fig. 14.3 from Kay, *Geol. Soc. Amer. Memoir* 48, p. 26, Fig. 9; reprinted with permission of the Geological Society of America.

Fig. 14.7A from Dewey and Burke, *Geology*, v. 2, p. 58, Fig. 1; reprinted with permission of the Geological Society of America.

Fig. 14.7B from K. Burke and J. F. Dewey, "Plume-generated triple junctions: Key indicators in applying plate tectonics to old rocks, "*J. Geol.*, v. 81, p. 420, Fig. 10; copyright © 1973, by permission of the University of Chicago Press.

Fig. 14.8 from *Phil. Trans. Royal Soc.* A, v. 267, pp. 323–329; by permission of the Royal Society.

Fig. 14.10 reprinted by permission of the American Association of Petroleum Geologists.

Fig. 14.11 from Karig and Sharman, *Geol. Soc. Amer. Bull.*, v. 86, p. 386, Fig. 7A; reprinted with permission of the Geological Society of America.

Fig. 14.14 from Dickinson et al., *Geol. Soc. Amer. Bull.*, v. 94, p. 223, Fig. 1; reprinted with permission of the Geological Society of America.

Fig. 14.15 from Dickinson et al., *Geol. Soc. Amer. Bull.*, v. 94, p. 232, Fig. 9; reprinted with permission of the Geological Society of America.

Index